RFID, Microwave Circuit, and Wireless Power Transfer Enabling 5/6G Communication

Augustine O. Nwajana
University of Greenwich, UK

IGI Global
Scientific Publishing
Publishing Tomorrow's Research Today

Vice President of Editorial	Melissa Wagner
Managing Editor of Acquisitions	Mikaela Felty
Managing Editor of Book Development	Jocelynn Hessler
Production Manager	Mike Brehm
Cover Design	Phillip Shickler

Published in the United States of America by
IGI Global Scientific Publishing
701 East Chocolate Avenue
Hershey, PA, 17033, USA
Tel: 717-533-8845
Fax: 717-533-8661
Website: https://www.igi-global.com E-mail: cust@igi-global.com

Library of Congress Cataloging-in-Publication Data

Names: Nwajana, Augustine Onyenwe, 1979- editor.
Title: RFID, microwave circuit, and wireless power transfer enabling 5/6G communication / edited by: Augustine Nwajana.
Description: Hershey PA : IGI Global Scientific Publishing, [2025] | Includes bibliographical references. | Summary: "This book aims to provide relevant theoretical frameworks and the latest empirical research findings in the area of RFID, microwave circuit, and wireless power transfer enabling 5/6G communication"-- Provided by publisher.
Identifiers: LCCN 2024057018 (print) | LCCN 2024057019 (ebook) | ISBN 9798369387993 (hardcover) | ISBN 9798369388006 (paperback) | ISBN 9798369388013 (ebook)
Subjects: LCSH: 5G mobile communication systems. | 6G mobile communication systems. | Radio frequency identification systems. | Microwave circuits. | Wireless power transmission.
Classification: LCC TK5103.25 .R48 2025 (print) | LCC TK5103.25 (ebook) | DDC 621.3845/6--dc23/eng/20250213
LC record available at https://lccn.loc.gov/2024057018
LC ebook record available at https://lccn.loc.gov/2024057019

British Cataloguing in Publication Data
A Cataloguing in Publication record for this book is available from the British Library.

In Memory of My Late Parents:
Mr Emmanuel Onwumbiko Nwajana [13/May/1932 – 07/June/2019] and
Mrs Cecilia Ogechukwu Nwajana [14/June/1951 – 03/September/2023]

Editorial Advisory Board

Table of Contents

Detailed Table of Contents

This chapter examines robust beamforming to increase power economy and reliability in magnetic-based wireless power transfer with many transmitters and receivers while accounting for MII imperfection. Transfer learning saves energy and improves efficiency in wireless communications by learning from prior tasks using correlation and similarity information. Since CSI procurement takes energy, its benefits gradually drop as power driven gadgets increase, yet spread. With system stability, channel state information techniques expand power insurance. ISWPT's integrated operation decreases system size, hardware cost, power consumption, and spectrum, enabling 6G wireless networks. Reconfigurable intelligent surfaces in wireless power transfer systems and designs can optimize energy efficiency by addressing multi-user circumstances and power distribution. Using simultaneous lightwave information and power transfer to combine EH and data transmission is fascinating and matures efficiency.

The RFID-Enabled Wireless Charging System for Electric Vehicles (EVs) represents a significant advancement in sustainable transportation infra- structure. By integrating Radio Frequency Identification (RFID) technology with wireless charging capabilities, the system not only enhances user experience but also promotes environmental sustainability. One key feature of the system is its security and efficiency. Charging is initiated only when a valid RFID card is swiped, ensuring that only authorized users can access the charging facility. This helps prevent unauthorized usage and ensures that the system operates smoothly and efficiently. The system allows EV batteries to be charged using solar panels, tapping into renewable energy sources and reducing reliance on non-renewable energy. This not only reduces the carbon footprint of EV charging but also promotes the use of clean energy in transportation. The wireless charging capability of the system is facilitated by two copper coils one on the charging pad and the other integrated into the EV.

A simple technique for transforming a single-band bandpass filter (BPF) into a dual-band BPF is presented in this chapter. The method starts with a 2nd-order single-band BPF design, giving rise to a 4th-order dual-band BPF after the proposed transformation. Both filters are implemented using compact U-shaped microstrip resonator for improved device miniaturization. The proposed work features a centre frequency of 1.4 GHz for the single-band BPF, with a 3.4% fractional bandwidth. The dual-band BPF operates at 1.35 and 1.45 GHz for the first and second bands, respectively. The design implementation employs the commercially available Rogers RT/Duroid 6010LM substrate, having a dissipation factor of 0.0023, dielectric constant of 10.7, and thickness of 1.27 mm. The practical responses of the prototype dual-band BPF indicate a good return loss of better than 18 dB across both bands, and an insertion loss of better than 0.1 dB. The design prototype achieved physical size of 0.23 x 0.18 λg. λg is the guided wavelength for the microstrip line impedance at the centre frequency of the filter.

There is growing demand for 5G as well as 6G technologies as per user's requirements and enhanced applications. As low latency and high data rates are key findings of 5G, in same way more improved bandwidth, maximum data enabled rate, and many new applications are basic achievable things from 6G systems. There are several types of microwave circuits used for such high frequency. RF/microwave circuits include antennas, filters, power amplifiers, phase shifters, power dividers, mixers, and multiplexers. Here, microwave antennas have been discussed under the microwave circuits. In the last few years, as per economic and social development which is greatly influenced by the advancements in mobile communication and technology, 5G technology has emerged as an important of the future 2020 generation which is already in use. After the development of fifth-generation technologies, researchers, scientists, and engineers are looking for wide bandwidth which should improve wireless systems and devices to provide better services and faster experience. Also, the development of 5G wireless network technology is the response to the crucial factors that lead to this demand because of its ability to provide extremely fast internet speed, high bandwidth, high performance, reduced latency, and high reliability and better gain. In this chapter, we will review the recent design with advancement in 5G antennas for beyond 5G applications and also we will give a few designs of antennae for 6G applications.

networks because the transmitting/receiving nodes may be travelling, the channels may have a complex framework, and coordination among multiple nodes is difficult. In spite of this, mm wave is significantly important in implementation of 5G/B5G communication networks. The novelty in the chapter is to suggesting a modern antenna design for mm wave application and link budget analysis so as to achieve flaw less communication with reduced call drop rate. In this chapter we are focusing on introduction and background. The chapter further will focus on mm wave in cellular mobile communication along with the applications in 5G/B5G wireless communication networks. The chapter will also deal with system architecture of 5G communication employing mm wave and antenna in mm wave communication will be discussed. Also various challenges associated with deployment of mm wave technology in 5G networks are discussed. At the last advantages and disadvantages of mm wave will be discussed in brief.

Mobile technology communication is evolving towards the seventh generation, the '6G' network. These networks will revolutionize society to the extent of connectedness and intelligence that we have not seen before. 6G will be important to implement the new generation smart cities by enhancing the digital communication technology with terahertz communication, ultra-low latency and artificial intelligence to operate the network system. This will result in smart cities that are fully interconnected to real time data as required for instance in self-drive cars, energy consumption, or security. 6G is found to be promising for URLLC, eXR and massive sensing at the edge necessary for Smart City instantiation. But there are issues such as technicality, data privacy, and regulations to consider. The present chapter contends using a 6G empowered smart city based on IoT, artificial intelligence, and edge computing for the increased efficiency of urban processes. These include dynamic network slicing to fulfill the latency and bandwidth requirement that is supported by Intelligent Resource management as well as Edge AI processing. The chapter therefore concludes by highlighting the need for joint efforts and forward planning to overcome hindrances in the 6G's delivery to support sustainable cities.

Smart industrial processes are impossible without the use of machinery. Majority of these machines comprise rotating machines. As supporting tools for developing maintenance strategies, a Data-driven approach for multifault diagnosis of Rotating Machines is currently used in smart industry. This Chapter presents Multisensor Data Fusion and its application in Predictive maintenance methodology for multifault diagnosis of Industrial Rotating Machines. The main objective of Multisensor Data Fusion, rather than the different sensors being applied independently, are optimally fused to take advantage of their respective strengths, by combining data from multiple sensors and related information to achieve specific inferences than could be achieved by a single sensor. This chapter presents a methodology using data extraction features domain parameters for multifault diagnosis of Rotating Machines. The analysis shows that Root Mean Square Amplitude increases as fault develops, Crest Factor is more for healthy bearing and less for faulty bearing, these can provide early warning of faults

In this article, a circular fractal array antenna is proposed for C-band and S-band applications. The antenna uses circular fractal geometry to achieve high gain across the S and C bands frequency spectrum. The antenna is fabricated using an FR-4 substrate and tested in anechoic chamber. The antenna design holds potential for multi-functional wireless systems and S and C band applications. The measurement results reveal that the designed antenna has two resonant frequencies at 3.5 GHz and 5.8 GHz and gain of around 3.1dB is noted. The size of proposed antenna is 28mm×28mm×1.6mm.

Intelligent Transportation Systems (ITS) represent a revolutionary integration of advanced information and communication technologies into transportation infrastructure and vehicles. This integration aims to significantly enhance the safety, efficiency, and sustainability of transportation networks across the globe. ITS is transformative, moving beyond mere enhancements of existing frameworks to leverage a broad spectrum of cutting-edge technologies, including real-time traffic monitoring, adaptive signal control, sophisticated vehicle-to-vehicle (V2V) and vehicle-to-infrastructure (V2I) communication systems, alongside advanced public transit solutions. This chapter offers an extensive overview of ITS, encapsulating its diverse technologies, widespread applications, and profound impact on global transportation systems. It introduces a paradigm shift in the conceptualization and management of transportation systems, addressing pressing urban challenges such as congestion, safety, and sustainability.

This chapter presents a Simultaneous Wireless Information and Power Transfer (SWIPT) base Cooperative Non-Orthogonal Multiple Access (CNOMA) system aided by Spatial Modulation (SM), designed by energy-efficient or "green" communication in beyond 5G (B5G) and 6G applications. The proposed system leverages NOMA to enhance spectral efficiency by allowing multiple users to share the same frequency and time resources, differentiated through power domain multiplexing. SWIPT is incorporated to enable near users to harvest energy from the received RF signals, which relay information to far users in a cooperative NOMA system. Spatial modulation reduces hardware complexity by activating a single antenna for data transmission, significantly enhancing energy efficiency while maintaining high data rates. This SM SWIPT NOMA framework addresses the challenges of power-constrained and spectrum-limited scenarios, providing a low-power, high-performance solution ideal for green communications. The system finds applications in smart cities, massive machine-type communication (mMTC) and vehicular networks where energy efficiency, coverage extension and user fairness are critical. Simulation results demonstrate the system's superior energy efficiency, spectral efficiency and performance gains compared to conventional techniques, making it a viable solution for sustainable communication in B5G and 6G networks. At a transmit power of 20 dBm, energy harvested in SM SWIPT NOMA is 135.12 joules and that of SM SWIPT OMA is 67.56 joules.

This chapter presents a comprehensive survey of edible antennas, exploring their design, development, and potential applications in emerging fields such as biomedical monitoring, healthcare, and environmental sensing. Edible antennas, crafted from biocompatible, biodegradable materials, offer unique advantages in applications where traditional antennas are unsuitable, especially in scenarios requiring ingestion or close human contact. The chapter begins with a detailed review of the state-of-the-art in edible antenna technology, discussing various fabrication methods, materials used, and performance metrics such as gain, efficiency, and bandwidth. Key considerations such as antenna miniaturization, impedance matching, and specific absorption rate (SAR) compliance are examined. Moreover, this survey delves into the integration of edible antennas with next-generation communication systems, including 5G/6G networks, which promise enhanced data rates, low latency, and reliable connectivity for real-time applications. The role of advanced materials such as metamaterials in improving antenna performance is also explored, along with future research directions and challenges such as signal attenuation in body tissues, energy harvesting, and sustainability. This survey concludes with insights into the future of edible antennas and their potential impact on personalized medicine, smart healthcare systems, and the Internet of Medical Things (IoMT).

Li-Fi technology uses visible light communication for data transmission and has the potential to transform wireless communication. The benefits offered by this technology have the potential to transform wireless communication, many industries, and improve communication in both indoor and outdoor environment. It provides significantly faster speeds and has better security standards compared to Wi-Fi. In this chapter, the authors discuss significant developments in Li-Fi, such as compatibility with 5G networks, support of IoT, security and privacy concerns. They have also presented an in-depth analysis of Li-Fi, including its potential benefits and drawbacks, as well as its range limitations, interoperability issues, and cost implications. This study aims to provide a comprehensive overview and contribute to the understanding

Preface

Future 5/6G technology is anticipated to address challenges such as low data rate and high latency in existing communication networks, particularly as the number of users overwhelms current 5G and 4G capabilities. Wireless communication has experienced remarkable advancements over the past two decades due to the continuous evolution of communication systems, largely driven by the expansion in modern communication infrastructures. Hence, the need for this edited collection of original research in this area of research.

RFID, Microwave Circuit, and Wireless Power Transfer Enabling 5/6G Communication provides relevant theoretical frameworks and the latest empirical research findings in RFID, microwave circuit, and wireless power transfer Enabling 5/6G Communication. It is written for professionals who want to improve their understanding of the strategic roles in the research area. RFID and wireless power transfer are modern technologies facilitated by recent advancements in microwave circuit design. These technologies have a wide range of applications in the state-of-the art 5G and future 6G networks. Some popular applications include indoor and outdoor tracking, sensing, operation of tag-attached objects, human bodies, etc. Wireless identification of people and physical objects has enabled them to become smartly connected. This has led to scientific breakthroughs in various fields of human endeavours including healthcare, health monitoring, disaster monitoring, logistics, social networking, smart environments, security services, etc. Though many applications rely on passive chipped tags, chip-less tags have been gaining popularity recently.

The book is organized in 15 chapters with a brief description of each chapter given as follows:

Chapter 1 covers recent advancements in wireless power transfer techniques for 6G Communication. The chapter examines robust beamforming techniques to increase power economy and reliability in magnetic-based wireless power transfer with many transmitters and receivers while accounting for MII imperfection. Transfer learning saves energy and improves efficiency in wireless communications by learning from prior tasks using correlation and similarity information. Since CSI procurement takes energy, its benefits gradually drop as power-driven gadgets increase and spread. With system stability, channel state information techniques expand power insurance. ISWPT's integrated operation decreases system size, hardware cost, power consumption, and spectrum, enabling 6G wireless networks. Reconfigurable intelligent surfaces in wireless power transfer systems and designs can optimize energy efficiency by addressing multi-user circumstances and power distribution. Using simultaneous lightwave information and power transfer to combine EH and data transmission is fascinating and matures efficiency.

Chapter 2 focuses on RFID-enabled wireless charging system for electric vehicles (EVs). This topic represents a significant advancement in sustainable transportation infrastructure. By integrating Radio Frequency Identification (RFID) technology with wireless charging capabilities, the system not only enhances user experience but also promotes environmental sustainability. One key feature of the system is its security and efficiency. Charging is initiated only when a valid RFID card is swiped, ensuring that only authorized users can access the charging facility. This helps prevent unauthorized usage and ensures that the system operates smoothly and efficiently. The system allows EV batteries to be charged using solar panels, tapping into renewable energy sources and reducing reliance on non-renewable energy. This not only reduces the carbon footprint of EV charging but also promotes the use of clean energy in transportation. The wireless charging capability of the system is facilitated by two copper coils one on the charging pad and the other integrated into the EV.

Chapter 3 investigates single- to dual-band bandpass filter (BPF) transformation for wireless communication systems. A simple technique for transforming a single-band BPF into a dual-band BPF is presented in this chapter. The method starts with a 2nd-order single-band BPF design, giving rise to a 4th-order dual-band BPF after the proposed transformation. Both filters are implemented using compact U-shaped microstrip resonator for improved device miniaturization. The proposed work features a centre frequency of 1.4 GHz for the single-band BPF, with a 3.4% fractional bandwidth. The dual-band BPF operates at 1.35 and 1.45 GHz for the first and second bands, respectively. The design implementation employs the commercially available Rogers RT/Duroid 6010LM substrate, having a dissipation factor of 0.0023, dielectric constant of 10.7, and thickness of 1.27 mm. The practical responses of the prototype dual-band BPF indicate a good return loss of better than 18 dB across both bands, and an insertion loss of better than 0.1 dB. The design prototype achieved physical size of 0.23 x 0.18 λg. λg is the guided wavelength for the microstrip line impedance at the centre frequency of the filter.

Chapter 4 presents microwave circuits enabling 5G/6G communication. There is growing demand for 5G as well as 6G technologies as per user's requirements and enhanced applications. As low latency and high data rates are key findings of 5G, in same way more improved bandwidth, maximum data enabled rate, and many new applications are basic achievable things from 6G systems. There are several types of microwave circuits used for such high frequency. RF/microwave circuits include antennas, filters, power amplifiers, phase shifters, power dividers, mixers, and multiplexers. Here, microwave antennas have been discussed under the microwave circuits. In the last few years, as per economic and social development which is greatly influenced by the advancements in mobile communication and technology, 5G technology has emerged as an important of the future 2020 generation which is already in use.

Chapter 5 investigates "The Role of Artificial Intelligence in 6G Networks, Architecture, Protocol, Transmission and Applications". The 6G wireless communication technology is the successor to the 5G cellular-based technique. It is aimed at the increased bandwidth and higher dimensions of data transmission with the measurement of less than one microsecond of latency communication. Artificial Intelligence (AI) is pivotal in establishing computational and intelligence infrastructure, data storing, transmission, and decision-making. With the increased frequencies of 6G, sampling rates will surge beyond those achievable with 5G (Fifth Generation). Access points will gain the capacity to serve multiple clients through orthogonal frequency-division multiple access concurrently. The expected spectrum of communication using 6G is up to terahertz (THz). Artificial Intelligence (AI) guides Quality of Service (QoS) in 6G communication. This chapter covers a high-level overview of the role of AI in 6G, architecture, protocols, transmission technologies, and applications.

Chapter 6 reports on "Enhancing Public Safety in Developing Countries Using a Mobile Application Solution: Nigeria as a Case Study". In today's world, the escalating global insecurity poses significant challenges for both developed and undeveloped countries. The high crime rates and shortcomings in reporting crimes are pressing issues, emphasizing just how crucial it is to ensure the safety of citizens. While law enforcement agencies play a vital role, relying solely on them is insufficient. Embracing modern technology and fostering active citizen participation are essential to address these concerns effectively. This research adopts an agile methodology to tackle the above concerns. The result of this approach is the development of a user-friendly mobile application that leverages cutting-edge technology. The primary objective of this application is to effectively map incidents of crime and provide real-time alerts, thereby significantly enhancing citizen security. With this, users are empowered to report criminal incidents directly from their smartphones, eliminating the need to physically approach law enforcement authorities.

Chapter 7 presents "Millimetre-Wave Communication for 5G/B5G Applications". Emerging millimeter wave (mm Wave) wireless communication systems reflect the culmination of more than a century of advancement in modern communications. Millimeter wave (mm Wave) communications have recently piqued the interest of researchers due to the enormous available bandwidth, which has the potential to result in rates of multiple Gbps (gigabit per second) per user. The novelty in the chapter is to suggest a modern antenna design for mm wave application and link budget analysis so as to achieve flaw less communication with reduced call drop rate. In this chapter we are focusing on introduction and background. The chapter further will focus on mm wave in cellular mobile communication along with the applications in 5G/B5G wireless communication networks. The chapter will also deal with system architecture of 5G communication employing mm wave and antenna in mm wave communication will be discussed.

Chapter 8 discusses "The Role of 6G in Empowering Smart Cities Enabling Ubiquitous Connectivity and Intelligent Infrastructure". Wireless communication is advancing, leading to 6G networks. These networks will reshape society with unprecedented connectivity and intelligence. 6G is key for future smart cities, integrating terahertz communication, ultra-low latency, and AI-driven networks. This will create hyper-connected cities where real-time data supports systems like autonomous vehicles, energy management, and public safety. 6G research shows its potential for ultra-reliable communication, immersive experiences like extended reality, and widespread sensing, crucial for smart city applications. However, challenges like technical issues, data privacy, and regulations need addressing. This chapter proposes a 6G-enabled smart city model using IoT, AI, and edge computing to improve urban operations. It explores dynamic network slicing, intelligent resource allocation, and edge AI processing to meet latency and bandwidth needs. The chapter concludes by emphasizing collaboration and strategic planning to overcome challenges, ensuring 6G technologies foster sustainable urban growth.

Chapter 9 investigates multisensor data fusion and its applications. Smart industrial processes are impossible without the use of machinery. Majority of these machines comprise rotating machines. As supporting tools for developing maintenance strategies, a Data-driven approach for multifault diagnosis of Rotating Machines is currently used in smart industry. This Chapter presents Multisensor Data Fusion and its application in Predictive maintenance methodology for multifault diagnosis of Industrial Rotating Machines. The main objective of Multisensor Data Fusion, rather than the different sensors being applied independently, are optimally fused to take advantage of their respective strengths, by combine data from multiple sensors and related information to achieve specific inferences than could be achieved by a single sensor. This chapter presents a methodology using data extraction features domain parameters for multifault diagnosis of Rotating Machines. The analysis shows that Root Mean Square Amplitude increases as fault develops, Crest Factor is more for healthy bearing and less for faulty bearing, these can provide early warning of faults.

Chapter 10 discusses "Circular Fractal Array Antenna for C and S-Band Applications". In this chapter, a circular fractal array antenna is proposed for C-band and S-band applications. The antenna uses circular fractal geometry to achieve high gain across the S and C bands frequency spectrum. The antenna is fabricated using an FR-4 substrate and tested in anechoic chamber. The antenna design holds potential for multi-functional wireless systems and S and C band applications. The measurement results reveal that the designed antenna has two resonant frequencies at 3.5 GHz and 5.8 GHz and gain of around 3.1dB is noted. The size of proposed antenna is 28mm×28mm×1.6mm.

Chapter 11 discusses "An ITS-G5 Antenna for Vehicle Communication". Intelligent Transportation Systems (ITS) represent a revolutionary integration of advanced information and communication technologies into transportation infrastructure and vehicles. This integration aims to significantly enhance the safety, efficiency, and sustainability of transportation networks across the globe. ITS is transformative, moving beyond mere enhancements of existing frameworks to leverage a broad spectrum of cutting-edge technologies, including real-time traffic monitoring, adaptive signal control, sophisticated vehicle-to-vehicle (V2V) and vehicle-to-infrastructure (V2I) communication systems, alongside advanced public transit solutions. This chapter offers an extensive overview of ITS, encapsulating its diverse technologies, widespread applications, and profound impact on global transportation systems. It introduces a paradigm shift in the conceptualization and management of transportation systems, addressing pressing urban challenges such as congestion, safety, and sustainability.

Chapter 12 focuses on "Energy-Efficient SWIPT aided NOMA with Spatial Modulation for Green Communication in Beyond 5G/6G Networks". The chapter introduces an energy-efficient system combining Simultaneous Wireless Information and Power Transfer (SWIPT), Cooperative Non-Orthogonal Multiple Access (CNOMA), and Spatial Modulation (SM) for beyond 5G and 6G networks. Designed for high data rates and low power use, it supports applications like smart cities, mMTC, vehicular networks. Non-Orthogonal Multiple Access (NOMA) enhances spectral efficiency by allowing multiple users to share frequency and time resources via power domain multiplexing. SWIPT enables near users to harvest RF energy to relay information to distant users, reducing infrastructure needs. Spatial Modulation (SM) minimizes hardware demands by activating one antenna per transmission, lowering power use while maintaining data rates. This SM SWIPT NOMA framework is optimized for power-limited, spectrum-constrained settings, aligning with B5G and 6G sustainability goals. Simulations demonstrate its efficiency: at 20 dBm, it achieves 135.12 joules of energy harvesting compared to 67.56 joules in SM SWIPT OMA, making it suitable for beyond 5G applications.

Chapter 13 investigates "Advancements in Edible Antennas: Design, Development, and Future Prospects in Biomedical and 5G/6G Applications". This chapter presents a comprehensive survey of edible antennas, focusing on their design, development, and applications in biomedical monitoring, healthcare, and environmental sensing. Made from biocompatible, biodegradable materials, edible antennas are ideal for scenarios where traditional antennas are unsuitable, especially when ingestion or close human contact is required. The chapter reviews the state-of-the-art in edible antenna technology, discussing fabrication methods, materials, and key performance metrics like gain, efficiency, and bandwidth. It also addresses challenges such as miniaturization, impedance matching, and SAR compliance. Integration with 5G/6G networks is explored, highlighting benefits like enhanced data rates, low latency, and reliable connectivity for real-time applications. The role of advanced materials such as metamaterials in improving performance is examined, along with future research directions and challenges, and potential impact of edible antennas on personalized medicine, smart healthcare, and IoMT are provided.

Chapter 14 presents "Li-Fi for Next Generation Wireless Communication Technical Insights and Innovations". Li-Fi technology uses Visible Light Communication for data transmission have the potential to transform wireless communication. The benefits offered by this technology, it has the potential to transform many industries and improve connectivity in both indoor and outdoor settings. It also provides significantly faster speeds and better security than standard Wi-Fi. This paper discusses significant developments in Li-Fi, such as its compatibility with 5G networks, support IoT, and how it improves the system of current wireless communication. In this chapter, we have presented an in-depth analysis of Li-Fi, including its potential benefits and drawbacks, as well as its range limitations, interoperability issues, and cost implications. The study aims to provide a comprehensive overview and contribute to the understanding and advancement of the technology by examining the basic ideas, technological advances, and issues of Li-Fi. The goal of this chapter is to improve our understanding and ability to use Li-Fi as a next generation technology in wireless communications.

Chapter 15 looks at an intelligent and optical security monitoring for 6G communications. The infrastructure of communication technologies in recent years has grown more complex and autonomous to meet societal requirements. Advanced technologies such as optical communication due to their enormous bandwidth and less latency capabilities have become the most sought-after candidates for 6G communications. The optical networks provide an agile and secure platform for effective data transmission. Thus, it is important to include cognitive and intelligent security monitoring for autonomous optical network management to improve reliability. Machine learning is a great tool that provides advancements in security monitoring in terms of accuracy and cost-effectiveness for optical network diagnosis and management. This chapter discusses the vulnerability of optical networks to various security threats in the physical layer along with countermeasures for network survivability. Several monitoring techniques using various performance parameters are also discussed for effective attack identification and localization to ensure robust optical network performance for 6G communications.

Augustine O. Nwajana
University of Greenwich, UK

Chapter 1
Wireless Power Transfer Techniques for 6G Communication

Tushar
https://orcid.org/0000-0001-7190-7687
United College of Engineering and Research, India

Nandita Pradhan
United College of Engineering and Research, India

Pooja Jaiswal
United College of Engineering and Research, India

ABSTRACT

This chapter examines robust beamforming to increase power economy and reliability in magnetic-based wireless power transfer with many transmitters and receivers while accounting for MII imperfection. Transfer learning saves energy and improves efficiency in wireless communications by learning from prior tasks using correlation and similarity information. Since CSI procurement takes energy, its benefits gradually drop as power driven gadgets increase, yet spread. With system stability, channel state information techniques expand power insurance. ISWPT's integrated operation decreases system size, hardware cost, power consumption, and spectrum, enabling 6G wireless networks. Reconfigurable intelligent surfaces in wireless power transfer systems and designs can optimize energy efficiency by addressing multi-user circumstances and power distribution. Using simultaneous lightwave information and power transfer to combine EH and data transmission is fascinating and matures efficiency.

1. INTRODUCTION

Future power usage systems will employ wireless power transfer, or WPT, as a wireless energy source. Sensor systems and the Internet of Things (IoT) will probably be used in Beyond 5G (B5G) and 6G communications. Furthermore, civilization 5.0, where cyberspace and the actual world are combined through digital transformation techniques is anticipated to be brought about by IoT and sensor

DOI: 10.4018/979-8-3693-8799-3.ch001

network technologies. "WPT methods have already been covered and reported. A WPT demonstration for a flying helicopter was described by Brown (Brown, 1984). The deployment of IoT and sensor networks across Beyond-5G (B5G)/6G networks and society 5.0 is made possible by the WPT system in Figure 1. The WPT system at 2.4 GHz, 5.7 GHz, and 920 MHz has already been established in Japan. However, 99.98% of all radio stations in Japan are located in the sub-6 frequency range, which hinders the spread of these WPT systems". As a result, it is crucial to raise WPT frequency in conjunction with data-transmission enhancement.

Figure 1. Data Power Transmission from 5G/6G BTS

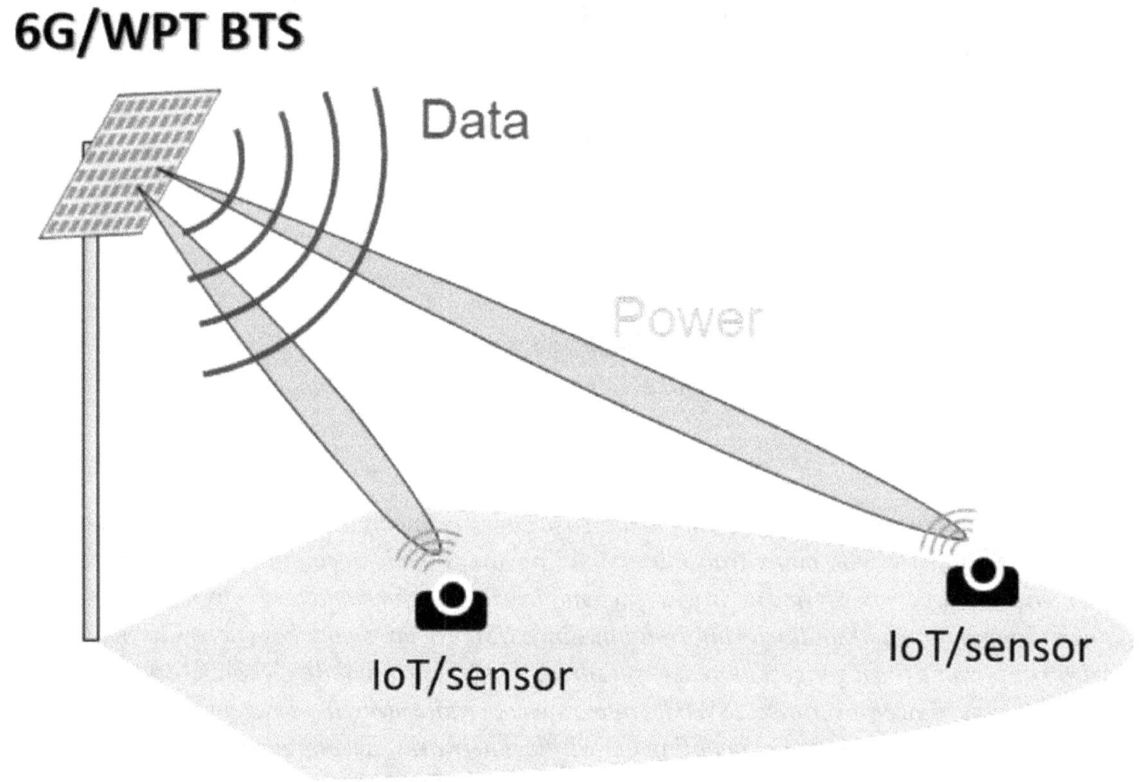

The beam-type WPT was designed to be utilized for IoT, sensors, and mobile or wearable device wireless power charging. Since 2013 (Report ITU-R, 2017), there has been discussion on ITU-R on the establishment of beam-WPT.

Figure 2. WPT and communication system based on B5G/6G BTS

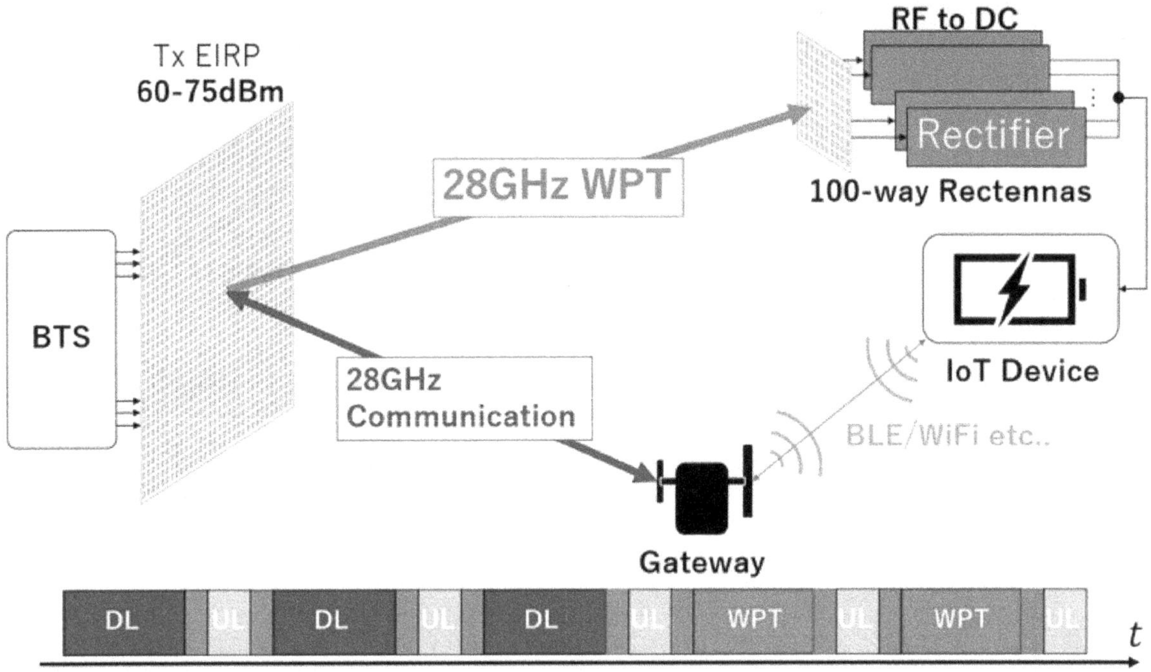

"Within the last decade, near field wireless power transfer (WPT) has been considered for energy-constrained smart devices in the Internet of Things (IoTs) as shown in Figure 2. Many near-field WPTs use inductive, capacitive, magneto dynamic and magnetic resonant coupling (MRC) (Yang et al., 2017; Sun et al., 2018; Liu et al., 2019; Lang, 2018). MRC-WPT provides safe, reliable, and interoperable wireless charging because it delivers power further and more efficiently (Yang et al., 2016)."

6G wireless communications will be introduced around 2027–2030 and will be 100 times faster than 5G (Union, 2015). "To overcome 5G's limits for new challenges, 6G wireless networks with additional enticing features are needed. 6G will combine network densification, fast throughput, positioning accuracy, energy efficiency, and massive connectivity. Generation 6G will follow previous trends (Zong et al., 2019). Learning from experience aids this. Transfer learning (TL) enables new tasks/domains learn from old ones to be faster and more efficient". TL may save energy and improve efficiency in numerous wireless communications sectors using task correlation and similarity information.

In this chapter, we discuss wireless powering topologies, challenges, and solutions. We emphasize Channel state information approaches' feasibility for driving several devices. Due to energy requirements, CSI procurement loses its benefits when power-driven devices increase, however distributed Channel state information methods expand power insurance coverage with system stability. Some EH systems employ solar, piezoelectric, wind, hydropower, or RF waves. This focuses on a Wireless Energy Transfer (WET) process for energizing EH devices via Transmissions. The right energy source is needed to gather electricity from environmental sources; however WET is Wi-Fi, inexpensive, energy-transmitting, and compact (Kamalinejad et al., 2015).

Integrating radar detection and wireless power transfer (ISWPT) into a single hardware platform is our novel approach. The ISWPT system that has been suggested offers several encouraging application scenarios, including intelligent vehicle networks (Liu et al., 2020). This allows for the simultaneous use of charge sensors and radar sensors. ISWPT provides a number of benefits over traditional independently designed systems with such integrated functioning, including:

(i) The integration saves spectrum resources and device costs, which is important for 6G networks in the future.
(ii) The radar beam often covers a variety of locations continuously, allowing nearby equipment to have a steady source of energy. In addition, radar power waste is decreased as a result of the radars' ability to gather energy from radar signals.

We study a unique ISWPT scenario where WPT and radar sensors are on one hardware platform. The multi-antenna transmitter produces one waveform to do both objectives. Radar sensing and WPT share antenna and transmit power resources, therefore their performance trades off. We quantify this performance-efficiency trade-off by enhancing the ISWPT system's transmitted signal beamforming vectors.

The following is a summary of this letter's primary contributions.

(i) As far as we are aware, we are the first to provide the idea of ISWPT, Sharing the same hardware platform, spectrum, power resources, and radar detection allows wireless power transfer. In particular, we examine a generic ISWPT system in which a multiple-antenna transmitter concurrently scans an area and provides energy to numerous energy receivers.
(ii) We analyze the performance trade-off between radar sensing and WPT by optimizing transmit beamforming vectors. The beamforming design problem is non-convex, but we use the SDR approach to solve it globally optimally. We also present a low-complexity sub-optimal beamforming solution to further simplify the ideal design.
(iii) Numerical findings demonstrate the performance trade-off between radar sensing and WPT. Numerical results support the unique ISWPT system's beamforming solutions' efficiency.

Reconfigurable intelligent surfaces (RISs) with passive reflecting features are being considered for cost-effective and energy-saving wireless communication networks (Du. Et al., 2021; Li, 2020; Cheng et al., 2022; Tang et al., 2021). The capacity, efficiency, and other performance of RIS in wireless information transfer have been thoroughly explored. Due to connection quality increase, RIS improves performance (Cheng et al., 2022). “Additionally, (Tang et al., 2021) found that RIS-aided wireless information transmission systems outperformed relay-aided systems due to decreased power usage.

The authors know of few RIS-aided WPT studies other than simultaneous wireless information and power transmission (Wu & Zhang, 2020; Liu et al., 2020). In (Wu & Zhang, 2020) and (Liu et al., 2020), weighted sum power and rate were assessed and maximized. The energy efficiency and quality of service of the RIS-aided simultaneous wireless information and power transfer system were examined in (Wu & Zhang, 2020) and (Liu et al., 2020)”. By adjusting the transmitter beamformer and RIS phase shift, (Pan et al., 2011) optimized the minimal gathered energy at the receivers under transmits power constraints. These studies on RIS-aided WPT systems analyze receiver energy rather than transfer efficiency. However, phase shift power consumption has not been considered in most of the research, which may have a major influence on efficiency and cannot be ignored, especially when the RIS contains many reflecting parts. Since RISs are frequently installed on high-rise building faces, where conventional energy supply is unavailable, their energy source presents a unique difficulty. The use of RIS in WPT is still in its infancy, and several difficulties need additional investigation.

"We present an absorb-then-reflect (ATR) WPT scheme and calculate the system-level power transfer efficiency (PTE) of a RIS-aided WPT system, taking into account RIS and BS antenna array power consumptions. We next compare the RIS scale's power transfer efficiency to the WPT system to assess RIS's feasibility and potential advantages". Significant contributions: Our RIS-aided WPT technology replaces joint beamforming with independent beamforming. Avoiding complicated joint beamforming improves framework efficiency. System performance is analyzed utilizing RIS-aided WPT framework. The PTE closed-form equation and tractable approximations are obtained. Further, the lowest RIS scale required to assure the RIS-aided system's performance advantage over the conventional system is examined. Our ART power transfer approach allows RIS-aided system self-sufficiency. We explored multiple-user power allocation and proposed two methods that combine energy requirement and status.

"The sixth generation (6G) of wireless networks enables Internet of Intelligence, allowing seamless and unlimited connection and complete synchronization of the physical, biological, and digital worlds for new indoor and outdoor applications (Lina, 2020). Mobile nodes with worldwide network coverage (10 Gb/s/m3), terrestrial, airborne, and undersea system integration, and native AI usage are predicted to grow rapidly in the 6G ecosystem. 6G is projected to improve energy efficiency, environmental awareness, security, resilience, digital inclusion, and electromagnetic field radiation awareness in addition to its technical elements". Limited wireless energy makes supplying a large and rising number of intelligent nodes problematic. Although hardware components allow this, future deployments must incorporate energy harvesting to prevent frequent battery changes to decrease human contact and material waste. Therefore, 6G networks should support wireless devices.

Integrating optical wireless communication (OWC) technology into 6G wireless networks for communication and power transfer may achieve the objectives (Lina, 2020). OWC-RF hybrid networking may also assist. Key causes are two:

(i) To enhance wireless network capabilities, additional frequency bands are necessary, and OWC systems are preferable due to their cheap cost, energy efficiency, and technical maturity of optical transceivers (Ghassemlooy et al., 2015).
(ii) Lightwave technology effectively transmits energy without stressing the oversaturated RF spectrum (Diamantoulakis et al., 2018).

LPT devices may transmit power and data like RF WPT (Diamantoulakis et al., 2018). SLIPT might combine OWC systems' capacity with LPT's efficiency and installation. Transmitter radiation pattern determines directed or non-directive SLIPT. Directive SLIPT works on land, air, and sea (Fakidis et al., 2021; Bashir & Alouini, 2022). Non-directive SLIPT works inside (Diamantoulakis et al., 2018). "Transceiver architecture, optical beam propagation, and design trade-offs for 6G SLIPT integration with VLC and FSO OWC technologies are covered in this chapter. SLIPT helps deploy umMTC and "connecting the unconnected" in urban, rural, and remote areas". Intelligent IoT devices and terrestrial, underwater, and aerial nodes like HAPs, UAVs, and small satellites are cost-, energy-, and environmentally-efficient. We conclude with possible obstacles and future research initiatives, including how SLIPT might be integrated with other 6G technologies like AI, reconfigurable intelligent surfaces (RISs), and quantum communications to achieve high energy efficiency and serve more complex use cases.

2. TECHNIQUES TO IMPROVE POWER DELIVERY VIA WIRELESS POWER TRANSFER SCHEME

The wireless power transfer strategy will be introduced in this book chapter to help meet the sustainable goals of 6G communication in order to improve power efficiency, stability, and dependability. The methods for enhancing power supply using wireless power transfer schemes are listed below.

2.1 Robust beamforming for the magnetic based wireless power transfer (WPT)

Power efficiency and reliability are improved by robust beamforming for magnetic-based wireless power transfer (WPT) with many transmitters and receivers. "We first propose a beamforming optimization issue to decrease transmit power while assuring reliable receiver power supplies, taking into account the magnetic mutual inductances information (MII) imperfection. Power efficiency of MRC-WPT systems, which manufacture high-quality electronic components and circuits, is improving (Mou et al., 2019)". Hardware limitations limit how much capacitance and resistance increase performance (Zhao et al., 2019). "In recent years, MIMO-inspired transmitter (TX) and/or receiver (RX) diversity has improved WPT system transmission efficiency and range (Zhao et al., 2020). MRC-WPT with several TXs and one RX has greater stable power transfer efficiency than single-TX (Tan et al., 2016). Multiple TXs can produce harmful magnetic fields through magnetic inductive coupling. To compensate for magnetic flux intensity loss at the receiver point, MRC-WPT systems need high transmit power, reducing power efficiency. The magnetic beamforming problem for MRC-WPT systems with numerous transmitters and/or receivers has been studied to overcome these concerns. By cooperatively regulating TX source currents, (Yang et al., 2017; Lang, 2018; Yang et al., 2016) established constructive magnetic field overlap at RX locations. The direction and strength of transmitter source currents were tweaked in (Yang et al., 2016) to increase MISO MRC-WPT system power output and efficiency. After that introduced a safe magnetic MISO beamforming method to maximize user power and limit unwanted power. In (Yang et al., 2017), the authors studied magnetic beamforming challenges for the MRC-WPT system with an arbitrary number of TXs and RXs using RX feedback mutual inductance information (MII). In (Park & Choi, 2020), transmitter source currents and receiver coil selections were adjusted to reduce transmit power while meeting total charged power and transmit power restrictions. All studies assumed complete knowledge of transceiver MIIs, despite the multiple beamforming algorithms reported to improve power efficiency for MRC-WPT systems under different transceiver power demands and restrictions. Due of restricted feedback (Zhao et al., 2019), model differences (Zhao et al., 2019; Guo & Ben, 2019), and privacy problems associated with MRC-WPT systems, the iterative Monte Carlo sampling and resampling search strategy may not effectively estimate or approximate the MII".

Figure 3. Example of considered system models comprising 5 TXs and 4 RXs

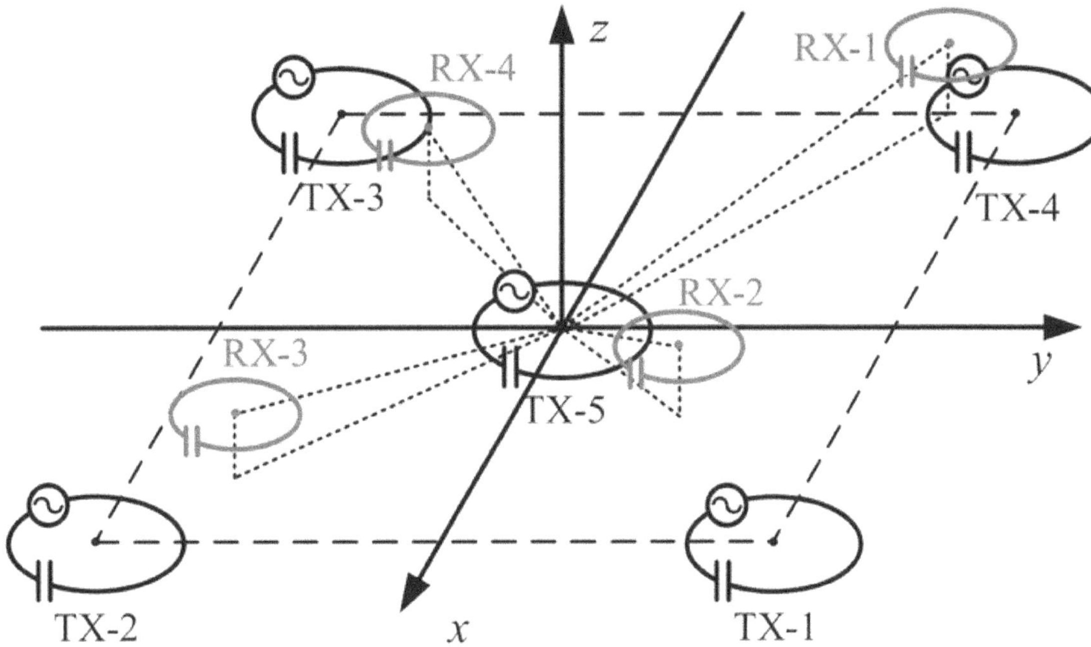

Figure 4. Equivalent circuit model for MIMO MRC-WPT systems

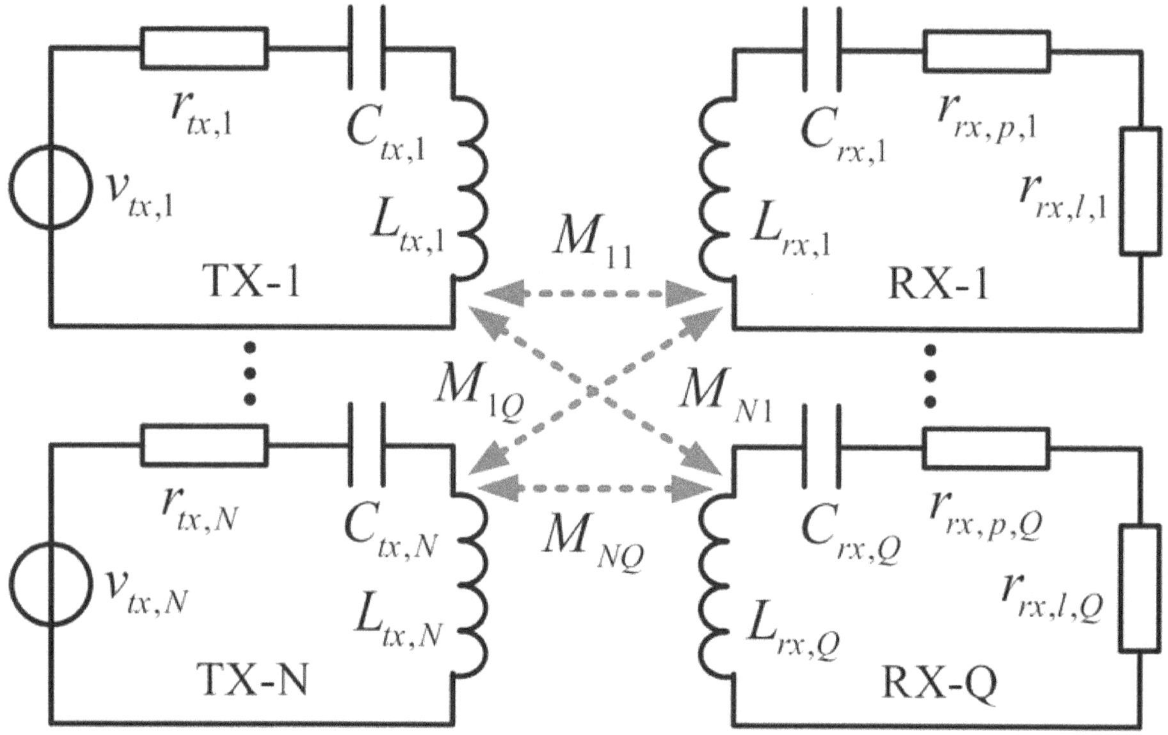

The robust beamforming issue for MISO MRC-WPT systems was investigated in (Sun et al., 2018), assuming norm-bounded error models for unintended receiver mutual inductances and a known MII for the intended user. "Due to inadequate sample or the high communication overhead from the necessity of the perfect MII regarding the intended receiver, these works are not suited for reliable and efficient MRC-WPT system implementation". Instead, they misallocate TX power. As shown in Figure 3, "this research investigates robust magnetic beamforming challenges to improve power efficiency and reliability of multiple TX and multiple RX MRC-WPT systems. We present an optimization problem for beamforming based on incomplete MII across transceiver coils to reduce transmit power while supplying power and efficiency to multiple-RXs simultaneously. The given issue has infinite restrictions and is severely nonconvex, therefore we separated it into two worst-case optimization problems: one for single RX and one for many RXs". For the single-RX case, the beamforming problem is a rank-one semidefinite programming (SDP). Since the problem's semidefinite relaxation (SDR) is tight, convex optimization methods like the CVX can solve it (Zhang et al., 2021). After that, SDR and randomness are used to determine the approximately optimal solution for multiple-RX situations. Numerical results support our beamforming methods. "Figure 4 depicts the MRC-WPT system with numerous TXs and RXs. The transmitter coils are fixed in the 2-D plane, whereas the reception coils are randomly distributed in the 3-D space above the transmitter array plane". Mutual inductances may convert the magnetic field created by multiple TXs at many RXs into voltage and current formulas.

Therefore, “the equivalent circuit model of the considered MIMO MRC-WPT system with N transmitter coils and Q receiver coils is shown in Figure 4. In this figure, $vtx,n(t) = vtx,ne^{j\omega t}$ and $itx,n(t) = itx,ne^{j\omega t}$ indicate the input sinusoidal voltage and supply current for the nth TX, respectively, where $n = 1, \ldots, N$. The time-varying magnetic flux generated by N TXs passes through the RXs and induces time-varying currents on them, where the induced current for the q-th RX is denoted by $irx,q(t) = irx\text{-},qe^{j\omega t}$ with $q = 1, \ldots, Q$”.

In Figure 4, rtx,n, Ltx,n, and Ctx,n are defined as the resistance, self-inductance, and capacitance for the nth TX, respectively

Following are the key contributions of this technique.

(i) We use norm-bounded error models to minimize transmit power while meeting deliverable power restrictions for multiple RXs, taking into account the imperfection of observed or predicted MII. We believe this is the first attempt to incorporate magnetic channel information error into the magnetic beamforming issue for magnetic-based WPT systems with multiple TXs and RXs.
(ii) The nonconvex magnetic beamforming optimization issue in the single-RX situation is securely approximated and turned to a worst-case optimization problem using the Cauchy-Schwarz inequality and S-procedure. We demonstrate that this problem's SDR is tight, allowing SDP to best solve it.
(iii) Slack variables are introduced to simplify the problem of many RXs. “Using the SDR and randomization-based approximation technique; a new linear programming (LP) issue that CVX can effectively solve is created to find the approximate optimal solutions for the original problem. Numerous numerical results confirm our theoretical analyses and show that our robust magnetic beamforming schemes can improve energy efficiency and reliability for magnetic-based WPT systems compared to the identical current scheme and the non-robust scheme”.

2.2 Transfer learning (TL)

In order to assist new activities become quicker and more efficient, transfer learning (TL) enables new tasks/domains to learn from experienced tasks/domains. With the correlation and similarity information between various jobs in numerous wireless communications domains, TL may assist reduce energy consumption and increase efficiency. This essay attempts to investigate two significant scientific difficulties using the aforementioned analysis in conjunction with 6G application and the benefits of TL (Wang et al., 2021).

(i) Why is TL necessary for 6G wireless communications?
(ii) In 6G wireless communications, how is TL used?

To the best of the author's knowledge, this is the first paper to specifically examine how TL is used in wireless communications and to investigate how TL will be implemented in upcoming 6G communication networks. In particular, the following is a summary of this article's primary contributions.

(i) The needs and anticipated technologies for the next generation of 6G development are outlined, along with five key research questions: high efficiency, seamless integration, innovative technologies, and secure green communication.

(ii) Analysis is conducted of common TL techniques used in wireless communications, covering both their general classification and basic logic. These techniques fall into two main categories: tool model TL (which includes classifier TL, deep network TL, adversarial TL, and reinforcement learning (RL) based TL) and data knowledge TL (which includes instance-based TL, feature-based TL, and transfer kernel learning).
(iii) Analysis of the literature pertaining to TL's application to wireless communications is provided; this serves as a crucial foundation and point of reference for TL services related to 6G in the future. We examine from the following angles: TL for a variety of applications, including wireless location, intrusion detection, content popularity prediction, cognitive radio network spectrum allocation, base station/access point (BS/AP) switching energy efficiency, and more.
(iv) Just as TL technology and the 6G era are mutually reinforcing, so is technology development and era demand. In this article, the issue is compared to a tree, with demand acting as the leaf and technology as the base. Demand will arise as technology becomes more widespread. Since these sophisticated TL techniques have not previously been used in the world of wireless communications, this article's interpretation will present prospects for future study on certain potential technical elements of TL that may be facilitated by 6G. Simultaneously, it is elucidated that TL might potentially aid in the settlement of some challenging 6G issues.
(v) With the gathering and public availability of large actual word datasets, sophisticated similarity analysis, and privacy protection throughout the TL process, the future obstacles and the problem that needs to be solved about the uses of TL in 6G wireless communications are suggested.

Figure 5 depicts the potential TL situations in 6G wireless communications. These examples are thought to represent only a small sample of the TL applications that will be available with 6G. Actually, a great deal of early study has been done to investigate these possibilities.

Figure 5. Possible scenarios of TL in 6G wireless communications

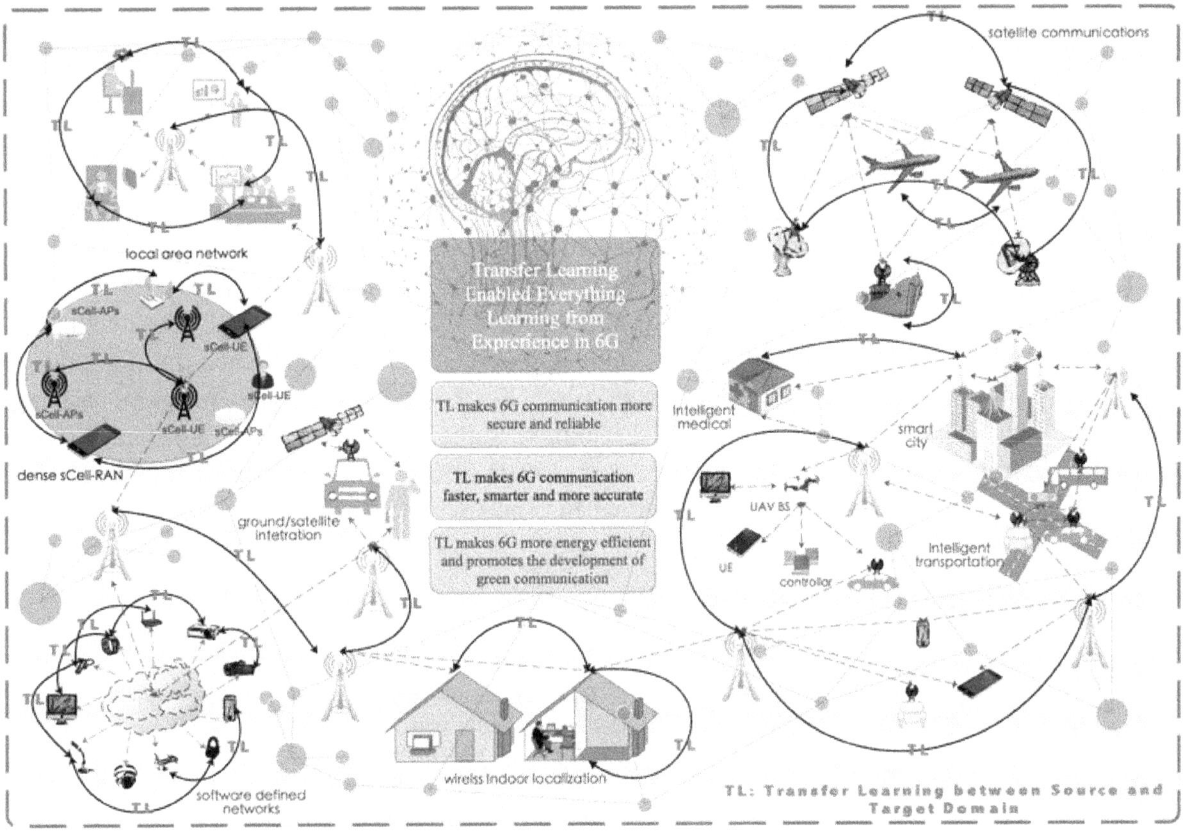

(a) *TL in Wireless Indoor Localization*

Indoor localization is a 6G use case. TL is used to Wi-Fi-based indoor localization system signal strength (RSS) data to validate its outstanding transmission value (Pan et al., 2011; Uguroglu & Carbonell, 2011). Future indoor wireless localization will be more effective owing to experience transfer between sites or localization operations at various times in the same location.

(b) *TL in Smart Societies*

The benefits of 6G will accelerate smart society, increasing smart medical, transportation, city, and UAV relay BS. Smart phones, remote control, autonomous vehicles, and 6G's extensive wireless connectivity will make society smarter. There will be lots of data to evaluate. The trained model must be sent to 6G smart societies. Modeling various users' needs is difficult even on the same project, necessitating model adjustment.

(c) *TL in Satellite-Ground Integrated Communications*

Transmission delay, outage likelihood, and asymmetry are high due to far space nodes and low channel quality. TL study assisted BSs/APs convert (Li et al., 2014; Sharma et al., 2017; Sharma et al., 2016; Sharma et al., 2019) and cache/recommendation system cold start (G. et al., 2014; Ba. Et al., 2015; Bharath et al., 2016; Wang et al. 2017). TL development to improve satellite ground integrated communications began with these attempts.

(d) *TL in 6G Wireless Communication Networks*

Stronger 6G wireless networks are needed. Software-defined networks, dense Cell-RAN, and local area networks require resource allocation, mutual coordination, and experience transfer between high-density and heterogeneous smart devices and users. Because scattered cognitive agents exchange experience, new subjects can join networks and acclimate to the radio environment faster in TL-based dynamic spectrum (Zhao et al., 2015). Radio resource management knowledge transfers for multihop backhaul networks encompass where, what, and when (Zhao & Grace, 2024).

Figure 6. Requirements and technologies of 6G

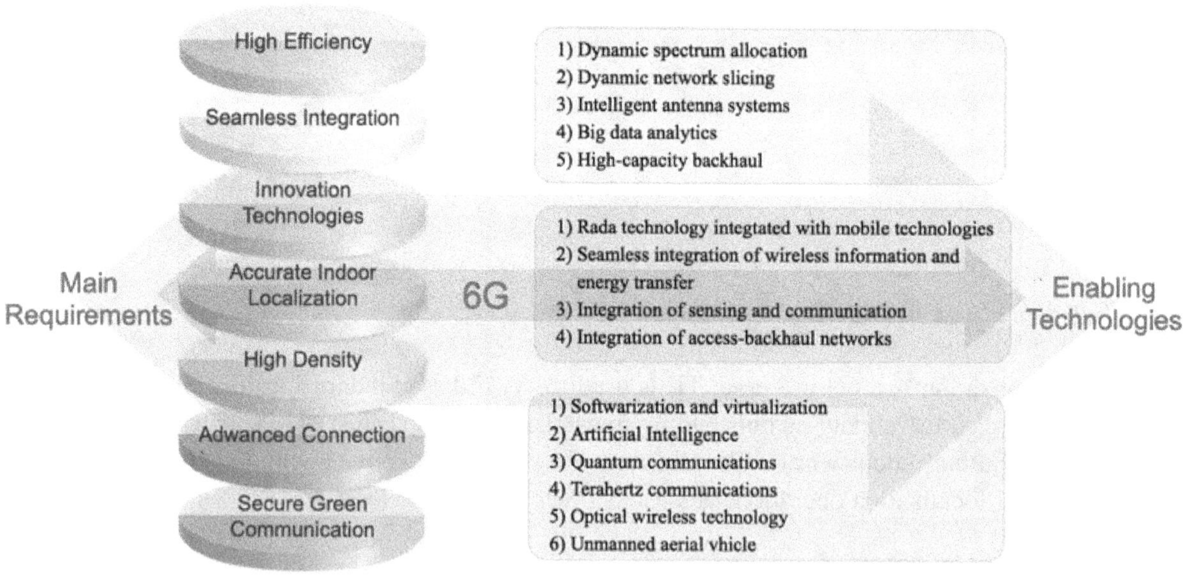

TL would improve 6G security, reliability, speed, smartness, accuracy, energy efficiency, and green communication.

2.2.1 Mutual Relationship between TL and 6G Communications

Thinking creatively to imagine new technologies is necessary to manage the expanding demand and consumption in light of the quantity, complexity, and performance requirements of communication systems. Reductions in learning time and energy usage are possible because TL is thought to have made further system efficiency possible in recent years. One good reason to consider TL as a possible technology for 6G wireless communications is its effective implementation in wireless communications, as previously discussed. In 6G communications, there are a number of issues that might be solved with the help of TL. Additionally, as Figure 7 illustrates, there may be a comparable reciprocal connection between them.

Figure 7. Tree illustration of mutual aid between TL and 6G

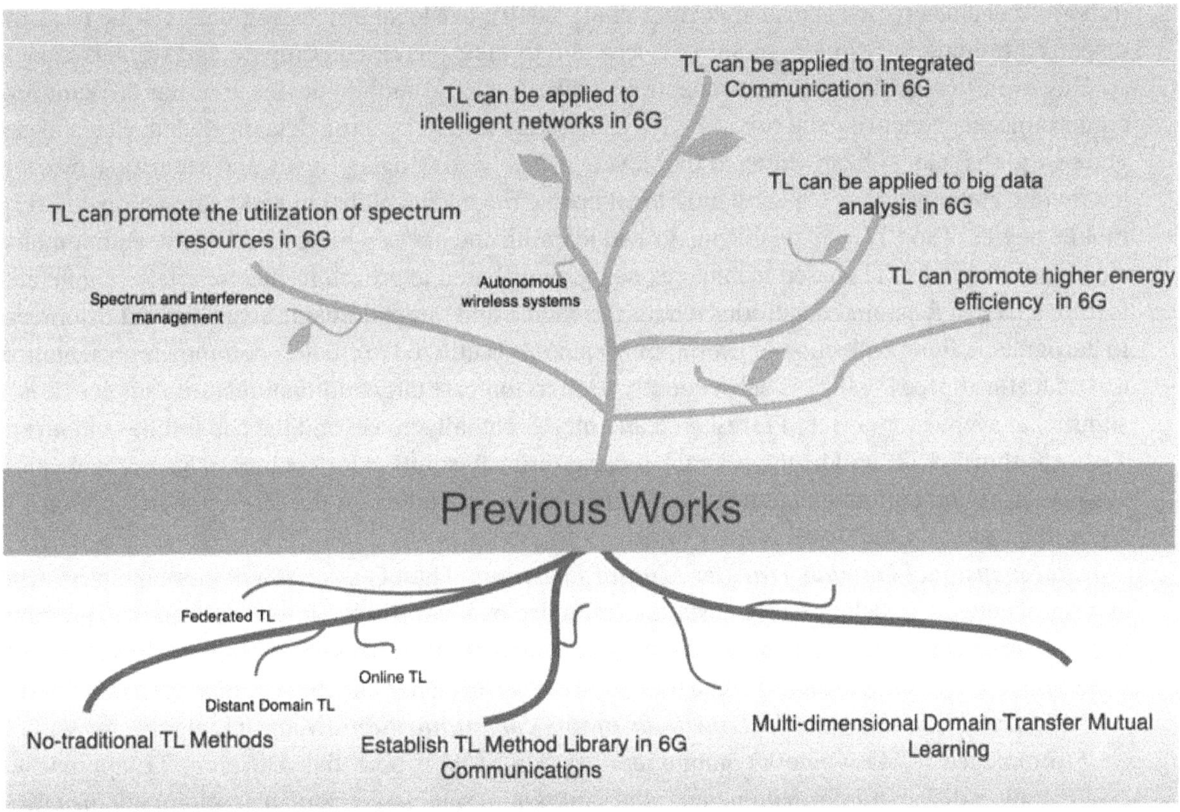

As 6G communication and TL evolve, sophisticated TL techniques will be more equipped to tackle challenging issues. These are a few potential avenues for advancement.

(i) ***Distant Domain TL (DDTL):*** In stark contrast to typical TL, DDTL (Tan et al., 2017) permits the source domain and the target domain to have no close link at all. In fact, they can be entirely distinct. Numerous heterogeneous communication technologies, including frequency bands, communication topologies, and service delivery, will be used in 6G. Furthermore, the mobile terminal's and the AP's hardware configurations will differ greatly. It will be difficult to integrate all communication technologies into a single platform as a result. By using DDTL to this vast heterogeneous platform, it will be possible to prevent the loss brought on by repetitive training in addition to tightly connecting heterogeneous devices to learn from one another.

(ii) ***Online TL:*** It is typically taken for granted that the data from the source and target domains are available and that a straight transfer may be completed. The actual application, however, is frequently not like this: possibly at first, just the data from the source domain and the destination domain are regularly delivered in little amounts. "Online transfer learning" is the term for this (Zhao et al., 2014). Many 6G applications, including dynamic network slicing, cell-free communication, and integration of sensing and communication, depend heavily on online real-time processing.

(iii) ***Federated TL (FTL):*** Federated learning (FL) includes FTL. With FL, a distributed machine learning technique, users may work together to jointly develop shared models while keeping their personal information private and save the gathered data locally on the device. One of FL's most important characteristics is privacy (Yang et al., 2019). 6G local processing data can be used for dispersed training on wireless mobile devices due to their increased compute and storage capacity. However, unreliable wireless connections make scattered mobile device training difficult and compromise data security and privacy. "FL technology retains training data on each device and develops a global model from scattered mobile devices, boosting data privacy and security. However, bandwidth and processing capacity limit local update model and global model aggregation on every mobile device. The FTL will enable integrated learning and privacy protection in vast, autonomous, close-knit networks. TL-based techniques can give a shared approach to the complete sample and feature space in federated conditions where the source and target domains share limited resources. In particular, a finite collection of overlapping samples is utilized to create a common representation across feature spaces, which is subsequently used to forecast target domain feature samples. It is a significant avenue for current FL research and may eventually go beyond the capabilities of current TL algorithms". FTL will be the primary security/privacy solution for cooperative communication systems in future communication networks, according to studies on the challenge of training FL algorithms across actual wireless networks.

(iv) ***Multidimensional Domain Transfer Mutual Learning:*** The 6G system will combine terrestrial and aerial networks, such as 3-D base stations provided by low-orbit satellites and drones, to promote vertical growth user communication. Three-dimensional network rises. An additional dimension has been added. Cross-dimensional TL can boost system performance and dimensional communication.

(v)Establish a library of TL methods for 6G communications. Many TL algorithms will be employed in 6G wireless communications due of their potential. Different TL approaches convey different information under the same source and target domains, improving the target domain differently. Researchers find manually selecting TL techniques laborious and imprecise. "A learning to transfer (L2T) technique (Ying et al., 2018) optimizes transfer learning (TL) from a source to a target domain by using past TL experiences to optimize what and how to transfer. Based on the above, we propose a transfer learning (TL) library for future wireless communication needs. We will next use the L2T algorithm to automatically match the library's best TL technique to environmental needs for adaptive TL in a complicated wireless communication environment".

2.3 Emphasis on the viability of Channel state information techniques

WET uses the far-field irradiative features of electromagnetic waves to power devices over shorter or relatively modest ranges. Due to recent improvements in MIMO innovation and the ongoing decrease in microwatts of device running energy, we expect a rise in WET applications in the upcoming years. We believe Interconnects will benefit from WET by making use of (Niyato et al., 2017):

(i) Wireless battery recharging, which makes servicing and fixing battery powered devices more easier.
(ii) A reduction in the equipment's final form factor.
(iii) An increase in the durability and reliability of the terminal device because of its interface design.
(iv) Increased energy efficiency and reduced carbon impact across the network.

WET and WIT readily coexist since wireless information transmission (WIT) is a key element of the Internet of Things (IoT) concept, as shown in Figs. 8 and 9. Consequently, there are two main architectural styles (Alsaba et al., 2018): wireless powered communication network (WPCN), in which WET occurs in the first phase's downstream and WIT occurs in the second phase.

Figures 8 and 9 show typical use cases for the aforementioned architectures in the IoT age; however, real networks may combine the aforementioned schemes (Alcaraz et al., 2017).

Figure 8. IoT architectures with WET applications

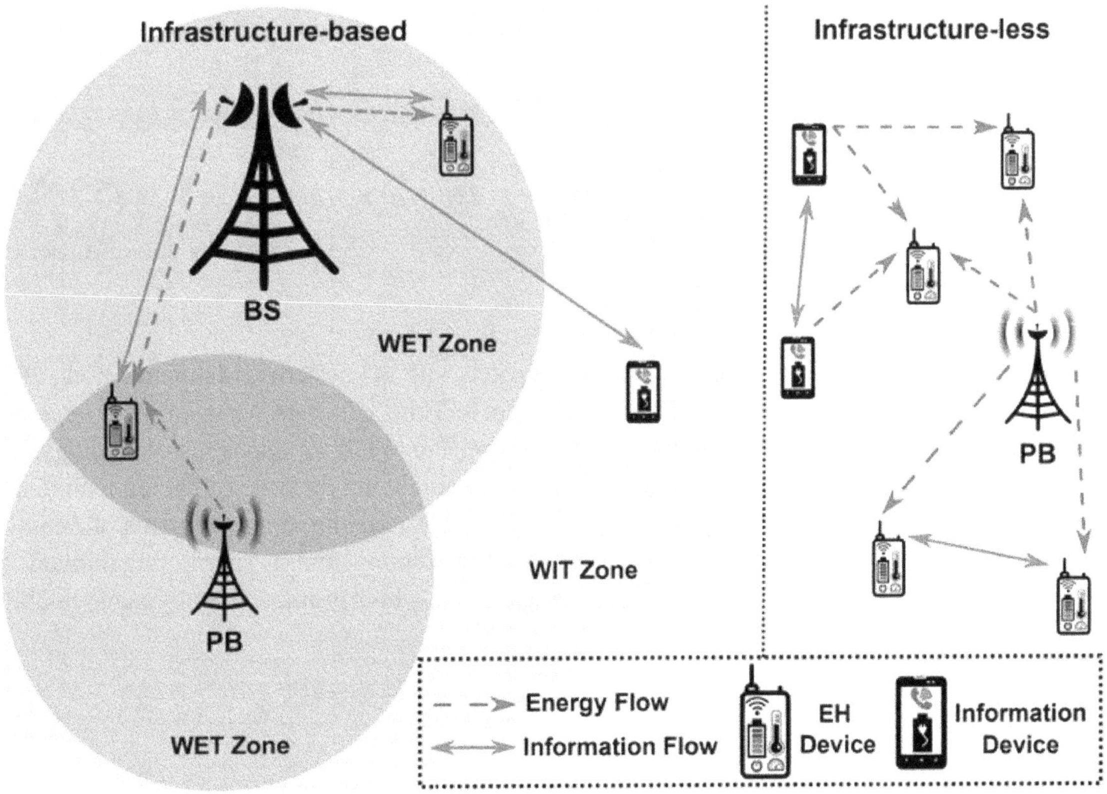

Figure 9. Out-of-band and in-band WET schemes. Advantages and disadvantages. Example of in-band WET using successive interference cancellation (SIC) for removing the interference caused by WET signals to the information decoding devices.

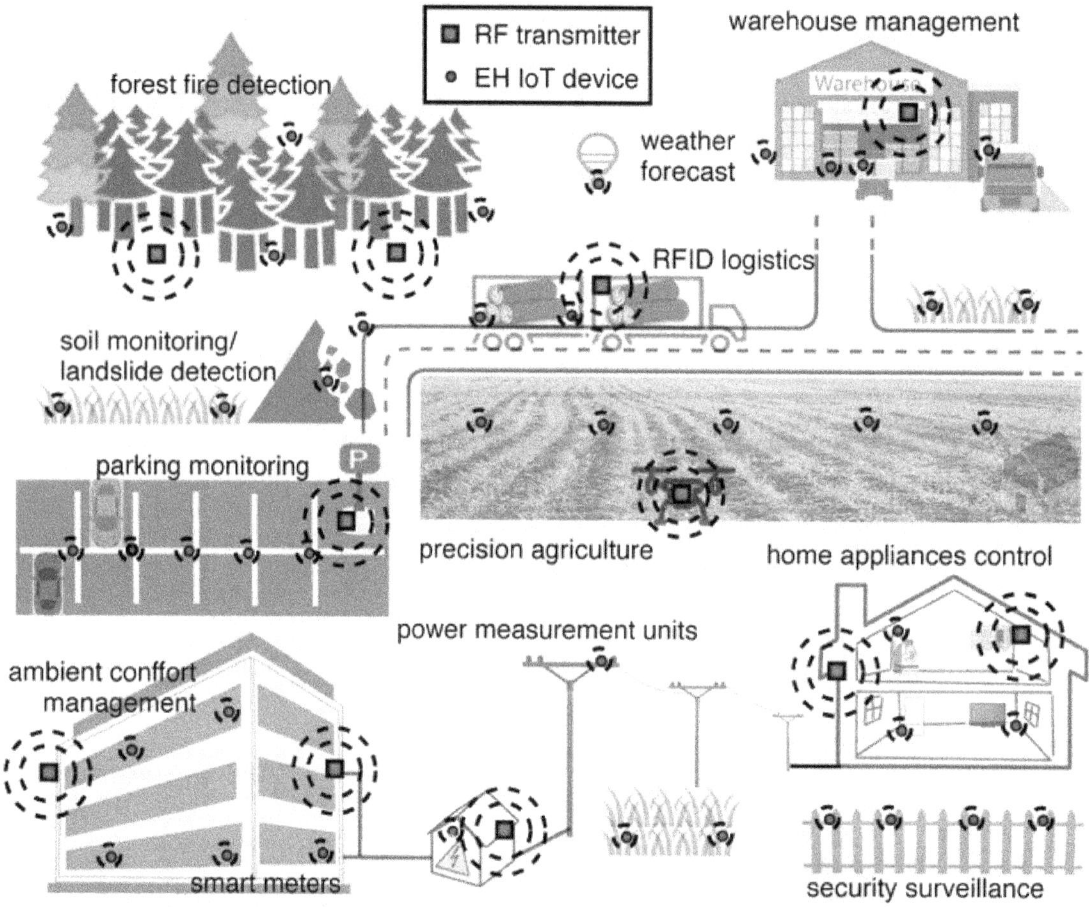

Up to 10 devices/m3 of huge connection are anticipated with 6G systems (Mahmood et al., 2020), the majority of which will be low-cost, low-power Internet of Things devices. Many of these large-scale IoT networks may be powered wirelessly and sustainably using WET, an appealing technology. Take note that WET promotes network sustainability through the following means: (i) making maintenance and servicing easier; (ii) making IoT devices more durable; (iii) assisting in the reduction of emissions across the network by addressing the issue of battery waste processing; and (iv) permitting miniaturized hardware implementations and supporting multi-user operation, which promotes deployment scalability.

(i) IoT WET Applications:

Owing to the above inherent advantages of RF EH and combined with the dwindle in devices' operating power and the developments in multiple input multiple-output (MIMO) technology, an increasing number of IoT WET applications is expected towards the 6G era. Note that optimizations for one application may be detrimental to another, sacrificing input power range for conversion efficiency, size for durability, or complexity for cost in ways suited only to a specific task (Soyata et al., 2016). Thus, applications requirements must be fully considered to ensure circuit trade-offs are given proper weight in the final product.

Figure 10. Selected WET-powered IoT use cases, e.g., monitoring/control systems in smart homes; transportation, healthcare, smart grid, weather forecast, emergency detection/response, soil monitoring and precision agriculture.

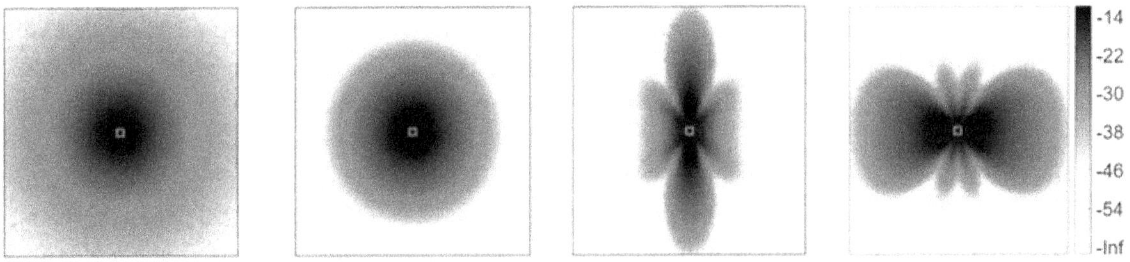

RFID is a crucial technology used in remote applications for inventory management, tracking, and identification. In comparison to conventional designs, the longevity, usefulness, and/or operation range of RFID tags also known as active or hybrid RFID tags are significantly increased by the incorporation of RF EH (Lu et al., 2015).

(a) ***Live labels:*** Information like price, flash sales, bar codes, and freshness index may be included on the labels of future items (found in stores and marketplaces). Simple customer comments may even be detected and forwarded to the retailer or producer. These days, tags with optical access points and light-based EH in ceilings are being studied (Katz & Brien, 2019). To prevent blockage, WET or hybrid designs could be utilized.

(b) ***Wireless sensor networks (WSNs):*** they have uses in the smart home, healthcare, business, and military, among other areas. One potential strategy to extend the life of WSNs5 is to use RF energy to replenish or eliminate the need for batteries. For example, writers in (None et al., 2016) suggested using WET to assist with building structural monitoring. In particular, far-field radiofrequency electro humidity (RF) sensors were created to measure a building's temperature, light, and humidity. The sensor node received 3.14, 2.88, 1.53, and 0.7 mW of power through air, wood, 2 inches of brick, and 2 inches of steel, respectively, at a distance of 1 m from a 3 W source. These results are adequate to power multiple cutting-edge sensors.

(c) ***Wake-up radios (WURs):*** A passive RF EH chip may still be utilized as a WUR that produces a wake-up pulse in response to a command from a nearby transmitter (Soyata et al., 2016). This is useful in situations when the device has higher energy requirements and cannot run only on RF energy sources. The lifespan of the device is undoubtedly increased by using a totally passive WUR, as the active part of the sensor will only be active for brief intervals of time and won't have to actively listen for orders when the device is idle.

(ii) Green WET:

There's no denying that WET encourages sustainability. The idea of a completely sustainable Internet of things, however, is not precisely consistent with the conventional understanding of WET-enabled networks, where PBs and hybrid BSs are dependably fueled by conventional energy sources, such as carbon-based energy. Rather, WET may be made more environmentally friendly by utilizing energy derived from renewable sources to power hybrid and PBs (Tran & Kaddoum, 2018; Liu & Ansari, 2019). There are two primary approaches to doing this: (a) Passively, in which case no locally accessible green energy sources are utilized. Either (b) actively, where the green energy is captured locally by the WET provider's equipment, or (c) at the industrial side, where it is harvested and delivered to the WET transmission infrastructure, requiring particular green contracts with the energy provider (Ghesla, 2017). Such energy may be used, for instance, by PBs linked to outside solar panels to wirelessly recharge interior RF-EH devices.

"The crucial issue of CSI acquisition in multi-user WET systems restricts the applicability of any research project that depends on flawless CSI. Figure 11 and 12. The more EH devices there are, the worse the situation gets. Scheduling methods may be required to minimize interference and collisions during training, consuming significant energy resources that are expensive for devices with limited energy resources. More training pilots could be required as the network expands in order to get the entire network CSI within each coherence time interval". Traditional training becomes ineffective or pricey in low-power huge IoT installations as the number of training pilots utilized for CSI acquisition increases.

Figure 11. Heat map of the average harvested energy in dBm under the CSI-free WET schemes: APS – EMW (left), AA – IS, SA (middle-left), AA – SSmin var (middle-right), AA – SSmax E (right).

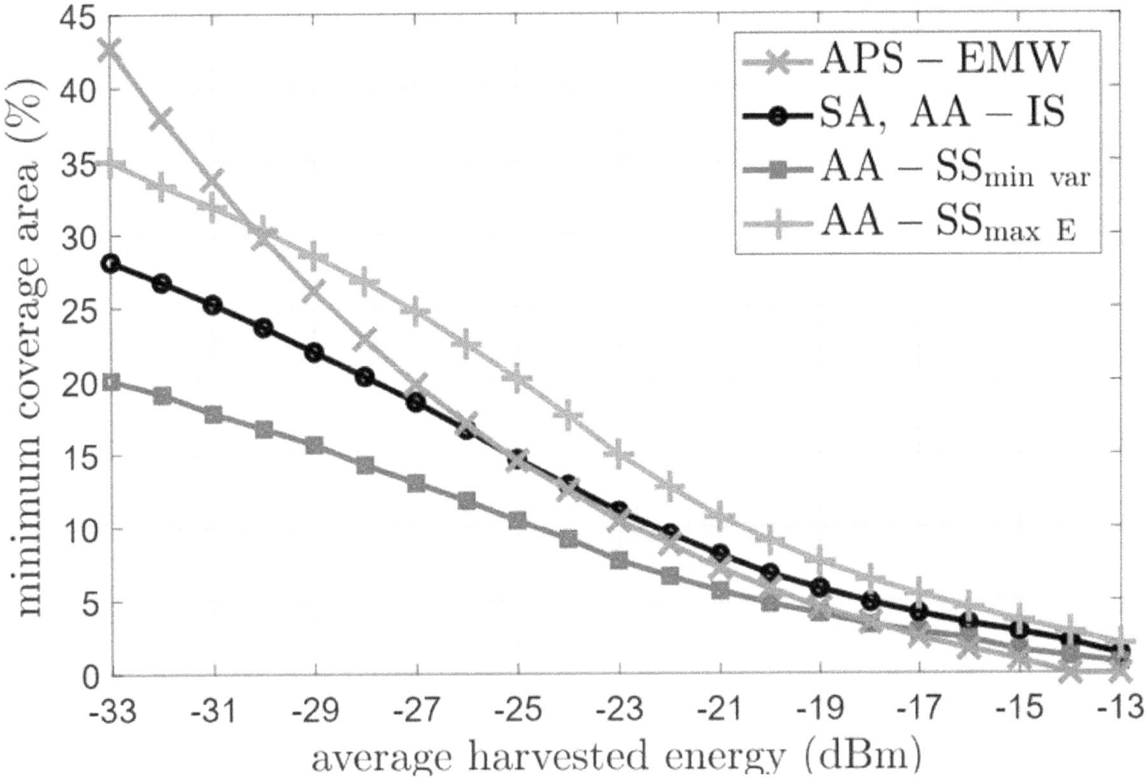

Figure 12. Area coverage for different average harvested energy requirements

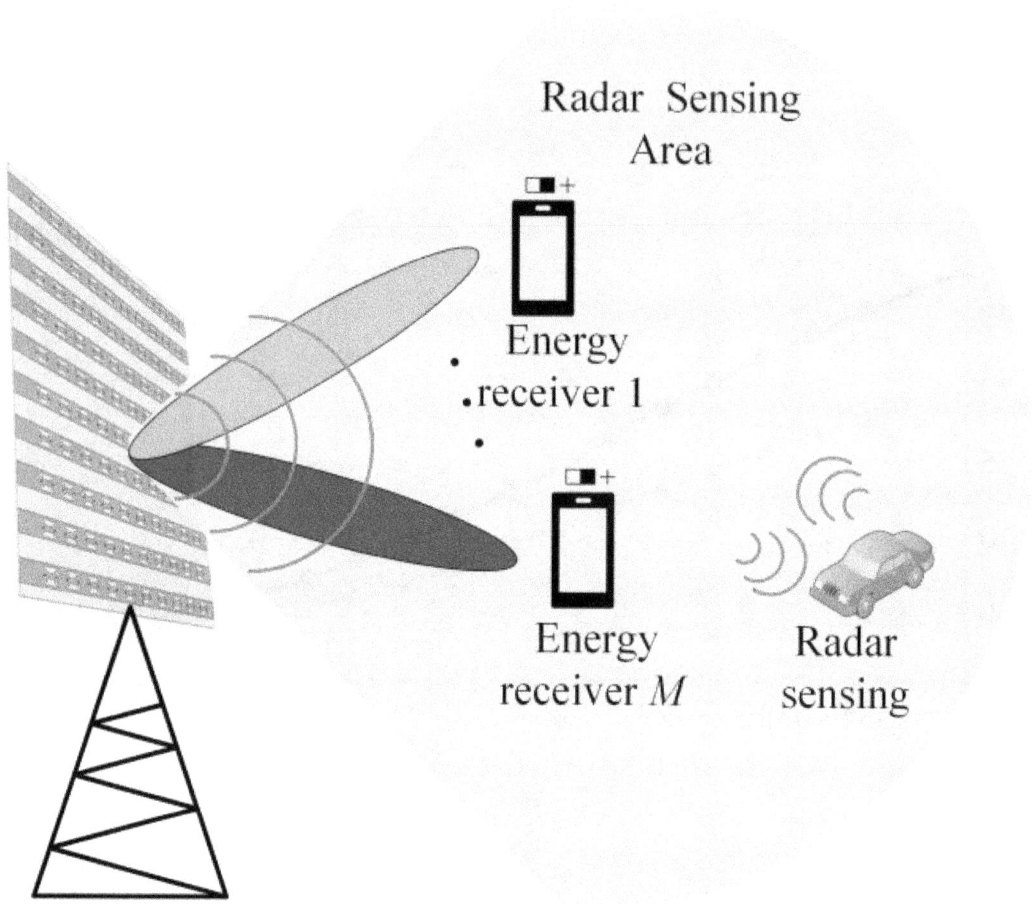

Wireless powering efficiency decays power-law with distance, which may aid. Since EH devices are farthest from the PB(s), their channels will dominate the beamforming design (Lopez et al., 2020). As shown in (Lopez et al., 2020), not doing CSI acquisition for users closer to the PB reduces channel training. The most remote and energy-constrained customers would still need to pay a lot on channel training. The performance of CSI-based systems gradually declines as user numbers increase (Lopez et al., 2019). In large-scale deployments, wireless signals' broadcast nature must be strategically used. This is especially true when powering several devices without CSI.

2.4 Concept of integrated sensing and wireless power transfer (ISWPT)

ISWPT integrates radar detection and wireless power transmission into one hardware platform. For future 6G wireless networks, ISWPT incorporates system size, hardware cost, power utilization, and spectrum conservation. Intelligent vehicle networks (Liu et al., 2020) can employ ISWPT to combine radar and charge sensors. ISWPT offers many benefits over traditional designs: "It conserves device cost and spectrum resources, which is vital for future 6G networks; the radar beam generally offers continuous coverage over a range of places, ensuring a reliable energy supply to equipment; and radar signal gathering decreases radar power waste". This novel ISWPT scenario combines radar detection and WPT on one hardware platform. One waveform is created by the multi-antenna transmitter for both purposes. Since radar sensing and WPT share antenna and transmit power, Figure 13 demonstrates a performance trade-off. We improve the ISWPT system's transmitted signal beamforming vectors to quantify this crucial performance trade-off.

Figure 13. Illustration of one ISWPT scenario, where one multi-antenna array simultaneously senses a range of area and charges multiple energy receivers

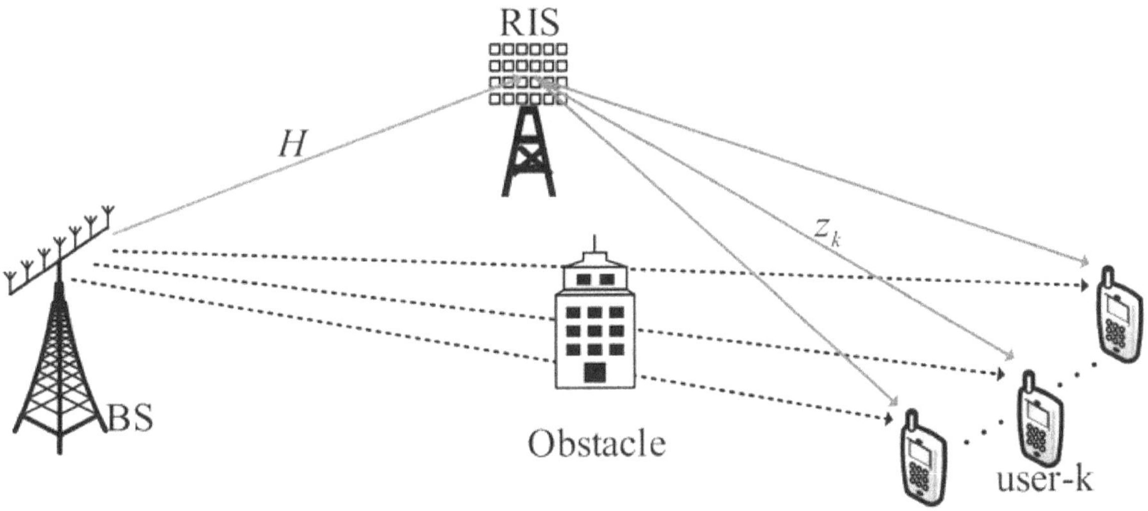

2.5 Reconfigurable intelligent surface (RIS) in the wireless power transfer (WPT) system and designs an efficient RIS-aided WPT scheme

Radiative wireless power transfer (WPT) using electromagnetic waves has garnered interest for increasing the battery life of energy-constrained terminal equipment such wireless sensor nodes (Psomas & Krikidis, 2019; Hu et al., 2021). A viable energy-efficient WPT system is needed because electromagnetic waves traveling in free space are severely attenuated (Yan et al., 2020; Zeng et al., 2017). Multiple antenna technology has been shown to improve energy transfer efficiency in radiative WPT systems (Zhang et al., 2021; Kashyap et al., 2015). "To assure energy outage probability, multiple antennas and beamforming can provide more energy to the receiver and use less energy at the transmitter (Zhang et al., 2021). WPT range may be improved using antennas. WPT systems with large antenna arrays improve transfer range and point-to-point efficiency (Huang & Clerckx, 2017)". Large antennas are impractical due to power consumption and radio-frequency chain hardware limitations (Yang et al., 2018).

A reconfigurable intelligent surface (RIS) with passive reflecting sections might save money and energy in future wireless communication networks (Du et al., 2021; Li et al., 2020; Cheng et al., 2022; Tang et al., 2021). RIS wireless information transmission capacity, efficiency, and other characteristics have been studied. RIS improves performance by improving connection quality (Huang et al., 2019). "In comparison, RIS-aided wireless information transfer beat relay-aided due to lower power usage (Bjornson et al., 2020).There are few RIS-aided WPT research other than simultaneous wireless information and power transfer, according to the authors. (Wu & Zhang, 2020a; Pan et al., 2020; Wu & Zhang, 2020b; Liu et al., 2020). (Wu & Zhang, 2020a) and (Pan et al., 2020) analyzed and maximized weighted-sum-power and rate. The energy efficiency and service quality of the RIS-aided simultaneous wireless information and power transmission system were studied. Under transmit power limits, the authors maximized receiver minimum collected energy by modifying the transmitter beamformer and RIS phase shift (Zhao et al., 2021). (Mohjazi et al., 2022) Checked the battery recharge time of a radio frequency energy harvesting node in a RIS-aided single input single-output WPT system with a single antenna at the power source and nodes".

(i) RIS-aided WPT studies focus on receiver energy rather than transfer efficiency. Most research has neglected phase shift power consumption, which may affect efficiency and cannot be ignored anymore, especially when the RIS includes several reflecting elements. Energy supply is tricky for the RIS since it is often put on high-rise building facades without lined energy supply. Thus, RIS in WPT is young and many aspects need further study. This paper presents an absorb-then-reflect (ATR) WPT scheme and analyzes the system-level power transfer efficiency (PTE) of a RIS-aided WPT system, taking into account RIS and BS antenna array power consumptions. The effect of the RIS scale on power transfer efficiency is compared to the normal WPT system to establish the feasibility and potential benefits of RIS in WPT. The main contributions: Our unique RIS-aided WPT system replaces cooperative beamforming with independent beamforming. The proposed framework is more efficient because it avoids complex joint beamforming.
(ii) We evaluate system performance using "the proposed RIS-aided WPT architecture". We develop the PTE and provide its closed-form equation and tractable approximations. Additionally, "the lowest RIS scale required to assure the RIS-aided system's performance advantage over the traditional system is assessed. We suggest ART power transfer for RIS-aided system self-sustainability".
(iii) We analyzed multi-user power allocation and proposed two methods that combine energy requirement and condition.

The study examines an efficient, self-sustaining RIS-aided WPT system.

Figure 14. System model

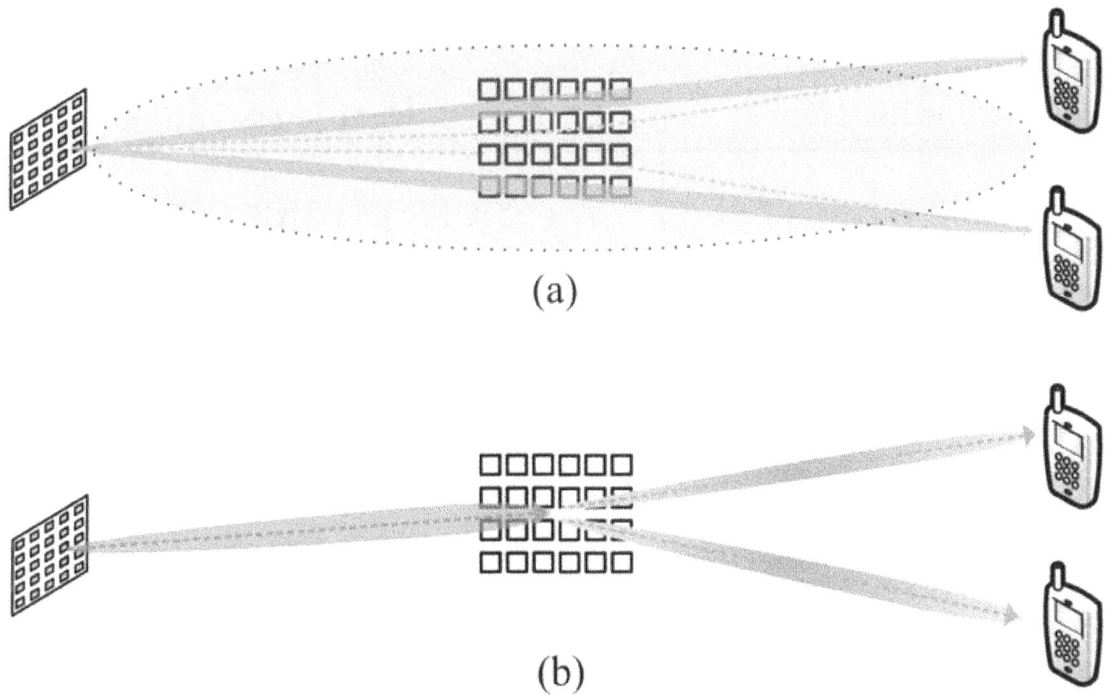

"The RIS plane is positioned between the base station (BS) and users in the WPT system to aid in power transmission. A uniform planar array (UPA) with M antennas is installed on the BS, and a UPA with N passive components makes up the RIS, as shown in Figure 14. N = Nx × Ny represents the number of rows and columns, respectively. Users with a single antenna that can gather energy from the RIS or the BS total K".

Figure 15. Proposed independent beamforming vs. the traditional joint beamforming. (a) Joint beam-forming. (b) Independent beamforming

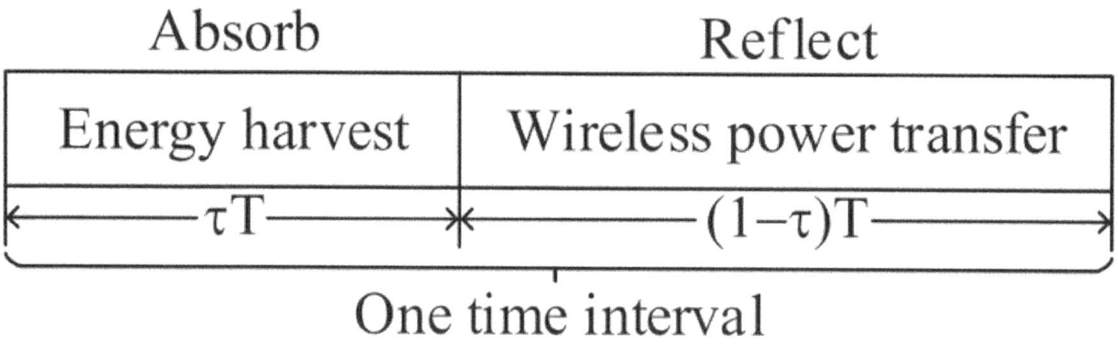

"Only line-of-sight propagation is studied in this study for these reasons: The WPT system has a limited valid distance and no rich scatters between the BS and users. The non-line-of-sight links have a signal strength 20 dB lower than the line-of-sight path (Lee et al., 2016; Rappaport et al., 2012)". The energy conversion unit has sensitivity above -10 dBm (Kamga & Aissa, 2019; Lu et al., 2015). Non-line-of-sight pathways are ignored for these three reasons.

The incident signal at the RIS is given by

yRIS =

$\sqrt{Pt}$

Hx + nRIS (1)

The received signal at the k-th user is given by

$y_k = z_k$

$\widehat{\Phi}$

$y_{RIS} + n_k$ (2)

The RIS may consist of a large number of elements, in this paper, the RIS power consumption will be taken into consideration. We denote

$P_{RIS} = NP_u$ (3)

Where P_u represents the power consumption of each passive element that depends on the RIS structure and phase resolution (the typical value of P_u is 0.3~8 mW) Thus, the power consumption of the RIS-aided WPT system can be formulated as

$$Ptotal_{RIS} = Pt + MP_a + NP_u \quad (4)$$

While in the traditional system, the power consumption does not include the RIS part, and is given by

$$Ptotal = P_t + MP_a \quad (5)$$

The difficulty of joint beamforming escalates with the scale of the RIS and the resolution of the phases due to its reliance on iterative optimization. This study presents an autonomous beamforming technique including two stages, as seen in Figure 15.

Figure 16. The time-switching based ATR scheme

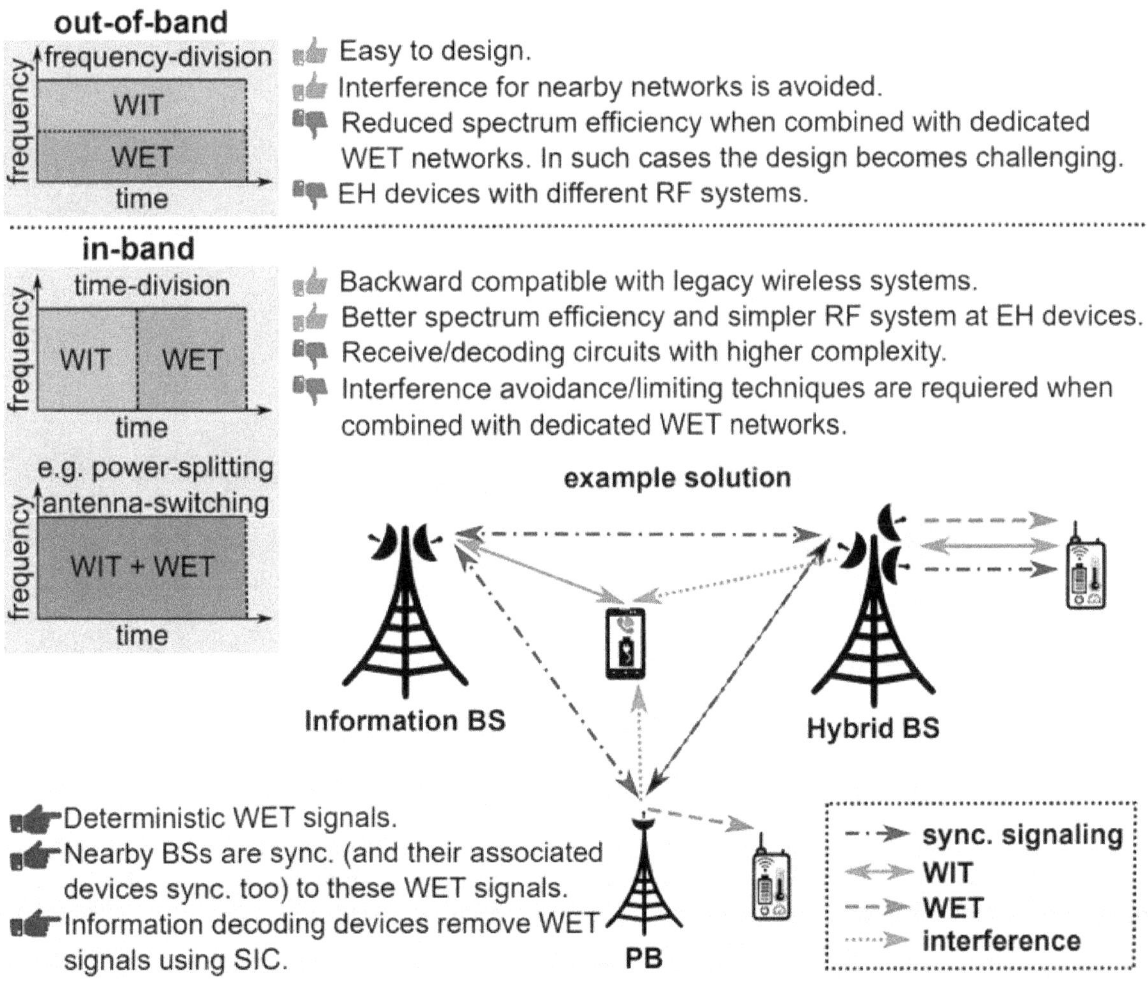

RIS power usage is considered. "External energy supply (EES) is used in RIS research. However, more energy sources will hinder RIS applications. In high-rise buildings without direct energy, the RIS will be mounted on walls or rooftops. This study suggests eliminating the EES via time-switching. Given that the RIS components may absorb incident electromagnetic waves (Wang et al., 2021; Renzo et al., 2020), a time-switching method for energy harvesting and power transmission at the RIS is implemented". The time interval is separated into two sections, as seen in Figure 16.In the first stage, the RIS harvests energy from the BS, where τ ($0 < \tau < 1$) denotes the time switching factor. In the second stage, the RIS forwards energy to the users.

Figure 17. A paradigm of power allocation schemes. (a) Power allocation.(b) Power state.

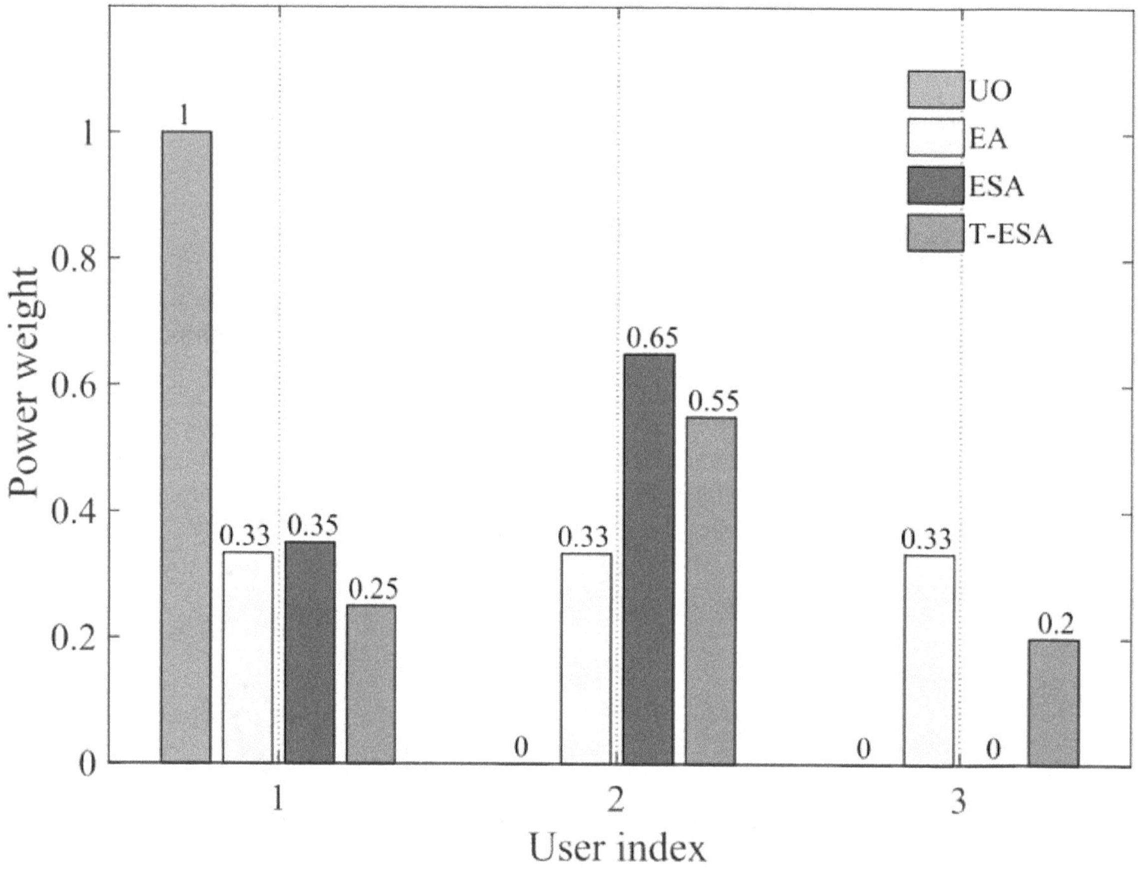

(a)

"The proposed RIS aided WPT system is practically valued, since the deployment of RIS plane is more convenient and flexible than the BS antenna array. A three-user scenario with the IES (0.55, 0.15, 0.3) is exemplified, namely $\varepsilon 1 = 0.55$, $\varepsilon 2 = 0.15$ and $\varepsilon 3 = 0.3$. The RIS scale is 5000 and the transmit power is 2 W".

"Figure 17 shows the power allocation weight and post-allocation using the suggested systems (Lu et al., 2015) and (Katz & Brien, 2019), which are energy state aware (ESA) and T-ESA. The utility optimization (UO) (Liu et al., 2020) and equal power (EA) allocation schemes are compared. This example sets the minimum T-ESA threshold at 0.2.The suggested ESA and T-ESA methods can better balance users' real-time energy states. In particular, the suggested systems allocate power first to user-2, the user with the worst energy status, and then to the user with the shortest distance, whereas the UO strategy allocates all power to the nearest user. The T-ESA technique also allocates energy to users who are neither closest to the RIS nor in the poorest energy condition but have energy levels below the threshold. This somewhat solves the short edge effect".

Figure 18. Energy utility after power allocation

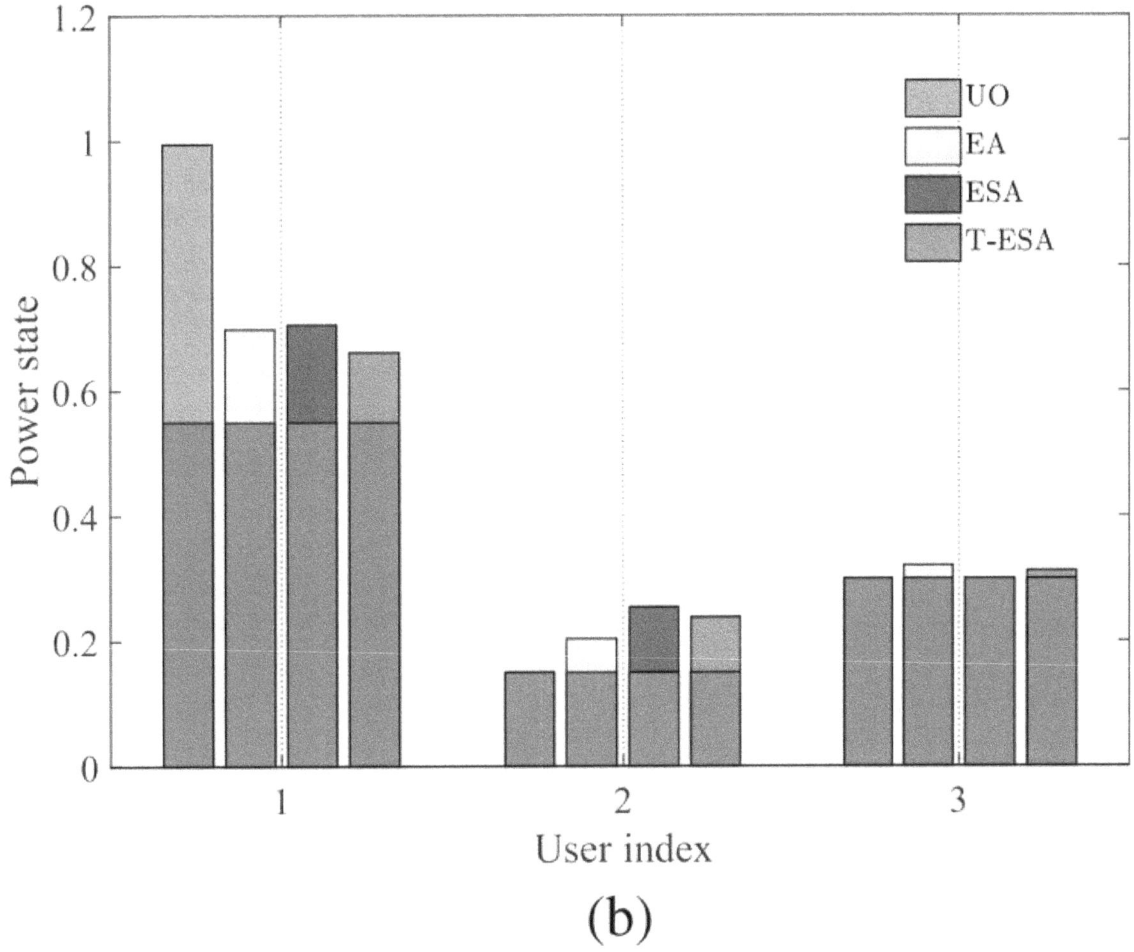

The system energy utilities are examined utilizing power allocation strategies, and Figure 18 displays the findings. It is noted that while the suggested T-ESA method performs somewhat worse, the proposed ESA power allocation strategy may, as predicted, produce the maximum power utility. But from the perspective of the energy utility, both ESA and T-ESA outperform the conventional power allocation techniques.

2.6 Simultaneous Lightwave Information and Power Transfer in 6G Networks

Combining data transmission and EH into a single method, simultaneous lightwave information and power transfer (SLIPT), is a promising strategy that makes use of optical transceivers' efficiency and technical maturity to guarantee both features. In this paper, we explore SLIPT as a promising technology to support 6G's sustainability objectives by going over optical beam propagation, SLIPT transceiver topologies, and key design trade-offs. Additionally, we classify SLIPT's applications based on their range from very short to long taking into account scenarios that occur indoors, underwater, and in the air. Finally, we highlight how well-suited SLIPT is for combining with other well-known technologies that are envisioned for use in 6G networks, such as quantum communications, artificial intelligence, and reconfigurable intelligent surfaces.

Transceiver Architecture, Optical Beam Propagation, and Design Trade-Offs

Depending on the beam-steering capabilities and directive or non-directive nature of light signals, several transceiver designs may be employed for SLIPT. Figure 19 depicts a generic SLIPT architecture.

Figure 19. SLIPT transceiver architecture. At the AP, an AC data-bearing signal and a DC bias to be utilized for EH are used to prepare the SLIPT signal for the downlink. The received signal's AC component is sent to the mobile equipment's digital baseband unit for decoding, while the DC component is used to charge the energy storage. An optical or radio frequency link can be used to implement the uplink. Trans-impedance amplifiers are referred to as TIAs.

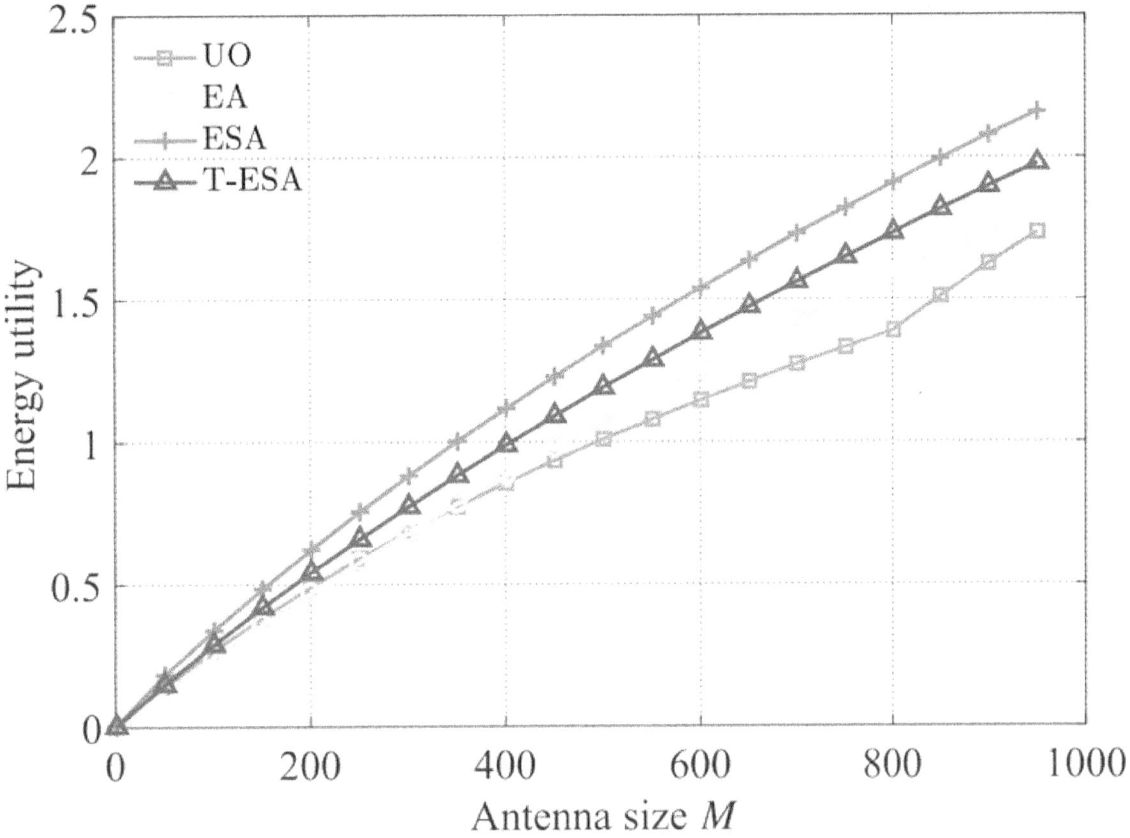

Directive SLIPT: Reduced geometric spreading losses allow light beam directivity to carry lightwave information and power between nodes (Fakidis et al., 2021). This offers great potential for underwater and aerial long-distance point-to-point communication and power transfer. This technology transmits via resonant lasers or narrow-beam LEDs. The transmitter's half-power angle may be limited to mrad to provide extremely directed light pulses. FSO receivers use a limited field-of-view (FoV) to collect the optical signal without interfering with ambient light-induced shot noise, whereas SLIPT systems with a larger FoV may leverage ambient light to boost energy collection. PD receivers can focus light into the active part of the PD using liquid convex lenses, which can change curvature. This area might be avalanche PD (APD) or positive-intrinsic-negative (PIN). APDs provide the high internal gain needed for underwater SLIPT (Palitharathna et al., 2022), but they also cause significant shot-noise. PIN PDs are good for low-complexity systems without a huge gain. In contrast, PV cells help identify EH and joint. By wavelength modification, GaAs-based PV cells may achieve 66% EH efficiency (Qaraqe et al., 2022). Due to PV cell transient length, data rates may be decreased. Directing SLIPT requires considering optical energy density, or eye safety. Eye safety restrictions increase with wavelength for collimated spatially coherent laser beams. A light intensity of 1 mW/cm2 is eye-safe for wavelengths around 850 nm and exposure periods under 10 ms. Space link applications can ignore this constraint.

Non-Directive SLIPT: Non-directive SLIPT helps in indoor multi-user VLC systems with user mobility when transmitter and receiver alignment is not attainable (Ma et al., 2019). For this, broad beam LEDs are employed to transmit light signals across a wide area. High power LEDs are needed because to the non-directive design, which may lower receiver irradiance. Recently, SLIPT systems have used energy-efficient organic LEDs to boost LPT efficiency (Nguyen et al., (2021). Micro-LEDs (mLEDs) are also used to produce dense multi-element transmitter architectures that evenly distribute lightwave power. Even though this design receives less optical power than directed SLIPT, PV cells catch energy because to their high effective area. Recent research has shown that organic and perovskite photovoltaic cells particularly printable carbon-based cells can be used as SLIPT receivers due to their ability to capture light in low-light interior environments and their printing versatility (Nguyen et al., (2021). However, PV cells' transient time limits bandwidth and may cause nonlinear effects. Pre-distortion can boost transmitter bandwidth. A hybrid design with a PV cell receiver for EH and a PD for information reception may be used.

Beam Steered SLIPT: Both directional and non directive SLIPT architectures fail under beam blockage, hence beam guided transceivers such multi-element-based, beam rotation-based, and RIS-based systems should be used. Multi-element SLIPT setups use many LEDs at the transmitter and PDs or PV cells at the reception. To give angle diversity and vast coverage, these parts are placed on a sphere, hemisphere, pyramid, or cone. LEDs at the transmitter may guide light toward users, while PD or PV cells at the receiver can modify light direction. In high-motion scenarios, beam steering improves performance. Retro reflectors enable resonant beam SLIPT, in which the air between the transmitter and receiver acts as a laser resonant cavity (Ijeh et al., 2022) . This design has beam-steering. Distributed access points (APs) and multi-element design can increase coverage and provide dependable blockage solutions. Beam steering helps undersea installations avoid node misalignment from ocean currents. In addition to multi-element deployment, RISs may direct light beams to improve SLIPT performance. Gimbal mechanisms, optical phased arrays (OPAs), microelectronic mirrors (MEMs), liquid crystals (LCs), etc. may rotate transmitters and receivers. Combining the above strategies can provide dependable solutions for many 6G applications.

RESULTS

Additionally, as can be shown in Figure 20, our suggested method achieves rising power efficiency vs. with g, whereas similar current and non-robust methods have constant and decreasing power efficiencies with g. As it becomes increasingly inaccurate, the non-robust technique is more likely to supply electricity to locations with the wrong MII. For instance, our suggested scheme's power efficiency is 69.6% for g = 0.35, compared to 40.3% and 31.9% for equivalent current and non-robust designs, respectively. In terms of numbers, compared to the same current and non-robust systems, our suggested robust method may increase the power efficiencies by 29.3% and 37.7%, respectively.

Figure 20. Comparison of the outage probabilities and power efficiencies versus error parameter g among three schemes for the case of multiple-RXs, where the total received power threshold is P = 24 W.

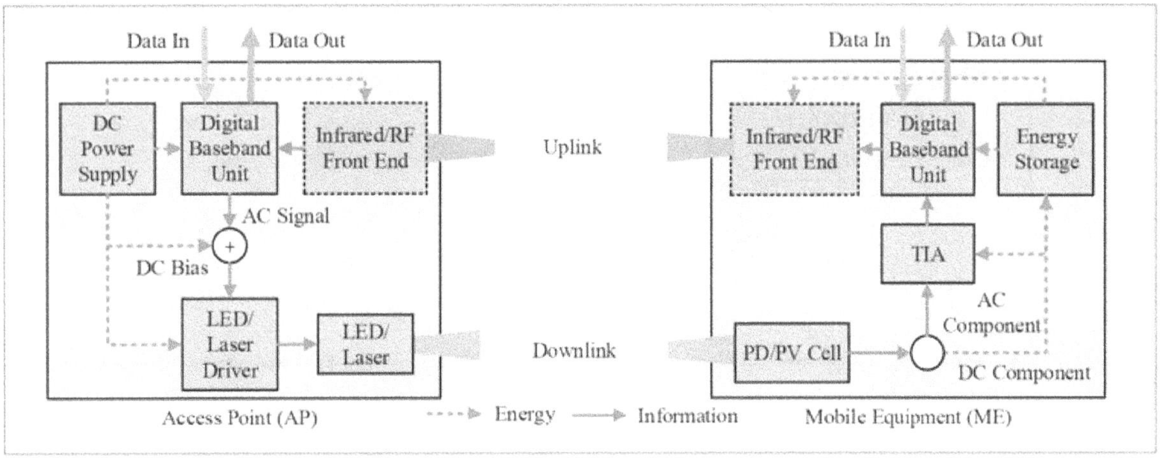

As the total received power constraints vary, Figure 20 shows "the outage probabilities and power efficiencies of the non-robust scheme, the same current scheme, and our proposed robust beamforming, where the power profile vector is [0.1 0.15 0.25 0.5] and the error parameter is g = 0.25. This figure shows that our technique can always offer non-outage power distribution to each RX, while the non-robust and identical current strategies have 0.72 and 0.11 approximate outage probabilities, respectively. As the power needed for four RXs increases, both our technique and the non-robust system improve power efficiency. The power thresholds for four RXs are 1 W, 1.5 W, 2.5 W, and 5 W, respectively, and our proposed system always has the highest power efficiency among three systems when P is more than 10 W. The results in Figure 21 demonstrate that our robust scheme may achieve higher power efficiency and reliable power distribution to each RX in MIMO MRC-WPT systems than the current scheme and the non-robust beamforming approach".

Figure 21. Comparison of the outage probabilities and power efficiencies versus power threshold among three schemes for the case of multiple-RXs when the error parameter g = 0.25.

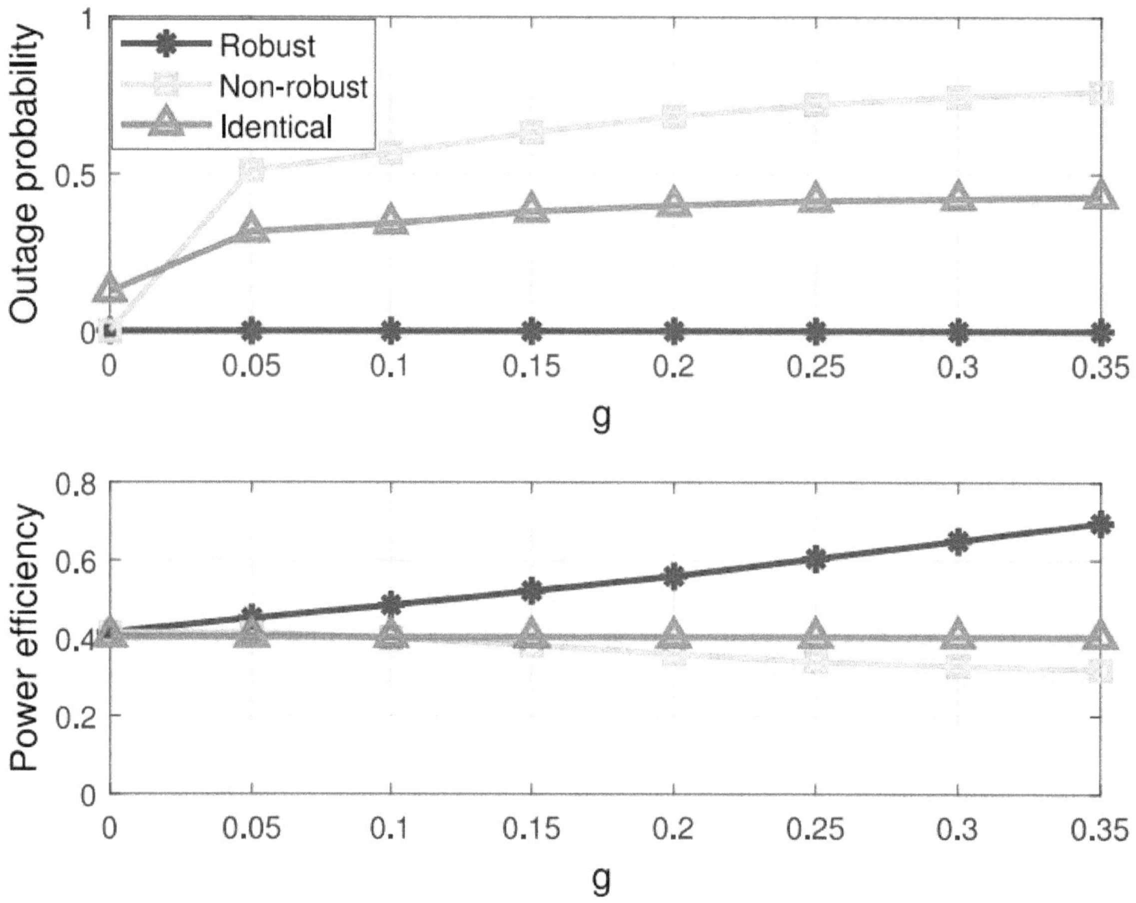

When the specifications and technology for 6G wireless communications were initially released, many of the requirements like high efficiency and high density were incompatible with one another, much as high density and a lot of data eventually resulted in a loss of energy efficiency. In addition to lessening this conflict, TL has an impact on nearly all 6G needs. The fundamental ideas and categorization of TL techniques were then presented.

We provided an in-depth analysis of the cutting-edge TL-based work in wireless communications, emphasizing the TL's exceptional contributions in six distinct application domains, including intrusion detection, indoor wireless localization, content popularity prediction, spectrum allocation, and BSs/APs switching energy. The foundation of next TL supporting adaptive wireless communications for 6G is formed by these works.

Two metrics indicate the likelihood of not harvesting energy within a coherent block time: (i) AHE and (ii) AEO probability per node. Unlike their CSI-based alternatives, whose efficiency rapidly decreases as the system grows denser, this use case shows how Channel state information approaches are independent of the number of devices in the network. The difference between Channel state information and CSI-based approaches is only noticeable for somewhat small systems; however, the real difference is significantly less because CSI collecting requires energy that is not considered. It is significant to highlight that, because we do not take into account the power resources needed for CSI acquisition, CSI-based results are optimistic. Due to the computational difficulties of identifying the optimal pre-coding feature vector as the number of participants and/or antennas increases, even the AA-CSI system may not be feasible to utilize in reality. Consequently, when N increases, the associated energy prices and CSI gathering procedures become problematic, limiting the positive effects of CSI-based systems. Moreover, the AA CSI approach outperformed the AA methodology only while powering a very small number of devices, and even then, the performance difference was not very significant for a single PB.

Figure 22. The objective function value of (10) versus different weight coefficient

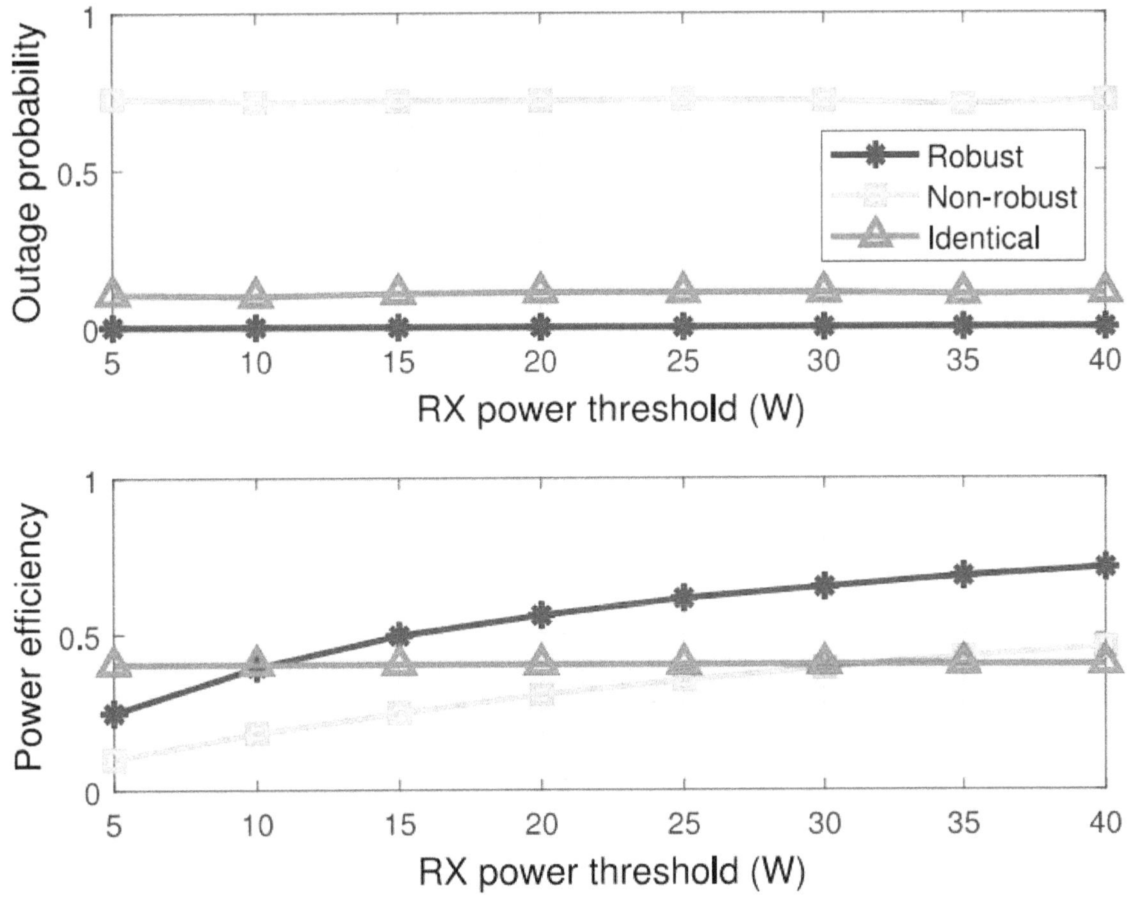

The performance of our suggested schemes is compared with two benchmark schemes (the SDR optimum solution and the Rank-1 approximation solution) in "Figure 22 under various weight coefficient ρ values in terms of minimizing the objective function value of issue (10). It is clear from Figure 2 that our suggested optimum solution performs similarly to the SDR optimal solution, suggesting that it is the globally optimal solution to issue (10)". Furthermore, our suggested ideal solution outperforms both the suggested low-complexity sub-optimal solution and the Rank-1 approximation solution, as predicted.

Figure 23. Radar beam pattern in different weight ρ: (a) ρ = 0.1; (b) ρ = 0.5

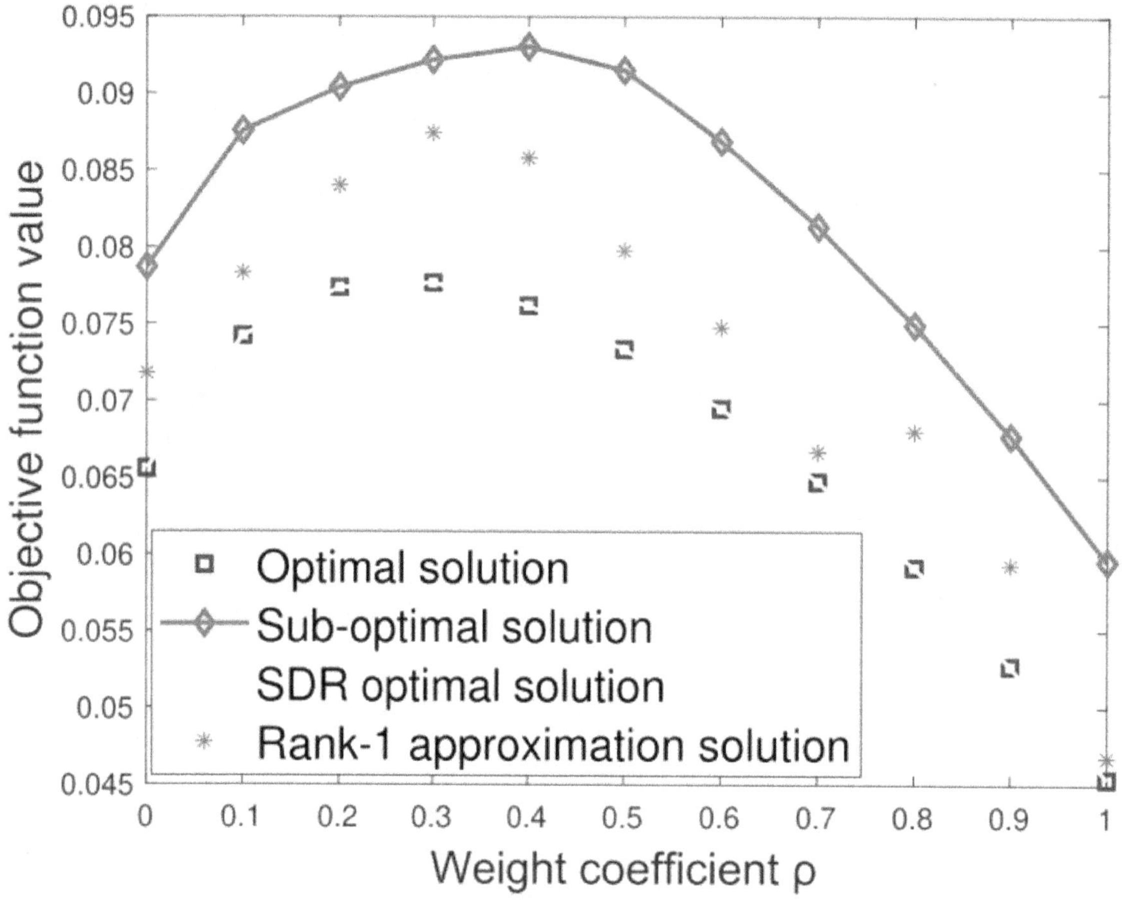

In Figure 23, "we evaluate the radar transmit beam patterns for the sub-optimal and optimum beamforming systems. Figures 23 (a) and 23 (b) depict transmit beam patterns for $\rho = 0.1$ and 0.5, respectively. If $\rho = 0$, the optimal beam for radar detection is the radar-only beam pattern described in equation (10). According to Figure 23, the optimal and sub-optimal beam patterns resemble the only radar optimal beamforming when the weighting factor is low, such as $\rho = 0.1$. However, when $\rho = 0.5$, the peak value of the main beams of the optimal and sub-optimal schemes decreases significantly compared to the only radar optimal beam pattern". It shows how WPT impacts radar sensing. The sub-optimal scheme's average beam pattern is identical to the optimum scheme's, proving our low-complexity design works.

Figure 24. Trade-off between radar beam pattern MSE and received power

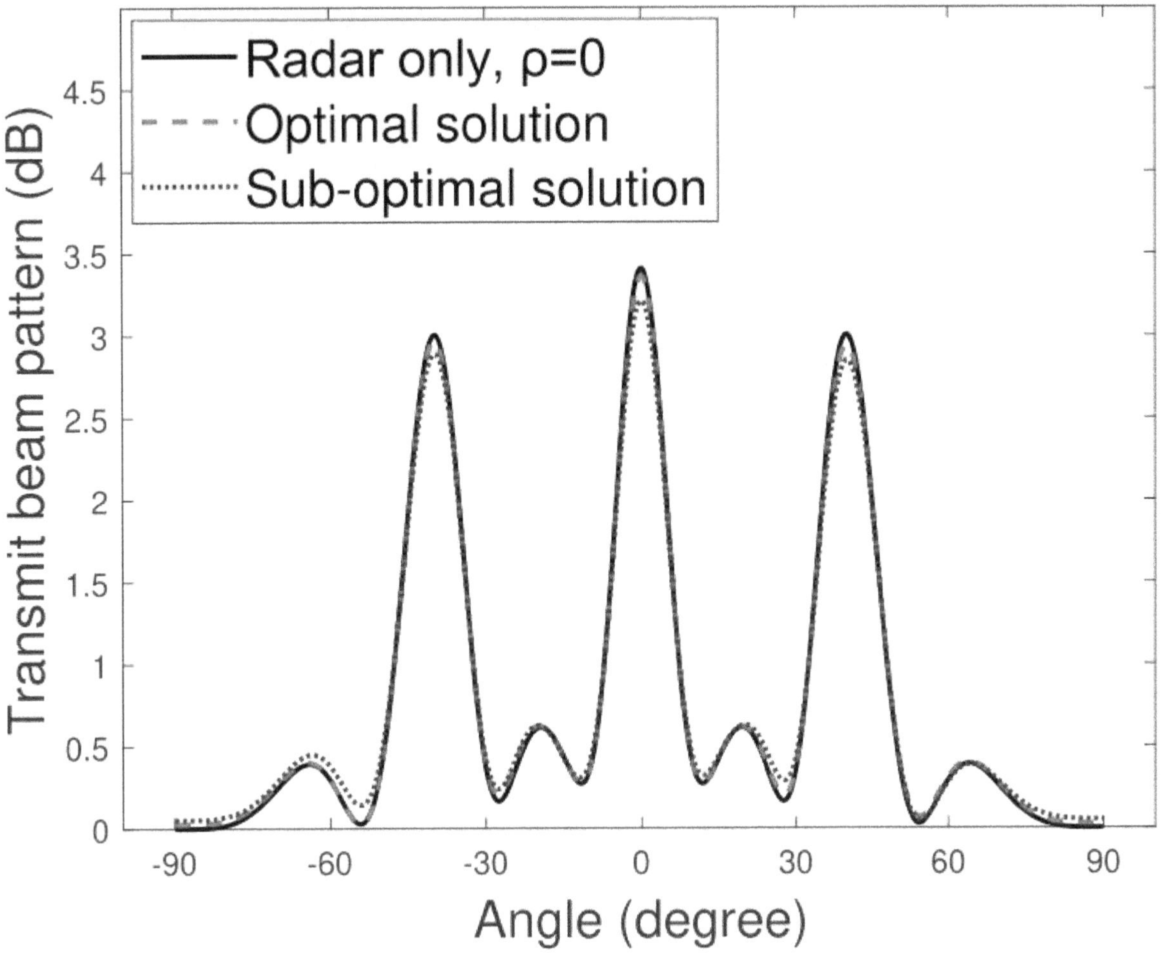

(a) $\rho = 0.1$

The trade-off between WPT and radar sensing for various beamforming methods is clearly seen in Figure 24, where the performance limitations of radar and WPT are denoted by the points "Radar only" and "WPT only," respectively. In the meanwhile, we can see that the MSE of the radar beam pattern for both optimum and sub-optimal designs rises when user received power grows. This suggests that improving user received power comes at the price of radar performance. Furthermore, it is evident that as received power rises, so does the trade-off performance difference between the optimal and sub-optimal schemes. This is due to the fact that MRT beamforming, which uses a suboptimal algorithm that ignores radar performance, requires greater transmit power as the needed received power rises. In addition, we compare the performance of our suggested optimal solution to the widely used one. According to our numerical results in Figure 22, our suggested optimal solution outperforms the rank-1 approximation solution in terms of beam pattern MSE while achieving the same performance as the SDR optimal solution.

The effectiveness of the suggested RIS-assisted WPT method is illustrated by the presentation of a few typical simulation outcomes. A multi-user scenario is simulated, with three users in the RIS-assisted WPT system (Figure 25(a)), and a benchmark traditional WPT system (Figure 25(b)) without RIS. D0 = 5 m, D1 = 7 m, D2 = 10 m, D3 = 12 m, D1 = 3 m, D2 = 5 m, and D3 = 8 m are the corresponding distances.

Figure 25. Simulated scenarios. (a) RIS aided WPT scenario. (b) Traditional WTP scenario.

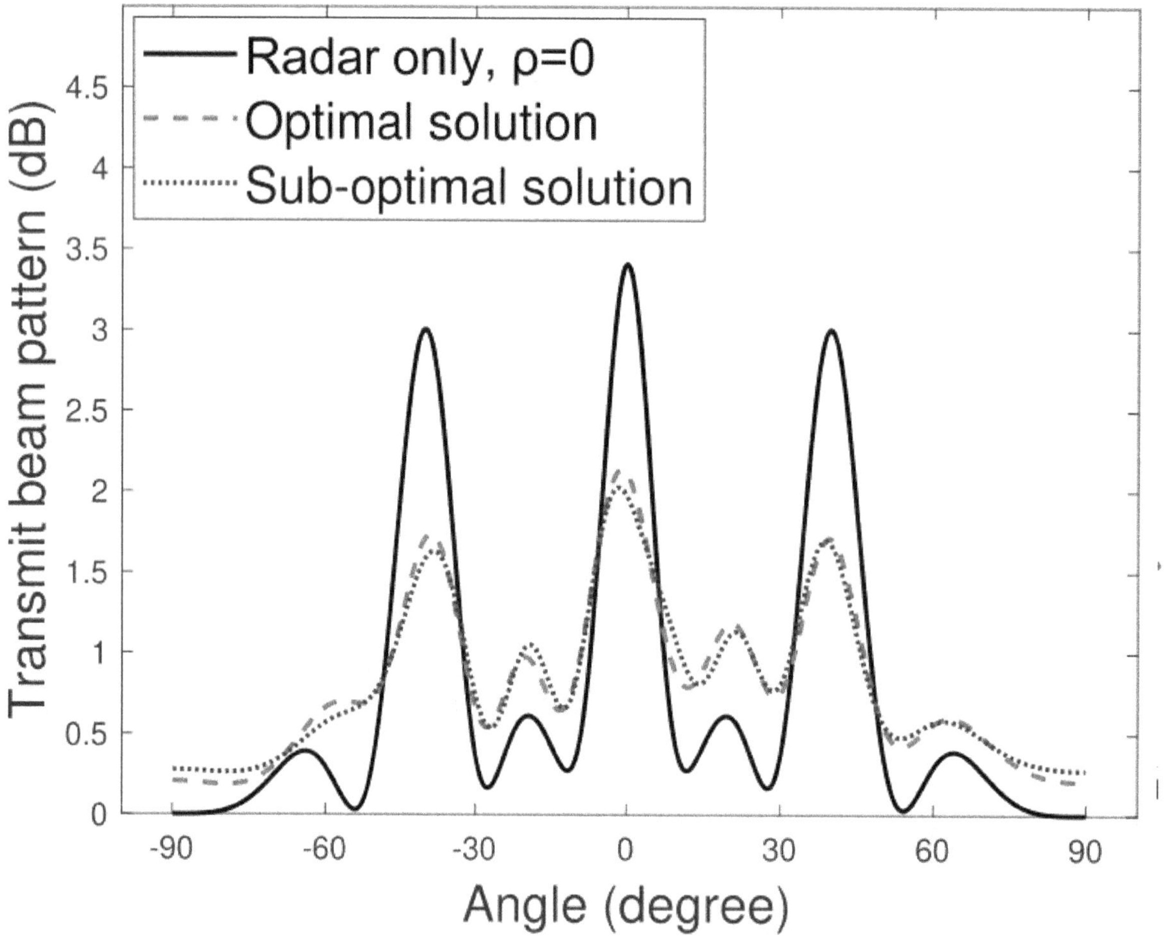

(b) $\rho = 0.5$

Figure 26. Harvested energy comparison between different WPT systems. (a) Comparison between RIS-aided system without direct links and traditional system. (b) Comparison between RIS-aided system with direct links and traditional system.

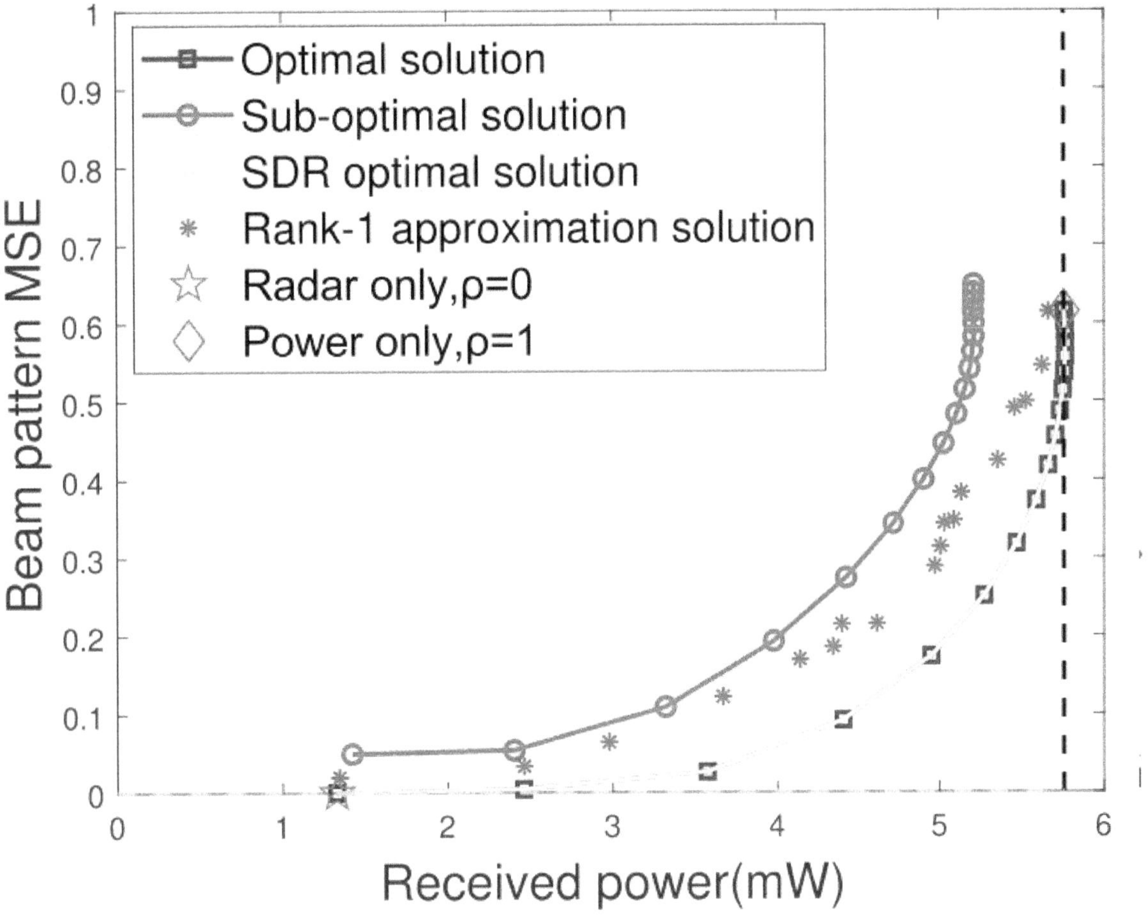

(b) $\rho = 0.5$

As anticipated, Figure 26(b) shows that the RIS-aided WPT scheme collects the same energy as the standard WPT system without RIS at RIS scale crossing point N1*. The suggested RIS-aided WPT approach will capture more energy than the conventional one when the RIS plane scale reaches N1*. The only distances that determine N1* are D0, dk, and Dk. Furthermore, it is independent of BS antenna count. In particular, N1*=237 in Figure 26(a) for M between 200 and 1000. Also exhibited is the gathered power with user interference power. As shown, the two RIS-aided system curves closely overlap, demonstrating that user interfering power has little performance benefit.

FUTURE OPPORTUNITIES AND CHALLENGES

Future research priorities in wireless communications will focus on finding ways to increase productivity while lowering expenses. Approaching this goal with the ability to learn from experience is crucial. New tasks are encouraged to learn from more experienced activities in order to become quicker and more effective. This process is known as transfer learning, or TL. In several wireless communication domains, TL's correlation and similarity data between various jobs may assist reduce energy consumption and boost productivity. It is therefore a highly useful issue to apply TL to future 6G communications.

In Figure 27 Data, similarity, and security three prospective application trends and problems of TL in 6G were examined (Wang et al. 2017). Applying TL to 6G will require careful consideration of these three challenges. It should give some thought and motivation for the use of TL in next 6G communications.

Figure 27. TL for Challenging Fields in 6G Communications

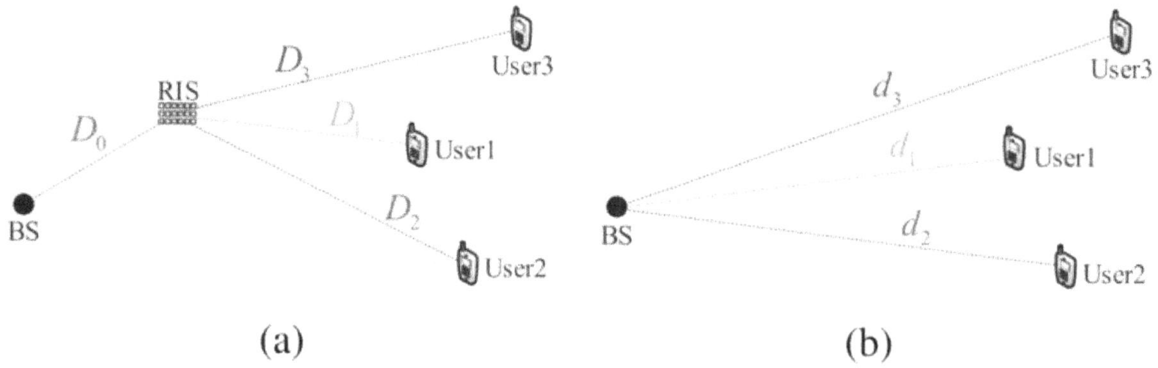

Security and Privacy:

Threat Identification and Mitigation in Real Time

Real-time network traffic analysis and security threat detection are possible. It may also be able to completely eradicate the problems. Additionally, machine learning models can recognize cyberattack trends, allowing for quick reactions to reduce risks (Paolini et al., 2023).

Privacy of Data

The enormous volumes of data generated by multiple devices and apps in 6G networks make data privacy a top priority. This problem is addressed by federated learning, which trains AI models across several devices without requiring data centralization. The raw data is kept private as each device processes its data locally and only transmits model updates to a central server. This method minimizes data transmission while lowering the risk of data breaches and illegal access. It is especially helpful in industries like healthcare and finance since it supports individualized models and aids in compliance with privacy requirements (Nguyen et al., 2021).

Safe Communication

In 6G networks, secure communication is essential for maintaining data confidentiality and integrity. By guaranteeing that only model updates, not raw data, are sent between devices and central servers, federated learning improves secure communication. By doing this, less sensitive data is transmitted across the network, which lowers the possibility of interception and illegal access. Federated learning offers a strong option for preserving secure communication in the quickly developing 6G landscape, despite obstacles such communication overhead and the difficulty of collecting model updates (Sirohi D et Al 2023).

Challenges in Wireless Power Transfer (WPT) techniques for 6G communication:

(i) ***Regulatory and Protocol Challenges:*** As 6G is still developing, laws and standards are still being set. The 3GPP has developed several groups to regulate ISTN, and SCN TC-SES is integrating UAV and satellite networks. Navigation is crucial to the RTM design since UAV location and guidance can improve data collection and reduce internal collisions. 6G is expected to be incorporated with LEO satellite networks to create an intelligent communication and navigation system for RTM-equipped UAVs.

Figure 28. Blockchain enabled secured UAV-RTM system

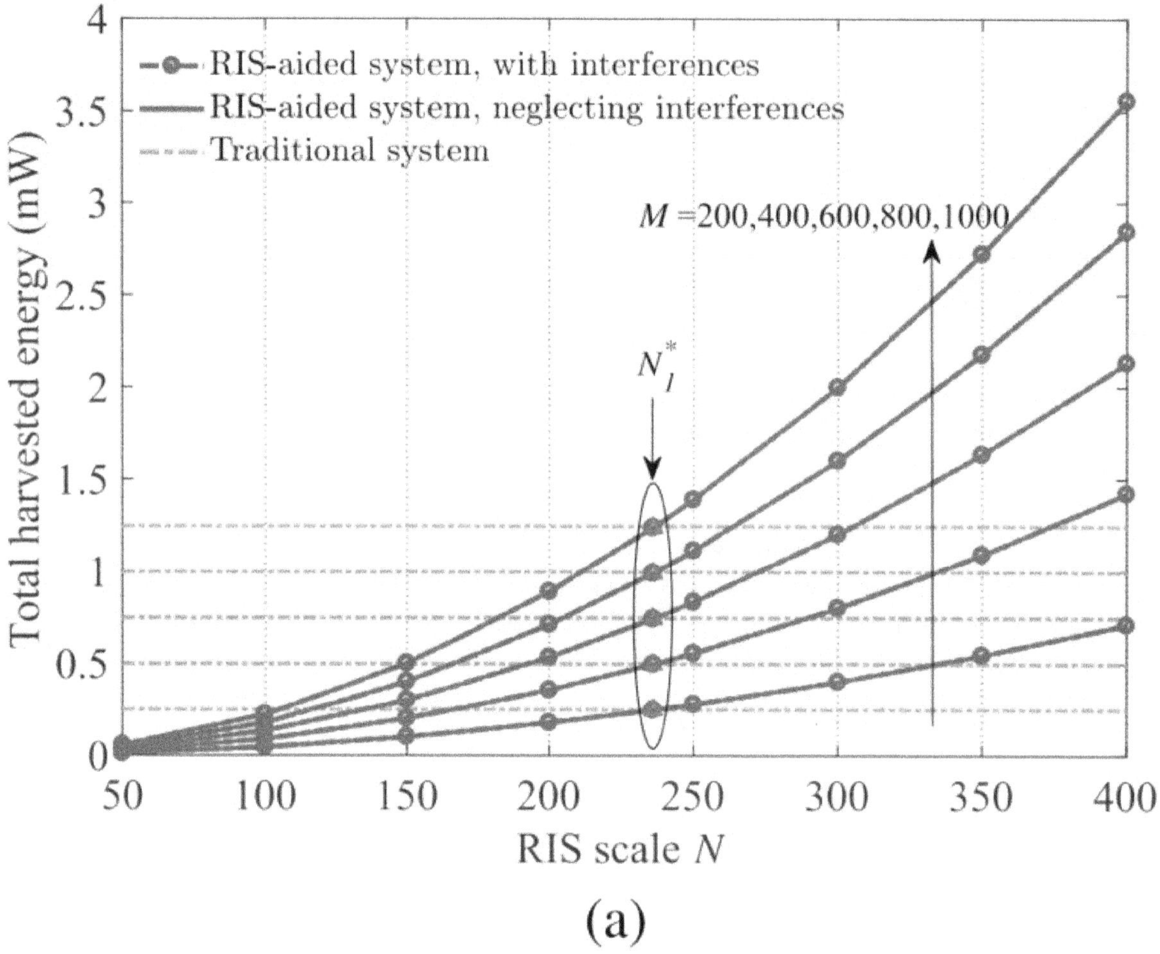

Since many devices will be connected in this UAV network, Figure 28 shows a Blockchain-enabled secured UAV-RTM system that protects data. By defending the system from jamming, spoofing, cyber-attacks, etc., 6G block chain technology will maximize security.

(ii) ***Technical Challenges:*** To seamlessly integrate AI into 6G networks, various technical difficulties must be overcome. These include developing advanced AI technologies to automate wireless network smart application information transmission. Satellite-based Non-Terrestrial Networks (NTN) has propagation delays and spectrum sharing issues. Regular feedback and increasing energy consumption are also issues in physical layer communication as antennas expand. In (Khalid et al., 2022; Zhang et al., 2022; Shah et al., 2024), AI-enabled integration of huge Multiple-Input-Multiple-Output (MIMO) techniques with Network Functions Virtualization (NFV) and Software-Defined Network (SDN) architectures is needed to efficiently serve customized services with various QoS needs. AI adoption in 6G networks must overcome these hurdles to improve performance and enable enhanced capabilities.

(a) *Low Efficiency over Distance:* RF and microwave techniques see significant efficiency decrease with increasing distance.
(b) *Interference Management:* WPT systems may disrupt 6G communication signals.
(c) *Energy Conversion Losses:* Rectification and conversion operations frequently cause energy loss.
(d) *Antenna Design:* Designing high-gain, efficient antennas for long-range power transfer presents design and manufacturing issues.

(iii) ***Economics Challenges:*** Wireless Power Transfer (WPT) systems for 6G connectivity provide many technical potential but also cost constraints. These issues affect WPT system development, deployment, and scaling. Economic challenges are listed below:

(a) High Initial Deployment Costs

WPT-enabled networks require significant upfront investment in energy beaming stations, rectennas, and advanced RF or laser transmitters. Innovative WPT technologies like millimeter-wave or terahertz energy transfer systems require significant R&D.

(b) Energy Efficiency and Losses

Long-distance wireless transmission losses raise operational expenses, making it less economically viable than traditional power delivery systems. Efficiency upgrades may be costly to meet environmental and safety criteria.

(c) Scalability Challenges

Multiple nodes are needed for large-scale 6G WPT deployments, raising per-unit expenses. Retrofitting or replacing components to integrate WPT with wireless systems is costly.

(d) Consumer Costs

WPT-compatible smartphones and IoT sensors cost more since they require rectennas. If efficiency difficulties persist, customers and providers may pay more for operational energy.

(e) Regulatory and Licensing Fees

Millimeter-wave or terahertz WPT spectrum licenses are expensive. Costs increase with electromagnetic exposure regulations.

(iv) ***Real World Applications:*** Wireless Power Transfer (WPT) for 6G communication could transform several sectors with unique uses. Examples of real-world use cases and their benefits:

(a) WPT permits wireless powering billions of IoT devices for monitoring traffic, air pollution, utilities, and infrastructure without regular battery replacements.

(b) Wireless charging at specific stations improves operational efficiency for electric buses, trains, and drones.
(c) WPT powers medical devices including pacemakers, insulin pumps, and sensors without invasive battery replacement operations.
(d) Health monitoring wearables with continuous wireless charging are more reliable and require less maintenance.
(e) With WPT, gadgets like smartphones, tablets, and laptops can charge wirelessly, providing a fully wireless experience.
(f) Home automation equipment such as smart speakers, security cameras, and appliances can be powered wirelessly.
(g) Dynamic wireless charging on roadways enables EVs to recharge while moving, decreasing the need for big battery packs.
(h) Allows self-driving vehicles to operate continuously without frequent stops.
(i) Wireless charging for industrial robots enhances factory automation efficiency and decreases downtime.
(j) WPT can power remote 6G base stations in places where regular power sources are impractical.
(k) Peer-to-peer WPT lets cell phones and computers exchange power wirelessly.
(l) Allows UAVs to operate continuously using distant recharge stations or satellite energy beams.
(m) WPT systems power essential communication devices and sensors in disaster-stricken areas with a degraded power grid.

(v) Analysis and challenges

Typical 6G antenna array transmitters can be changed for WPT. Therefore, while high-frequency architectures provide several implementation issues, these are mostly not specific to WPT transmitters. Thus, we focus on the key characteristics anticipated to be substantially linked with radiating WPT for 6G IoE networks.

(a) Extended exposure to RF, microwave, or laser radiation may pose health risks. WPT infrastructure is expensive to build and maintain, especially for big installations.
(b) Creating effective energy distribution mechanisms for millions of 6G devices is a challenge.
(c) Ensure cost-effectiveness of WPT for low-energy devices, such as sensors or IoT nodes.
(d) Urban areas with moving items or barriers can hinder WPT efficiency.

Further developments in ultra low power circuits and their incorporation into a wide range of materials will greatly expand the possibilities of WET technology in the context of the 6G future. In real-world applications, time-varying wireless channels and the causal link between the present WET process and future WIT make real-time information/energy scheduling a difficult task. Consequently, a formal proposal for a cluster-based EB scheme will be made in a later study, along with its connection with DAS for additional enhancements.

When compared to traditional independently designed systems, ISWPT's integrated functioning offers the following benefits: (i) The integration saves spectrum resources and device costs, which is important for 6G networks in the future. (ii) The radar beam typically covers a wide region continuously, facilitating a steady energy supply to nearby equipment. In addition, radar power waste is decreased by collecting energy from radar signals.

SLIPT and AI: Standard SLIPT resource management can be computationally difficult due to sophisticated transceiver designs and non-linear effect equalization and compensating techniques. ML architectures trained to extract optimum resource allocation parameters save computing effort (Le et al., 2021). In prediction-based "lightweight" ML architectures, recurrent neural networks can alleviate the problems of typical machine learning systems, such as computational complexity that delays training and insufficient training data. In dynamic, unpredictable, and less informational applications, deep learning, reinforcement learning, and transfer learning may increase accuracy. Decentralized federated learning (FL) may train ML models locally and coordinate parameter improvements in large networks. Mobile data processing can revolutionize dispersed AI, hence SLIPT-assisted machine learning might be used. LPT reduces mobile device data transfer to the AP throughout the radio frequency spectrum in FL-based wireless networks (Tran et al., 2020). The FL paradigm may train statistical models, power local devices with SLIPT, and upload data to the central server utilizing various access methods by carefully specifying a time allocation parameter. The techniques may be applied to mobile edge computing, UAVs, and more. A block-by-block artificial neural network (ANN) employing the sensors' average optical power and an initial block that anticipates human movement-induced obstacles can be employed. Then, many ANN blocks may optimize the beamforming matrix, time allocation parameter, and average transmits power.

RIS-Assisted SLIPT: SLIPT and most OWC connections depend on LoS availability, which may limit real-world use. After seeing their application in RF systems, RISs have been studied for lightwave systems to improve coverage and connectivity, especially for mobile users. Metasurfaces and mirror-based RISs in OWC systems guide light toward the photodiode (PD). They can be wall-mounted or transceiver-integrated (Abumarshoud et al., 2021). RISs can enhance UAV communication. Aerial platforms can host RISs for changeable position and many devices. Localization mechanisms should be added to these systems to maximize RIS potential. Device position is crucial for beam steering.

SLIPT for Quantum Information Transfer: Quantum cryptography, QKD protocols, superdense coding, and quantum communications may increase security. Recently, optical wireless quantum communication networks were experimentally confirmed for satellite and terrestrial communications (Grillo et al., 2021).

CONCLUSION

In this chapter, we formulated and studied the following points:

(i) In MRC-WPT systems with several TXs and RXs, magnetic beamforming must account for MII flaws to increase power transfer efficiency and reproducibility. We divide this nonconvex QCQP with indefinite limitations into two robust optimization problems reflecting the worst-case single- and multiple-RX scenarios to solve it. In MIMO MRC-WPT systems, robust magnetic beamforming lowered outage risk while increasing power transfer efficiency for every RX.
(ii) TL will help 6G researches take use of wireless communications' flexibility, energy efficiency, speed, and scalability.
(iii) Using DAS with a good channel status information plan may decrease peak power failures and boost power coverage for all EH equipment.

(iv) We examined a novel ISWPT configuration combining wireless power transfer and radar detection on a single hardware platform. We characterized the wireless power transfer-radar detection trade-off using optimal transmitted beamforming vectors.

(v) We can achieve a self-sustaining WPT system with improved PTE and energy utility by using a well-developed RIS plane. We can utilize RIS-assisted WPT for smart home and low power sensor node charging.

(vi) SLIPT was developed to address interest in offering EH capabilities and information transmission simultaneously for energy-efficient 6G networks. SLIPT can support many applications and enable next-generation networking despite its limited adoption. We discussed various SLIPT designs, prospective applications, practical issues, and design trade-offs.

Abbreviations

AI - Artificial Intelligence
APD - Avalanche Photodiode
APs - Access Points
ATR - Absorb Then Reflect
BS_S - Base Station
Cell PAN - Cell Personal Area Network
CSI - Channel Source Information
CVX - Convex Optimization
DDTL - Distance Domain Transfer Learning
EES - External Energy Supply
EH - Energy Harvesting
ESA - Energy State Aware
FL - Federated Learning
FoV - Front of View
FSO - Free Space Optics
FTL - Federated Transfer Learning
HAP_S - High Altitude Platforms
IoT - Internet of Things
ISWPT - Integrated Sensing and Wireless Power Transfer
L2T - Learn to Transfer
LPT - Lightware Power Transfer
MEM_S - Microelectronic Mirrors
MII - Mutual Inductance Information
MIMO - Multiple Input Multiple Output
MISO - Multiple Input Serial Outputs
MRC - Magnetic Resonant Coupling
MRC-WPT - Magnetic Resonant Coupling-Wireless power Transfer
OPA_S - Optical Phase Arrays
OWC - Optical Wireless Communication
PIN - Positive Intrinsic Negative
PTE - Power Transfer Efficiency

PV - Photovoltaic
QCQP - Quadratically Constrained Quadratic Programming
RF - Radio Frequency
RFID - Radio Frequency Identification
RSS - Received Signal Strength
RX - Receiver
SDP - Semi definite Programming
SLIPT - Simultaneous Lightware Information and Power Transfer
T-ESA - Total Energy State Aware
TL - Transfer Learning
TX - Transmitter
UAV_S - Unmanned Aerial Vehicles
umMTC - Ultra Massive Machine type Communication
UO - Utility optimization
UPA - Uniform Planar Array
VLC - Visible light Communication
VLC - Visible Light Communication
WET - Wireless Energy Transfer
WIT - Wireless Information Transfer
WPCN - Wireless Powered Communication Network
WPT - Wireless Power Transfer
WSN_S - Wireless Sensor Networks
WUR_S - Wake up Radios
IoE - Internet of Everything
AAN - Autonomous Aerial Networks
RTM - Road Traffic Management

REFERENCES

Bariah, L., Mohjazi, L., Muhaidat, S., Sofotasios, P. C., Kurt, G. K., Yanikomeroglu, H., & Dobre, O. A. (2020). A prospective look: Key enabling technologies, applications and open research topics in 6G networks. *IEEE Access : Practical Innovations, Open Solutions*, 8, 174792–174820.

Abumarshoud, H., Mohjazi, L., Dobre, O. A., Di Renzo, M., Imran, M. A., & Haas, H. (2021). LiFi Through Reconfigurable Intelligent Surfaces: A New Frontier for 6G? *IEEE Vehicular Technology Magazine*, •••, 2–11. DOI: 10.1109/MVT.2021.3121647

Alcaraz, L., & Alves, H. (2017). Ultrareliable Short-Packet Communications With Wireless Energy Transfer. *IEEE Signal Processing Letters*, 24(4), 387–391. DOI: 10.1109/LSP.2017.2667620

Alsaba, Y., Rahim, S. K. A., & Leow, C. Y. (2018). Beamforming in Wireless Energy Harvesting Communications Systems: A Survey. *IEEE Communications Surveys and Tutorials*, 20(2), 1329–1360. DOI: 10.1109/COMST.2018.2797886

Bashir, M. S., & Alouini, M.-S. (2022). Energy Optimization of a Laser-Powered Hovering-UAV Relay in Optical Wireless Backhaul. *IEEE Transactions on Wireless Communications*, •••, 1–1. DOI: 10.1109/TWC.2022.3216797

Bharath, B. N., Nagananda, K. G., & Vincent Poor, H. (2016). A Learning-Based Approach to Caching in Heterogenous Small Cell Networks. *IEEE Transactions on Communications*, 64(4), 1674–1686. DOI: 10.1109/TCOMM.2016.2536728

Bjornson, E., Ozdogan, O., & Larsson, E. G. (2020). Intelligent Reflecting Surface Versus Decode-and-Forward: How Large Surfaces are Needed to Beat Relaying? *IEEE Wireless Communications Letters*, 9(2), 244–248. DOI: 10.1109/LWC.2019.2950624

Brown, W. C. (1984, September). The History of Power Transmission by Radio Waves. *IEEE Transactions on Microwave Theory and Techniques*, 32(9), 1230–1242. DOI: 10.1109/TMTT.1984.1132833

Cheng, Y., Peng, W., Huang, C., Alexandropoulos, G. C., Yuen, C., & Debbah, M. (2022). RIS-Aided Wireless Communications: Extra Degrees of Freedom via Rotation and Location Optimization. *IEEE Transactions on Wireless Communications*, 21(8), 1–1. DOI: 10.1109/TWC.2022.3151702

Cheng, Y., Peng, W., Huang, C., Alexandropoulos, G. C., Yuen, C., & Debbah, M. (2022). RIS-Aided Wireless Communications: Extra Degrees of Freedom via Rotation and Location Optimization. *IEEE Transactions on Wireless Communications*, 21(8), 1–1. DOI: 10.1109/TWC.2022.3151702

Diamantoulakis, P. D., Karagiannidis, G. K., & Ding, Z. (2018). Simultaneous Lightwave Information and Power Transfer (SLIPT). *IEEE Transactions on Green Communications and Networking*, 2(3), 1–1. DOI: 10.1109/TGCN.2018.2818325

Du, L., Zhang, W., Ma, J., & Tang, Y. (2021). Reconfigurable Intelligent Surfaces for Energy Efficiency in Multicast Transmissions. *IEEE Transactions on Vehicular Technology*, 70(6), 6266–6271. DOI: 10.1109/TVT.2021.3080302

Du, L., Zhang, W., Ma, J., & Tang, Y. (2021). Reconfigurable Intelligent Surfaces for Energy Efficiency in Multicast Transmissions. *IEEE Transactions on Vehicular Technology*, 70(6), 6266–6271. DOI: 10.1109/TVT.2021.3080302

Fakidis, J., Videv, S., Kucera, S., Claussen, H., & Haas, H. (2021). On the Design of Optical Energy Harvesting and Storage Systems for Outdoor Small Cells. *ICC 2021 -. IEEE International Conference on Communications : [proceedings]. IEEE International Conference on Communications*, 15, 1–6. DOI: 10.1109/ICC42927.2021.9500505

Fakidis, J., Videv, S., Kucera, S., Claussen, H., & Haas, H. (2021). On the Design of Optical Energy Harvesting and Storage Systems for Outdoor Small Cells. *ICC 2021 -. IEEE International Conference on Communications : [proceedings]. IEEE International Conference on Communications*, 15, 1–6. DOI: 10.1109/ICC42927.2021.9500505

None Ruisi Ge, None Hong Pan, None Zhibin Lin, None Na Gong, None Jinhui Wang, & None Xiaowei Chen. (2016). *RF-powered battery-less Wireless Sensor Network in structural monitoring*. DOI: 10.1109/EIT.2016.7535298

Ghassemlooy, Z., Arnon, S., Uysal, M., Xu, Z., & Cheng, J. (2015). Emerging Optical Wireless Communications-Advances and Challenges. *IEEE Journal on Selected Areas in Communications*, 33(9), 1738–1749. DOI: 10.1109/JSAC.2015.2458511

Ghesla, C. (2017). Defaults in green electricity markets: Preference match not guaranteed. *Journal of the Association of Environmental and Resource Economists*, 4(S1), S37–S84. DOI: 10.1086/691977

Grillo, M., Dowhuszko, A. A., Khalighi, M.-A., & Hamalainen, J. (2021). Resource allocation in a Quantum Key Distribution Network with LEO and GEO trusted-repeaters. *2021 17th International Symposium on Wireless Communication Systems (ISWCS)*. DOI: 10.1109/ISWCS49558.2021.9562139

Guo, H., & Ben, B. (2019). *Reinforcement Learning-Enabled Reliable Wireless Sensor Networks in Dynamic Underground Environments. ArXiv*. Cornell University., DOI: 10.1109/MILCOM47813.2019.9021051

Hu, Y. H., Yuan, X., Zhang, G., & Schmeink, A. (2021). Sustainable Wireless Sensor Networks With UAV-Enabled Wireless Power Transfer. *IEEE Transactions on Vehicular Technology*, 70(8), 8050–8064. DOI: 10.1109/TVT.2021.3090849

Huang, C., Zappone, A., Alexandropoulos, G. C., Debbah, M., & Yuen, C. (2019). Reconfigurable intelligent surfaces for energy efficiency in wireless communication. *IEEE Transactions on Wireless Communications*, 18(8), 4157–4170. DOI: 10.1109/TWC.2019.2922609

Huang, C., Zappone, A., Alexandropoulos, G. C., Debbah, M., & Yuen, C. (2019). Reconfigurable intelligent surfaces for energy efficiency in wireless communication. *IEEE Transactions on Wireless Communications*, 18(8), 4157–4170. DOI: 10.1109/TWC.2019.2922609

Huang, Y., & Clerckx, B. (2017). Large-Scale Multiantenna Multisine Wireless Power Transfer. *IEEE Transactions on Signal Processing*, 65(21), 5812–5827. DOI: 10.1109/TSP.2017.2739112

Ijeh, I. C., Khalighi, M. A., Elamassie, M., Hranilovic, S., & Uysal, M. (2022). Outage probability analysis of a vertical underwater wireless optical link subject to oceanic turbulence and pointing errors. *Journal of Optical Communications and Networking*, 14(6), 439. DOI: 10.1364/JOCN.454191

ITU-R. Report ITU-R, SM.2303-2. (2017). "Wireless power transmission using technologies other than radio frequency beam," 2017.

Kamalinejad, P., Mahapatra, C., Sheng, Z., Mirabbasi, S., Leung, V. C. M., & Guan, Y. L. (2015). Wireless energy harvesting for the Internet of Things. *IEEE Communications Magazine*, 53(6), 102–108. DOI: 10.1109/MCOM.2015.7120024

Kamga, G. N., & Aissa, S. (2019). Wireless Power Transfer in mmWave Massive MIMO Systems With/Without Rain Attenuation. *IEEE Transactions on Communications*, 67(1), 176–189. DOI: 10.1109/TCOMM.2018.2799217

Kashyap, S., Bjornson, E., & Larsson, E. G. (2015). *On the feasibility of wireless energy transfer using massive antenna arrays in Rician channels. KTH Publication Database DiVA*. KTH Royal Institute of Technology., DOI: 10.1109/SPAWC.2015.7226997

Katz, M., & O'Brien, D. (2019). Exploiting novel concepts for visible light communications: From light-based IoT to living surfaces. *Optik (Stuttgart)*, 195, 163176. DOI: 10.1016/j.ijleo.2019.163176

Khalid, R., Naeem, M., & Ejaz, W. (2022). Autonomous Aerial Networks with Wireless PowerTransfer: Resource Optimization, Standardization, and Challenges. *IEEE Communications Standards Magazine*, 6(3), 24–31. DOI: 10.1109/MCOMSTD.0001.2200022

Lang, H. D. (2018). Optimization of wireless power transfer systems with multiple transmitters and receivers. *Ph.D. dissertation, Dept. Elect. Eng., Toronto Univ., Toronto, Canada.*

Le, T. D. Georges Kaddoum, Tran, H.V., & Chadi Abou-Rjeily. (2021). Evolution Strategies for Lightwave Power Transfer Networks. *IEEE Wireless Communications Letters, 10*(11), 2572–2576. DOI: 10.1109/LWC.2021.3107731

Lee, G., Sung, Y., & Seo, J. (2016). Randomly-Directional Beamforming in Millimeter-Wave Multiuser MISO Downlink. *IEEE Transactions on Wireless Communications*, 15(2), 1086–1100. DOI: 10.1109/TWC.2015.2483493

Li, D. (2020). Ergodic Capacity of Intelligent Reflecting Surface-Assisted Communication Systems With Phase Errors. *IEEE Communications Letters*, 24(8), 1646–1650. DOI: 10.1109/LCOMM.2020.2997027

Li, D. (2020). Ergodic Capacity of Intelligent Reflecting Surface-Assisted Communication Systems With Phase Errors. *IEEE Communications Letters*, 24(8), 1646–1650. DOI: 10.1109/LCOMM.2020.2997027

Li, R., Zhao, Z., Chen, X., Palicot, J., & Zhang, H. (2014). TACT: A Transfer Actor-Critic Learning Framework for Energy Saving in Cellular Radio Access Networks. *IEEE Transactions on Wireless Communications*, 13(4), 2000–2011. DOI: 10.1109/TWC.2014.022014.130840

Liu, F., Yuan, W., Masouros, C., & Yuan, J. (2020). Radar-Assisted Predictive Beamforming for Vehicular Links: Communication Served by Sensing. *IEEE Transactions on Wireless Communications*, 19(11), 7704–7719. DOI: 10.1109/TWC.2020.3015735

Liu, F., Yuan, W., Masouros, C., & Yuan, J. (2020). Radar-Assisted Predictive Beamforming for Vehicular Links: Communication Served by Sensing. *IEEE Transactions on Wireless Communications*, 19(11), 7704–7719. DOI: 10.1109/TWC.2020.3015735

Liu, G., Sun, Z., & Jiang, T. (2019). Joint time and energy allocation for QoS-aware throughput maximization in MIMO-based wireless powered underground sensor networks. *IEEE Transactions on Communications*, 67(2), 1400–1412. DOI: 10.1109/TCOMM.2018.2874990

Liu, J., Xiong, K., Lu, Y., Wing, D., Zhong, Z., & Han, Z. (2020). Energy Efficiency in Secure IRS-Aided SWIPT. *IEEE Wireless Communications Letters*, 9(11), 1884–1888. DOI: 10.1109/LWC.2020.3006837

Liu, J., Xiong, K., Lu, Y., Wing, D., Zhong, Z., & Han, Z. (2020). Energy Efficiency in Secure IRS-Aided SWIPT. *IEEE Wireless Communications Letters*, 9(11), 1884–1888. DOI: 10.1109/LWC.2020.3006837

Liu, X., & Ansari, N. (2019). Toward Green IoT: Energy Solutions and Key Challenges. *IEEE Communications Magazine*, 57(3), 104–110. DOI: 10.1109/MCOM.2019.1800175

Lopez, A., Alves, H., Souza, R. D., & Montejo-Sanchez, S. (2019). Statistical Analysis of Multiple Antenna Strategies for Wireless Energy Transfer. *IEEE Transactions on Communications*, 67(10), 7245–7262. DOI: 10.1109/TCOMM.2019.2928542

Lopez, O. L. A., Mahmood, N. H., Alves, H., Lima, C. M., & Latva-aho, M. (2020). Ultra-Low Latency, Low Energy, and Massiveness in the 6G Era via Efficient CSIT-Limited Scheme. *IEEE Communications Magazine*, 58(11), 56–61. DOI: 10.1109/MCOM.001.2000425

Lu, X., Wang, P., Niyato, D., Kim, D. I., & Han, Z. (2015). Wireless Networks With RF Energy Harvesting: A Contemporary Survey. *IEEE Communications Surveys and Tutorials*, 17(2), 757–789. DOI: 10.1109/COMST.2014.2368999

Lu, X., Wang, P., Niyato, D., Kim, D. I., & Han, Z. (2015). Wireless Networks With RF Energy Harvesting: A Contemporary Survey. *IEEE Communications Surveys and Tutorials*, 17(2), 757–789. DOI: 10.1109/COMST.2014.2368999

Ma, S., Zhang, F., Li, H., Zhou, F., Wang, Y., & Li, S. (2019). Simultaneous Lightwave Information and Power Transfer in Visible Light Communication Systems. *IEEE Transactions on Wireless Communications*, 18(12), 5818–5830. DOI: 10.1109/TWC.2019.2939242

Mahmood, N. H., L´opez, O. L. A., Park, O. S., Moerman, I., Mikhaylov, K., Mercier, E., Munari, A., Clazzer, F., B¨ocker, S., & Bartz (Eds.), H. (2020). White paper on critical and massive machine type communication towards 6G. *6G Research Vision.*

Mohjazi, L., Muhaidat, S., Abbasi, Q. H., Imran, M. A., Dobre, O. A., & Renzo, M. D. (2022). Battery Recharging Time Models for Reconfigurable Intelligent Surfaces-Assisted Wireless Power Transfer Systems. *IEEE Transactions on Green Communications and Networking*, 6(2), 1173–1185. DOI: 10.1109/TGCN.2021.3120834

Mou, X., Gladwin, D. T., Zhao, R., & Sun, H. (2019). Survey on magnetic resonant coupling wireless power transfer technology for electric vehicle charging. *IET Power Electronics*, 12(12), 3005–3020. DOI: 10.1049/iet-pel.2019.0529

Nguyen, D. H., Matsushima, T., Qin, C., & Adachi, C. (2021). Toward Thing-to-Thing Optical Wireless Power Transfer: Metal Halide Perovskite Transceiver as an Enabler. *Frontiers in Energy Research*, 9, 679125. DOI: 10.3389/fenrg.2021.679125

Nguyen, V. L., Lin, P. C., Cheng, B. C., Hwang, R. H., & Lin, Y. D. (2021). Security and privacy for 6G: A survey on prospective technologies and challenges. *IEEE Communications Surveys and Tutorials*, 23(4), 2384–2428. DOI: 10.1109/COMST.2021.3108618

Niyato, D., Kim, D. I., Maso, M., & Han, Z. (2017). Wireless Powered Communication Networks: Research Directions and Technological Approaches. *IEEE Wireless Communications*, 24(6), 88–97. DOI: 10.1109/MWC.2017.1600116

Palitharathna, K. W. S., Suraweera, H. A., Godaliyadda, R. I., Herath, V. R., & Ding, Z. (2022). Lightwave Power Transfer in Full-Duplex NOMA Underwater Optical Wireless Communication Systems. *IEEE Communications Letters*, 26(3), 622–626. DOI: 10.1109/LCOMM.2021.3140117

Pan, C., Ren, H., Wang, K., Elkashlan, M., Nallanathan, A., Wang, J., & Hanzo, L. (2020). Intelligent Reflecting Surface Aided MIMO Broadcasting for Simultaneous Wireless Information and Power Transfer. *IEEE Journal on Selected Areas in Communications*, 38(8), 1719–1734. DOI: 10.1109/JSAC.2020.3000802

Pan, S. J., Tsang, I. W., Kwok, J. T., & Yang, Q. (2011). Domain Adaptation via Transfer Component Analysis. *IEEE Transactions on Neural Networks*, 22(2), 199–210. DOI: 10.1109/TNN.2010.2091281 PMID: 21095864

Paolini, E., Valcarenghi, L., Maggiani, L., & Andriolli, N. (2023). Real-time clustering based on deep embeddings for threat detection in 6G networks. *IEEE Access : Practical Innovations, Open Solutions*, 11, 115827–115835. DOI: 10.1109/ACCESS.2023.3325721

Park, S., & Choi, W. (2020). Transmitter Current Control and Receiver Coil Selection in Magnetic MIMO Power Transfer Systems. *IEEE Wireless Communications Letters*, 9(10), 1782–1785. DOI: 10.1109/LWC.2020.3004815

Psomas, C., & Krikidis, I. (2019). Energy beamforming in wireless powered mm-Wave sensor networks. *IEEE Journal on Selected Areas in Communications*, 37(2), 424–438. DOI: 10.1109/JSAC.2018.2872381

Qaraqe, M., Usman, M., Serbes, A., Ansari, I. S., & Alouini, M.-S. (2022). Power Hotspots in Space: Powering CubeSats via Inter-Satellite Optical Wireless Power Transfer. *IEEE Internet of Things Magazine*, 5(3), 180–185. DOI: 10.1109/IOTM.001.001.2200071

Rappaport, T. S., Eshar Ben-Dor, Murdock, J., & Qiao, Y. (2012). 38 GHz and 60 GHz angle-dependent propagation for cellular & peer-to-peer wireless communications. *International Conference on Communications*.DOI: 10.1109/ICC.2012.6363891

Renzo, M., Zappone, A., Debbah, M., Alouini, M.-S., Yuen, C., Rosny, J., & Tretyakov, S. (2020). Smart Radio Environments Empowered by Reconfigurable Intelligent Surfaces: How It Works, State of Research, and The Road Ahead. *IEEE Journal on Selected Areas in Communications*, 38(11), 2450–2525. Advance online publication. DOI: 10.1109/JSAC.2020.3007211

Sadman, A. S., Kar, A., Emon, A. R., & Mubin, M. A. (2024, April). An Overview of the Architecture and Challenges of RTM System by Integrating 6G Technology with UAV Network. In 2024 3rd International Conference on Advancement in Electrical and Electronic Engineering (ICAEEE) (pp. 1-6). IEEE.

Sharma, S., Darak, S. J., & Srivastava, A. (2017). Energy saving in heterogeneous cellular network via transfer reinforcement learning based policy. *Proc. Int. Conf. Commun. Syst. Netw.*, 397–398. DOI: 10.1109/COMSNETS.2017.7945411

Sharma, S., Darak, S. J., & Srivastava, A. (2019). Transfer reinforcement learning based framework for energy savings in cellular base station network. *Proc. Asia Pacific Radio Sci. Conf.*, 1–4. G., E. B., Bennis, M., & Debbah, M. (2014). Anticipatory caching in small cell networks: A transfer learning approach. *Proc. Kuvs Workshop Anticipatory Netw.*, 1–3. Ba¸ E., g, stu., Bennis, M., & Debbah, M. (2015). A transfer learning approach for cache-enabled wireless networks. *Proc. Int. Symp. Model. Optim. Mobile, Ad-Hoc, Wireless Netw.*, 161–166. DOI: 10.23919/URSIAP-RASC.2019.8738418

Sharma, S., Darak, S. J., Srivastava, A., & Zhang, H. (2016). A transfer learning framework for energy efficient Wi-Fi networks and performance analysis using real data. *IEEE International Conference on Advanced Networks and Telecommunications Systems (ANTS).* DOI: 10.1109/ANTS.2016.7947853

Sirohi, D., Kumar, N., Rana, P. S., Tanwar, S., Iqbal, R., & Hijjii, M. (2023). Federated learning for 6G-enabled secure communication systems: A comprehensive survey. *Artificial Intelligence Review*, 56(10), 11297–11389. DOI: 10.1007/s10462-023-10417-3 PMID: 37362891

Soyata, T., Copeland, L., & Heinzelman, W. (2016). RF Energy Harvesting for Embedded Systems: A Survey of Tradeoffs and Methodology. *IEEE Circuits and Systems Magazine*, 16(1), 22–57. DOI: 10.1109/MCAS.2015.2510198

Sun, H., Zhu, F., Lin, H., & Gao, F. (2018). Robust Magnetic Resonant Beamforming for Secured Wireless Power Transfer. *IEEE Signal Processing Letters*, 25(8), 1226–1230. DOI: 10.1109/LSP.2018.2849586

Tan, B., Zhang, Y., Pan, S. J., & Yang, Q. (2017). Distant domain transfer learning. *Proc. Nat. Conf. Artif. Intell.*, 2604–2610.

Tan, L., Guo, J. X., Huang, X., & Wen, F. (2016). Output power stabilisation of wireless power transfer system with multiple transmitters. *IET Power Electronics*, 9(7), 1374–1380. DOI: 10.1049/iet-pel.2015.0577

Tang, W., Chen, M. Z., Chen, X., Dai, J. Y., Han, Y., Di Renzo, M., Zeng, Y., Jin, S., Cheng, Q., & Cui, T. J. (2021). Wireless Communications With Reconfigurable Intelligent Surface: Path Loss Modeling and Experimental Measurement. *IEEE Transactions on Wireless Communications*, 20(1), 421–439. DOI: 10.1109/TWC.2020.3024887

Tang, W., Chen, M. Z., Chen, X., Dai, J. Y., Han, Y., Di Renzo, M., Zeng, Y., Jin, S., Cheng, Q., & Cui, T. J. (2021). Wireless Communications With Reconfigurable Intelligent Surface: Path Loss Modeling and Experimental Measurement. *IEEE Transactions on Wireless Communications*, 20(1), 421–439. DOI: 10.1109/TWC.2020.3024887

Tran, H.-V., & Kaddoum, G. (2018). RF Wireless Power Transfer: Regreening Future Networks. *IEEE Potentials*, 37(2), 35–41. DOI: 10.1109/MPOT.2017.2749118

Tran, H. V., Kaddoum, G., Elgala, H., Abou-Rjeily, C., & Kaushal, H. (2020). Lightwave Power Transfer for Federated Learning-Based Wireless Networks. *IEEE Communications Letters*, 24(7), 1472–1476. DOI: 10.1109/LCOMM.2020.2985698

Uguroglu, S., & Carbonell, J. (2011). Feature selection for transfer learning. *Proc. Joint Eur. Conf. Mach. Learn. Knowl. Discovery Databases,* 430–442.

Union, I. (2015). IMT traffic estimates for the years 2020 to 2030. *ITU-R Radiocommun. Sector ITU, Geneva, Switzerland, Tech. Rep. ITU-R M*, 2370.

Wang, H., Zhang, Z., Zhu, B., Dang, J., Wu, L., Wang, L., Zhang, K., Zhang, Y., & Li, G. Y. (2021). Performance Analysis of Multi-Branch Reconfigurable Intelligent Surfaces-Assisted Optical Wireless Communication System in Environment With Obstacles. *IEEE Transactions on Vehicular Technology*, 70(10), 9986–10001. DOI: 10.1109/TVT.2021.3092722

Wang, M., Lin, Y., Tian, Q., & Si, G. (2021). Transfer Learning Promotes 6G Wireless Communications: Recent Advances and Future Challenges. *IEEE Transactions on Reliability*, 70(2), 790–807. DOI: 10.1109/TR.2021.3062045

Wang, Y., Chen, Y., Dai, H., Huang, Y., & Yang, L. (2017). *A learning-based approach for proactive caching in wireless communication networks*. DOI: 10.1109/WCSP.2017.8170984

Wu, Q., & Zhang, R. (2020). Joint Active and Passive Beamforming Optimization for Intelligent Reflecting Surface Assisted SWIPT Under QoS Constraints. *IEEE Journal on Selected Areas in Communications*, 38(8), 1735–1748. DOI: 10.1109/JSAC.2020.3000807

Wu, Q., & Zhang, R. (2020a). Weighted Sum Power Maximization for Intelligent Reflecting Surface Aided SWIPT. *IEEE Wireless Communications Letters*, 9(5), 586–590. DOI: 10.1109/LWC.2019.2961656

Wu, Q., & Zhang, R. (2020b). Joint Active and Passive Beamforming Optimization for Intelligent Reflecting Surface Assisted SWIPT Under QoS Constraints. *IEEE Journal on Selected Areas in Communications*, 38(8), 1735–1748. DOI: 10.1109/JSAC.2020.3000807

Yan, H., Chen, Y., & Yang, S.-H. (2020). UAV-Enabled Wireless Power Transfer With Base Station Charging and UAV Power Consumption. *IEEE Transactions on Vehicular Technology*, 69(11), 12883–12896. DOI: 10.1109/TVT.2020.3015246

Yang, G., Reza, M., & Zhang, R. (2017). Magnetic MIMO Signal Processing and Optimization for Wireless Power Transfer. *IEEE Transactions on Signal Processing*, 65(11), 2860–2874. DOI: 10.1109/TSP.2017.2673816

Yang, G., Vedady, M. R., & Zhang, R. (2016). *Magnetic beamforming for wireless power transfer. ArXiv*. Cornell University., DOI: 10.1109/ICASSP.2016.7472415

Yang, L., Zeng, Y., & Zhang, R. (2018). Wireless Power Transfer With Hybrid Beamforming: How Many RF Chains Do We Need? *IEEE Transactions on Wireless Communications*, 17(10), 6972–6984. DOI: 10.1109/TWC.2018.2865313

Yang, Q., Liu, Y., Chen, T., & Tong, Y. (2019). Federated machine learning: Concept and applications. *ACM Transactions on Intelligent Systems and Technology*, 10(2), 1–12. DOI: 10.1145/3298981

Ying, W., Zhang, Y., Huang, J., & Yang, Q. (2018). Transfer learning via learning to transfer. *Proc. Int. Conf. Mach. Learn.*, 5072–5081.

Zeng, Y., Clerckx, B., & Zhang, R. (2017). Communications and signals design for wireless power transmission. *IEEE Transactions on Communications*, 65(5), 2264–2290. DOI: 10.1109/TCOMM.2017.2676103

Zhang, H., Shlezinger, N., Guidi, F., Dardari, D., Imani, M. F., & Eldar, Y. C. (2022). Near-Field Wireless Power Transfer for 6G Internet of Everything Mobile Networks: Opportunities and Challenges. *IEEE Communications Magazine*, 60(3), 12–18. DOI: 10.1109/MCOM.001.2100702

Zhang, Y., Chen, D., & Jiang, T. (2021). Robust Beamforming Design for Magnetic MIMO Wireless Power Transfer Systems. *IEEE Transactions on Signal Processing*, 69, 5066–5077. DOI: 10.1109/TSP.2021.3108686

Zhang, Y., Chen, D., & Jiang, T. (2021). Robust Beamforming Design for Magnetic MIMO Wireless Power Transfer Systems. *IEEE Transactions on Signal Processing*, 69, 5066–5077. DOI: 10.1109/TSP.2021.3108686

Zhao, L., Wang, Z., & Wang, X. (2021). Wireless Power Transfer Empowered by Reconfigurable Intelligent Surfaces. *IEEE Systems Journal*, 15(2), 2121–2124. DOI: 10.1109/JSYST.2020.2985121

Zhao, P., Hoi, S. C., Wang, J., & Li, B. (2014). Online transfer learning. *Artificial Intelligence*, 216, 76–102. DOI: 10.1016/j.artint.2014.06.003

Zhao, Q., & Grace, D. (2024). *Agent transfer learning for cognitive resource management on multi-hop backhaul networks*. Ieee.org. https://ieeexplore.ieee.org/document/6633554,1-10

Zhao, Q., Grace, D., & Clarke, T. (2015). Transfer learning and cooperation management: Balancing the quality of service and information exchange overhead in cognitive radio networks. *Transactions on Emerging Telecommunications Technologies*, 26(2), 290–301. DOI: 10.1002/ett.2913

Zhao, Y., Huang, J., Li, X., & Xu, C. (2020). Optimal node placement for magnetic relay and MIMO wireless power transfer network. *Proc. 15th Int. Conf. Wireless Algorithms Syst. Appl.*, 743–754. DOI: 10.1007/978-3-030-59016-1_61

Zhao, Y., Li, X., Ji, Y., & Xu, C.-Z. (2019). Random Energy Beamforming for Magnetic MIMO Wireless Power Transfer System. *IEEE Internet of Things Journal*, 7(3), 1773–1787. DOI: 10.1109/JIOT.2019.2962699

Zong, B., Fan, C., Wang, X., Duan, X., Wang, B., & Wang, J. (2019). 6G Technologies: Key Drivers, Core Requirements, System Architectures, and Enabling Technologies. *IEEE Vehicular Technology Magazine*, 14(3), 18–27. DOI: 10.1109/MVT.2019.2921398

Chapter 2
RFID-Enabled Wireless Charging System for Electric Vehicles

Sarthak Bisht
SRM Institute of Science and Technology, India

Tia Mittal
SRM Institute of Science and Technology, India

Karan Bhambhani
SRM Institute of Science and Technology, India

T. Y. J. Naga Malleswari
SRM Institute of Science and Technology, India

ABSTRACT

The RFID-Enabled Wireless Charging System for Electric Vehicles (EVs) represents a significant advancement in sustainable transportation infra- structure. By integrating Radio Frequency Identification (RFID) technology with wireless charging capabilities, the system not only enhances user experience but also promotes environmental sustainability. One key feature of the system is its security and efficiency. Charging is initiated only when a valid RFID card is swiped, ensuring that only authorized users can access the charging facility. This helps prevent unauthorized usage and ensures that the system operates smoothly and efficiently. The system allows EV batteries to be charged using solar panels, tapping into renewable energy sources and reducing reliance on non-renewable energy. This not only reduces the carbon footprint of EV charging but also promotes the use of clean energy in transportation. The wireless charging capability of the system is facilitated by two copper coils one on the charging pad and the other integrated into the EV.

1. INTRODUCTION

The rise of electric vehicles (EVs) as a disruptive technology in transportation and energy has ushered in a new era of energy efficiency, environmental sustainability, and technology outcomes. Electric cars, powered by electricity stored in onboard batteries, are a great alternative to traditional internal

DOI: 10.4018/979-8-3693-8799-3.ch002

combustion engine cars and bring many benefits to people and the world. Global interest in electric vehicles continues to grow due to concerns about air pollution, climate change, and energy security, with demand for electric vehicles becoming increasingly effective. This study expands on the motivation behind the development of consumer electronics in trucks, addressing key concepts such as benefits, environmental impact, economics, technological advancement, and policy considerations. The transition to electric vehicles represents a revolution in transportation, and electric vehicles will play an important role in reducing household emissions, reducing pollution, and solving transportation problems. The energy efficiency of electric vehicles, combined with advances in battery technology and renewable energy connections, enables them to achieve significant benefits in transportation, being clean and green. However, widespread adoption of EVs depends on energy efficiency and availability, demonstrating an urgent need for consumer electronics that can meet these needs, differentiate between EV owners, and facilitate connected charging.

Electrification is another important reason why consumers are paying for electric vehicles. Optimizing the charging process to reduce energy consumption, increase charging, and extend battery life is crucial for the long-term support of electric vehicles. Smart payment algorithms, smart grid integration, and robust load management strategies play an important role in achieving energy savings while balancing grid stability, security (Muralikrishnan et al., 2020), and demand-side management goals. In addition, integrating renewable energy sources such as solar and wind energy (Vinayak Rangrao Patil, 2020) into electric vehicle charging further increases the sustainability of electric vehicles. Environmental factors also emphasize the importance of user-friendly electric vehicle charging systems. By encouraging the use of electric vehicles and reducing dependence on fossil fuel-powered vehicles, consumers are spending money to help improve air quality, reduce noise, and protect the environment. Analysis of the electrical life cycle, including supply, production, vehicle operation, and end-of-life repair, highlights the importance of the need to manage the business. Integrating safety standards into the design, operation, and management of the charging system makes the electric vehicle a valuable contribution to global efforts to reduce climate change and achieve carbon neutrality targets.

Business and consumer preferences are driving the evolution of consumer electronics. Understanding the needs, preferences, and behaviors of electric vehicle owners and potential customers is essential to creating solutions that meet customer needs, increasing and improving the overall user experience. Factors such as affordability, charging options, charging speed, interoperability, network coverage, and convenience will influence consumers' decisions and behaviors to use electric vehicles. Therefore, promote collaboration between business partners, policy makers, service providers, technology developers, and end users to create energy- and user-centric charging infrastructure to adapt to the changing market of electric vehicles. Technological advances in electric vehicle charging, including charging technology, wireless charging solutions, vehicle-to-grid integration (V2G), and smart payment features, help improve the customer experience. Fast charging can quickly top up electric car batteries, shorten charging time, and make it more convenient for users. Wireless charging solutions simplify the charging process and make it easier for users by eliminating the need for physical cables and connectors. V2G integration allows electric vehicles to interact with the grid, provide grid services, improve energy efficiency, and support bidirectional energy use. Smart payment can instantly monitor, control, and optimize the payment process based on the grid, energy costs, customer preferences, and power outages, thereby improving overall performance and user satisfaction. Policy decisions and administrative procedures also play an important role in the payment process. Encouraging the widespread use of electronic devices for charging vehicles. Government grants, subsidies, tax credits, and regulations encourage investment in cooperatives, promote cooperative models, encourage public-private partnerships, and assist and support the creation of new solutions. Policy leaders on decarbonizing transportation, promoting clean energy, and promoting sustainable urban living will increase the use of renewable energy, access to electricity, and economic growth.

Dimitriadou et al. (2023) Wireless charging is possible by galvanic isolation between the input side and the output circuitry. Feng, H.et al (2020). Due to the fast development of power electronics and improved semiconductor materials, wireless charging is widely used to charge home appliances and various portable wearable devices Rigogiannis et al (2022).

EV wireless charging has many pros as follows:

- Easy, safe, user-friendly charging with less maintenance cost, as there are no mechanical parts.
- No risk of using cables.
- Increase the lifespan of charging infrastructure by installing the charging transmitter underground, avoiding exposure to dangerous environmental conditions.
- Avoids vandalism that is theft of charging cable, other component removal, etc.

The cons of EV wireless charging are as follows:

- Insufficient infrastructure for necessary installation. Having static and dynamic wireless charging stations in highways will facilitate new building infrastructure. Ahmad et al. (2017)
- The current electrical infrastructure. The topologies such as vehicle to grid (V2G) will solve the problem.
- Transferring power without loss and precise alignment of transmitting and receiving coils to inductive coupling.
- Power pads mismatch Zhao, J. et al. (2016)

RFID is a wireless technology to identify and track the objects using radio waves through embedded tags automatically. RFID technology plays a vital role in the electric vehicle ecosystem. Secure authentication of EV owners before accessing charging stations, maintaining EV owner profiles, their preferences, and charging history; authorized EV owners can charge in any of the networked charging stations; seamless payment integration by pre-registered payment methods for automatic billing; offering various subscription plans and their payment automatically; enabling interoperability, allowing users to charge vehicles anywhere; personalized charging; enhancing user experience; fleet management are the advantages of RFID-enabled EV charging. Convenience, scalability, interoperability, and security are the main features of RFID in the EV charging domain.

Interoperability and Standardization:

Diverse vehicle architectures such as different manufacturers, communication protocols, and voltage levels leads to interoperability issues and challenges the integration of RFID technology in wireless charging systems. Failure happens if there is a misalignment between the charging infrastructure and vehicle. (Zhao, Q et.al, 2019) Compatibility issues between EVs brands and chargers due to lack of standardized data format. SAE J2954 for wireless charging and ISO/IEC 18000 for RFID is the existing standard that provide basic guidelines. These standards need to be revised considering EV specific requirements for a comprehensive framework. Uniform communication protocols are difficult for secure communication between vehicle and charging station including standardization of encryption for data security and compatibility issues. Interoperable certification processes ensure interoperability and improve consumer confidence (Kim, J,et.al, 2021), (Chen, X, et.al, 2020). The challenges faced with EVs are charging station availability, EV cost, battery capacity, impact on the power grid. To address these challenges (Ishtiak Mahmud, et al, 2023) considered user behaviour, economic culture, travel demand, psychological outlook of Malaysians and current government policies, towards EV adoption.

The alternative charging technologies are mentioned in the table.

Table 1. Various charging technologies

Charging technology	Overview	Strength	Weakness
Plug-in Charging	Direct charging via cable	High efficiency, low cost, more usage	Manual plugging and unplugging, limited durability
Conductive charging	Via charging pads	High efficiency, low cost, more usage	More maintenance
Dynamic charging	Charging when vehicle in motion	No stop required for charging, ideal for public transportation	Highcost infrastructure, vehicle speed
RFID enabled wireless charging	Use of electro-magnetic fields	Fully automated, Enhance security	Lower efficiency, High installation and maintenance cost, Limited charging speed

Cybersecurity

Data Interception which is a privacy breach by tapping RFID data during transmission, Cloning and spoofing through unauthorized charging by imitating legitimate vehicles by counterfeiting RFID tags, Denial of service and malware injections are some of the cyber security threats in RFID enabled systems (Wang, H.,et.al, 2020). Securing data using advanced encryption protocols, mutual authentication of RFID readers and tags, Firewalls and Intrusion detection systems (Zhang, Y.,et.al, 2022)to protect backend architecture, regular software updates are few mitigation strategies (Li, X.,et.al, 2021). The main security concerns of RFID technology are limited computational capacity of RFID tags for encryption making the tags passive and unsecured wireless communication by eavesdropping.

Trends in Global EV Adoption Rates

In 2023, 4% of rise from 2019 and global EV sales reached to 14% making China, Europe and US as leading markets. 27 million crossed in 2023 doubling since 2020. By 2030 there will be rise of 30% in EV sales (IEA, 2023).

Government Policies for EV promotion:

$7.5 million was allocated by the government of US for EV charging infrastructure. Europe proposed 55 programs to support RFID Enabled systems and standardized charging technologies. ISO 15118 mandates for interoperability standards between EVs and charging infrastructure compatible with RFID systems (Park, J., et.al, 2022). EV manufacturers getting many tax benefits. China and Germany started pilot projects for RFID enabled wireless charging to increase EV adaption and reduce carbon emissions (Chen, Q., et.al, 2020).

In summary, the motivation behind the development of user-friendly electric vehicle charging is multifaceted, including profits, environmental impacts, market conditions, technological advances, consumer preferences, and policy considerations. By addressing these incentives and challenges, stakeholders in the EV ecosystem can create a sustainable and accessible payment system that supports the transition to a clean environment and greener and better transportation in the future. The transportation section is facing numerous challenges as it is dependent on the availability of resources, fossil fuels, and environmental conditions. R. Das et al. (2020). (Collaboration, innovative solutions, and effective collaboration are key to realizing the full potential of EV charging customers and unlocking the benefits of EV charging. It is offered voluntarily to people, society, and the world. Due to the convenience, security, and lower cost features, RFID-enabled wireless charging is suitable to charge EVs.

EV charging infrastructure is one of the practical barriers for EV sales. Infrastructure cost, power grid limitations, space constraints, interoperability, and environmental impact are some of the hurdles to scaling the EV charging infrastructure. Technological innovations in ultra-fast chargers, strategic planning in the deployment of infrastructure, deploying energy storage systems, and introducing incentives through government policies may give positive solutions.

Case study: RFID-enabled wireless charging for taxis was first introduced in Oslo, Norway, in 2024. The RFID system provides authentication and a simplified billing system. Operational efficiency was increased. The availability of vehicles increased for passengers. Carbon emissions were reduced. System durability enhanced. There are some challenges, such as high initial costs, too faced in this pilot project.

Energy sources, operational efficiency, and infrastructure are to be considered from EV production, use, and disposal to evaluate the environmental impact of EVs and their charging systems. Fossil fuels give higher emissions of 500-800 g CO_2e/kWh, whereas renewable energy gives zero emissions. Advanced charging stations can reduce life cycle emissions, providing sustainable transportation solutions. To address pollution gas emissions, renewable energy resources (RER) playing vital role in increase in EV adoption (Anis Ur Rehman, et.al, 2024). RERs are introduced in many services such as PV systems, EV batteries, wind farms etc. Grid tired solar based EV charging stations getting wider recognition among all renewable energy integrations due to the minimal emissions and cost-effective energy production (W. Zhong, et.al 2018). Solar powered charging stations have many advantages such as enhanced efficiency, efficient scheduling and EV-charging related activities (B.N. Alhasnawi,et.al, 2021)

2. LITERATURE SURVEY

Chung, Ching-Yen et al (2013) reported an RFID mesh network design for smart charging for electric vehicles. In this context, as the number of electric vehicles (EVs) on the road increases, electric charging plays an important role in meeting the needs of both players in the region and EV users. Ching's paper presents a network RFID system for user identification and payment authorization as part of a smart payment system that enables tracking and control of payments. Zigbee-based mesh RFID provides an effective solution to identify and authorize vehicle payments between EV drivers and ensure EV payments occur efficiently while adhering to project boundaries and meeting expectations. J. Joyce Jacob et al. (2023) reported a good idea for wireless payment of electric vehicles using RFID along with good route planning. The problem of carbon dioxide emitted by gas-powered vehicles into the environment is one of the main factors that has led to the rise and popularity of electric vehicles (EVs) in today's smart cities in recent years. In this context, electric vehicles provide an environmentally friendly environment for modern smart cities. However, there are many problems (smart payment, route planning, communication, etc.) in the use of electric vehicles in today's smart cities. Therefore, new technologies and solutions are needed to increase the performance of existing equipment. In response to these problems, this project proposes a new wireless charging technology for electric vehicles using RFID tags and provides optimal charging through an optimized schedule. There is a good charge when charging electric vehicles.

(S. Karthikeyan et al., 2021) reported a solar energy-based fast charging system for electric vehicles. Among these, photovoltaic energy production appears to be the most efficient energy production, green and clean energy used today. (K. Parvin et al. 2022). Photovoltaic cells convert the energy produced from the sun into electricity with the help of solar cells inside. Here we are building a solar-powered fast charging station for electric cars. Charging LEV batteries must be done clearly. Here we created a small and useful garage for electric cars. Here, the ARDUINO UNO serves as the heart of the entire circuit. This clears all circuits and continues operating. Therefore, ARDUINO recognizes the physical card detected by RFID and then flags the data to Arduino for further processing. He then pointed the Arduino and started charging the EV from the electric field. Many factors, such as cost, compactness, and efficiency, are considered important requirements of the system concept. Arduino 8 restarts the timer, showing the charging time. Outdoors, this can be used to intelligently charge electric vehicles. Design challenges and future work are also discussed in detail. Depends on card payment. This will provide enough power for the electric car. (Cai, Kunhai et al., 2017) introduced a six-degree-of-freedom positioning system based on robust RFID for UAVs. Although unmanned aerial vehicles (UAVs) are mostly used outdoors, there is also interest in using UAVs indoors. Positioning a drone precisely in an indoor environment without Global Positioning System (GPS) service is an interesting and difficult task. In this project, we demonstrate RFUAV, a drone system enhanced with radio frequency identification (RFID) that provides 6 degrees of freedom (6-DoF) behavior for drones. For RFUAV, three or more ultra-high frequency (UHF) RFID tags are attached to the drone and interrogated by a commercial off-the-shelf (COTS) RFID reader with multiple antennas. Based on the measurement level of the RFID tag field, the Bayesian filter-based algorithm RFUAV's RFID tracker is used to track the tag's position in the global management system. RFUAV's pose estimator calculates the 6-DoF bit pose of the UAV based on the tag position. We tested the performance of RFUAV in a representative model-rich environment, achieving a position error of 0.04 m and an orientation error of 2.5 degrees. With the development of electric vehicles, many studies are being carried out on the layout and technology of public charging stations. started. Global business management system. RFUAV's pose estimator calculates the 6-DoF bit pose of the UAV based on the tag position. We tested the performance of RFUAV in a representative, sample-rich environment and achieved a position error of 0.04 m and a direction error of 2.5 degrees. With the development of electric vehicles, many studies have begun to be carried out on the management and identification of public parking lots. The system and technology were designed for electric car users. Renewable energy connected to a solar panel has the necessary features for a grid car in offices, schools, homes, etc., developed an RFID-based security access system using the frequently used AVR microcontroller. Referring to this article, we examined the connection between RFID and AVR.

EV Battery

Due to the great benefits, they provide, electric vehicles (EVs) are expected to replace internal combustion engines (ICEs) in transportation. Nowadays, electric batteries need to be charged wirelessly whenever possible. This study provides a comprehensive review of various electric vehicle wireless charging methods. There are two ways to distribute power wirelessly to charge EV batteries: dynamic EV charging and EV dynamic charging. Static wireless EV battery charging technology uses capacitive and inductive methods for power transfer; however, only the inductive method is used in dynamic wireless EV battery charging technology. This work provides a comprehensive evaluation of these systems, focusing on various switches and controllers for electronic charging, magnetic coupling inductor core types, and wireless power transfer (WPT) systems. In addition, this article also discusses the design considerations of the static wireless electric vehicle battery charging system and examines its balance. The report also describes the challenges and future developments in wireless charging of EV batteries. Electric vehicles (EVs) have attracted great attention recently due to their long battery life and low fuel consumption. As many devices can now be connected with the increasing Internet of Things (IoT), limited battery capacity and the lack of charging or battery replacement are the two biggest problems with electric cars. Creating the necessary infrastructure and having a reliable battery management system (BMS) that can estimate additional costs are two solutions. For some electric vehicles, battery replacement may be an option. A lower number of EV charging stations is a major problem for the wide adoption of EVs. K. Alanne, (2019), N. Moghadasi, et al. (2022).

Many researchers focus on EV scheduling for the efficient use of charging stations. A. AbuElru et al (2020) present a new approach to improve the economic feasibility of low-carbon emission prices by the optimization of EV charging schedules and EV integration in China's carbon market. X. Ding et al. (2023). To increase the net profit with the presence of EVs in the smart grid and to efficiently manage energy, it develops a model using historical data and deep learning methods. E.A. Rene et al. (2023) propose a new approach for optimization of the placement of plug-in EV charging stations in distributed networks with the help of particle swarm optimization and hybrid genetic algorithm.

3. PROPOSED WORK

The system of electric vehicle wireless charging using RFID (Radio-Frequency Identification) technology represents a significant breakthrough in the realm of electric mobility, offering a streamlined and user-friendly approach to charging EVs. At its core, this system operates on the principles of electromagnetic induction, where a charging pad generates a magnetic field that transfers energy wirelessly to the vehicle's battery through an induction coil. The key components of this system are the charging pad and the RFID tag. The charging pad, typically installed on the ground or at designated charging stations, generates the magnetic field required for wireless energy transfer. On the other hand, the RFID tag, affixed to the EV, contains crucial information such as the vehicle's identification number, battery capacity, and charging requirements. When an EV equipped with an RFID tag enters the proximity of the charging pad, the RFID reader within the pad authenticates the vehicle and initiates the charging process automatically.

The authentication and charging initiation processes are seamless and efficient, eliminating the need for physical cables or connectors. This not only simplifies the charging experience for EV owners but also reduces wear and tear on charging equipment. Additionally, the RFID tag allows the system to monitor and optimize the charging process based on the vehicle's specific requirements. For instance, the system can adjust the charging rate and duration to optimize battery life and charging efficiency, thereby enhancing the overall performance and longevity of the EV's battery system. One of the key advantages of the RFID-based wireless charging system is its convenience and ease of use. EV owners no longer need to manually connect charging cables or worry about compatibility issues with different charging connectors. Instead, the system seamlessly initiates the charging process as soon as the vehicle enters the range of the charging pad, offering a hassle-free and user-centric charging experience. This convenience factor is particularly beneficial for increasing EV adoption rates and encouraging more consumers to switch to electric vehicles. Moreover, the RFID-based wireless charging system contributes to a cleaner and more sustainable transportation ecosystem. By eliminating the use of traditional fuel-powered vehicles and reducing reliance on fossil fuels, EVs powered by wireless charging systems help mitigate air pollution, reduce greenhouse gas emissions, and promote environmental sustainability. Additionally, the system's ability to integrate renewable energy sources further enhances its eco-friendly credentials, aligning with global efforts to combat climate change and transition towards cleaner energy solutions. While the RFID-based wireless charging system offers numerous benefits, it also faces certain challenges and considerations that need to be addressed for widespread adoption. One such challenge is the need for infrastructure development to expand the coverage of charging pads and create a robust charging network. Collaborative efforts between government agencies, private companies, and EV manufacturers are essential to accelerate infrastructure deployment and ensure seamless accessibility for EV owners across various locations. Additionally, interoperability issues and standardization efforts play a crucial role in enhancing compatibility between different charging systems and EV models, ensuring a cohesive and user-friendly charging experience for all EV users.

In conclusion, the system of electric vehicle wireless charging using RFID technology represents a transformative innovation in the EV industry, offering a convenient, efficient, and sustainable method of charging EVs. With ongoing advancements and collaborative initiatives, this system has the potential to revolutionize the way we charge (Shun-Neng Yang et al., 2011) electric vehicles, driving widespread adoption, reducing environmental impact, and shaping the future of transportation towards a cleaner and greener paradigm.

4. SYSTEM ARCHITECTURE

Figure 1. Architecture Diagram

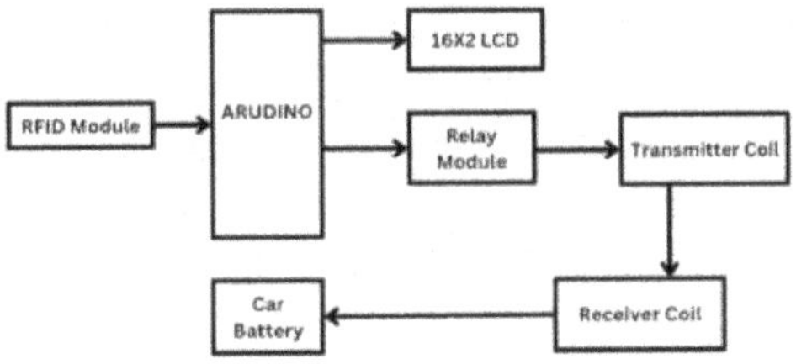

Figure 1 has several important features. The RFID module is used for authentication and communication with the Arduino Uno, which acts as a control system. 16x2 LCD provides feedback to users. The relay module controls the flow of energy to the transmitter coil, which transmits the energy wirelessly. The receiving coil in the electric car receives this energy and charges the car's battery. The system uses an Arduino Uno to control authentication, power flow, and user interaction to enable safe and efficient wireless charging of electric vehicles.

Figure 2. Circuit Diagram

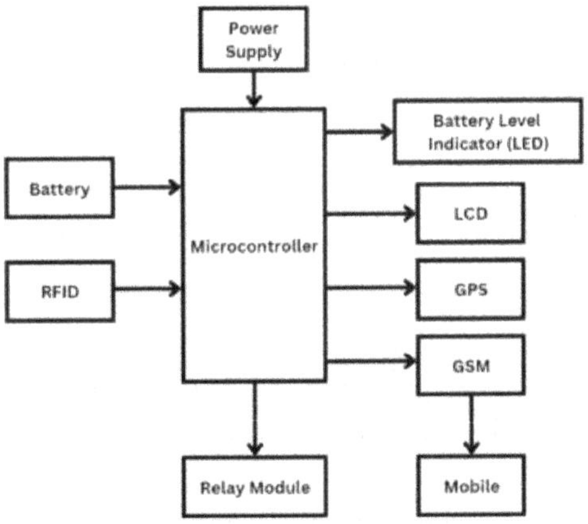

The circuit diagram in Figure 2 includes the battery, RFID module, microcontroller, relay module, power supply, battery indicator (LED), LCD, GPS, GSM module, and mobile device showing the rise of light. The battery powers the system, and the RFID module authenticates the user through the microcontroller. The microcontroller controls relay modules for device control, LCD interfaces for user input, and communication with GPS and GSM modules. The battery level indicator LED provides a good indication of battery, while the GSM module communicates with mobile devices for remote monitoring or control. This combination makes it versatile and efficient, providing a variety of functions from user authentication to remote control.

Jumper cables are used to connect the RFID MFRC522 module, LCD 16x2 display and relay module to the Arduino Uno. Connect the relay module to MOSFET Z44 to control wireless charging. Use a solar panel to charge the batteries of your Arduino Uno and other devices. Finally, use the adapter to power the setup. Control logic can be created using Arduino Uno to control users via RFID, display information on the LCD, and control devices via relay devices. Testing should ensure that all components are working properly and the Arduino Uno can be controlled as intended. In this system, the battery that uses electrical power sends it to the (Muralikrishnan, P, et.al, 2020) wireless power transfer receiver coil using the electrical power obtained from the wireless power transfer on the road. In addition, electricity is obtained from existing electrical energy, Solar panels (Zhao, J.,et.al,.2016). Here, the movement of the car is controlled by the Arduino board. RFID reader connected to Arduino. After parking at the wireless charging station, the user slides the RFID tag into the RFID reader. The principle of wireless car charging is inductive connection. When the user scans the RFID, according to the code integrated into the Arduino, the Arduino ensures that the power to be sent to the transmitter sends electricity from the transmitter, and the power to be sent to the vehicle from the receiver coil on the right. Inductive coupling. Inductive coupling between two coils is the principle of wireless transmission and wireless communication.

5. RESULTS

The analysis of the proposed RFID-base system for electric system for electric vehicles (EVs) encompasses various dimensions critical to understanding its potential results and implications. One key aspect is the user experience and convenience offered by the system. By eliminating the need for physical cables or connectors, the system stream- lines the charging process and enhances user satisfaction. EV owners can simply park their vehicles over the charging pad, and the system automatically initiates the charging process, making it intuitive and hassle-free. This convenience factor is expected to positively impact EV adoption rates, as users are more likely to embrace electric mobility when presented with seamless and user-friendly charging solutions. Moreover, the charging efficiency and optimization capabilities of the system play a pivotal role in its overall performance. The system's ability to adjust charging parameters such as charging rate, duration, and battery health based on data from the RFID tag ensures optimal energy transfer and minimizes energy losses. This optimization leads to faster charging times, improved battery life, and enhanced overall performance of EVs. Analyzing the system's charging efficiency metrics, such as energy transfer efficiency and charging speed, can provide insights into its effectiveness in meeting EV owners' needs and expectations. From an environmental perspective, the RFID-based wireless charging system has significant implications for sustainability and emissions reduction. By eliminating the use of traditional fuel-powered vehicles and reducing reliance on fossil fuels, EVs powered by wireless charging systems contribute to mitigating air pollution and reducing greenhouse gas emissions MD.S. Adnan Khan et al (2019), M.A. Judge et al (2022)

Furthermore, the system's integration with renewable energy sources further enhances its eco-friendly credentials, aligning with global efforts to combat climate change and promote cleaner energy solutions. Conducting a comprehensive environmental impact assessment, including lifecycle analysis and emissions reduction estimates, can shed light on the system's sustainability benefits and environmental footprint compared to traditional charging methods. Cost-effectiveness is another crucial aspect that influences the feasibility and adoption of the RFID-based wireless charging system. Evaluating the system's initial installation costs, maintenance expenses, energy savings, and potential revenue streams from charging services is essential to assess its overall financial viability and return on in- vestment (ROI). Factors such as reduced maintenance costs (due to fewer moving parts and wear), increased EV adoption rates due to enhanced user experience, and potential revenue opportunities from offering charging services can contribute positively to the system's cost-effectiveness and long-term sustainability. Conducting a thorough cost- benefit analysis and ROI assessment would provide valuable insights into the economic feasibility and business case for deploying the RFID-based wireless charging system. Scalability and future developments are also critical considerations for the system's analysis. As EV adoption rates continue to rise, scalability becomes essential to meet growing demand and ensure seamless charging experiences for a larger user base. Assessing the system's scalability potential, compatibility with evolving EV technologies, and adaptability to future market trends and regulatory requirements is crucial for its long-term success and relevance in the electric mobility ecosystem. Addition of charging stations with RFID enabled, with varying power needs, allowing incremental upgrades, connected with smart grids, leveraging renewable energy resources in urban, sub urban and rural areas can accelerate the EV adaption (Park, C. et.al, 2022). Roadway infrastructure for dynamic charging, embedded coil systems in parking spaces, minimal space for charging stations in dense areas are some of the factors to be considered to improve EV adaption (Wu, T., et.al,2020) Rate of EVs increased when there are incentives from government and public-private partnerships in foundational infrastructure. Public and semi-private charging stations at dense population areas such as malls, residential complexes and workplaces addresses assess the diverse settings. Additionally, exploring potential advancements such as faster charging technologies, enhanced energy management systems, and integration with smart grid technologies can provide insights into the system's evolution and continuous improvement. In conclusion, the analysis of the proposed RFID-based wireless charging system for EVs encompasses key dimensions such as user experience, charging efficiency, environmental impact, cost-effectiveness, scalability, and future developments. By examining these aspects comprehensively, stakeholders can gain valuable insights into the system's potential results, implications, and areas for optimization, driving informed decision-making and strategic planning for the deployment and adoption of this innovative charging technology in the electric mobility sector.

Figure 3. Waiting for Identification

In Figure 3, the charging station is in standby mode, waiting for RFID identification of an electric vehicle (EV) to start charging. When the electric vehicle's RFID card communicates with the payment station's RFID reader, the system verifies the vehicle's identity information, ensuring the payment process begins smoothly.

Figure 4. Invalid RFID Credentials

Figure 4 shows an example where the use or non-use of RFID may result in the user's credentials not being verified. Therefore, the payment process does not start because the system requires a valid RFID card for authentication before starting the payment process.

Figure 5. Valid RFID Credentials

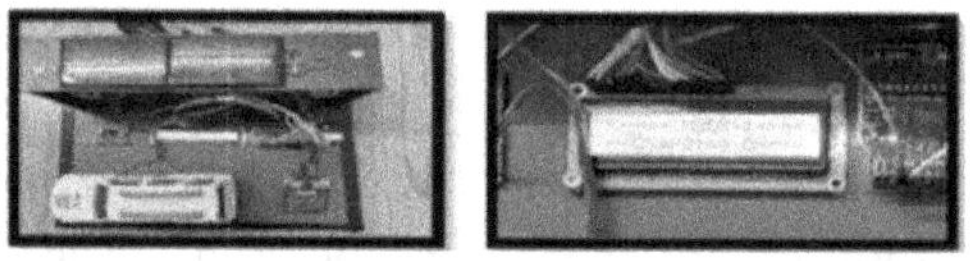

Figure 5 shows the charging process of the RFID-based wireless charging system. When a valid RFID tag is issued, the system verifies the user's credentials to ensure payment authorization. This authentication process verifies the user's identity and eligibility for paid services. Once tested, the system will initiate the charging process and allow the electric vehicle to start receiving power wirelessly. Seamless integration of RFID technology enables safe and efficient charging, increases customer convenience and encourages the adoption of wireless charging solutions in vehicle electrical. The evolution of the RFID-based wireless charging system from Figure 3 to Figure 5 shows that the process is seamless and efficient for electric vehicle (EV) charging. In Figure 3, the toll booth is waiting for the RFID identification of the electric vehicle before the payment can begin, indicating that the system is ready to start payment after authentication. Integration of RFID technology enables the transition to a temporary payment system when the vehicle's electronic RFID card communicates with the payment centre's RFID reader, ensuring the vehicle's identity is verified and the payment process begins smoothly. Figure 4 describes the situation where when an invalid RFID card is detected, the system will reject the user's certificate and then stop the payment process. This ensures the safety and security of payments, highlighting the importance of good RFID tags for authentication. This authentication mechanism not only verifies the user's identity but also their eligibility to receive payment services, providing ease of use and efficiency. The robustness and efficiency of the charging system enable easy, safe and efficient charging of electric vehicles. Seamless integration of RFID technology increases customer convenience, supports the use of wireless charging solutions and highlights the importance of personal identification and process evidence in enabling the electric vehicle to charge electricity.

6. CONCLUSION

The development and implementation of an electric vehicle (EV) wireless charging system using Radio-Frequency Identification (RFID) technology signify a significant advancement in the realm of electric mobility infrastructure. This project showcases the potential of RFID-based solutions to revolutionize the EV charging landscape by offering a seamless, user-friendly, and efficient charging experience. The integration of RFID technology into the charging system introduces several key advantages. Firstly, RFID enables secure and reliable user authentication, ensuring that only authorized users can access the charging facilities. This authentication process can be further enhanced with encryption protocols and multi-factor authentication methods to bolster security measures and protect against unauthorized access. Moreover, the RFID-based wireless charging system eliminates the hassle of physical cables and connectors, streamlining the charging process and enhancing user convenience. EV owners can simply park their vehicles over the charging pad, and the system automatically initiates the charging process based on the authenticated user's profile and charging preferences. This level of automation not only saves time and effort for users but also contributes to a more seamless and integrated charging infrastructure.

REFERENCES

AbuElrub, F., Hamed, F., & Saadeh, O. (2020, December). Hamed, O. Saadeh, Microgrid integrated electric vehicle charging algorithm with photovoltaic generation. *Journal of Energy Storage*, 32, 101858. DOI: 10.1016/j.est.2020.101858

Adnan Khan, M. D. S., Kadir, K. M., Mahmood, K. S., Ibne Alam, M. I., Kamal, A., & Al Bashir, M. M. (2019, December). Technical investigation on V2G, S2V, and V2I for next generation smart city planning. *J. Electronic Sci. Technol.*, 17(4), 100010. DOI: 10.1016/j.jnlest.2020.100010

Ahmad, A., Alam, M. S., Varshney, Y., & Khan, R. H. A state of the Art review on Wireless Power Transfer a step towards sustainable mobility. In Proceedings of the 2017 14th IEEE India Council International Conference (INDICON), Roorkee, India, 15–17 December 2017; pp. 1-Bijan. (2017). Design and control of a 6-degree-of-freedom precision positioning system. DOI: 10.1109/INDICON.2017.8488032

Alanne, K., & Liimatainen, H. (2019). Design implications of the electrification of passenger vehicle stock on renewable energy integration in Finnish apartment buildings. *Sustainable Cities and Society*, 47, 101507.

Alhasnawi, B. N., & Jasim, B. H. (2021, August). A new internet of things enabled trust distributed demand side management system. *Sustainable Energy Technologies and Assessments*, 46, 101272. DOI: 10.1016/j.seta.2021.101272

Rehman, A. U., Alblushi, I. G. M., Khalid, H. M., Said, Z., Iqbal, A., & Muyeen, S. M. (2024). Techno economics and energy dynamics of a solar powered smart charging infrastructure for electric vehicles with advanced IoT based monitoring and RFID based security. *Sustainable Futures : An Applied Journal of Technology, Environment and Society*, 8, 100376.

Chen, Q., & Wang, R. (2020). Policy incentives and their role in shaping EV infrastructure innovation. *Transportation Research Part D, Transport and Environment.*

Chen, X., Li, R., & Zhao, M. (2020). *Standardization gaps in EV wireless charging technologies*. Renewable Energy Journal.

Das, R., Wang, Y., Putrus, G., Kotter, R., Marzband, M., Herteleer, B., & Warmerdam, J. (2020, January). multi-objective techno-economic-environmental optimisation of electric vehicle for energy services. *Applied Energy*, 257, 113965. DOI: 10.1016/j.apenergy.2019.113965

Dimitriadou, K., Rigogiannis, N., Fountoukidis, S., Kotarela, F., Kyritsis, A., & Papanikolaou, N. (2023). Current Trends in Electric Vehicle Charging Infrastructure; Opportunities and Challenges in Wireless Charging Integration. *Energies*, 16(4), 2057. DOI: 10.3390/en16042057

Ding, X., Gan, Q., & Shaker, M. P. (2023, April). Optimal management of parking lots as a big data for electric vehicles using internet of things and Long–Short term Memory. *Energy*, 268, 126613. DOI: 10.1016/j.energy.2023.126613

Feng, H., Tavakoli, R., Pantic, Z., & Onar, O. C. (2020). Advances in High-Power Wireless Charging Systems: Overview and Design Considerations. *IEEE Transactions on Transportation Electrification*, 6(3), 886–919. DOI: 10.1109/TTE.2020.3012543

IEA (2023). "Global EV Outlook 2023: Scaling up the transition to electric mobility."

Joyce Jacob, J., Abinaya S, Divya Priya R, Poonam Khatarkar, Abdullaeva Barno and G. Sathish kumar, (2023) E3S Web Conf., 399 01010 DOI: https://doi.org/.DOI: 10.1051/e3sconf/202339901010

Judge, M. A., Khan, A., Manzoor, A., & Khattak, H. A. (2022, May). Overview of smart grid implementation: Frameworks, impact, performance and challenges. *Journal of Energy Storage*, 49, 104056. DOI: 10.1016/j.est.2022.104056

Karthikeyan, S., Bragruthshibu, H., Logesh, R., Srinivasan, K., & Tarjanbabu, S. "Solar Based Fast Tag Charger for Electrical Vehicle," 2021 7th International Conference on Advanced Computing and Communication Systems (ICACCS), Coimbatore, India, 2021, pp. 776-779. DOI: 10.1109/ICACCS51430.2021.9441940

Kim, J., Park, S., & Lee, H. (2021). Blockchain-based standardization for RFID wireless charging. *IEEE Transactions on Industrial Electronics.*

Li, X., Zhang, L., & Wu, D. (2021). Intrusion detection for RFID-enabled EV systems: A deep learning approach. *Computers & Security.*

Mahmud, I., & Medha, M. B. M. Hasanuzzaman, Global challenges of electric vehicle charging systems and its future prospects: A review, Research in Transportation Business & Management, Volume 49, 2023,101011, ISSN 2210-5395, https://doi.org/DOI: 10.1016/j.rtbm.2023.101011

Moghadasi, N., Collier, Z. A., Koch, A., Slutzky, D. L., Polmateer, T. L., Manasco, M. C., & Lambert, J. H. (2022, September). Trust and security of electric vehicle-to-grid systems and hardware supply chains. *Reliability Engineering & System Safety*, 225, 108565. DOI: 10.1016/j.ress.2022.108565

Muralikrishnan, P., & Kalaivani, M. (2020). IOT Based Electric Vehicle Charging Station Using Arduino Uno. *International Journal of Advanced Science and Technology*, 29, 4101–4106.

Park, C., Kim, J., & Shin, S. (2022). Urban planning strategies for RFID-enabled wireless EV charging. *Transportation Research Part D, Transport and Environment.*

Park, J., & Kim, S. (2022). Standardization and policy frameworks for RFID in EV infrastructure. *IEEE Access : Practical Innovations, Open Solutions.*

Parvin, K., Hannan, M. A., Hui Mun, L., Hossain Lipu, M. S., Abdolrasol, M. G. M., Jern Ker, P., Muttaqi, K. M., & Dong, Z. Y. (2022, October). The future energy internet for utility energy service and demand- side management in smart grid: Current practices, challenges and future directions. *Sustainable Energy Technologies and Assessments*, 53, 102648. DOI: 10.1016/j.seta.2022.102648

Rene, E. A., Tounsi Fokui, W. S., & Nembou Kouonchie, P. K. (2023, June). Optimal allocation of plug-in electric vehicle charging stations in the distribution network with distributed generation. *Green Energy Intell. Transport.*, 100094(3), 100094. Advance online publication. DOI: 10.1016/j.geits.2023.100094

Rigogiannis, N., Kotsidimou, A., Arzanas, I., Kyritsi, C., & Papanikolaou, N. (2022, August). Comparative Performance Study of Hybrid Si/SiC Insulated-Gate Bipolar Transistors. In 2022 IEEE 7th Forum on Research and Technologies for Society and Industry Innovation (RTSI) (pp. 191-197). IEEE.

Wang, H., & Chen, Z. (2020). Threat modeling and defense mechanisms for RFID-based IoT systems. *Journal of Cybersecurity*.

Wu, T., & Zhang, Y. (2020). Dynamic wireless EV charging: Technical and economic perspectives. *Journal of Advanced Transportation*.

Yang, S.-N., Cheng, W.-S., Hsu, Y.-C., Gan, C.-H., & Lin, Y.-B. (2011, December). Charge scheduling of electric vehicles in highways. *Mathematical and Computer Modelling*, •••, 6.

Zhang, Y., He, S., & Yang, Z. (2022). Security challenges in RFID-based electric vehicle charging systems. *IEEE Transactions on Industrial Informatics*.

Zhao, J., Cai, T., Duan, S., Feng, H., Chen, C., & Zhang, X. (2016). A General Design Method of Primary Compensation Network for Dynamic WPT System Maintaining Stable Transmission Power. *IEEE Transactions on Power Electronics*, 31, 8343–8358. DOI: 10.1109/TPEL.2016.2516023

Zhao, Q., & Ma, Y. (2019). Interoperability frameworks for next-generation EVs. *Energy Conversion and Management*.

Zhong, W., Yu, R., Xie, S., Zhang, Y., & Yau, D. K. Y. (2018, July). On stability and robustness of demand response in V2G mobile energy networks. *IEEE Transactions on Smart Grid*, 9(4), 3203–3212. DOI: 10.1109/TSG.2016.2629378

Chapter 3

Investigating Single- to Dual-Band Bandpass Filter Transformation for Wireless Communication Systems

Anup Raju Vasistha
University of Greenwich, UK

Chandramouli H. Mahadevaswamy
University of Greenwich, UK

Solomon H. Ebenuwa
https://orcid.org/0000-0001-5780-4817
University of Greenwich, UK

Augustine O. Nwajana
https://orcid.org/0000-0001-6591-5269
University of Greenwich, UK

ABSTRACT

A simple technique for transforming a single-band bandpass filter (BPF) into a dual-band BPF is presented in this chapter. The method starts with a 2nd-order single-band BPF design, giving rise to a 4th-order dual-band BPF after the proposed transformation. Both filters are implemented using compact U-shaped microstrip resonator for improved device miniaturization. The proposed work features a centre frequency of 1.4 GHz for the single-band BPF, with a 3.4% fractional bandwidth. The dual-band BPF operates at 1.35 and 1.45 GHz for the first and second bands, respectively. The design implementation employs the commercially available Rogers RT/Duroid 6010LM substrate, having a dissipation factor of 0.0023, dielectric constant of 10.7, and thickness of 1.27 mm. The practical responses of the prototype dual-band BPF indicate a good return loss of better than 18 dB across both bands, and an insertion loss of better than 0.1 dB. The design prototype achieved physical size of 0.23 x 0.18 λg. λg is the guided wavelength for the microstrip line impedance at the centre frequency of the filter.

DOI: 10.4018/979-8-3693-8799-3.ch003

1. INTRODUCTION

In the rapidly evolving landscape of communication technology, the demand for smaller and more efficient devices has become increasingly prevalent. This necessity has spurred research and development efforts towards enhancing the performance of primary devices, particularly filters, which play a critical role in signal transmission and reception. Filters are passive components designed to selectively transmit or receive signals within specified frequency bands (Li et al., 2023; Nwajana, 2021; Yan J. et al., 2023; Nwajana et al., 2021; Mu et al., 2023; Nwajana et al., 2017). They find applications across a myriad of wireless communication systems. One remarkable advancement in this domain is the exploration of dual-band BPF as a single component in communication systems (Wei et al., 2023; Alazemi, 2023; Kumar et al., 2021; Li, & Xiao, 2023). This cutting-edge approach aimed at achieving compactness and reduced losses in the system. Dual-band BPFs, realised by employing dual- and multi-mode resonators (Liu Q. et al., 2023), have gained attention for their ability to significantly reduce the physical size of the filter device while enhancing overall performance. Resonators play a pivotal role in the investigation of BPF design techniques. Diverse resonator structures including hairpin (Nwajana, 2020; Shankar et al., 2023; Liu H. et al. 2023), quarter-wavelength (Ma et al., 2022), open-loop ring (Pradhan et al., 2023; Liu et al., 2020), waveguide (Xiong et al., 2023), substrate integrated waveguide (Nwajana & Obi, 2022), and folded-arms square open-loop (Nwajana, & Paul, 2024), have been employed by researchers to successfully achieve BPF components.

In this chapter, a simple technique for transforming a single-band BPF into a dual-band BPF is presented. Microstrip U-shaped resonator structure (Mahadevaswamy et al. 2024) is employed in the implementation of the proposed filter due to its simplicity and relative compact size, when compared to many other popular microstrip resonator structures. The proposed method can be extended to the design of any multi-band BPF and implemented using any resonator of choice. To realise the microstrip U-shaped resonator structure, the conventional microstrip half-wavelength resonator structure shown in Figure 1 (a) is folded into a U-shape structure with dimensions relative to the standard guided wavelength as shown in Figure 1 (b).

The rest of the chapter is organized by firstly reviewing related works in section 2. The theoretical configuration of the proposed dual-band BPF is covered in section 3 while the practical implementation is the focus of section 4. In section 5, the results and discussion are presented while the conclusion is drawn in section 6.

Figure 1. Microstrip Resonators: (a) Half-wavelength configuration; (b) U-shaped configuration.

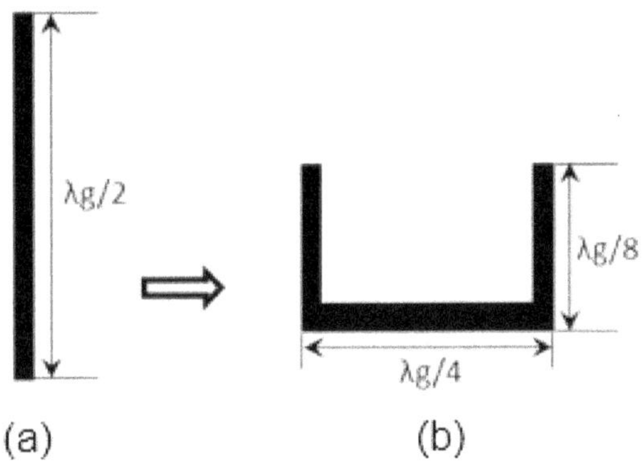

2. RELATED WORK

This section covers an overview of related works in dual-band BPF design and implementation. Due to the growing interest in the design of dual-band BPFs, several authors have proposed various techniques for implementing this type of filter. Zhang et al. (2024) proposed a multilayer packaged dual-band BPF based on square substrate integrated waveguide (SIW) technology. The design exploited multiple pairs of orthogonal degenerated modes in the square SIW cavities in the implementation of two single-band and one dual-band BPFs. The proposed technique is valid but involves the complexities of dealing with multilayered device manufacturing. Gu & Dong (2022) reported a dual-band BPF based on spiral stub-loaded dual-mode SIW cavities. The two passbands generated in the proposed design were formed by the two modes of the dual-mode SIW cavities. The technique is valid but when compared to microstrip, SIW devices are relatively bulky and hence take up larger footprint in any system they are found. Yeo & Nwajana (2013) proposed a technique that employs the microstrip dual-mode square patch resonators in the implementation of a dual-band BPF. The achieved filter is of compact size considering the 50% reduction in the number of resonators used in the design implementation. The design technique is valid, but the resultant filter achieved a measured return loss of about 8.5 dB which is considered poor when compared to the reference 20 dB standard. Yan Z. et al. (2023) investigated a switchable dual-band BPF design method with proposed application in 5G signal detector. Though the proposed design seems valid, the results achieved are very poor with a return loss of worse than 7 dB across the two passbands of the filter. Yang, Zhu, & Wang (2023) proposed a multi-mode resonator (MMR) overlapping mode control technique to realize dual-band BPF based on the SIW implementation. The design was achieved by a pair of via perturbation, with two kinds of slotlines introduced to control the overlapping modes. The issue with the proposed design is the effectiveness of the two slotlines in controlling the overlapping modes. The practical filter achieved using the proposed technique would need manual tuning to achieve optimum performance. Xu et al. (2022) reported the design of a dual-band BPF using dual-mode dielectric waveguide resonators. The results of the design are encouraging but as it is widely known, waveguide devices are very bulky and difficult to integrate with other devices in an existing system. Lee, Yoon, & Kim (2022) proposed a dual-band BPF based on a stepped impedance resonator (SIR) with meander line and folded structure. The use of the SIR in the design ensures that the proposed filter can be adjusted to reject spurious bands. However, based on the results presented in the paper, it is unclear if the intension of the proposed work was attained.

The dual-band BPF proposed in this paper is simply achieved from the transformation of a single-band BPF and implemented using microstrip technology for compactness and ease of integration with other devices in any communication system.

3. THEORETICAL FILTER CONFIGURATION

The procedure for achieving a single-band BPF begins with the transformation of the standard normalised prototype Chebyshev lowpass filter element-values of g0 = 1.0, g1 = 0.6648, g2 = 0.5445, and g3 = 1.2210 to realise a second order channel filter at the desired centre frequency of 1.4 GHz, fractional bandwidth of 3.4%, and the input/output characteristic impedance (Z0) of 50 Ohms. The single-band BPF transformation is based on the principles reported in Hong & Lancaster (2004), and the final circuit configuration is shown in Figure 2(a). The single-band BPF is then converted into the corresponding dual-band BPF configuration shown in Figure 2(b). The transformation from the single-band to the dual-band BPF is achieved by simply extending the circuit and retaining a similar mutual coupling inductor-only network between each pair of resonators as shown in Figure 2(b). The numerical values for the capacitive and the inductive parameters indicated in Figure 2 are C = 44.4564 pF, L = 0.2907 nH, L01 = 5.6841 nH, L12 = 5.1442 nH. The numerical values are all based on the filter design formulations previously reported in Nwajana & Obi (2023).

Figure 2. BPF circuit configurations: (a) Single-band; (b) Dual-band.

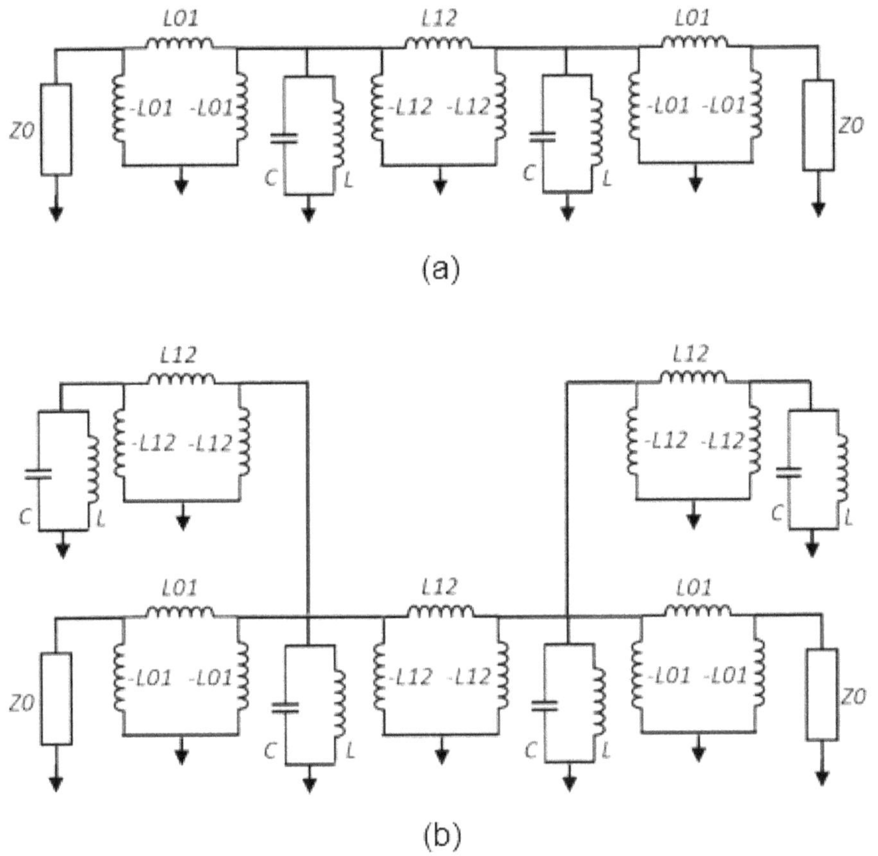

4. PRACTICAL FILTER CONFIGURATION

The microstrip U-shaped resonator structure reported in Saeed & Nwajana (2023) and Mahadevaswamy et al. (2024) is employed in the implementation of the proposed prototype filters due to its simplicity and relative compact size, when compared to other numerous popular microstrip resonator structures. The width and guided wavelength for designing the U-shaped resonator are determined from Nwajana & Obi (2023). Advanced Design System (ADS) EM Simulation Software is used to construct and simulate the layout of both filter models, with the coupling arrangements shown in Figure 3, forming the design foundation. The layouts were created on Rogers RT/Duroid 6010LM substrate with a dielectric constant of 10.7, a thickness of 1.27 mm, and a loss tangent of 0.0023. The microstrip U-shaped resonator is designed and constructed to resonate at the filter centre frequency of 1.4 GHz as shown in Figure 4.

Figure 3. BPF coupling arrangements: (a) Single-band; (b) Dual-band.

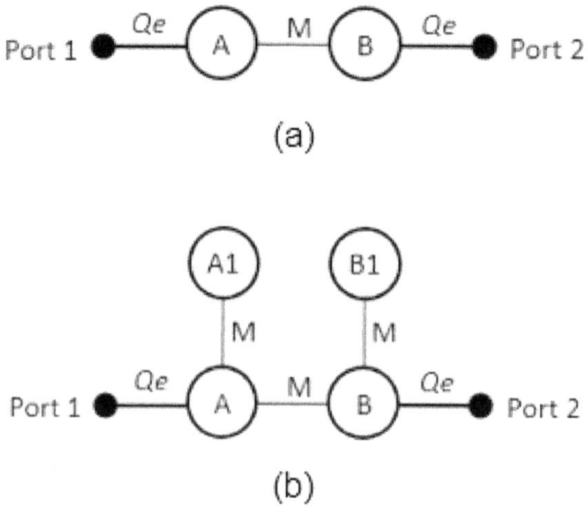

$$M = \frac{FBW}{\sqrt{g_1 g_2}} = 0.0565 \quad (1)$$

$$Q_e = \frac{g_0 g_1}{FBW} = 19.5529 \quad (2)$$

Figure 4. Practical dimensions (in mm) of the proposed U-shaped resonator at 1.4 GHz and the EM simulation response.

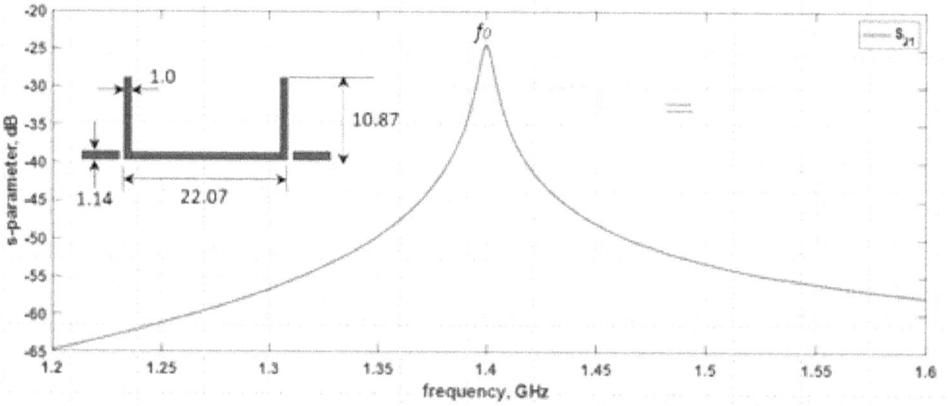

Equations (1) and (2) proposed in Hong & Lancaster (2004) were employed in the determination of the mutual coupling coefficient, M, between two adjacent resonators (A and B, A and A1, B and B1); and the external quality factor, Qe, between the input port and the first filter resonator (that is, port 1 and A), and the output port and the last filter resonator (that is, port 2 and B), respectively.

The practical layouts of the proposed single- and dual-band BPFs are established using the full-wave Keysight ADS EM simulator which simulated the mutual coupling coefficients and the external quality factor against practical structures as reported in Hong & Lancaster (2004). The design curves for extracting the mutual coupling coefficients and the external quality factor are shown in Figures 5 and 6, respectively. The resonator pairs shown in Figures 5 and 6 were synchronously tuned since they are of equal dimensions, that is, the resonators are resonating at the same frequency. To extract the mutual coupling coefficients from Figures 5, the separation distance, s, is continuously adjusted and simulated until the practical mutual coupling coefficient value, calculated using Equation (3), matches the theoretical value determined from Equations (1). To extract the external quality factor from Figure 6, the tapping distance, t, is continuously adjusted and simulated until the practical external quality factor value, calculated using Equation (4), matches the theoretical value determined from Equation (2). The physical dimensions of the filter layouts, and the corresponding images of the fabricated filter devices are shown in Figures 7 and 8, respectively.

$$k = {(f_2^2 - f_1^2)} \Big/ {(f_2^2 + f_1^2)} \quad (3)$$

$$Q_{ext} = {f_0} \Big/ {(f_2 - f_1)} \quad (4)$$

Figure 5. Microstrip layout and EM simulation response for extracting the mutual coupling coefficient.

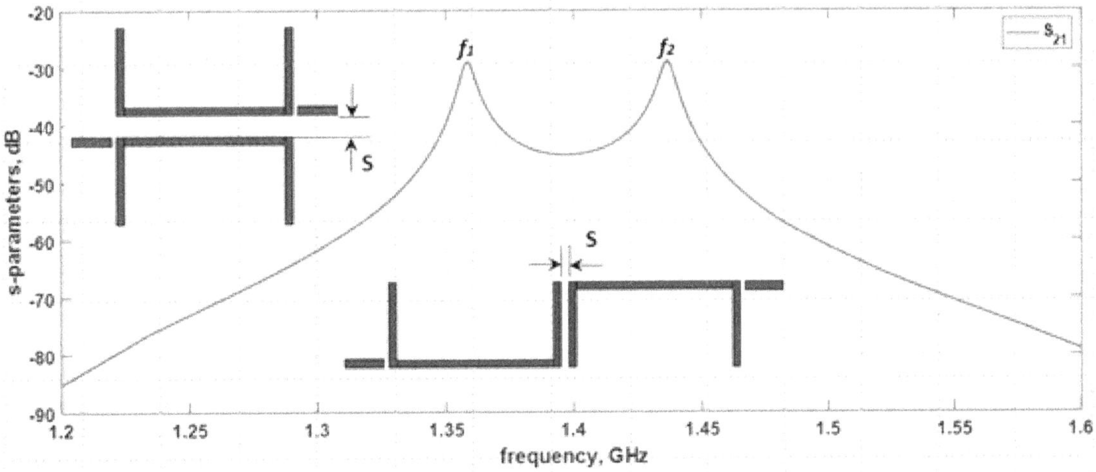

Figure 6. Microstrip layout and EM simulation response for extracting the external quality factor.

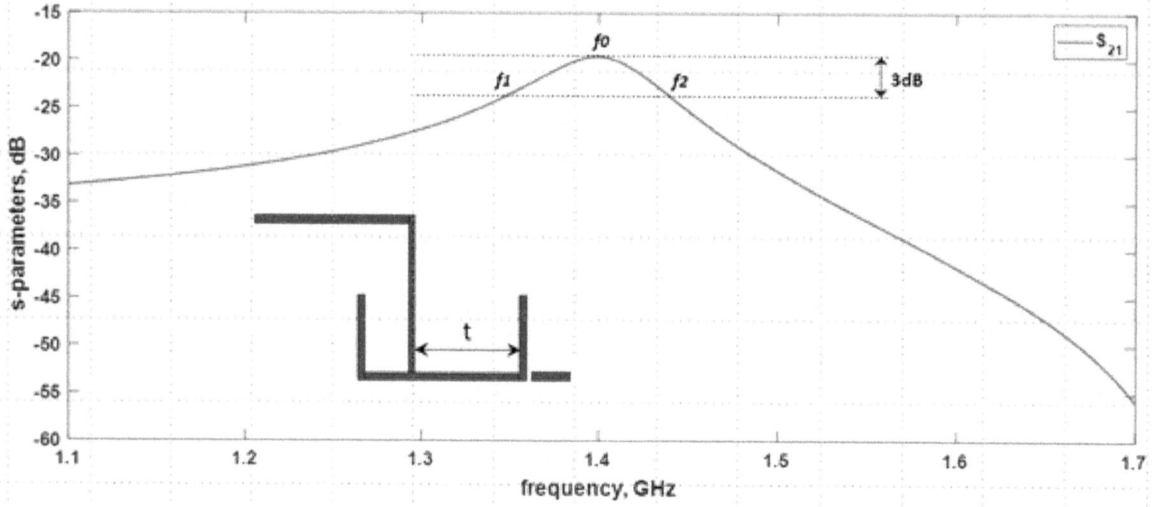

Figure 7. Microstrip BPF layouts with dimensions in mm: (a) Single-band; (b) Dual-band.

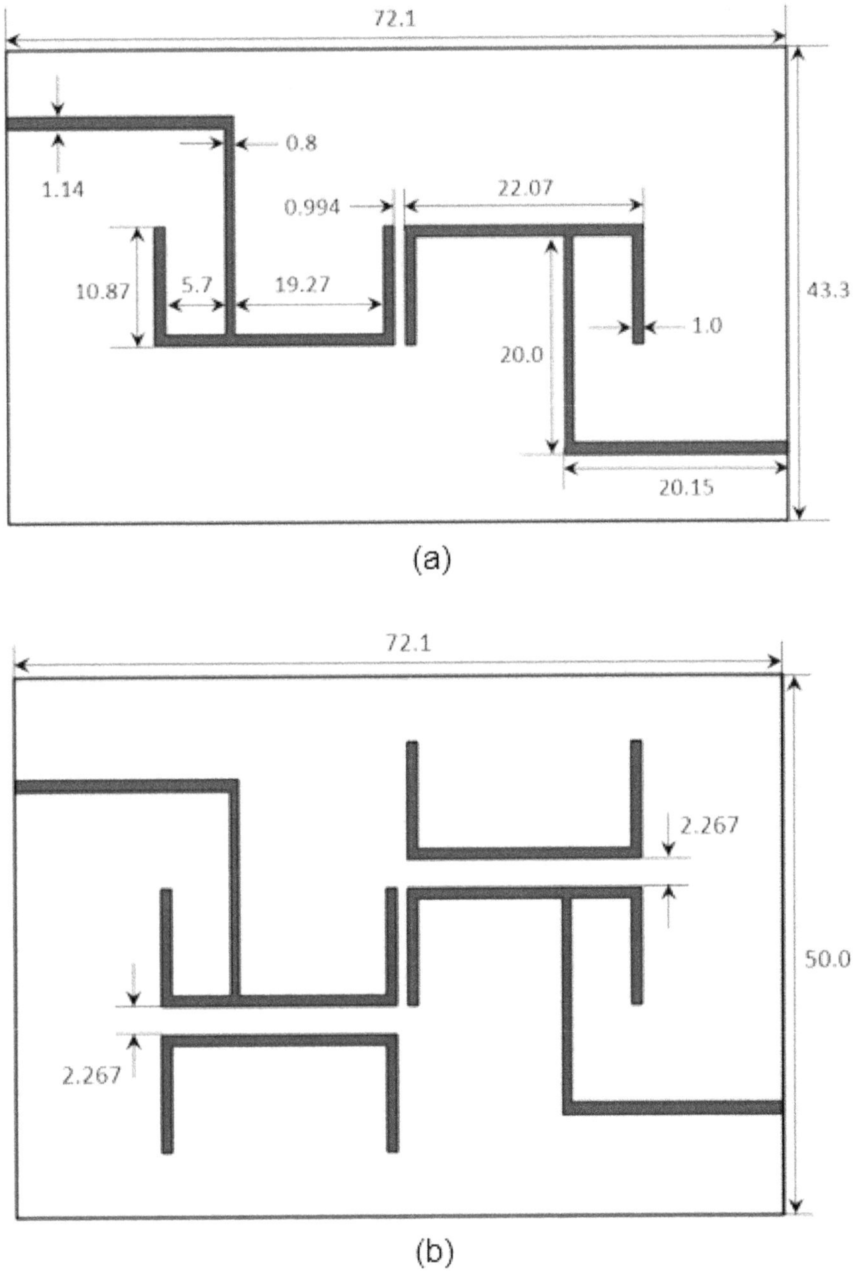

Figure 8. Fabricated BPF image: (a) Single-band; (b) Dual-band.

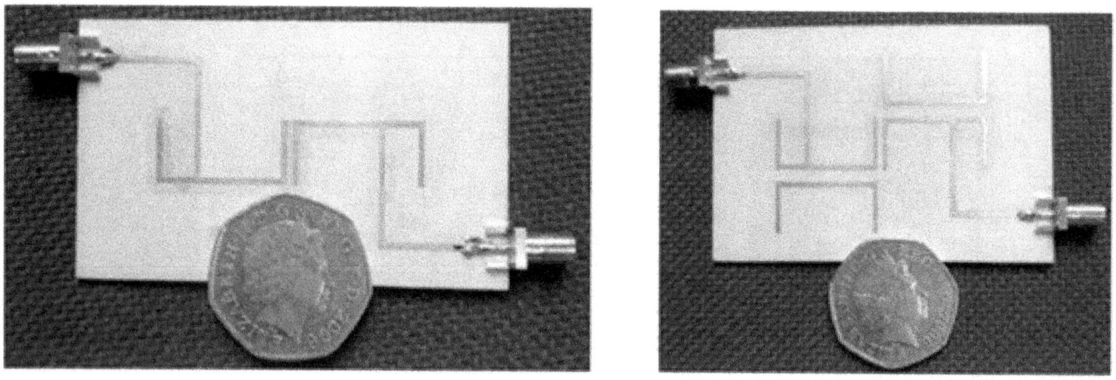

(a) (b)

5. RESULTS AND DISCUSSION

This section captures and discusses both the theoretical and practical results for the proposed single- and dual-band BPFs. The dual-band BPF implemented using the microstrip U-shaped resonator has been developed from the transformation of a single-band BPF. The results for the single- and dual-band BPFs are presented in Figure 9 for ease of analysis and comparison. The results show good agreements between the corresponding theoretical and practical responses. Looking at Figure 9(a), the single-band BPF transmits at the specified centre frequency of 1.4 GHz, with an insertion and return losses of 0.1 dB and 20 dB, respectively. In Figure 9(b), the dual-band BPF transmits at 1.35 GHz and 1.45 GHz, with insertion losses of better than 0.1 dB across both bands. The return loss achieved in the first band is better than 24 dB, while that of the second band is better than 18 dB. The proposed dual-band BPF results are compared to the current state-of-the-art and captured in Table 1. The performance comparison reinforces the design's competitive edge over similar designs reported in literature. The designs achieved a physical size of 0.23 λg x 0.15 λg for the single-band BPF, and 0.23 λg x 0.18 λg for the dual-band BPF. It is important to note that the reported results assume copper conductor with cladding thickness of 35 micron (μm) and conductivity of 5.8 x 10^7 S/m. Thickness variation of the substrate material used in the design, and metal surface roughness were not considered in this research work.

Figure 9. Theoretical and practical BPF results comparison: (a) Single-band; (b) Dual-band.

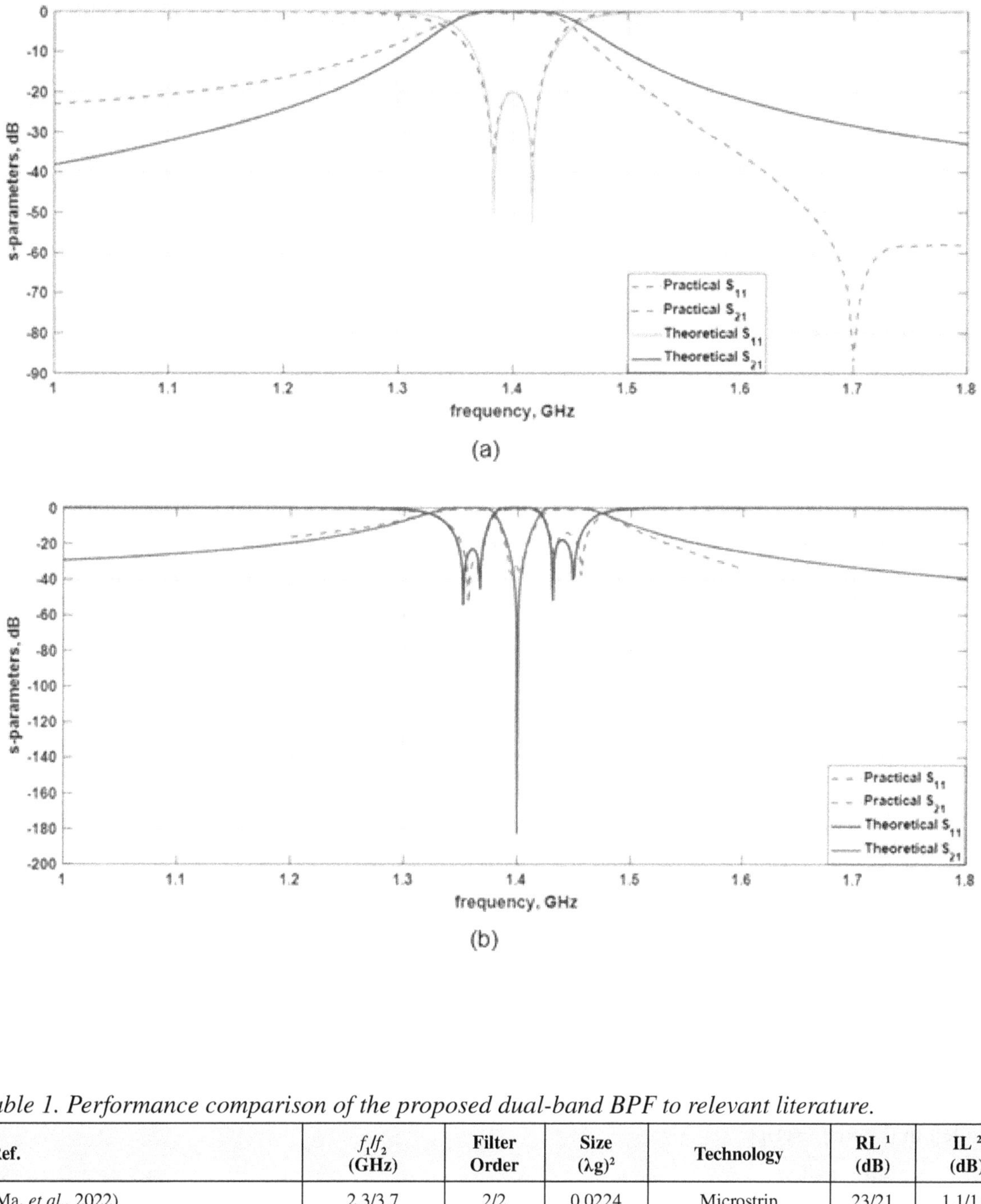

Table 1. Performance comparison of the proposed dual-band BPF to relevant literature.

Ref.	f_1/f_2 (GHz)	Filter Order	Size $(\lambda g)^2$	Technology	RL [1] (dB)	IL [2] (dB)
(Ma, *et al.*, 2022)	2.3/3.7	2/2	0.0224	Microstrip	23/21	1.1/1.9
(Gu, & Dong, 2023)	7.8/12.3	2/2	0.0156	Hybrid	15/16	1.3/1.5
(Wei, *et al.*, 2023)	0.9/2.5	2/2	0.0130	Microstrip	20/12	0.3/0.9
(Lee, *et al.*, 2022)	2.9/4.4	2/2	0.0308	Microstrip	17/17	0.1/0.1

continued on following page

Table 1. Continued

Ref.	f_1/f_2 (GHz)	Filter Order	Size $(\lambda g)^2$	Technology	RL [1] (dB)	IL [2] (dB)
Pradhan, *et al.*, 2023)	1.6/4.7	2/2	0.0410	SIW	14/21	1.1/1.2
(Li, & Xiao, 2023)	1.1/3.4	2/3	0.0221	Hybrid	14/14	0.4/1.3
(Xu, *et al.*, 2022)	2.9/3.5	2/2	0.0306	Waveguide	17/18	1.0/1.5
(Yang, *et al.*, 2023)	42.8/60.5	4/6	1.8696	SIW	18/12	1.9/1.4
This work	1.35/1.45	2/2	0.0410	Microstrip	24/18	0.1/0.1

[1.] Return loss. [2.] Insertion loss.

6. CONCLUSIONS

A dual-band BPF achieved from the transformation of a single-band BPF has been developed and realized using microstrip U-shaped resonator technology. The proposed dual-band BPF operates at 1.35 and 1.45 GHz centre frequencies. The realization and fabrication are carried out on Rogers RT/Duroid 6010LM substrate with a dielectric constant of 10.7, a thickness of 1.27 mm and a loss tangent of 0.0023. The theoretical and practical results of the proposed dual-band BPF show good agreement with a return loss of better than 18 dB across the two bands, with an insertion loss of better than 0.1 dB on each band. The minor deviations between the theoretical and practical results of the dual-band BPF may be traced to the surface roughness and thickness variation of the substrate material, which are not considered in the simulation. The fabricated dual-band BPF device achieved a physical size of 0.23 λg x 0.18 λg. Looking at the performance comparison in Table 1, it is easy to conclude that the proposed dual-band BPF competes well with the current state-of-the-art as it has the best insertion loss, and a very good return loss as recorded in Table 1. The proposed dual-band BPF device finds application in modern multifunctional communication systems where it is used for isolating small sections of a frequency spectrum within a wider spectrum.

Author contributions: Conceptualization, A.O.N.; data curation, C.H.M.; formal analysis, C.H.M.; funding acquisition, S.H.E.; investigation, A.R.V.; methodology, A.R.V.; project administration, A.O.N.; Resources, A.O.N.; software, C.H.M.; supervision, A.O.N.; validation, S.H.E.; visualization, C.H.M.; writing—original draft preparation, A.R.V.; writing—review and editing, S.H.E. All authors have read and agreed to the published version of the manuscript.

Funding: This research received no external funding.

Competing interests: The authors declare no competing of interest.

Data availability statement: No additional data is applicable besides the data/information already included in the manuscript.

REFERENCES

Alazemi, A. J. (2023). Dual-band and wideband bandpass filters using coupled lines and tri-stepped impedance stubs. *Micromachines*, 14(6), 1254. DOI: 10.3390/mi14061254 PMID: 37374838

Gu, L., & Dong, Y. (2023). Compact dual-band bandpass filter with high selectivity based on spiral-stub-loaded dual-mode QMSIW resonators. *Microwave and Optical Technology Letters*, 65(2), 559–566. DOI: 10.1002/mop.33541

Hong, J. S. G., & Lancaster, M. J. (2004). *Microstrip filters for RF/microwave applications*. John Wiley & Sons.

Kumar, K. V. P., Velidi, V. K., Althuwayb, A. A., & Rao, T. R. (2021). Microstrip dual-band bandpass filter with wide bandwidth using paper substrate. *IEEE Microwave and Wireless Components Letters*, 31(7), 833–836. DOI: 10.1109/LMWC.2021.3077879

Lee, T. H., Yoon, K. C., & Kim, K. G. (2022). Miniaturized Dual-Band Bandpass Filter Using T-Shaped Line Based on Stepped Impedance Resonator with Meander Line and Folded Structure. *Electronics (Basel)*, 11(2), 219. DOI: 10.3390/electronics11020219

Li, X. F., & Xiao, J. K. (2023). Dual-Band Bandpass Filter Based on Suspended Coplanar Waveguide-Microstrip Hybrid. IEEE Transactions on Circuits and Systems I: Regular Papers.

Li, Y. C., Wu, D. S., Zhang, T. Z., & Xue, Q. (2023). Dual-band dual-channel bandpass filters using high quality factor dielectric resonators. IEEE Transactions on Circuits and Systems II: Express Briefs.

Liu, H., Kuang, J., Cao, Y., & Zhang, Y. (2023). Flexible Design of Dual-Band Common-Mode Filters Using Hairpin Ring Resonators and Defected Ground Slots. *IEEE Transactions on Electromagnetic Compatibility*, 65(5), 1360–1370. DOI: 10.1109/TEMC.2023.3304532

Liu, H., Wang, Z., Hu, S., Xu, H. X., & Ren, B. (2020). Design of tri-band balanced filter with wideband common-mode suppression and upper stopband using square ring loaded resonator. *IEEE Transactions on Circuits and Wystems. II, Express Briefs*, 67(10), 1760–1764. DOI: 10.1109/TCSII.2019.2949772

Liu, Q., Zhang, D. W., Gong, K., Qian, H., Qu, W. Z., & An, N. (2023). Single-and dual-band bandpass filters based on multiple-mode folded substrate-integrated waveguide cavities. *IEEE Transactions on Microwave Theory and Techniques*, 71(12), 5335–5345. DOI: 10.1109/TMTT.2023.3276058

Ma, M., You, F., Qian, T., Shen, C., Qin, R., Wu, T., & He, S. (2022). A wide stopband dual-band band-pass filter based on asymmetrical parallel-coupled transmission line resonator. *IEEE Transactions on Microwave Theory and Techniques*, 70(6), 3213–3223. DOI: 10.1109/TMTT.2022.3168024

Mahadevaswamy, C. H., Vasistha, A. R., & Nwajana, A. O. (2024, July). Dual-band bandpass filter derived from the transformation of a single-band bandpass filter. In 2024 24th International Microwave and Radar Conference (MIKON) (pp. 200-203). IEEE. DOI: 10.23919/MIKON60251.2024.10633986

Mu, R., Wu, Y., Pan, L., Zhao, W., & Wang, W. (2023). A Miniaturized Low-Loss Switchable Single-and Dual-Band Bandpass Filter. *International Journal of RF and Microwave Computer-Aided Engineering*, 2023, 2023. DOI: 10.1155/2023/9025980

Nwajana, A. O. (2020). Circuit modelling of bandpass/channel filter with microstrip implementation. *Indonesian Journal of Electrical Engineering, and Informatics*, 8(4), 696–705.

Nwajana, A. O. (2021). A step-by-step approach to bandpass/channel filter design. [IJECME]. *International Journal of Electronics, Communications, and Measurement Engineering*, 10(2), 1–14. DOI: 10.4018/IJECME.2021070101

Nwajana, A. O., Dainkeh, A., & Yeo, K. S. (2017). Substrate integrated waveguide (SIW) bandpass filter with novel microstrip-CPW-SIW input coupling. *Journal of Microwaves, Optoelectronics and Electromagnetic Applications*, 16(2), 393–402. DOI: 10.1590/2179-10742017v16i2793

Nwajana, A. O., & Obi, E. R. (2022). A review on SIW and its applications to microwave components. *Electronics (Basel)*, 11(7), 1160. DOI: 10.3390/electronics11071160

Nwajana, A. O., & Obi, E. R. (2023). Application of compact folded-arms square open-loop resonator to bandpass filter design. *Micromachines*, 14(2), 320. DOI: 10.3390/mi14020320 PMID: 36838019

Nwajana, A. O., Obi, E. R., Ijemaru, G. K., Oleka, E. U., & Anthony, D. C. (2021). Fundamentals of RF/Microwave bandpass filter design. In *Handbook of research on 5G networks and advancements in computing, electronics, and electrical engineering* (pp. 149–164). IGI Global. DOI: 10.4018/978-1-7998-6992-4.ch005

Nwajana, A. O., & Paul, R. (2024). Three-Way Equal Filtering Power Divider for Modern Communication Systems. *Engineering Reports*, •••, 13036. DOI: 10.1002/eng2.13036

Nwajana, A. O., & Paul, R. (2024, July). 3-way equal filtering power divider using compact folded-arms square open-Loop resonator. In 2024 25th International Microwave and Radar Conference (MIKON) (pp. 191-195). IEEE.

Pradhan, N. C., Koziel, S., Barik, R. K., Pietrenko-Dabrowska, A., & Karthikeyan, S. S. (2023). Miniaturized dual-band SIW-based bandpass filters using open-loop ring resonators. *Electronics (Basel)*, 12(18), 3974. DOI: 10.3390/electronics12183974

Saeed, M. A., & Nwajana, A. (2023, September). U-Shaped terahertz microstrip patch antenna for 6G future communications. In 2023 7th International Electromagnetic Compatibility Conference (EMC Turkiye) (pp. 1-4). IEEE.

Shankar, E., Kumar, K. V. P., & Velidi, V. K. (2023, April). A miniaturized dual narrow-band bandpass filter using a hairpin resonator and coupled lines. In *2023 International Conference on Recent Advances in Electrical, Electronics, Ubiquitous Communication, and Computational Intelligence (RAEEUCCI)* (pp. 1-4). IEEE. DOI: 10.1109/RAEEUCCI57140.2023.10134052

Wei, G. Y., Wang, Y. X., Liu, J., & Li, H. P. (2023). Design of a Planar Compact Dual-Band Bandpass Filter with Multiple Transmission Zeros Using a Stub-Loaded Structure. *Progress in Electromagnetics Research Letters*, 109, 109. DOI: 10.2528/PIERL22122603

Xiong, S., Guan, X., Ren, B., Wan, S., & Zhang, X. (2023, November). Dual-Band Bandpass Filter Based on Two Pairs of Degenerate Modes of a Regular Hexagonal Prism Dielectric Waveguide Resonator. In 2023 Cross Strait Radio Science and Wireless Technology Conference (CSRSWTC) (pp. 01-03). IEEE. DOI: 10.1109/CSRSWTC60855.2023.10426903

Xu, Z., Wu, Y., Dong, Q., & Wang, W. (2022). Miniaturized dual-band filter using dual-mode dielectric waveguide resonator. *IEEE Microwave and Wireless Components Letters*, 32(12), 1411–1414. DOI: 10.1109/LMWC.2022.3193424

Yan, J. M., Kang, B. J., Yang, Y., & Cao, L. (2023). Single-and Dual-Band Bandpass Filters Based on a Novel Microstrip Loop-Type Resonators Loaded with Shorted Stubs. *Progress in Electromagnetics Research Letters*, 113, 113. DOI: 10.2528/PIERL23083105

Yan, Z., Zhang, X., He, X., Ye, Y., Hou, B., Hu, C., & Wen, W. (2023). A high performance dual-band switchable bandpass filter and its application in 5G signal detector. *IEEE Transactions on Components, Packaging, and Manufacturing Technology*, 13(8), 1242–1253. DOI: 10.1109/TCPMT.2023.3302729

Yang, X. L., Zhu, X. W., & Wang, X. (2023). Dual-band substrate integrated waveguide filters based on multi-mode resonator overlapping mode control. IEEE Transactions on Circuits and Systems II: Express Briefs.

Yeo, K. S. K., & Nwajana, A. O. (2013). A novel microstrip dual-band bandpass filter using dual-mode square patch resonators. *Progress In Electromagnetics Research*, C, 36, 233–247.

Zhang, G., Zhou, X., Hu, J., Tam, K. W., Li, D., Zhang, Z., Ma, T., & Yang, J. (2024). Multilayer Packaged Single-and Dual-Band Dual-Channel Bandpass Filters Based on Substrate Integrated Waveguide Cavities. *IEEE Transactions on Components, Packaging, and Manufacturing Technology*, 14(1), 114–121. DOI: 10.1109/TCPMT.2023.3330170

Chapter 4
Microwave Circuits Enabling 5G/6G Communication

Smrity Dwivedi
https://orcid.org/0000-0001-5160-6617
Indian Institute of Technology, Varanasi, India

ABSTRACT

There is growing demand for 5G as well as 6G technologies as per user's requirements and enhanced applications. As low latency and high data rates are key findings of 5G, in same way more improved bandwidth, maximum data enabled rate, and many new applications are basic achievable things from 6G systems. There are several types of microwave circuits used for such high frequency. RF/microwave circuits include antennas, filters, power amplifiers, phase shifters, power dividers, mixers, and multiplexers. Here, microwave antennas have been discussed under the microwave circuits. In the last few years, as per economic and social development which is greatly influenced by the advancements in mobile communication and technology, 5G technology has emerged as an important of the future 2020 generation which is already in use. After the development of fifth-generation technologies, researchers, scientists, and engineers are looking for wide bandwidth which should improve wireless systems and devices to provide better services and faster experience. Also, the development of 5G wireless network technology is the response to the crucial factors that lead to this demand because of its ability to provide extremely fast internet speed, high bandwidth, high performance, reduced latency, and high reliability and better gain. In this chapter, we will review the recent design with advancement in 5G antennas for beyond 5G applications and also we will give a few designs of antennae for 6G applications.

1. INTRODUCTION

Microwave circuits play a crucial role in enabling 5G and future 6G communication systems. These systems rely on high-frequency electromagnetic waves, typically in the microwave and millimeter-wave ranges, for high-speed data transmission and connectivity. Here's a detailed look at how microwave circuits are essential in these next-generation communication technologies:

1. Microwave Frequency Bands for 5G and 6G
    - **5G Frequency Bands**:

DOI: 10.4018/979-8-3693-8799-3.ch004

- 5G operates in several frequency bands, including sub-6 GHz and millimeter-wave (24 GHz to 100 GHz), often referred to as the "5G New Radio" (NR) spectrum.
- Sub-6 GHz bands provide good coverage and penetration, while millimeter-wave frequencies enable high-speed data transfer, though they have shorter ranges and are more susceptible to obstacles.

- **6G Frequency Bands**:
    - 6G will likely push the boundaries further into terahertz frequencies (100 GHz to 1 THz) and integrate ultra-high frequency bands for even faster speeds, lower latency, and more reliable communication.

2. Microwave Circuits and Components in 5G/6G

Microwave circuits enable essential functions in the transmission and reception of signals in the RF (radio frequency) spectrum. Key components include:

- **Power Amplifiers**: These amplify weak signals for transmission, and high-efficiency power amplifiers are critical for maximizing coverage and capacity in 5G/6G networks. Gallium nitride (GaN) and gallium arsenide (GaAs) are commonly used semiconductor materials in these amplifiers.
- **Low-Noise Amplifiers (LNAs)**: LNAs are used to amplify weak signals received by the antennas. In 5G/6G, their performance needs to be optimized for low noise levels, as they directly affect the overall system's sensitivity and ability to handle high-frequency signals.
- **Filters**: Filters are used to select specific frequency bands and reject unwanted signals (such as noise or interference). In the millimeter-wave and terahertz range, filters must be precise, compact, and able to handle high-frequency signals without introducing significant losses.
- **Mixers**: Mixers are used to shift the frequency of a signal, often from RF to intermediate frequencies (IF) for processing. These are crucial in systems that need to operate over a wide range of frequencies, such as 5G and 6G.
- **Antennas**: Advanced antenna technologies, like massive MIMO (Multiple Input, Multiple Output), use large arrays of antennas to improve data throughput and coverage. Microwave circuits manage the signals fed to and from these antennas, ensuring high-efficiency transmission and reception.

3.Key Microwave Technologies Enabling 5G/6G

- **Phased Array Antennas**: Phased array antennas allow for the electronic steering of beams, enabling adaptive beamforming and enhancing the performance of MIMO systems. These antennas are vital for 5G and 6G systems as they provide faster data transmission and improved reliability.
- **Small Cell and Beamforming**: Small cells, which are low-power, short-range base stations, help to overcome the propagation limitations of high-frequency signals in 5G/6G networks. Beamforming technologies, using arrays of microwave circuits, allow the focusing of radio signals in specific directions to enhance coverage and capacity in dense urban environments.
- **Beam Steering**: With beamforming, microwave circuits enable dynamic adjustment of the direction of the signal beams, allowing devices to receive stronger signals, improving both data rates and network efficiency. This is essential for high mobility in 5G/6G.

4.The Role of Microwave Circuits in 5G/6G Applications

- **Ultra-Reliable Low-Latency Communication (URLLC)**: Microwave circuits enable low-latency, high-reliability communication, essential for applications like autonomous vehicles and industrial automation.
- **Enhanced Mobile Broadband (eMBB)**: High-speed, large-capacity data transmission in areas like augmented reality (AR) and virtual reality (VR) depends on advanced microwave circuits operating in the millimeter-wave bands.
- **Massive Machine-Type Communication (mMTC)**: 5G and 6G will support billions of connected devices, which will require efficient microwave circuit designs for low-power, long-range communication.
- **Integrated Sensing and Communication (ISAC)**: Future 6G networks will likely integrate sensing capabilities alongside communication, utilizing microwave circuits for real-time sensing applications, such as location tracking, environmental monitoring, and gesture recognition.

Scientists, researchers and experts are seeking to develop and advance in new technologies and new areas such as fifth communication, based on the needs and requirements of wide and fast communication. For decades, the emergence of fourth-generation technology has provided users with high-speed connections, good bandwidth and high data; but now the bandwidth needs to be increased and based. As expected, fifth technology has also emerged among others. has been developed and improved according to the design. The design of the antenna is one of the important aspects of the fifth machine. Scientists are working hard to create the best, most suitable models, low cost, high performance. There are many types of designs, including rear antennas, graphene antennas for terahertz applications, array antennas for high direct and gain, and MIMO antennas to improve bandwidth and above. There are also many materials used to create fifth-generation antennas and above-fifth-generation antennas. Among them, metamaterials, graphene, gold, SiO2, polyimide and other materials can be used for fifth-generation and higher antenna power and terahertz applications. Using some of the above design features, some features and designs are presented in the next section. As mentioned above, its main features are used to achieve wide bandwidth and high performance in fifth-generation and above applications. Due to the rapid development of wireless communication and the high demand to combine and integrate multiple wireless models into a single efficient platform, it is extremely important that the antenna's operating frequency, radiation pattern and polarization can be adjusted and changed as needed. . . Reconfigurable antennas change and adjust their operating frequency, impedance bandwidth, polarization and electrical configuration according to the usage and performance of each application. They can easily emit different types of radiation at different frequencies and polarizations. Nowadays, achieving the required functionality of antenna enhancements and combining them into a complete system in a cost-effective and efficient manner is a difficult task for antenna designers. As we move from the 5th generation to the 6th generation, these analyses will be very useful in creating smart 6th generation antennas. Another difficult task is to transform the antenna into a reconfigurable antenna by using different technologies to change the internal structure and parameters of the antenna. For each antenna operation in the environment, many factors such as better gain, good performance, stable power structure and perfect impedance matching need to be taken into account. To achieve these features throughout the operation, antenna designers need to pay attention to issues such as the characteristics of the antenna need to be changed (such as frequency, electrical pattern or polarization) and how the antenna content is adjusted to restore the product you need because it is important and a reconfiguration process that maximizes efficiency while minimizing the negative impact on antenna performance. Reconfigurable antennas provide the same functionality as single multi-purpose antennas.

As mentioned above, researchers are working on the fifth generation (5G) technology to achieve wide and good bandwidth, reliable operation, high channel capacity instead of more information and high data, and thus have created many ways that can be used for wireless communication. In addition, these new technologies help to create connections between machines, devices and all devices to reduce latency, improve network performance and use the entire spectrum, and the working capacity was good for him. The fifth generation will also help to bring about major changes in business, science and medicine as the industry grows and new technologies with high data are created. For the successful deployment of the fifth generation technology, multi-channel communication is important and necessary to provide more network services and improve capacity according to the need. MIMO is an important technology that uses multiple transmitting and receiving antennas to transmit and receive large amounts of data over the same radio. It is used to improve and increase the strength of the signal circulating in the radio channel. MIMO technology has created new and innovative wireless models such as wireless LAN, wifi-MAX, long-term transmission, fifth generation, beyond the fifth generation and a new generation with more information, better capabilities and less support. Application to meet the needs of users with the help of different types and varieties of antennas. The negative characteristics of a perfect and well-designed antenna can greatly improve the transmission of communication signals. Therefore, high-precision, perfect MIMO antenna design and product selection, design, simulation, analysis and optimization techniques are of particular interest to the antenna communication community to improve the performance of wireless communication systems through the improvement and reliability of transmission, Speed, data rate and channel capacity. In the past and recent years, through the efforts and work of antenna designers, an important process has been found in the antenna design of planar MIMO antenna. Millimeter wave and terahertz (THz) (frequency level for 6G) electromagnetic (EM) waves enable wide-bandwidth and low-latency telecommunications and communications, information, and security (J. Lambrechts et al.,2016). Terahertz applications are also evolving in ways that will fundamentally change and reshape communications, art, and information (A. Leitenstrofer et al. 2023). Physics imposes strict limits on the propagation of high-frequency signals. The transceiver cannot learn and predict the limitations of physics. Replacing legacy transceivers with dynamic adaptive systems brings the potential for instantaneous reconfigurability and performance. Advances in artificial intelligence (AI) and machine learning (ML) show that computational algorithms can be combined with human intelligence to enable automation and optimization in the transceiver and propagation channels are often used by passive components.

Dynamic configurations using AI can generate feedback and adaptation will improve performance. Focusing on 5G and 6G communications, this article focuses on the concept of

Intelligent Integrated System (IIS). IIS is based on re-engineered hardware and "smart" hardware. Advances in Reconfigurable Smart Surfaces (RIS) (S. Venkatesh et al. 2023), (S. Venkatesh et al. 2023), (Y. Wang et al, 2023) allow the modification and modification of metasurfaces to adapt to changes in exposure. More than a century ago, in the 1890s, the ability to send signals wirelessly using electromagnetic waves was first discovered in the famous wireless telegraphy experiments performed by Nobel Prize winner G. Marconi.

Figure 1. Conceptual illustration of the 6G communication network that encompasses the 5G network (J. A. del Peral-Rosado et al. 2018).

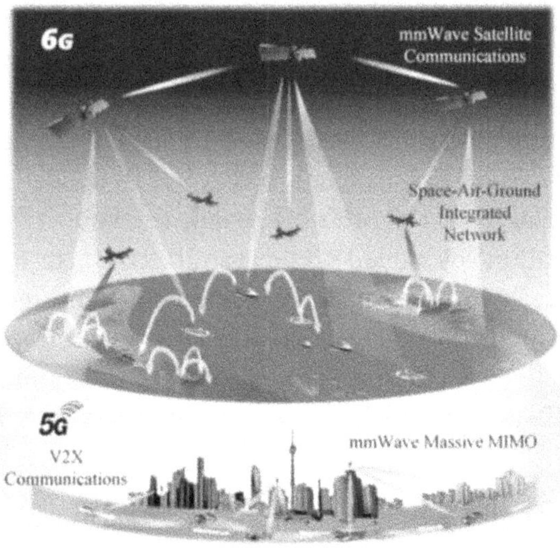

It took nearly 80 years to turn this into a technology. A business application where people could connect quickly. Since then, mobile communication technology has developed rapidly due to the development of communication theory and communication, microelectronics and integration, microwave circuits and antennas. A new generation has emerged almost every decade, starting in the 1980s (J. A. del Peral-Rosado et al. 2018). The first generation (1G) mobile communication is based on analog communication using frequency division multiplexing access (FDMA). It only allows voice signal transmission with limited and unbalanced spatial coverage (T. Farley, 2005).

The second generation (2G) uses digital communication using time-division multiplexing access (TDMA) and code-division multiplexing access (CDMA). Broader support and support for messaging between users. The third generation (3G) uses a variant of advanced CDMA technology and supports many feature-rich services, including the first to support multimedia data transmission. By leveraging orthogonal frequency-division multiplexing (OFDM) and multiple-input multiple-output (MIMO) technology, the fourth generation (4G) has been developed, including 3.9G Long-Term Evolution (LTE) and 4G LTE advanced. The turning point of the 4G era is the worldwide popularization of smart phones, changing people's lifestyles and the connections between people. In terms of the spectrum created for several generations of mobile communication, we can make two observations. First, many frequency bands are gradually released to obtain a larger channel bandwidth that can meet the demand for more information value. Second, all radio frequencies are below 4GHz, which is mainly due to two reasons: 1) electromagnetic waves below 4GHz are not easily affected by interference and weather changes; over the communication link. 4G mobile network infrastructure cannot meet the demand for high-speed telecommunications. Reliance on (6G). 5G was deployed in 2019 and is about to be commercialized on a large scale. The International Telecommunication Union (ITU) defines three main demands for the new 5G Radio (NR): Enhanced Mobile Broadband (eMBB), Massive Machine Type Communications (mMTC), and ultra-reliable low-latency low-latency communication (URLLC). 5G is expected to support multi-gigabits per second (Gb/s) data rates, millisecond-level latency and traffic rates, as well as providing major improvements in spectrum, power, and operating costs. To meet these requirements, many network services and hardware technologies have been developed, including ultra-dense connections, full-spectrum access, massive MIMO, and full-duplex. First, the sub-6 GHz spectrum is already crowded with shared trunks for cellular communications, satellite and airborne communications, and wireless local area networks (WLANs). In contrast, there are many unlicensed frequency bands in the millimeter wave frequency from 6 GHz to 300 GHz; there is a lot of usable spectrum. Second, at the same relative bandwidth, the actual bandwidth of the millimeter wave frequency is larger than the bandwidth of the low microwave frequency. The Third Partnership Project (3GPP) divided the 5G New Radio (NR) into the 410–7125 MHz FR1 band and the 7125 MHz FR2 band, also known as the millimeter wave band. In addition to the narrow frequency bands around 3.5 GHz and 4.9 GHz, many countries have also released several millimeter wave frequency bands for 5G NR communications, including the Ka band, Q band, and even the E band. Transition to 2030 and beyond due to the rapid development of new technologies such as Virtual Reality, Vehicle-to-X Network, Drone Network, Medium Earth Orbit (MEO) and Low Orbit - Satellite networks in Earth Orbit (LEO) and ocean data networks will make 5G communications inadequate.

2. LITERATURE REVIEWS FOR THE DESIGN OF ANTENNA:-

Part of the electromagnetic spectrum of the microwave and infrared ranges is defined as the terahertz band with spectral variation such as 0.1 THz - 10 THz (S. Das et al., 2021). Due to the lack of data on the terahertz frequency, the full use of this terahertz band in applications is still very limited. However, in recent years, the best communication models have provided users with more and faster information. In order to achieve good connectivity in continuous wireless communication, it increases the working frequency and wide bandwidth requirement. In this case, only the wide terahertz frequency band spectrum (0.1-10 THz) can meet the requirements of many devices to continuously communicate with high reliability systems, devices and devices from high terabit data. The advantages of terahertz communication are based on achieving high communication speed and reception in communication. On the other hand, the transmission and reception links mainly depend on the characteristics of terahertz antenna design. Microstrip patch antenna (MPA) in terahertz field (Ch Krishna et al., 2017) 2021), (K.V. Babu et al., 2023) It is in high demand due to its many advantages such as thin design, light weight, compact size, easy-to-make model. The effect, gain, bandwidth, efficiency, etc. are all very low. However, the overall performance of the microstrip antenna can be improved by confining the electromagnetic wave by combining the photonic band gap (PBG) and electromagnetic band gap models in planar antenna geometries. 2008). In the PBG design, photonic crystals are embedded in the dielectric substrate to suppress unnecessary frequency bands and create a resistive effect on the surface waves propagating through the dielectric substrate, improving the characteristics of the patch antenna. Often the channel is affected by problems such as fading, which degrades the performance of terahertz wireless communication. As a solution, many electronic devices can be used to solve the fading problem to improve the performance of terahertz communication. MIMO antenna technology can reduce fading and increase the ability to accommodate wide bandwidths. MIMO antennas are considered as the core components in terahertz systems and provide better signal reliability and efficiency as well as higher data rates. Researchers have gradually introduced various designs of single antenna, array antenna configurations and MIMO terahertz patch antenna, such as photonic crystal terahertz patch antenna arrays (Benlakehal

ME, 2022), periodic and aperiodic 1 × 2 patch array photonic antenna crystal (Benlakehal ME, 2022), flexible terahertz antenna based on photonic crystal and silicon materials (EddineTemmar MN et al., 2021) Photonic crystal terahertz antenna based on polyimide substrate (Hocini A et al., People, 2021) 2019), PBG substrate based terahertz antenna using segmented object orientation technology (Khezzar D et al., 2021), simple rectangular terahertz microstrip patch antenna for tracking application (R. H. Mahmud et al., 2020), rhoternahertz A season (Krishna et al., 2021), graphene based terahertz patch antenna with glass substrate (R. Goyal et al.). 2018) S. Ananda et al. 2014), Simple rectangular antenna based on graphene nanoribbons for terahertz applications, coaxial-fed polyimide-based fractal MIMO antenna (K. V. Babu et al. 2014) 2022), multilayer terahertz rectangular patch antenna array (A. Azarbar et al., 2014), trapezoidal microstrip patch antenna by embedding photonic crystals in the base layer (A. Singh et al., 2015), synthetic photonic bandgap substrate based on terahertz microstrip binary particle swarm optimization algorithm (MNE Tenmar, 2019), tree-like microscale wideband MIMO specifically designed for terahertz systems (K. V. Babu, 2022), U-shaped and elliptical slot loaded hexagonal terahertz patch antenna (S.M. Shamim et al., 2021), M cladding loaded rectangular patch antenna. Younis et al., 2013), simple rectangular patch antenna based on graphene (S. M. Shamim et al., 2021), embedded feed microstrip antenna with loaded 1D, 2D, 3D photonic crystal substrates (R. K. Kushwaha et al., 2021). 2021), dual-polarized graphene-based patch antenna (M. Shalini et al., 2019), PBG integrated patch antenna with frequency selective surface (FSS) and photonic band transmission (PBG) (R. K. Kushwha, 2018) Microstrip antenna (A. Neeb et al. 2014), parallel connected array antenna with PBG and SRR (R K Kushwaha et al. 2018), hexagonal pattern MIMO antenna array (M. Singh et al. 2021), multilayer terahertz array antenna model (Vettikalladi.12 H.et) 2019), microstrip MIMO antenna for terahertz short-range wireless communication system (T Okan, 2021), 2 × 2 patch antenna array with PBG substrate (ME Benlakehel et al., 2022), PBG substrate with homogeneous and periodic 1 × 2 patch antenna array (ME Benlakehel et al. 2022), 2 × 2 MIMO antenna with photonic crystal substrate (ME. Benlakehel et al. 2022), CPW-Fed MIMO uses parasitic element array antenna (K. V. Basu et al. 2022), layered antennas supported by a PBG substrate with a metal ring (I. Ahmad et al. 2021), simple microstrip antennas based on DGS and PBG (S. Ulla et al., 2019), a simple patch antenna with 5x5 type PBG structure (SK Tripathi et al., 2019), graphene-based patch antenna with PBG and various designs with wide bandwidth and increasing frequency. Terahertz antennas are currently used in high-speed wireless communication equipment. THz MIMO antennas are used to improve separation and provide high gain with high frequencies to meet the high speed and data needs of today's world and users. Terahertz MIMO antennas can be designed using materials that improve their parameters. Metamaterials are artificial materials with different microwave and optical structures that improve their properties due to their weak refraction. Metamaterial MIMO antennas have attracted researchers due to their improved performance, which can improve gain, bandwidth, separation, etc. Metamaterial loaded antennas were developed by Patel and Kosta as early as 2013 (SK Patel et al. 2012), but these antennas were implemented as single antennas. In this work, split-ring resonator, an important building block in metamaterial design, was used. Broadband and high gain antennas are achieved by integrating metamaterials into patch and ground plane. Defects in the ground plane are designed to increase the gain and bandwidth of the metamaterial antenna (R. S. Keerthi et al., 2021). The metamaterial antenna was designed and a good response was obtained by using optimization method to improve the negative (EI Kenawy et al. 2022). The nonlinear behavior of metamaterial antennas helps to increase the bandwidth of ISM and RFID applications. This antenna is designed using low cost FR4 material and is suitable for many applications due to its low cost and affordable price (Najumunnisa, M. et al. 2017). A single terahertz antenna was designed and simulated using a metamaterial simulator based on the finite element method (A. T. Devpriya et al., 2019). The antenna design has low bandwidth but can be further improved with other technologies in the future. One of the antennas that can be considered for wide range, multi-band and high speed antennas is MIMO antenna. MIMO antennas can be designed with metamaterials for fifth generation and terahertz communication to meet the high speed and high data requirements of today's world. application form. Tunability is achieved by changing the chemical potential of the formed graphene (P. Das et al., 2017). Miniaturization of THz MIMO antennas is also possible by integrating metamaterials into them. In this design, split ring resonator (SRR)-C shape is used to achieve miniaturization and broadband results (KK Naik et al., 2017). 2021). Metamaterial-based terahertz MIMO antennas have been designed using metamaterials such as open-loop resonators, thin wires, etc. In some studies on antenna development, the ground layer is also negative (R. Kumari et al. 2022). Split-ring resonators are also used in substrate materials to increase efficiency. Compared to ordinary patch-based split-ring resonators, this structure is a bit more difficult to fabricate because the split-ring resonator needs to be embedded in the substrate to be effective (Koutsoupidou, M. et al. 2014) and can be used in IoT applications. Metamaterial components are used to improve the gain and bandwidth parameters and enable their implementation in 6G high-speed devices (S. A. Khaleel et al. 2022).

3. ANTENNAS DESIGN FOR 5G AND 6G ERA

3.1 Antenna Design for Fifth Generation Technologies

In last few years, as per economic and social development which is greatly influenced by the advancements in the field of mobile communication and technology, 5G technology has emerged as an important of the future 2020 generation which is already in use. 5G technology is an enhance technology with evolutionary and revolutionary services for users. It is the recent generation of technology which is providing us ultra-high data rates, very low latency, more capacity, and good quality of service. It is worth mentioning that 5G technology has opened unleash new opportunities to leapfrog traditional barriers to development. As 5G technology supports IoT also, it gives easy major societal transformation in the fields of education, industry, healthcare, and other social sectors. 5G technology is now unlocked an extensive IoT ecosystem wherein many devices will be connected and by maintaining a trade-off between latency, cost, and speed a network can suffice the communication needs as per requirement. As 5G has come, the major 3GPP standards undergo continual change as it investigates an organized release of new functionality and is responsible for new releases of standards as per the planned schedules. It has stated three different usage scenarios of 5G communications which are as follows, (1) Enhanced Mobile Broadband (eMBB): gives ultra high speed indoor and outdoor connection and supports good and uniform quality of service at the edge of the cell, on highways, in aircraft, and train. Also, it provides high data rates upto 20 Gbps in the indoor area and 2 Gbps in the outdoor area. (2) Massive Machine Type Communications (mMTC): It helps to support IoT to interconnect a very large number of devices in which a single base station can support 10000 or more devices for different applications like smart power grids, smart cities, sensors. (3) Ultra-reliable and Low Latency Communications (uRLLC): It is used for the requirement such as low latency (below 1 ms) and low packet loss (1 in 10000 packets). At some instances it supports remote medical surgery, safety in transportation, and wireless control of manufacturing process. As given in figure 2, 5G technology is used by eight specification requirements namely frequency bands, mobility, data rate, forward error correction, access technology, latency, spectral efficiency, and connection density taking into account the connection reliability (H. C. Huang, 2018), (W. Wang et al. 2017). The use of shorter frequencies for 5G networks is the reason that why 5G can be faster from previous network. The 5G technology supports beam division multiple access and filters bank multicarrier. The highly directive beams of signals help to achieve space division multiple access which can be termed as BDMA. It can handle many users in 5G systems thereby increasing the system capacity very fast. In FBMC, a bank of filters is being used which provides better spectral efficiency. The low latency feature of 5G technology is driving new ways of using high quality video in real time for user interface. The use of artificial intelligence (AI) along with video analytics is more likely to turn high definition camera streams into actionable information. One of the pivotal parts of the 5G device is an antenna which is required to work at an enhanced gain, bandwidth with lesser radiation losses. So, antenna design for 5G devices becomes very crucial while maintaining the above mentioned parameters for 5G communication as per requirement and use. There are several work on this topic have already been done and still has been continuing to design the best antennas in low cost with high gain and bandwidth. So, in this chapter, we have tried to explore few a5G antennas in a holistic way that were proposed in recent years considering their performance enhancement techniques. The chapter also aims to direct the researcher for further advancement in the 5G antenna design as per their applications and to take it for 6G applications (Sumit Kumar et al. 2020).

Figure 2. Specifications of 5G technology (Sumit Kumar et al. 2020).

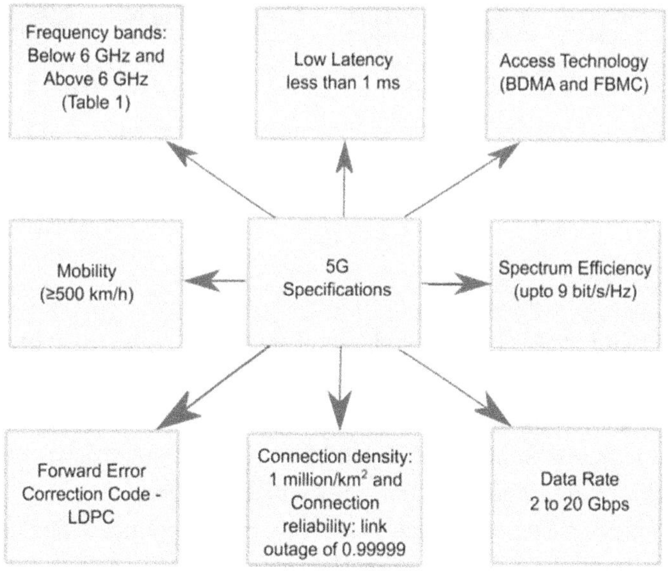

3.2 Antenna Design for beyond 5G Technologies

The microstrip patch antenna having with high demand and their compactness as well as ease of fabrication are widely used (Sarthak Singhal, 2019) (Lin Guo, 2014). Many universities has already published many research items related to 6G technologies, its advantages and few designs but that time there was no use and even they were not properly written Samsung research, 2020).Then many research institute in collaboration with industries, are working and has produces few important results in this area (K. B. Letaeif et al. 2019). This new technology for beyond 5G will basically work in terahertz frequency range which is already defined for range 0.1 and 10 THz, and then it is divided into two more categories according to frequency range as given in Figure 1. (B. Duan, 2020), (L. Chi et al. 2020). Here, from 0.1–0.3 THz is known as a sub-TeraHertz region and next part is known as pure terahertz band from later from 0.3–10 THz (M. K. Imara et al. 2020).

Figure 3. Range of frequency in beyond 5G and sixth generations (I. Fitri et al. 2017).

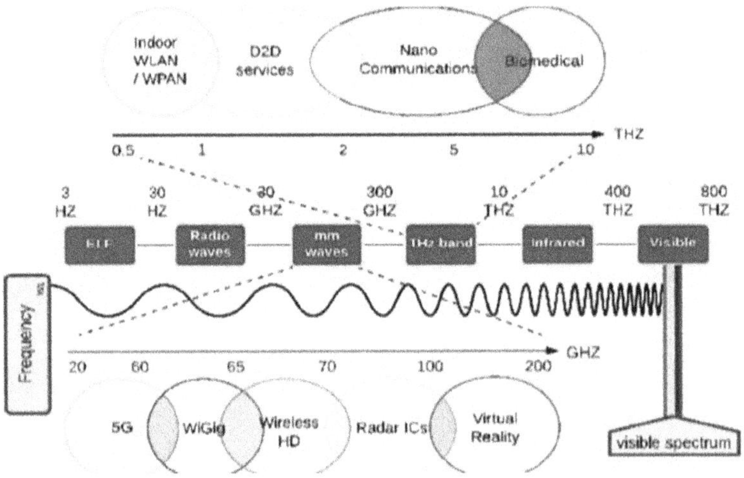

This high frequency band is also known as the terahertz gap just because of lack of materials which could respond to these high frequencies. This frequency band may also be known as Tera waves . As per Figure 2 which represents the frequency range in sixth generation wireless technology and system. As previously reported by many researchers that beyond 5G are the smart system giving secure information and it will run using artificial intelligence technology which will connect every smart device with smart intelligence system. In today research area, there are many types antenna which are being used according to design and application for high frequency like reflect array antenna are giving better performance with a new substrate material like polymide at high speed and useful in many applications such as spectroscopy in research and biomedical field. Also, few more design like elliptical pattern antenna which has high gain and good radiation pattern with ease of fabrication and can be used for any distance whether it is short or long. Due to increase demand of changing frequency, polarization and radiation pattern, reconfigurable antennas have been introduced with many smart material like graphene, metamaterials, metasurfaces etc giving MIMO applications with less loss and high gain with enhanced bandwidth at high frequency. Depending on distance covered, fading effect and introduction of solid state devices with these antenna are creating new interest between researchers as dielectric resonator antenna, MIMO antenna with cmos technology are the best suited for getting proper dispersion and less fading effect. Nano technology is also being used to get high bandwidth and less isolation effects (I. Fitri et al. 2017), (M. kokkonen et al. 2020), (Zhuo Wei Miao et al. 2018), (Binbin Cheng et al. 2018).

Hence, to develop future generation beyond 5G, frequency is increased into THz region and for that frequency, fabrication is very complex as it has very small size even in micrometer which is very difficult to make and measure as well. For that purpose, very good materials like graphene, polymide, metasurface, gold etc are required, and also well-equipped instruments are needed to perform complex fabrication and its measurement such as high frequency vector network analyzers. This frequency demands small dimensions as per wavelength and also few more additional components in some cases and costly substrate material and for the radiating elements. In other word, this frequency region will face free space loss and molecular absorption as sky wave and ground wave communication system simultaneously. As per high frequency, few materials like perfectly electric conductors show the high loss and cannot work properly at this frequency. Even some dielectric materials as mentioned above, can work at this high frequency but some improvement and advancement are required so that they can properly work without having ohmic loss and dielectric loss. Even few new technology are evolved such as use of surface Plasmon or SPP which is having less loss and behaving like slow wave structures at required and desired modes as they are also use as wave (Shivani Chandra et al., 2019), (Shivani Chandra et al. 2021), (Smrity Dwivedi, 2022), (Shivani Chandra et al. 2020), (Smrity Dwivedi, 2022), (Smrity Dwivedi, 2023), (L. Dussopt et al. 2003), (P. J. Rainville et al. 1992), (L. R. Tan et al, 2013), (G. H. Huff et al. 2007), (Rakhi Patil et al. 2022), (Sudhir Ingle Uria et al. 2022), (Krishna Dharavathu et al. 2022), (Peng Yang, 2020), (Abd Khadum Ali et al, 2022), (V. P. Eduardo et al. 2022).

4. PRACTICAL DESIGN OF ANTENNA FOR 5G, BEYOND 5G AND 6G SYSTEMS

(a) This is author own design and in this structure, design and simulation of high frequency antennas for 5G and its scope for extension of 5G have been proposed and investigated. As new era has come in which fifth generation system antennas are being used and several research on this topic still going on to improve its performance up to maximum data rate for the users as their demand of wide bandwidth and high rate of data is increasing as time passes. Here, in this paper, microwave frequency region 3 to 6 GHz has been explored and used for fifth generation systems. FR4 material has been used as substrate formation having capability of operating at both low and high frequency and also copper has been used for base material for design triangular shape array 1x4 structures to achieve high gain. Structure is simulated with the help of CST microwave studio complete ground. Results obtained from these designs are 7.18dBi, 7.22dBi, 7.24dBi and 7.2dBi for all four ports excited separately or simultaneously. The S11 parameter is -19.64dB for frequency 4.135GHz. Similar procedure has been used to simulate the single triangular structure with complete ground with gain of 7.16dBi. This comparison shows about structure durability and enhancement for 5G design. To improve the quality of work many research papers have been read and used to compare the results. Also, all advantages and disadvantages have been discussed by using the results. Here, the bandwidth is increased from 0.29 GHz to 0.32 GHz. Here, a simple structure has been discussed without using any slots and simple ground structure has been taken and discussed in coming sections with simple feeding.

Figure 4. Design of antenna (a), S parameter of antenna (c), Radiation pattern (c).

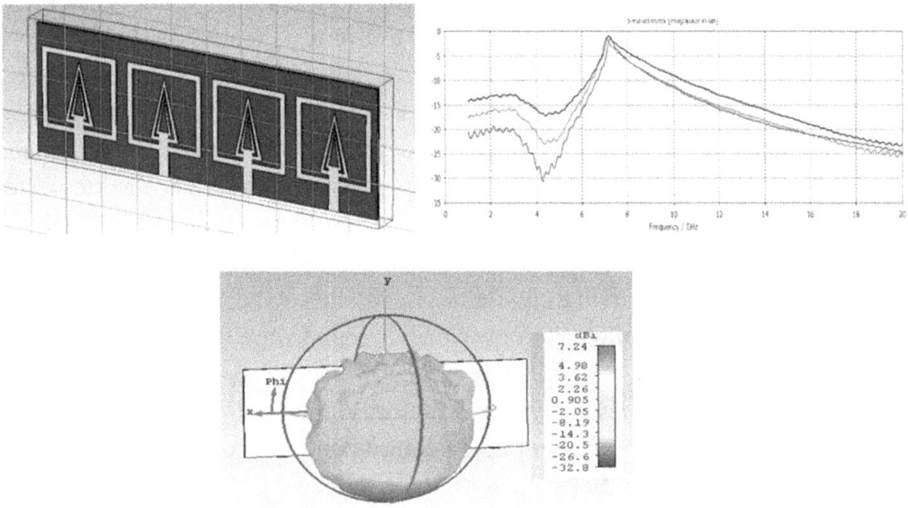

Table 1. Proposed structures comparison results

Structure Taken	Gain (dBi)
Triangular structure with complete ground	7.85
array structure (Port 1)	7.18
Array structure (Port 2)	7.22
array structure (Port 3)	7.24

(b) Author has done design and simulation of high frequency antennas for 5G and beyond 5G have been proposed and investigated. This new era of 5G and beyond 5G are useful to connect most of the available devices and create a proper balance between systems and users as per demand of technology and requirement. For that purpose, a large bandwidth is supposed to be required with better gain. Hence, microwave frequency region 8 to 10 GHz has been explored for 5G and also 7 to 20 GHz is being explored for beyond 5G that is why both possibilities have been taken here. FR4 has been used as substrate material having capability of operating at both low and high frequency and also copper has been used for base material for design triangular shape array MIMO structure to achieve high gain. Structure is simulated with the help of CST microwave studio complete ground. Results obtained from these designs are -14.5dBi, -13.9dBi, 14.3dBi, 13.3dBi for all four ports excited separately or simultaneously. The S11 parameter is -72.82dB for frequency 9.3086GHz. Similar procedure has been used to simulate the single triangular structure with partial ground with gain of 10.5dBi. Few literatures survey has been discussed with evolution of beyond 5G technology. Also merit and demerit have been discussed for proposed structures. Bandwidth is increased from 0.29 GHz to 0.32 GHz. As many designed have already been discussed in section I and II, the new thing which is being introduced in this paper, a simple design, easy fabrication, low loss as it has no slots, available materials with simple feeding techniques (S. Dwivedi, 2024).

Figure 5. Design of antenna (a), S parameter of antenna (c), Radiation pattern (c).

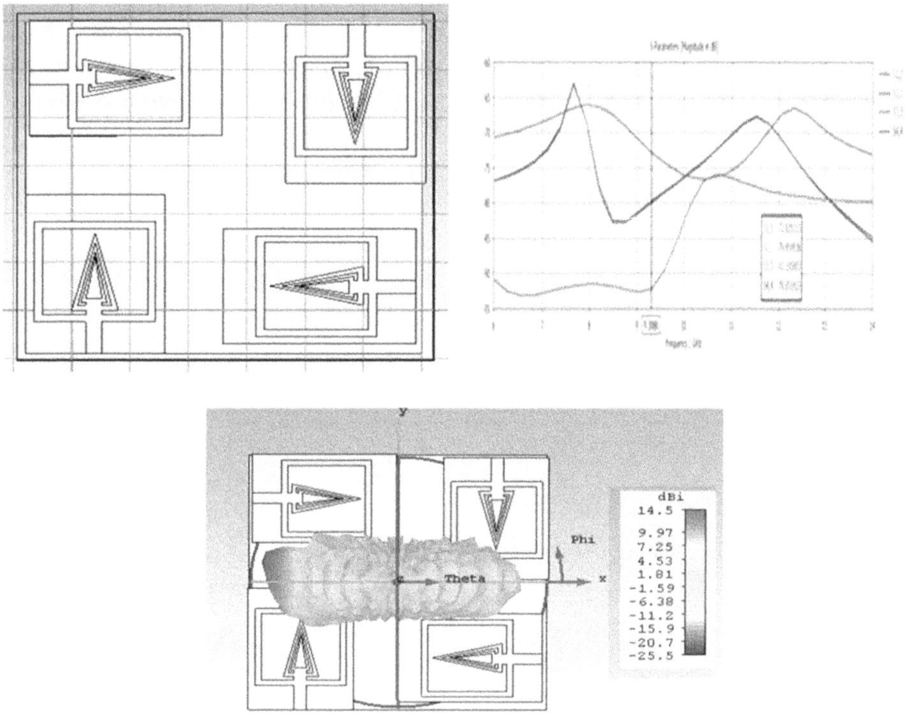

Table 2. Comparison table of proposed structures

Proposed structure	**Gain (dBi)**
Triangular structure with partial ground	10.5
MIMO structure (Port 1)	14.5
MIMO structure (Port 2)	13.9
MIMO structure (Port 3)	14.3
MIMO structure (Port 4)	13.3

(c)This is author own paper in which I designed and simulated of THz antennas for beyond 5G has been proposed and investigated. This new era of beyond 5G is useful to connect most of the devices and create a proper balance between systems and users as per demand of technology. For that purpose, a large bandwidth is supposed to be required with better gain. Hence, a new frequency region that is at THz band is utilized. Graphene has been taken as base material having capability of operating at high frequency and also to support graphene, polymide and silicon dioxide have been taken as substrate materials along with layering of air and ground as well. Few literatures survey has been discussed with evolution of terahertz Bandwidth is increased from 0.3 THz to 0.6 THz and the gain is increased 3.02 dB to 10.09 dB for the frequency range from 0.1 THz to 1 THz. a technique of gain improvement has been applied with different material layering on the same structure and simulated which is giving the simulated bandwidth greater than previous that is 52.9 GHz and higher simulated gain achieved with 10.9 dB (Smrity Dwivedi, 2023).

Table 3. Comparison table of proposed structures

Proposed structure	Frequency (THz)	Bandwidth (GHz)	Gain (dB)
Basic structure	0.45	30.4	4.21
First structure	0.45	35.6	8.48
Second structure	0.45	37	8.49
Third structure	0.45	32	8.05
Fourth structure	0.45	52.9	10.9

Figure 6. The radiating patch and ground plane are represented by grey color, while substrate is represented by purple color and PEC with white color.

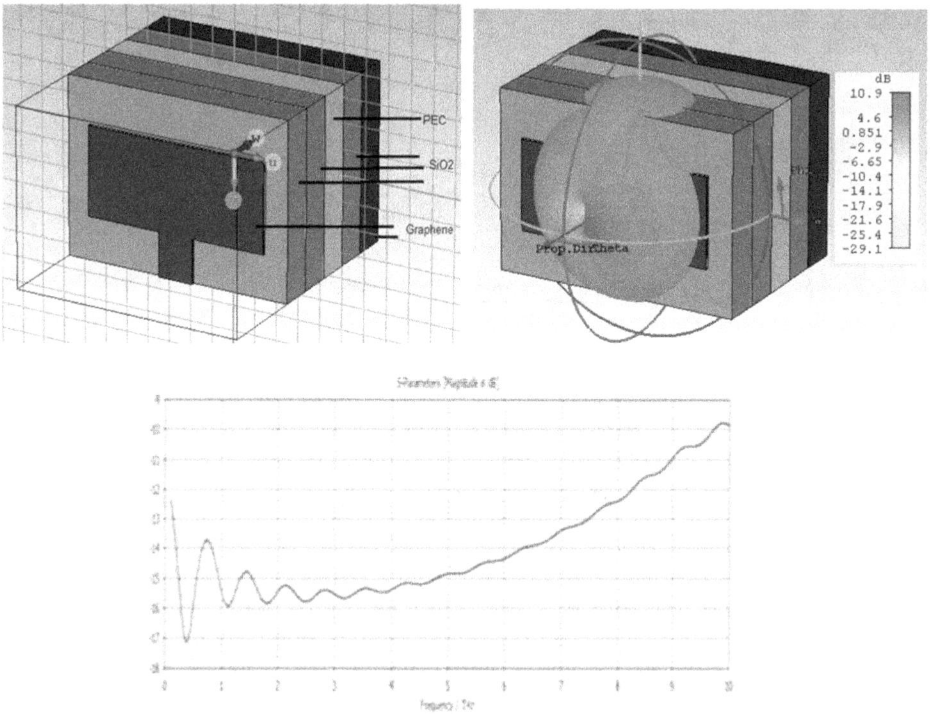

(d)This is author own work in which author has taken inserted H- shaped antenna with partial ground with three major designs and their comparison at terahertz frequency for beyond 5G applications. This new era of beyond 5G is useful to connect most of the devices and create a proper balance between systems and users as per demand of technology. Antenna design is proposed with normal rectangular patch and two more modifications have been done in same patch and it is simulated with the help of CST microwave studio for its important parameters such as bandwidth and gain. For the three structures first design, second design and third design bandwidth is 100 GHz, 250 GHz, 400 GHz with gain 13.5 dBi, 14.4 dBi, 14.8 dBi respectively. A simple PEC material as a radiating element and polyamide as substrate have been taken. Partial ground has been taken with microstrip line feedline (Smrity Dwivedi, 2023), (Utkarsh Gupta et al., 2021),(Smrity Dwivedi, 2023), (Smrity Dwivedi, 2023).

Table 4. Comparison table of proposed structures

Proposed structure	**Bandwidth (GHz)**	**Gain (dBi)**
First Design	100	13.5
Second Design	250	14.4
Third Design	400	14.8

Figure 7. Final structure Antenna with inserted H- shaped.

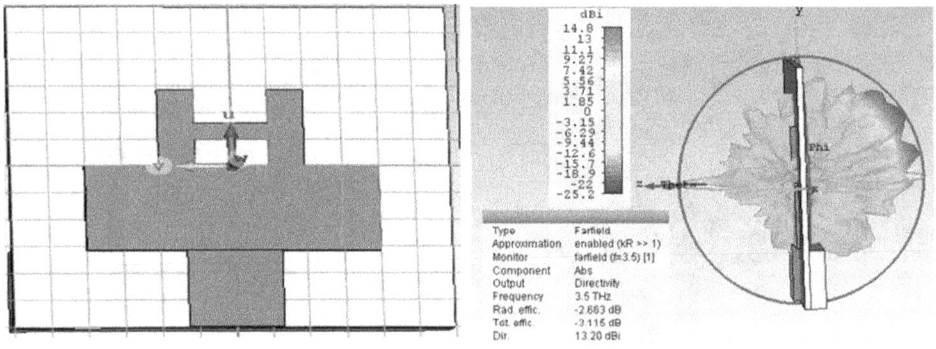

(e)This is author own work in which author has proposed design and simulation of THz antennas for beyond 5G. This new era of beyond 5G is useful to connect most of the devices and create a proper balance between systems and users as per demand of technology. For that purpose, requirement of high bandwidth with maximum gain is one primary constraint for improving performance because of that high frequency regime is used in this paper. Hence, a high frequency has been selected to design the antenna, without using the material like graphene, gold etc. A simple copper and roger have been taken for patch and substrate respectively. Here, array has been adopted which is giving MIMO characteristics which is best suited for high gain extra wide band terahertz frequency. Terahertz MIMO antenna used here, having 2-elements and 3- elements arrays. Antenna structures have been simulated using commercial software. The gain of 3- element arrays is 24.5 dBi and the gain of 2-element arrays is 22.1 dBi. Few literatures survey has been discussed with evolution of terahertz. Bandwidth is increased from 0.2 THz for 2-element arrays and 0.45 THz for 3- element arrays for the frequency range from 2 THz to 6 THz.

Table 5. Designed structures comparison.

Proposed Structure	Frequency (THz)	Bandwidth (THz)	Gain (dBi)
2-elements array	2.74	0.2	22.1
3-elements array	5.3502	0.45	24.5

Figure 8. The patch and substrate are mentioned in above figure with front view of 3- elements MIMO antenna.

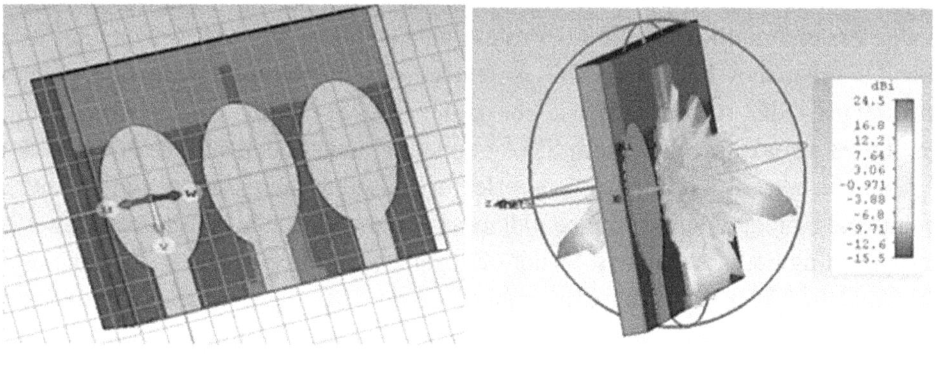

(f) Author has simulated an elliptical leaf pattern six leaves THz antenna in this paper with wide band applications for beyond 5G communication systems. Antenna design is proposed with two elliptical leaves and increased up to six leaves to enhance the performance. Bandwidth and gain of antenna are measured for two leaves, four leaves and six leaves are 300 GHz, 350 GHz, and 500 GHz with 24.9 dBi, 25 dBi and 25.2 dBi gain respectively. Author has got three different gains for three frequencies which is giving three wide bandwidth which is need of beyond 5G applications. Maximum gain 25.2dBi at 500 GHz bandwidth has been obtained for six leaves elements.

Table 6. Proposed structures design comparison

Designed structure	**Bandwidth (GHz)**	**Gain (dBi)**
Two leaves pattern	300	24.9
Four leaves pattern	350	25
Six leaves pattern	500	25.2

Figure 9. Design of antenna with six elliptical leaves and radiation pattern.

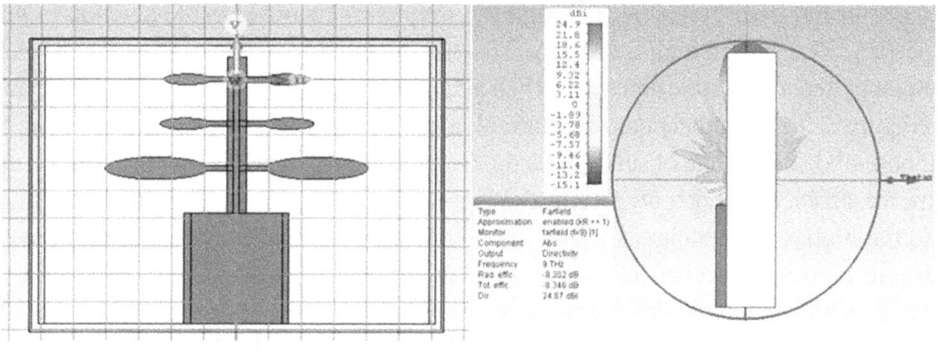

(g) Present paper is especially for the analysis of frequency reconfiguration antenna which removes the problem of achieving multifunctional systems using single antenna at different frequency according to user requirements with enhance performance in low volume. At high frequency from millimetre to terahertz range, data rate and efficient systems are the most important part at less transmission loss to get reliable wireless communication. In order to use an antenna at different frequencies for a particular frequency band, a Frequency Reconfigurable Antenna design is proposed using dual patch of gold material structure. It uses SiO_2 whose dielectric constant 3.6 (60×60 μm^2) as a substrate with a thickness of 1.575 μm. It uses a coaxial feeding with SMA connector and a metal surface at the back of substrate of the same size which act as a reflector. This reflector is used to increase the gain and other characteristics of the antenna by reducing the surface radiation. A varactor diode is used between two patches as a switch for connecting both the patches. Structure resonates at three different frequencies which is 5.6 THz, 6.2 THz, and 14.6 THz with a maximum gain of 6.44 dBi, 7.75 dBi, and 8.54 dBi respectively.

Figure 10. Reconfigurable antenna with switch, S11 parameter, radiation pattern

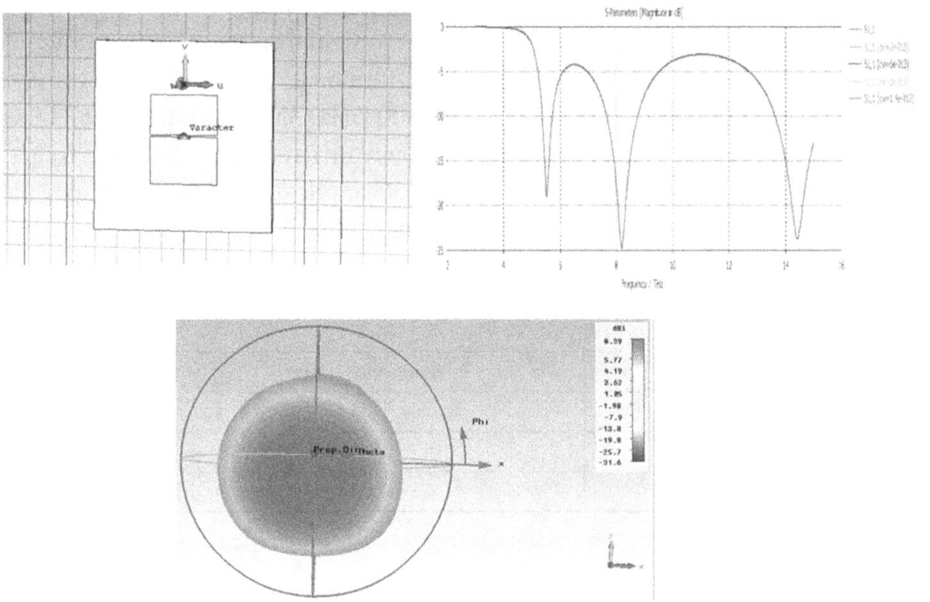

(h) Here, author has used graphene to design the frequency reconfigurable antenna as graphene has unique properties to work with high frequency. This unique tunable property of graphene can be used to design a frequency reconfigurable antenna. When a bias voltage is applied to the designed antenna array as shown in Figure 8, the chemical potential of graphene is varied from 0.4 eV to 0.8 eV. With the increase of chemical potential and change of surface conductivity the dramatical change in antenna properties are observed. Increased chemical potential gives the increase in return loss and shifts of antenna resonance frequency to the higher frequencies. To analyze the property of frequency reconfigurable antenna, the proposed structure is simulated in CST software and the result is shown in figure 16. Chemical potential of all the four patch elements of 1×4 antenna array are changed simultaneously from 0.4eV to 0.8eV. Obtained simulated result is summarized in table 1. Shift in frequency is observed from 0.64 THz to 0.74

THz with change of chemical potential from 0.4eV to 0.8eV. It is clear that the achieved frequency band which is covered by the proposed frequency reconfigurable antenna is 0.64-0.74 THz which gives the bandwidth of 100 GHz which is high enough. Hence, with such reconfigurable antennas we can easily remove the drawback of narrow band antennas.

Figure 11. Orthographic view of 1×4 graphene antenna array

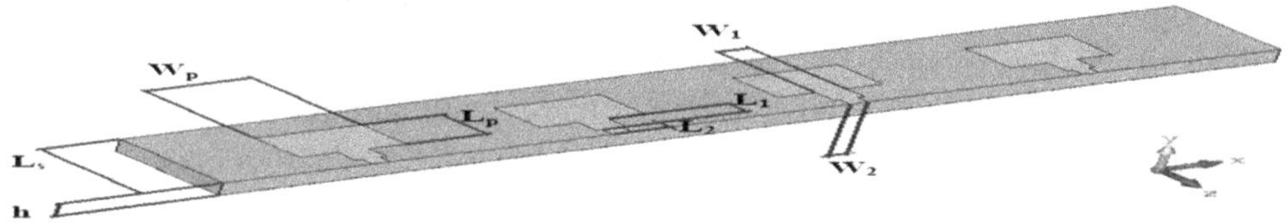

Figure 12. Effect of variation of graphene chemical potential on resonating frequency.

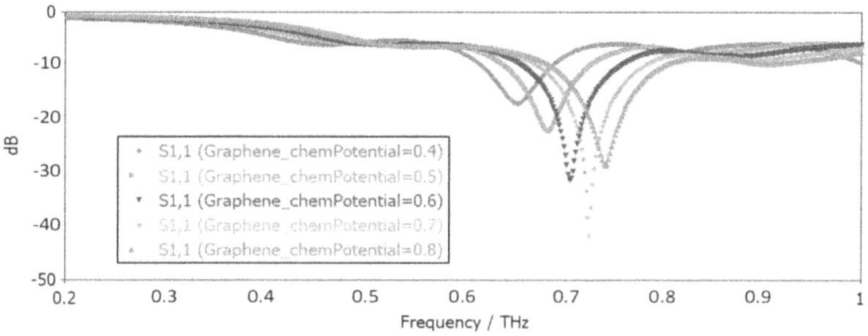

Table 7. Performance parameters of frequency reconfigurable antenna

Graphene Chemical Potential (eV)	Frequency (THz)
0.4	0.64
0.5	0.69
0.6	0.70
0.7	0.71
0.8	0.74

5. CONCLUSION

Communications for many future 6G applications. The next generation of 6G wireless communications must overcome many challenges in the terahertz frequency band. This article provides a detailed overview of the most important antenna challenges in the terahertz band. We specifically focus on antennas for ultra-high-speed terahertz wireless communications in 6G technology. The antenna design challenges to meet 6G antenna specifications are identified. The rigorous data analysis of Ultra-5G antennas has been completed and successfully analyzed and presented. There are many different types of antennas designed outside of 5G, and we discussed their benefits to help people better understand advanced antennas. Each effort has different goals and therefore different guidelines for frequency selection, antenna types, equipment, manufacturing, and development. In Chapter 3, the authors present their work covering a frequency range from GHz to THz. Many designs including inverse antennas, MIMO antennas, differential antennas, antennas made of different materials such as gold, graphene, metamaterials, SiO2, etc. Also different frequencies from 5GHz to 10 THz were selected for the design. It was found that the maximum value was approximately 24.5 dBi by design.

REFERENCES

Ahmad I, Ullah S, Ullah S, Habib U, Ahmad S, Ghaffar A, Alibakhshikenari M, Khan S, Limiti E (2021) Design and analysis of a photonic crystal based planar antenna for THz applications. Electronics 10(16):1941. ronic s1016 1941.DOI: 10. 3390/ elect

Ananda, S., Sriram Kumara, D., Wub, R. J., & Chavali, M. (2014). Graphene nanoribbon-based terahertz antenna on polyimide substrate. *Optik (Stuttgart)*, 125(19), 5546–5549. DOI: 10.1016/j.ijleo.2014.06.085

Azarbar, A., Masouleh, M. S., & Behbahani, A. K. (2014). A new terahertz microstrip rectangular patch array antenna. *Int J Electromagn Appl*, 4, 25–29.

Babu, K. V.. (2023). *Design and analysis of fractal-based THz antenna with co-axial feeding technique for wireless applications.Recent advances in graphene nanophotonics*. Springer Nature Switzerland.

Babu, K. V., Das, S., Sree, G. N. J., Madhav, B. T. P., Patel, S. K. K., & Parmar, J. (2022). Design and optimization of micro-sized wideband fractal MIMO antenna based on characteristic analysis of graphene for terahertz applications. *Optical and Quantum Electronics*, 54(5), 281. DOI: 10.1007/s11082-022-03671-2

Benlakehal ME et al (2022) Design and analysis of novel microstrip patch antenna array based on photonic crystal in THz. Opt Quantum Electron 54(5):1-16.2

Benlakehal, M. E., Hocini, A., & Khedrouche, D.. (2022). Design and analysis of a 1 × 2 microstrip patch antenna array based on photonic crystals with a graphene load in THz. *Journal of Optics*. Advance online publication. DOI: 10.1007/s12596-022-01006-8

Benlakehal, M. E., Hocini, A., Khedrouche, D., Temmar, M. N., & Denidni, T. A. (2022). Design and analysis of a 1 × 2 microstrip patch antenna array based on periodic and aperiodic photonic crystals in terahertz. *Optical and Quantum Electronics*, 54(10), 672. DOI: 10.1007/s11082-022-04076-x

Benlakehal, M. E., Hocini, A., Khedrouche, D., Temmar, M. N., & Denidni, T. A. (2022). Design and analysis of MIMO system for THz communication using terahertz patch antenna array based on photonic crystals with graphene. *Optical and Quantum Electronics*, 54(11), 693. DOI: 10.1007/s11082-022-04081-0

Benlakehal, M. E., Hocini, A., Khedrouche, D., Temmar, M. N. E., & Denidni, T. A. (2022). Design and analysis of a 2 × 2 microstrip ratch antenna array based on periodic and non-periodic photonic crystals substrate in THz. *Optical and Quantum Electronics*, 54(3), 190. DOI: 10.1007/s11082-022-03563-5

Chandra, S., & Dwivedi, S. (2019, March). Comparative analysis of reconfigurable patch antenna array for different liquid crystal substrates. In *2019 URSI Asia-Pacific Radio Science Conference (AP-RASC)* (pp. 1-4). IEEE.

Chandra, S., & Dwivedi, S. (2020, February). Simulation analysis of high gain and low loss antenna with metallic ring and spoof SPP transmission line. In 2020 7th International Conference on Signal Processing and Integrated Networks (SPIN) (pp. 328-331). IEEE.

Chandra, S., & Dwivedi, S. (2021, April). Graphene based Radiation Pattern reconfigurable antenna. In *International Conference for convergence in Technology* (pp. 2-4).

Cheng, B., Cui, Z., Lu, B., Qin, Y., Liu, Q., Chen, P., He, Y., Jiang, J., He, X., Deng, X., Zhang, J., & Zhu, L. (2018). 340-GHz 3-D imaging radar with 4Tx-16Rx MIMO array. *IEEE Transactions on Terahertz Science and Technology*, 8(5), 509–519. DOI: 10.1109/TTHZ.2018.2853551

Chi, L., Weng, Z., Qi, Y., & Drewniak, J. L. (2020). A 60 GHz PCB Wideband Antenna-in-Package for 5G/6G Applications. *IEEE Antennas and Wireless Propagation Letters*, 1225(c), 1–1. DOI: 10.1109/LAWP.2020.3006873

Das, P., Singh, A. K., & Mandal, K. (2022). Metamaterial loaded highly isolated tunable polarisation diversity MIMO antennas for THz applications. *Optical and Quantum Electronics*, 54(4), 250. Advance online publication. DOI: 10.1007/s11082-022-03641-8

Das, S., Anveshkumar, N., Dutta, J., & Biswas, A. (Eds.). (2021). *Advances in terahertz technology and its applications*. Springer., DOI: 10.1007/978-981-16-5731-3

del Peral-Rosado, J. A., Raulefs, R., López-Salcedo, J. A., & Seco-Granados, G. "Survey of cellular mobile radio localization methods: From 1G to 5G," *IEEE Commun. Surveys Tuts.*, vol. 20, no. 2, pp. 1124–1148, 2nd Quart. 2018. DOI: 10.1109/COMST.2017.2785181

Devapriya, A. T., & Robinson, S. Investigation on metamaterial antenna for terahertz applications. J. Microw.Optoelectron. Electromagn. Appl. (2019). DOI: 10.1590/2179-10742019v18i31577

Duan, B. Y. (2020). Evolution and innovation of antenna systems for beyond 5G and 6G. *Frontiers of Information Technology & Electronic Engineering*, 21, 1–3.

Dwivedi, S. (2022). *ReconfigurableArray antenna*. Intech Open., DOI: 10.5772/intechopen.106343

Dwivedi, S. (2022). Design and Analysis of Metamaterial Waveguide Antenna for broadband applications. In *ICT Analysis and Applications: Proceedings of ICT4SD 2022* (pp. 417–423). Springer Nature Singapore.

Smrity Dwivedi, "Metamaterials-Based Antenna for 5G and Beyond, "DOI: , Metamaterial Technology and Intelligent Metasurfaces for Wireless Communication Systems, IGI global publishers of timely knowledge, 2023.DOI: 10.4018/978-1-6684-8287-2.ch001

Dwivedi, S. (2023). *Metamaterials-Based Antenna for 5G and Beyond*. IGI Global., DOI: 10.4018/978-1-6684-8287-2.ch001

Dwivedi, S. "Design Demonstration ofFrequency Reconfiguration Antenna using Varactor diode with SiO ", 2023 International Conference on Computer,Electronics & Electrical Engineering & their Applications (IC2E3), 2023.

Dwivedi, S. "Proposed Design for Beyond 5G Antenna for Upgraded Applications with review," , 2023 *14th International Conference on Computing Communication and Networking Technologies (ICCCNT)*. DOI: 10.1109/ICCCNT56998.2023.10307737

Dwivedi, S. (2023, December). Terahertz Elliptical Leaf Pattern Antenna for Beyond 5G Applications. In 2023 10th IEEE Uttar Pradesh Section International Conference on Electrical, Electronics and Computer Engineering (UPCON) (Vol. 10, pp. 340-343). IEEE.

Dwivedi, S. (2023, August). Partial Ground-Inserted H-shaped Antenna for Beyond 5G Applications. In International Conference on ICT for Sustainable Development (pp. 313-321). Singapore: Springer Nature Singapore.

Dwivedi, S. (2024, March). Design of triangular MIMO antenna for 5G and beyond. In 2024 11th International Conference on Signal Processing and Integrated Networks (SPIN) (pp. 40-44). IEEE.

Dwivedi, S. (2024). Metamaterial Antennas for Wireless Communication Systems. In Electromagnetic Wave Control Techniques of Metasurfaces and Metamaterials (pp. 32-60). IGI Global.

Dwivedi, S. (2024). Antenna Design for Beyond 5G: 5G and 6G Applications With Different Materials and Techniques Including Metamaterials and Metasurfaces. In Electromagnetic Wave Control Techniques of Metasurfaces and Metamaterials (pp. 106-140). IGI Global.

El-Kenawy, E. S. M., Ibrahim, A., Mirjalili, S., Zhang, Y. D., Elnazer, S., & Zaki, R. M. (2022). Optimized ensemblealgorithm for predicting metamaterial antenna parameters. *Computers, Materials & Continua*, 2022, 023884. DOI: 10. 32604/ cmc

Emara, M. K., Stuhec-Leonard, S. K., Tomura, T., Hirokawa, J., & Gupta, S. (2020). Laser-Drilled All- Dielectric Huygens' Transmit-Arrays as 120 GHz Band Beamformers. *IEEE Access : Practical Innovations, Open Solutions*, 8, 153815–153825. DOI: 10.1109/ACCESS.2020.3018297

Farley, T. (2005). Mobile telephone history. *Telektronikk*, 101(3–4), 22–34.

Fitri, I., & Akbart, A. A. "A new gridded parasitic patch stacked microstrip antenna for enhanced wide bandwidth in 60 GHz Band," in 2017 International Conference on Broadband Communication, Wireless Sensors and Powering, BCWSP 2017, 2018, vol. 2018-Janua, pp. 1–4, DOI: 10.1109/BCWSP.2017.8272571

Goyal, R., & Vishwakarma, D. K. (2018). Design of a graphene-based patch antenna on glass substrate for high-speed terahertz communications. *Microwave and Optical Technology Letters*, 60(7), 1594–1600. DOI: 10.1002/mop.31216

Guo, L., Huang, F., & Tang, X. (2014). A novel integrated MEMS helix antenna for terahertz applications. *Optik (Stuttgart)*, 125(1), 101–103. DOI: 10.1016/j.ijleo.2013.06.016

Gupta, U., & Dwivedi, S. "Frequency Reconfigurable Antenna Using Metamaterial Split Ring Resonators for Smart Applications", *2021 International Conference on Electrical, Communication, and Computer Engineering (ICECCE)*, 2021. DOI: 10.1109/ICECCE52056.2021.9514067

Hocini, A., Temmar, M. N., Khedrouche, D., & Zamani, M. (2019). Novel approach for the design and analysis of a terahertz microstrip antenna based on photonic crystals. *Photonics and Nanostructures*, 36, 100723. DOI: 10.1016/j.photonics.2019.100723

Holloway, D., Dienstfrey, A., Kuester, E. F., O'Hara, J. F., Azad, A. K., & Taylor, A. J. (2009). A Discussion on the Interpretation and Characterization of Metafilms/Metasurfaces: The Two Dimensional Equivalent of Metamaterials. *Metamaterials (Amsterdam)*, 3(2), 100–112. DOI: 10.1016/j.metmat.2009.08.001

Holloway, K., Kuester, E. F., Baker-Jarvis, J., & Kabos, P. (2003). A Double Negative (DNG) Composite Medium Composed of Magneto-Dielectric Spherical Particles Embedded in a Matrix. *IEEE Transactions on Antennas and Propagation*, 51(10), 2596–2603. DOI: 10.1109/TAP.2003.817563

Huang, H. C. (2018, March). Overview of antenna designs and considerations in 5G cellular phones. In 2018 International Workshop on Antenna Technology (iWAT) (pp. 1-4). IEEE.

Keerthi, R. S., Dhabliya, D., Elangovan, P., Borodin, K., Parmar, J., & Patel, S. K. (2021). Tunable high-gain and multiband microstrip antenna based on liquid/copper split-ring resonator superstrates for C/X band communication. *Physica B, Condensed Matter*, 618, 413203.

Khaleel, S. A., Hamad, E. K., Parchin, N. O., & Saleh, M. B. (2022). MTM-inspired graphene-based THz MIMO antenna configurations using characteristic mode analysis for 6G/IoT applications. *Electronics (Basel)*, 11(14), 2152.

Khezzar, D., Khedrouche, D., & Denidni, T. A. (2021). New design of a broadband PBG-based antenna for THz band applications. *Photonics and Nanostructures*, 46, 100947. DOI: 10.1016/j.photonics.2021.100947

M. Kokkonen, S. Myllymaki, and H. Jantunen, "3×3 dipole lens antenna at 300 GHz with different permittivity lenses," in 2nd 6G Wireless Summit 2020: Gain Edge for the 6G Era, 6G SUMMIT 2020, 2020, pp. 4–7, .DOI: 10.1109/6GSUMMIT49458.2020.9083825

Koutsoupidou, M., Karanasiou, I. S., & Uzunoglu, N. (2014). Substrate constructed by an array of split ring resonators for a THz planar antenna. *Journal of Computational Electronics*, 13(3), 593–598. Advance online publication. DOI: 10.1007/s10825-014-0575-y

Krishna, ChM., Das, S., Lakrit, S., Lavadiya, S., Madhav, B. T. P., & Sorathiya, V. (2021). Design and analysis of a super wideband (0.09–30.14 THz) graphene-based log periodic dipole array antenna for terahertz applications. *Optik (Stuttgart)*, 247, 167991. DOI: 10.1016/j.ijleo.2021.167991

Krishna, C. M., Das, S., Nella, A., Lakrit, S., & Madhav, B. T. P. (2021). A micro-sized rhombus shaped THz antenna for high-speed short-range wireless communication applications. *Plasmonics*, 16(6), 2167–2177. DOI: 10.1007/s11468-021-01472-z

Kumar, S., Dixit, A. S., Malekar, R. R., Raut, H. D., & Shevada, L. K. (2020). Fifth generation antennas: A comprehensive review of design and performance enhancement techniques. *IEEE Access : Practical Innovations, Open Solutions*, 8, 163568–163593.

Kumari, R., Tomar, V. K., & Sharma, A. (2022). Miniaturization and performance enhancement of super wide band four element MIMO antenna using DNG metamaterial for THz applications. *Optical and Quantum Electronics*, 54(9), 577. Advance online publication. DOI: 10.1007/s11082-022-04011-0

Kushwaha, R. K., & Karuppanan, P. (2021). Investigation and design of microstrip patch antenna employed on PCs substrates in THz regime. *Aust J Electr Electron Eng*, 18(2), 118–125. DOI: 10.1080/1448837X.2021.1936779

Kushwaha, R. K., Karuppanan, P., & Malviya, L. D. (2018). Design and analysis of novel microstrip patch antenna on photonic crystal in THz. *Physica B, Condensed Matter*, 545, 107–112. DOI: 10.1016/j.physb.2018.05.045

Kushwaha, R. K., Karuppanan, P., & Srivastava, Y. (2018). Proximity feed multiband patch antenna array with SRR and PBG for THz applications. *Optik (Stuttgart)*, 175, 78–86. DOI: 10.1016/j.ijleo.2018.08.139

Lambrechts, J., & Sinha, S. (2016). *Microsensing Networks for Sustainable Cities* (Vol. 18). Springer., DOI: 10.1007/978-3-319-28358-6

Leitenstorfer, A., Moskalenko, A. S., Kampfrath, T., Kono, J., Castro-Camus, E., Peng, K., Qureshi, N., Turchinovich, D., Tanaka, K., Markelz, A. G., Havenith, M., Hough, C., Joyce, H. J., Padilla, W. J., Zhou, B., Kim, K.-Y., Zhang, X.-C., Jepsen, P. U., Dhillon, S., & Cunningham, J. (2023, June). The 2023 terahertz science and technology roadmap. *Journal of Physics. D, Applied Physics*, 56(22), 223001. Advance online publication. DOI: 10.1088/1361-6463/acbe4c

Letaief, K. B., Chen, W., Shi, Y., Zhang, J., & Zhang, Y. J. A. (2019, August). The Roadmap to 6G: AI Empowered Wireless Networks. *IEEE Communications Magazine*, 57(8), 84–90. DOI: 10.1109/MCOM.2019.1900271

Mahmud, R. H. (2020). Terahertz microstrip patch antennas for the surveillance applications. *Kurdistan Journal of Applied Research*, 5(1), 16–27.

Miao, Z.-W., Hao, Z.-C., Wang, Y., Jin, B.-B., Wu, J.-B., & Hong, W. (2018). A 400-GHz High-Gain Quartz-Based Single Layered Folded Reflectarray Antenna for Terahertz Applications. *IEEE Transactions on Terahertz Science and Technology*, 9(1), 78–88. DOI: 10.1109/TTHZ.2018.2883215

Naik, K. K., Suman, M., & Rao, E. K. (2021). Design of complementary split ring resonators on elliptical patch antenna with enhanced gain for terahertz applications. *Optik (Stuttgart)*, 243, 167434.

Najumunnisa, M., Sastry, A. S. C., Madhav, B. T. P., Das, S., Hussain, N., Ali, S. S., & Aslam, M. (2022). A metamaterial inspired AMC backed dual band antenna for ISM and RFID applications. *Sensors (Basel)*, 22(20), 8065. Advance online publication. DOI: 10.3390/s22208065 PMID: 36298414

Nejati, A., Sadeghzadeh, R. A., & Geran, F. (2014). Effect of photonic crystal and frequency selective surface implementation on gain enhancement in the microstrip patch antenna at terahertz frequency. *Physica B, Condensed Matter*, 449, 113–120. DOI: 10.1016/j.physb.2014.05.014

Okan, T. (2021). High efficiency unslotted ultra-wideband microstrip antenna for sub-terahertz short range wireless communication systems. *Optik (Stuttgart)*, 242, 166859. DOI: 10.1016/j.ijleo.2021.166859

Patel, S. K., & Kosta, Y. P. (2012). Meandered multiband metamaterial square microstrip patch antenna design. Waves Rand. *Complex Media*, 2012, 723837. DOI: 10. 1080/ 17455 030

Roslan, N. H., Awang, A. H., & Hizan, H. M. (2008) The effect of photonic crystal parameters on the terahertz photonic crystal cavities microstrip antenna performances. In: *Proceedings of the 2018 IEEE International RF and Microwave Conference (RFM)*, Penang, Malaysia, 17–19 December 2018, pp 147–150.

Shalini, M., & Madhan, M. (2019). Design and analysis of a dualpolarized graphene based microstrip patch antenna for terahertz applications. *Optik (Stuttgart)*, 194, 163050. DOI: 10.1016/j.ijleo.2019.163050

Shamim, S. M., Das, S., Hossain, M. A., & Madhav, B. T. P. (2021). Investigations on Graphene-Based Ultra-Wideband (UWB) microstrip patch antennas for Terahertz (THz) Applications. *Plasmonics*, 16(5), 1623–1631. DOI: 10.1007/s11468-021-01423-8

Shamim, S. M., Uddin, M. S., Hasan, M., & Samad, M. (2021). Design and implementation of miniaturized wideband microstrip patch antenna for high-speed terahertz applications. *Journal of Computational Electronics*, 20(1), 604–610. DOI: 10.1007/s10825-020-01587-2

Sharma, M., Sharma, K., Dwivedi, S., Singh, D. K., Gill, R., & Kumar, N. (2023, December). A Multiple Millimeter-Wave 5G MIMO Antenna Including n257/n258/n259/n260/n261 FR2-Bands with High Diversity Performance. In *2023 International Conference on Power Energy, Environment & Intelligent Control (PEEIC)* (pp. 279-284). IEEE.

Singh, A., & Singh, S. (2015). A trapezoidal microstrip patch antenna on photonic crystal substrate for high-speed THz applications. *Photonics and Nanostructures*, 14, 52–62. DOI: 10.1016/j.photonics.2015.01.003

Singh, M., Singh, S., & Islam, M. T. (2021). *Highly efficient ultra-wideband MIMO patch antenna array for short range THz applications, emerging trends in terahertz engineering and system technologies devices, materials, imaging, data acquisition and processing*., DOI: 10.1007/978-981-15-9766-4

Singhal, S. (2019). Elliptical ring terahertz fractal antenna. *Optik (Stuttgart)*, 194, 163129. Advance online publication. DOI: 10.1016/j.ijleo.2019.163129

Sivukhin, D. V. (1957). The Energy of Electromagnetic Fields in Dispersive Media. *Opt. Spektrosk.*, 3, 308–312.

Smith, D. R., Padilla, W. J., Vier, D. C., Nemat-Nasser, S. C., & Schultz, S. (2000). Composite Medium with Simultaneously Negative Permeability and Permittivity. *Physical Review Letters*, 84(18), 4184–4186. DOI: 10.1103/PhysRevLett.84.4184 PMID: 10990641

Temmar, M. N., Hocini, A., Khedrouche, D., & Denidni, T. A. (2020). Enhanced flexible terahertz microstrip antenna based on modified silicon-air photonic crystal. *Optik (Stuttgart)*, 217, 164897. DOI: 10.1016/j.ijleo.2020.164897

Temmar, M. N. E., Hocini, A., Khedrouche, D., & Denidni, T. A. (2021). Analysis and design of MIMO indoor communication system using terahertz patch antenna based on photonic crystal with graphene. *Photonics and Nanostructures*, 43, 100867. DOI: 10.1016/j.photonics.2020.100867

Temmar, M. N. E., Hocini, A., Khedrouche, D., & Zamani, M. (2019). Analysis and design of a terahertz microstrip antenna based on a synthesized photonic bandgap substrate using BPSO. *Journal of Computational Electronics*, 18(1), 231–240. DOI: 10.1007/s10825-019-01301-x

Tripathi, S. K., & Kumar, A. (2019). High gain miniaturised photonic band gap terahertz antenna for size-limited applications. *Aust J Electr Electron Eng*, 16(2), 74–80. DOI: 10.1080/1448837X.2019.1602944

Ullah S, Ruan C, Haq TU, Zhang X (2019) High performance THz patch antenna using photonic band gap and defected ground structure. J Electromagn Waves Appl 33(15):1943–1954. . 2019. 16549 29.DOI: 10. 1080/ 09205 071

Vasu Babu, K., Das, S., Varshney, G., Sree, G. N. J., & Madhav, B. T. P. (2022). A micro-scaled graphene-based treeshaped wideband printed MIMO antenna for terahertz applications. *Journal of Computational Electronics*, 21(1), 289–303. DOI: 10.1007/s10825-021-01831-3

Vasu Babu, K., Sree, G. N. J., Kumari, S. V., & Das, S. (2022). Design and analysis of a CPW-fed fractal MIMO THz antenna using an array of parasitic elements. In Das, S., Nella, A., & Patel, S. K. (Eds.), *Terahertz devices, circuits and systems*. Springer., DOI: 10.1007/978-981-19-4105-4_4

Venkatesh, S., Lu, X., Saeidi, H., & Sengupta, K. (2022, August). A programmable terahertz metasurface with circuit-coupled meta-elements in silicon chips: Creating low-cost, large-scale, reconfigurable terahertz metasurfaces. *IEEE Antennas & Propagation Magazine*, 64(4), 110–122. DOI: 10.1109/MAP.2022.3176588

Venkatesh, S., Sturm, D., Lu, X., Lang, R. J., & Sengupta, K. (2022, October). Origami microwave imaging array: Metasurface tiles on a shape-morphing surface for reconfigurable computational imaging. *Advancement of Science*, 9(28), 2105016. Advance online publication. DOI: 10.1002/advs.202105016 PMID: 35896946

Vettikalladi H, Sethi WT, Abas AFB, Ko W, Alkanhal MA, Himdi M (2019) Sub-THz antenna for high-speed wireless communication systems. Int J Antennas Propag 2019, Article ID 9573647, 9 pages. .DOI: 10.1155/2019/9573647

Wang, C., Chen, Y., & Yang, S. (2018, November). Dual-band dual-polarized antenna array with flat-top and sharp cutoff radiation patterns for 2G/3G/LTE cellular bands. *IEEE Transactions on Antennas and Propagation*, 66(11), 5907–5917. DOI: 10.1109/TAP.2018.2866596

Wang, Y., Zhang, W., Liu, C., Sun, J., & Wang, C.-X. ''Reconfigurable intelligent surface for NLOS integrated sensing and communications,'' in *Proc. IEEE/CIC Int. Conf. Commun. China (ICCC)*, IEEE, Aug. 2022, pp. 708–712, DOI: 10.1109/ICCC55456.2022.9880639

Wen, L.-H., Gao, S., Mao, C.-X., Luo, Q., Hu, W., Yin, Y., & Yang, X. (2018). A wideband dual-polarized antenna using shorted dipoles. *IEEE Access : Practical Innovations, Open Solutions*, 6, 39725–39733. DOI: 10.1109/ACCESS.2018.2855425

Yang, D., Liu, S., & Zhao, Z. (2017, May). A broadband dual-polarized printed dipole antenna with low cross-polarization and high isolation for base station applications. *Microwave and Optical Technology Letters*, 59(5), 1107–1111. DOI: 10.1002/mop.30467

Younssi, M., Jaoujal, A., Yaccoub, M. D., El Moussaoui, A., & Aknin, N. (2013). Study of a microstrip antenna with and without superstrate for terahertz frequency. *International Journal of Innovation and Applied Studies*, 2(4), 369–371.

Chapter 5
The Role of Artificial Intelligence in 6G Networks, Architecture, Protocol, Transmission, and Applications

W. Aldrin Joan Pandian
https://orcid.org/0009-0000-0501-9614
VIT Bhopal University, India

Palak Mangal
VIT Bhopal University, India

D. Lakshmi
https://orcid.org/0000-0003-4018-1208
VIT Bhopal University, India

I. Jasmine Selvakumari Jeya
VIT Bhopal University, India

ABSTRACT

The 6G wireless communication technology is the successor to the 5G cellular-based technique. It is aimed at the increased bandwidth and higher dimensions of data transmission with the measurement of less than one microsecond of latency communication. Artificial Intelligence (AI) is pivotal in establishing computational and intelligence infrastructure, data storing, transmission, and decision-making. With the increased frequencies of 6G, sampling rates will surge beyond those achievable with 5G (Fifth Generation). Access points will gain the capacity to serve multiple clients through orthogonal frequency-division multiple access concurrently. The expected spectrum of communication using 6G is up to terahertz (THz). Artificial Intelligence (AI) guides Quality of Service (QoS) in 6G communication. This chapter covers a high-level overview of the role of AI in 6G, architecture, protocols, transmission technologies, and applications.

DOI: 10.4018/979-8-3693-8799-3.ch005

1.0 INTRODUCTION

Wireless communication refers to the transmission of data or information over a distance without the use of physical connections or wires. This type of communication utilizes electromagnetic waves, such as radio frequencies, microwaves, and infrared signals, to convey information between devices. Figure 1 shows the evolution of wireless communication.

Figure 1. Evolution of Wireless Communication

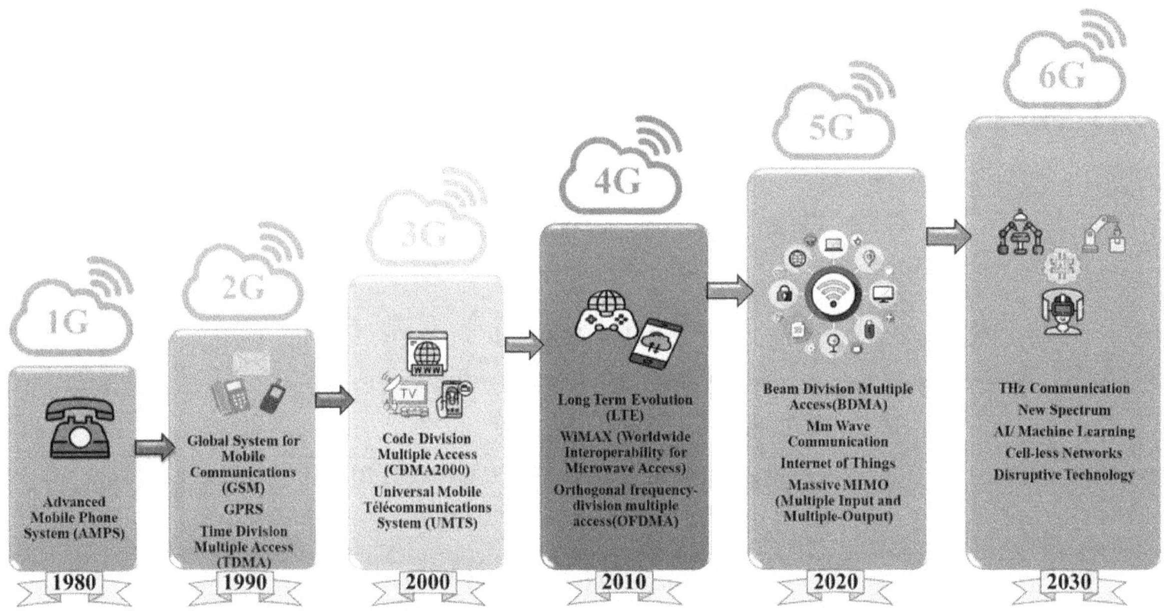

In the early 1980s wireless mobile technology was started with the first generation (1G). 1G was used for low data transfer, especially for voice/audio communication with a speed of 2.4Kbps.

In the early 1990s, the second generation (2G) wireless technology was introduced. Time Division Multiple Access (TDMA) is a cellular telecommunication technology that was developed to transmit high-quality audio and text communication with a transmission speed of 64Kbps.

In the early 2000s, the third generation (2G) wireless technology was introduced. Code Division Multiple Access (CDMA) is a cellular telecommunication technology that was developed to transmit high-quality audio and text communication with a transmission speed of 64Kbps.

In the early 2000s, the third generation (3G) wireless technology was introduced. Code Division Multiple Access (CDMA) is a cellular telecommunication technology that was developed to transmit high-quality audio and text communication with a transmission speed of 64Kbps.

In the early 2010s, the fourth generation (4G) wireless technology was introduced. Orthogonal Frequency Division Multiple Access (OFMDA) is a cellular telecommunication technology that was developed to transmit high-quality audio and text communication with a transmission speed of 1Gbps.

In the early 2020s, the fifth generation (5G) wireless technology was introduced. Beam Division Multiple Access (BDMA) is a cellular telecommunication technology that was developed to transmit multiple inputs and multiple outputs with a transmission speed of 1Gbps to 10Gbps.

In the 2030s, it is expected that sixth-generation (6G) wireless technology will be introduced. This technology aims to revolutionize connectivity by moving towards a cellular-less network structure. 6G is being developed to support multiple-input and multiple-output (MIMO) systems with an ambitious goal of achieving transmission speeds of 1 Tbps. These advancements will significantly enhance data transmission rates, connectivity, and overall network performance, paving the way for new applications and services (Chataut et al., 2024). Artificial intelligence (AI) is expected to play a crucial role in shaping the development and deployment of sixth-generation (6G) wireless technology.

6G communication technology is set to transform numerous technologies and game-changers in a range of industrial applications. Artificial intelligence drives the entire spatial framework of 6G architecture. In the forthcoming 6G networks, mobile edge computing will be inherently integrated, unlike the necessity to incorporate it into current 5G networks. The fusion of edge and core computing will be more seamless, forming a unified communication and computation infrastructure framework by the onset of 6G network deployment. This strategy promises numerous advantages as 6G technology comes into play, such as enhanced access to AI capabilities and bolstered support for advanced mobile devices and systems. 6G is expected to be available for commercial use in 2030 (Ahammed et al., 2023).

6G core technologies include AI, Distributed Radio Access Networks (RAN), Cloud Infrastructure, Sensors, Quantum Computing, new material based on liquid crystals, High-Performance Computing, photonic and electronic-based integration solutions, integrated sensing, localization, and communication (Nayak & Patgiri, 2022).

6G networks aim for "Intelligent Connectivity," "Deep Connectivity," "Holographic Connectivity," and "Ubiquitous Connectivity." Key technologies include terahertz communication, AI-based wireless, and integrated Space-Air-Ground-Sea networks (Yajun et al., 2019).

2.0 ARTIFICIAL INTELLIGENCE IN 6G NETWORKS

2.1 Introduction to AI in 6G Networks

Artificial Intelligence is so important in the 6G network because it will be used to enhance various aspects of network operations such as Management, Security, etc.

2.2 AI-Driven Network Performance Optimization:

2.2.1 Network Management:

Artificial Intelligence in management can be used to continuously monitor network conditions and real-time adjustments can be made to the parameters to optimize performance. Using this we can also be able to adjust the bandwidth, power levels and even other parameters to have well-optimized connectivity and performance (Umoga et al., 2024).

2.2.2 Predictive Maintenance and Self-Healing:

Predictive Maintenance is the main feature that is attained with the help of Artificial Intelligence. When we use AI in a 6G network we can also get the help of predictive Maintenance (Umoga et al., 2024). The AI will have the ability to predict potential network failures and will take all kinds of measures to prevent them from having them. Because of this, we can reduce the downtime and also can improve the reliability of the network. Predictive Maintenance can learn the patterns of the issues and can annihilate them.

2.3 Enhanced Security and Privacy

2.3.1 Real-time Threat Detection and Mitigation:

Analyzing network traffic and detecting security threats can be done in real time. It can also be able to annihilate the issues. The Machine learning models can also understand the patterns of cyberattacks and will enable prompt responses to mitigate the risks. **(Paolini et al., 2023)**

2.3.2 Data Privacy:

In 6G networks, ensuring data privacy is a major concern because of the vast amounts of data produced by numerous devices and applications. Federated learning tackles this issue by training AI models on multiple devices without needing to centralize the data. Each device processes its data locally and only sends model updates to a central server, keeping the raw data private. This approach reduces the risk of data breaches and unauthorized access while also minimizing data transmission. It helps comply with privacy regulations and supports personalized models, making it particularly useful in fields like healthcare and finance. (Nguyen et al., 2021)

2.3.3 Secure Communication:

Secure communication in 6G networks is crucial to protecting the integrity and confidentiality of data. Federated learning enhances secure communication by ensuring that only model updates, rather than raw data, are transmitted between devices and central servers. This minimizes the amount of sensitive information that is sent over the network, reducing the likelihood of interception and unauthorized access. Although it faces challenges like communication overhead and the complexity of aggregating model updates, federated learning provides a robust solution for maintaining secure communication in the rapidly advancing 6G landscape. (Sirohi D et al., 2023)

2.4 AI in Resource and Spectrum Management

2.4.1 Dynamic Resource Allocation:

The AI algorithm can manage the network resource by analyzing the real-time demand and usage patterns so that they can dynamically allocate the required bandwidth, processing power, and even other resources. By doing that, AI helps in optimizing the use of the available resources and also ensuring that network performance remains high. The benefits of this are less network congestion, and minimum latency and also enhance the overall efficiency of the network. The user experience will be faster and more reliable connections as a result. (Yang et al., 2020)

2.4.2 Spectrum Optimization:

It can also control the AI radio spectrum, which is the frequencies used for communication. It can find out the underutilized frequencies and relocate them to the areas of high demand. By using dynamic management, it helps us to use the spectrum more efficiently as possible. Reducing the interference of different network users and maximizing the network capacity. Because of this more devices can be connected simultaneously without compromising the performance (Shi et al, 2023).

2.5 AI for Edge Computing and IoT Integration

2.5.1 Edge Intelligence:

In the 6G landscape, AI is being integrated with advanced communication technologies to boost network performance and capabilities. Techniques like federated learning, decentralized learning, split learning, distributed reinforcement learning, and trustworthy learning models are used to make 6G networks smarter and more efficient. However, there are challenges. One major issue is how edge AI interacts with the complex metaverse framework, which brings significant technical and operational difficulties. Additionally, incorporating edge AI into virtual environments adds more layers of complexity that must be overcome to fully benefit from edge computing in the 6G era (Nivetha et al., 2024).

2.5.2 IoT Device Management:

The combination of advanced AI techniques and 6G communication technologies has huge potential to transform various industries through effective IoT device management. AI is crucial for managing large-scale IoT deployments by monitoring device health, predicting maintenance needs, and optimizing operations to prevent failures. This leads to a more robust and scalable IoT network, supporting diverse applications from smart homes to industrial IoT solutions. By using edge Intelligence and 6G, industries can achieve better network efficiency, enhanced data processing, and increased connectivity, paving the way for more innovative and intelligent applications (Shen et al., 2022).

2.6 AI-Driven Quality of Service (QoS) and User Experience

2.6.1 Personalized Services:

Technology advancements have paved the way for highly personalized digital services. With the integration of AI powered by 6G, services can now cater to individual preferences and behaviors. AI-driven Quality of Service (QoS) optimization in 6G networks, like the EvoNetSlice framework, dynamically allocates resources based on traffic patterns and user behavior. This approach not only boosts network efficiency but also adapts to changing QoS needs, greatly enhancing the user experience. EvoNetSlice optimizes 6G network slicing for resource allocation and QoS management, effectively balancing conflicting goals to ensure optimal network performance. (Yang et al., 2022)

2.6.2 Adaptive Applications:

The use of edge cloud metaverse and mobile Edge Computing (ECM) is transforming how applications engage with users. Decentralized metaverse architectures allow applications to be more flexible and responsive to user needs. However, challenges such as optimizing 6G network slicing for resource allocation and QoS management, as well as balancing conflicting objectives for efficient network performance, must be addressed to fully exploit the potential of adaptive applications. The EvoNetSlice framework tackles these issues by integrating dynamic resource allocation and real-time QoS, achieving effective multi-objective optimization to balance various network goals. (Venkata et al, 2024)

2.7 Challenges and Future Directions in AI for 6G

2.7.1 Technical Challenges:

The seamless integration of Artificial Intelligence (AI) into 6G networks faces several significant technical challenges that must be overcome. These include the development of advanced AI technologies to automate information delivery between smart applications in wireless networks. Additionally, satellite-based Non-Terrestrial Networks (NTN) pose unique difficulties, such as long propagation delays and spectrum sharing complexities. The increase in the number of antennas also presents challenges in physical layer communication, including frequent feedback and higher energy consumption. Moreover, efficiently supporting customized services with diverse Quality of Service (QoS) requirements necessitates AI-enabled integration of massive Multiple-Input-Multiple-Output (MIMO) techniques with Network Functions Virtualization (NFV) and Software-Defined Network (SDN) architectures. Overcoming these challenges is essential for the successful deployment of AI in 6G networks, enhancing performance and enabling advanced capabilities. (Chataut, R et al., 2024)

2.7.2 Ethical and Regulatory Issues:

The integration of Artificial Intelligence (AI) into 6G networks brings about several technical challenges and points to promising future research directions. Some of the key technical challenges include dealing with long propagation delays, high Doppler shifts, and complex beam and resource allocation in Non-Terrestrial Networks (NTN). Additionally, optimizing the interplay between communication and sensing functions through AI techniques is essential. It's also important to align industry standards to ensure AI can be widely adopted in mobile communication systems.

All the research is focusing on how AI can enhance intelligent decision-making, automation, and predictive capabilities in wireless networks. This includes developing AI-driven solutions for satellite-based NTN and creating AI-enabled semantic communication systems for next-generation networks. Ethical and regulatory issues are also crucial when integrating AI into 6G networks. Data privacy and security are top priorities, especially with the increased use of AI, as protecting user data from breaches and unauthorized access is a major concern. AI systems must also address biases that could lead to unfair treatment of certain groups, ensuring fairness and equity in 6G applications. Compliance with global and regional regulations, such as GDPR, is vital, requiring a careful balance between following rules and fostering innovation. Additionally, the ethical implications of AI, including surveillance, autonomy, and decision-making, must be considered to build user trust and acceptance. Ensuring the responsible and secure implementation of AI technologies in future wireless systems is key to their successful integration (abdulraqeb et al., 2024).

2.8 Future Research Directions:

Future research directions for AI in 6G networks cover several areas. One major focus is developing advanced AI algorithms that can handle the complexity and scale of 6G networks, particularly those designed for real-time processing, scalability, and energy efficiency. Additionally, research into edge AI and federated learning is crucial because these methods enhance data privacy and reduce latency by processing data closer to the source—vital for applications requiring immediate decision-making. Integrating quantum computing with AI is another exciting avenue, as it could significantly increase computational power, enabling more efficient data processing and analysis in 6G networks.

AI for network security is another essential area of research. Developing techniques to improve threat detection and mitigation, including the prediction and prevention of cyber-attacks, is paramount. Equally important are sustainable AI practices, which focus on creating energy-efficient algorithms and methods to minimize the environmental impact of 6G networks. This involves optimizing AI workloads and developing eco-friendly AI technologies. Furthermore, improving collaboration between humans and AI systems is crucial for maximizing the benefits of AI in 6G networks. This includes creating intuitive interfaces and ensuring transparency in AI decision-making processes.

Exploring innovative technologies such as non-terrestrial networks (NTN) and leveraging AI for intelligent networking to enhance performance and resource management are also key research areas. Implementing AI-enabled architectures for smart resource management and service provisioning aims to address challenges like scalability, real-time decision-making, energy efficiency, and privacy concerns while maximizing network performance through AI-enhanced mobile edge computing, intelligent mobility, and smart spectrum management.

The shift towards AI-native 6G networks underscores the need for goal-oriented and semantic communications to optimize bandwidth, latency, energy consumption, and electromagnetic field radiation, paving the way for more efficient wireless services. These research directions emphasize the critical role of AI in shaping the future of 6G networks by enabling intelligent networking, efficient resource utilization, and improved user experiences (R, Deepa et al., 2024).

3.0 ARCHITECTURE OF 6G

3.1 Introduction to 6G Network Architecture

3.1.1 Overview of 6G architecture

The 6G architecture represents a revolutionary shift in how networks are designed, optimized, and operated, with an emphasis on ultra-high speed, low latency, and ubiquitous connectivity. Unlike its predecessors, 6G heavily integrates AI and machine learning at every layer of the network, making the architecture more intelligent, adaptive, and autonomous. AI will manage network resources dynamically, optimize traffic flow, predict system needs, and enhance security without constant human intervention. A key feature of 6G architecture is the decentralization of network components, where core network functions are distributed across various nodes and closer to the edge, reducing latency and improving responsiveness. This decentralized model is supported by edge computing, where data processing happens closer to the user, reducing the need for data to travel long distances (Khan etal ., 2020). The architecture also incorporates new communication technologies like terahertz waves, quantum computing, and software-defined networking (SDN), providing a flexible, scalable, and programmable network infrastructure. Additionally, energy efficiency is a critical focus, with 6G networks designed to minimize power consumption through intelligent resource management and the integration of sustainable technologies. Overall, 6G architecture is designed to meet the demands of future applications like immersive augmented reality (AR), real-time holographic communication, massive IoT (Internet of Things), and URLLC, creating a seamless and intelligent network ecosystem. (Dogra et al., 2020).

Figure 2. Simplified Architecture of 6G Networks with AI Integration

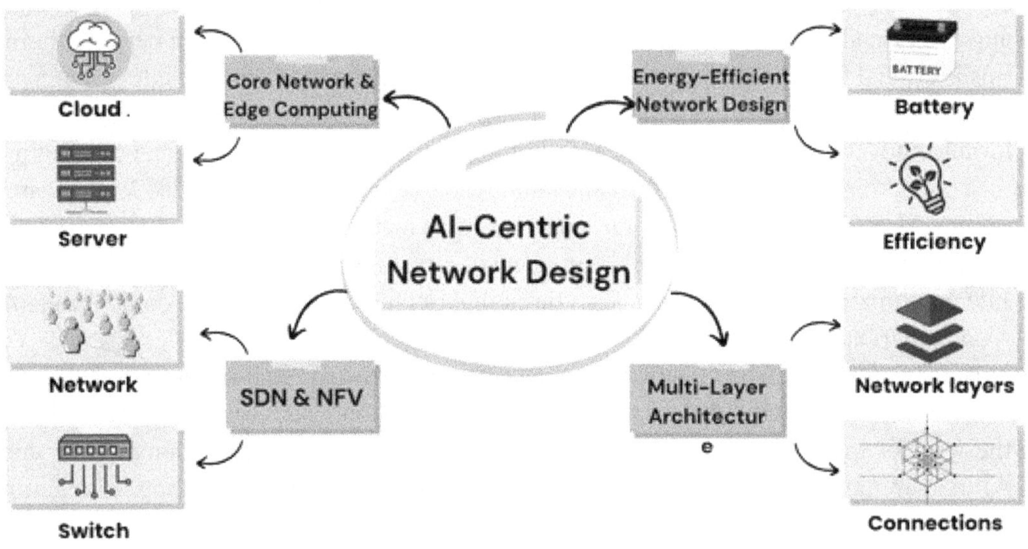

3.1.2 Key differentiators from 5G architecture

The architecture of 6G networks introduces several significant advancements over 5G, with the primary differentiator being the integration of AI at the core of the network. While 5G employs AI for specific optimizations, 6G will rely on AI for fully autonomous network management, enabling real-time self-optimization, self-configuration, and self-healing capabilities. This makes 6G networks highly adaptive and intelligent, significantly reducing the need for human intervention. Another key difference is the use of higher frequency spectrums, such as terahertz (THz) waves, compared to the millimeter waves used in 5G. These higher frequencies provide much faster data rates, supporting speeds of up to 1 terabit per second (Tbps), and enabling advanced applications like real-time holographic communication and ultra-immersive augmented reality (AR). Additionally, 6G embraces a more decentralized architecture, with computing power distributed closer to the edge of the network (Li et al., 2021). This approach, known as edge computing, reduces latency by processing data locally rather than relying on distant centralized servers, which is more common in 5G. Overall, 6G focuses on creating a more flexible, efficient, and scalable network that can support the data-intensive, real-time applications of the future.

3.2 AI-Centric Network Design

3.2.1 How AI fundamentally shapes and optimizes 6G network architecture?

In 6G network architecture, AI plays a central role in shaping and optimizing the entire network, making it more intelligent, adaptive, and efficient than any previous generation. Unlike traditional networks that rely on predefined rules and human intervention, AI allows 6G networks to learn from data, predict network conditions, and autonomously make decisions to optimize performance in real time. This fundamental shift enables AI to manage key aspects such as traffic flow, resource allocation, and power consumption dynamically, based on the network's current state and future predictions. For example, AI algorithms can predict traffic surges in certain areas or detect potential network congestion and automatically reroute data to underutilized parts of the network, ensuring smoother service and reduced latency (Ziegler et al., 2020). Similarly, AI enables Self-Optimizing Networks (SONs), which continuously monitor and fine-tune network parameters without manual oversight, adapting to changes in user behavior, device mobility, and environmental factors. This real-time optimization significantly enhances the efficiency and reliability of the network, supporting the diverse and demanding applications 6G is designed to handle, such as holographic communication, immersive augmented reality (AR), and massive IoT.

AI also enhances energy efficiency by intelligently adjusting power levels based on network traffic, reducing energy consumption in low-traffic periods while ramping up resources when needed (Alhammadi et al., 2024). Through continuous learning and adaptation, AI can optimize how 6G handles vast amounts of data, improving overall network performance while reducing operational costs and environmental impact. In summary, AI's deep integration into 6G architecture allows for a flexible, adaptive, and proactive network that continuously evolves and optimizes itself.

3.2.2 Key AI-driven features for scalability, flexibility, and efficiency

In 6G networks, AI-driven features are critical to achieving scalability, flexibility, and efficiency, addressing the growing demands for high-speed, low-latency, and data-intensive applications. These features allow 6G networks to dynamically scale to accommodate varying levels of traffic, adjust to changing user requirements, and operate efficiently across diverse environments. One key feature is AI-based resource management, which ensures efficient use of network resources like bandwidth, power, and processing capacity. AI continuously analyzes network traffic patterns and user behavior to dynamically allocate resources where they are most needed. This allows the network to scale up during periods of high demand and scale down during low usage, optimizing resource usage and reducing operational costs.

Flexibility in 6G is driven by AI's ability to support network slicing, a process where multiple virtual networks are created on the same physical infrastructure. AI can intelligently manage these slices, ensuring each one is tailored to specific services or applications—whether it's URLLC for autonomous vehicles or enhanced mobile broadband (eMBB) for high-definition streaming. This allows 6G networks to adapt flexibly to different use cases without requiring changes to the physical network itself. In terms of efficiency, AI enhances energy conservation and operational efficiency. AI algorithms continuously monitor network conditions and user mobility, optimizing signal strength and power levels based on real-time data (Moro-Visconti et al., 2023). This helps reduce energy consumption while ensuring high performance. Moreover, AI can automate network maintenance by identifying and addressing potential issues before they become critical, reducing downtime and improving overall network reliability. These AI-driven features make 6G networks highly scalable to meet increasing demands, flexible enough to serve diverse applications, and efficient in terms of both energy use and resource management, ensuring a sustainable and cost-effective network infrastructure.

3.3 Core Network and Edge Computing in 6G

3.3.1 Role of a distributed core network and edge computing in reducing latency and enhancing performance

In 6G networks, the distributed core network and edge computing play a pivotal role in reducing latency and enhancing overall performance by bringing data processing and network functions closer to the end user. Unlike traditional centralized core networks where all data must travel to and from central servers, 6G employs a distributed core architecture, where computing resources are spread across multiple, smaller data centers or nodes located closer to the network edge. This decentralization significantly reduces the distance that data needs to travel, which in turn minimizes latency (Mansouri et.al., 2021). For example, applications that require real-time responses, such as augmented reality (AR), virtual reality (VR), or autonomous driving, benefit from having the processing tasks performed near the edge of the network rather than in a distant centralized server. This proximity enables ultra-low-latency communication, often in the range of milliseconds, which is critical for applications requiring real-time interactivity.

Edge computing further enhances performance by processing and storing data locally at the edge of the network, near where it is generated. Instead of sending all data to the cloud or central servers for processing, edge computing allows devices such as smartphones, IoT sensors, and autonomous vehicles to perform computations locally. This not only reduces latency but also alleviates network congestion by offloading data traffic from the core network. By distributing the computational workload between the core network and the edge, 6G networks are able to deliver higher speeds, more reliable connections, and enhanced performance for mission-critical applications (Hamdan et al., 2020). Moreover, edge computing can filter, analyze, and manage data before it reaches the core network, ensuring that only relevant information is transmitted, reducing the overall data load. This architecture also enhances the scalability of the network, allowing it to handle the immense data traffic expected from future applications like smart cities, massive IoT, and immersive multimedia experiences.

3.3.2 Architectural shifts from centralized to decentralized systems

The transition from centralized to decentralized systems in 6G architecture marks a fundamental shift in how networks are designed and operated. In previous generations like 4G and 5G, network architectures were primarily centralized, with most data processing, storage, and decision-making happening in large, centralized data centers or core networks. While this approach worked well for many applications, it often introduced significant latency and bottlenecks as all data had to be transmitted to and from these centralized locations, regardless of where it originated. In 6G, the architecture shifts to a more decentralized system, where computing, storage, and network resources are distributed across multiple nodes closer to the end users, forming a more localized, intelligent infrastructure. This decentralization reduces the physical distance that data must travel, which drastically lowers latency and enhances the speed and efficiency of data processing (Dib et al., 2020). A key element of this shift is edge computing, where local nodes or "edge" devices handle much of the data processing and analysis near the point of data generation (e.g., IoT devices, smartphones, sensors). This reduces the reliance on sending data back and forth to centralized servers. For example, in applications like autonomous driving or augmented reality, immediate, real-time responses are crucial, and the decentralized nature of 6G ensures that processing happens locally, avoiding delays caused by sending data to far-off cloud servers.

Another advantage of this decentralized approach is scalability. With more computing power distributed across the network, 6G can handle a massive number of connected devices, particularly with the rise of the Internet of Things (IoT), which will involve billions of connected sensors and devices. This flexibility ensures that the network can dynamically allocate resources where they are most needed, preventing network congestion and overloading of central servers. Furthermore, decentralized systems enhance network resilience. If a particular node or edge server experiences an issue, other nodes can take over, providing redundancy and ensuring continuous service. This makes 6G networks more robust and fault-tolerant compared to centralized models, where a failure in a core node could disrupt services across a large region.

3.4 Software-Defined Networking (SDN) and Network Function Virtualization (NFV)

3.4.1 How SDN and NFV enable flexibility, programmability, and virtualization of network resources

In 6G networks, Software-Defined Networking (SDN) and Network Function Virtualization (NFV) bring immense flexibility by allowing network operators to decouple software from hardware, enabling more dynamic control over network resources. SDN achieves this by separating the control plane (where decisions are made) from the data plane (which handles the actual data transfer), allowing centralized control of the entire network through software. This means network administrator can reprogram and reconfigure network resources in real-time without needing to modify or upgrade physical infrastructure (Kenner et al.. 2024). The ability to quickly adjust traffic flows, reroute data, or allocate bandwidth makes

SDN crucial for supporting the scalability and agility required by 6G applications, such as URLLC and mMTC.

Meanwhile, NFV enhances virtualization by transf00orming traditional hardware-based network functions—such as routers, firewalls, and load balancers—into software that can run on standard servers. This virtualization allows multiple network functions to run on the same hardware, making the network more resource-efficient and adaptable to varying demands. By leveraging virtualized functions, 6G networks can easily scale and customize services through network slicing, where isolated virtual segments of the network are allocated to specific applications (Kaur et al.. 2020). For instance, an ultra-low-latency slice can support autonomous vehicle communications, while another slice can be optimized for high-bandwidth applications like immersive AR/VR. Together, SDN and NFV enable 6G networks to be programmable, scalable, and highly efficient, delivering tailored, on-demand network services with minimal physical infrastructure changes.

3.4.2 Their integration into 6G architecture for enhanced performance

The integration of Software-Defined Networking (SDN) and Network Function Virtualization (NFV) into the 6G architecture is a key enabler of enhanced network performance, flexibility, and scalability. By embedding these technologies into the core design of 6G, the network can operate in a more agile and responsive manner, addressing the needs of diverse, data-intensive applications and services. SDN allows for centralized, software-based control of the network, enabling real-time adjustments and optimizations across the entire system. This means that network resources can be allocated dynamically based on current demands, ensuring that critical applications like autonomous vehicles or remote surgeries receive the necessary bandwidth and low-latency connections.

Similarly, NFV enhances the efficiency of the 6G network by virtualizing essential network functions, allowing them to be deployed and managed in software on general-purpose hardware. This eliminates the need for specialized equipment, making the network more cost-effective and easier to scale (Ziegler et al., 2020). NFV also enables on-demand provisioning of network resources, so services can be rapidly deployed, scaled, or adjusted to meet fluctuating user requirements. When integrated with SDN, these virtualized functions can be dynamically managed and optimized in real-time, ensuring that the network can handle high traffic loads without compromising on performance. Together, SDN and NFV help create a more programmable, flexible, and efficient 6G network that can meet the performance needs of next-generation technologies while providing seamless, high-quality service across a wide range of applications.

3.5 Energy-Efficient and Sustainable Network Design

3.5.1 Importance of energy efficiency in 6G architecture

In the 6G architecture, energy efficiency is of paramount importance due to the exponential growth in data traffic, connected devices, and resource-intensive applications that require vast computational power and continuous connectivity. As 6G networks aim to support emerging technologies like massive Internet of Things (IoT), real-time holographic communication, and URLLC, they must be designed to handle these demands without significantly increasing energy consumption. Without a focus on energy efficiency, the environmental impact and operational costs of running such extensive networks would be unsustainable.

Energy efficiency in 6G is critical not only to minimize carbon footprints and reduce operational costs for network providers, but also to extend the battery life of devices and ensure the sustainability of the entire ecosystem. By integrating AI-driven optimization, edge computing, and smart resource allocation, 6G networks can intelligently manage energy usage by dynamically adjusting power levels based on real-time demand, traffic conditions, and device usage (Sodhro et al.,2020). This approach ensures that energy is consumed only where and when it's needed, contributing to greener and more sustainable network operations. Additionally, energy-efficient designs are essential to meet the global push towards carbon-neutral and sustainable technologies, aligning with the goals of reducing greenhouse gas emissions and achieving environmental sustainability in telecommunications.

3.5.2 Role of AI and intelligent algorithms in minimizing energy consumption

AI and intelligent algorithms play a pivotal role in minimizing energy consumption in 6G networks by enabling smart, real-time management of network resources. AI's ability to continuously monitor, analyze, and predict network conditions allows for more efficient use of power and computational resources, reducing unnecessary energy consumption. Through predictive algorithms, AI can forecast traffic patterns, user behavior, and device mobility, allowing the network to dynamically adjust energy usage based on real-time demand. For example, during periods of low network activity, AI can automatically reduce the power supply to idle base stations or devices, effectively putting them into a low-power or sleep mode. Similarly, when traffic surges occur, AI can rapidly reallocate resources, boosting energy where needed while avoiding over-provisioning (Farzaneh et al., 2021) . AI also enables intelligent load balancing, ensuring that data is routed through the most energy-efficient paths, reducing the strain on specific network nodes and improving overall efficiency.

Additionally, AI-driven resource allocation algorithms can optimize energy consumption by distributing tasks between the core network and edge computing nodes, where processing closer to the user can reduce latency and avoid long-distance data transmissions, which consume more power. By continuously learning and adapting to changing conditions, AI helps to strike a balance between maintaining high performance and minimizing energy usage, making 6G networks both energy-efficient and sustainable without compromising on service quality. This AI-driven optimization is essential for managing the massive scale of 6G, which will support billions of connected devices and high-performance applications, all while keeping energy consumption at sustainable levels.

4.0 6G PROTOCOLS AND TRANSMISSION

4.1 Introduction to 6G Protocols

4.1.1 Protocol Evolution from 5G to 6G

The change from 5G to 6G protocols showcases a noteworthy improvement in the domain of wireless communication, stressing heightened levels of intelligence, efficiency, and adaptability within the network infrastructure. The primary objective of 6G is to deliver exceptionally high data transmission speeds and real-time communication capabilities across a wide-ranging expanse, focusing on the integration of cutting-edge technologies to support novel applications and improved energy efficiency. The incorporation of block chain technology is anticipated to play a pivotal role in enhancing both the security measures and optimizing the overall performance of the 6G network. The utilization of testbeds will play a critical role in the validation process of these innovative technologies within authentic operational environments, underscoring the significance of practical experimentation as we progress towards the realization of 6G networks. Moreover, the forthcoming 6G landscape is envisioned to be characterized by a more open and diverse ecosystem, necessitating the development of inventive security frameworks such as software-defined Zero Trust Architecture (ZTA) to ensure robust access control mechanisms and effectively combat malicious cyber activities. These advancements in security measures will be essential in safeguarding the integrity and confidentiality of data transmission within the 6G network, thereby reinforcing the trustworthiness and resilience of the overall communication infrastructure. As such, the evolution towards 6G networks heralds an era of unprecedented connectivity and technological sophistication, underpinned by a multifaceted approach to security and performance optimization. (Singh et al., 2024)

4.1.2 Key Objectives and Requirements of 6G Protocols

The evolution in 6G protocols brings about notable distinctions compared to 5G. Whereas 5G predominantly improved Quality of Service (QoS) and offered assurances for industrial use cases, the upcoming 6G era is poised to transform applications with state-of-the-art technologies such as post-quantum cryptography, artificial intelligence, machine learning, and edge computing. Anticipated to function at elevated frequency spectrums, 6G networks will provide exceptionally minimal latency and achieve data transmission speeds reaching up to 1000 Gbps, representing a significant leap beyond the capabilities of 5G networks (Wang et al., 2023). Furthermore, 6G is set to integrate innovative security measures including advanced authentication mechanisms, encryption techniques, access control policies, and communication strategies to establish resilient privacy and trust frameworks within future network infrastructures. In addition, 6G is designed to address issues concerning security, privacy, energy efficiency, and roaming authentication through the implementation of cutting-edge protocols like on-demand anonymous access and streamlined batch authentication processes. The enhancement of time synchronization, a critical component for a myriad of applications, will also be a focal point in the development of 6G networks.

4.2 Advanced Modulation and Coding Techniques

4.2.1 New Modulation Schemes for 6G

In the domain of 6G communication networks, there is a growing interest in exploring various advanced modulation techniques aimed at improving both spectral and energy efficiency. Techniques such as sparsity modulation, including spatial modulation and index modulation, as well as low peak-to-average power ratio (PAPR) schemes like $(1 + D)\ \pi/2(N + 1)$ BPSK, have been the focus of attention due to their demonstrated enhancements in performance metrics (Li et al., 2023). Moreover, the combination of hierarchical modulation (HM) with bit-interleaved coded modulation (BICM) and iterative decoding (ID) is currently under scrutiny for its ability to provide unequal error protection, exhibiting superior efficacy when compared to conventional methods. In addition to this, there is ongoing research into the integration of maximum distance separable (MDS) codes with Orthogonal Frequency Division Multiplexing (OFDM) and Reconfigurable Intelligent Surfaces (RIS), with the aim of achieving robust data transmission across intricate wireless channels while improving the Bit Error Rate (BER) performance. Sparse Code Multiple Access (SCMA) has also emerged as a promising technique for 6G networks, offering solutions to challenges related to codebook design and decoding efficiency, thus addressing critical issues in next-generation wireless communication systems.

4.2.2 Error Correction and Coding Strategies

To achieve URLLC in 6G networks, advanced error correction and coding techniques are essential. Innovations like Low-Density Parity-Check (LDPC) codes and polar codes are being fine-tuned to meet the rigorous demands of 6G. Researchers are also exploring the use of machine learning to dynamically adjust coding parameters based on changing channel conditions. Additionally, combining Forward Error Correction (FEC) with Automatic Repeat Request (ARQ) mechanisms is being improved to boost the reliability and robustness of data transmission. These strategies are vital for ensuring dependable communication in the varied and complex scenarios anticipated for 6G (Rowshan et al., 2024).

4.2.3 Impact on Transmission Efficiency

The development of advanced modulation and coding techniques for 6G is greatly improving transmission efficiency. These innovations are aimed at enhancing both spectral and energy efficiency, which are crucial to meet the high-performance demands of future networks. Techniques like spatial modulation and index modulation, along with low PAPR schemes such as $(1 + D)\ \pi/2(N + 1)$ BPSK, are being designed to boost system performance by reducing power consumption and increasing data rates. Hierarchical modulation (HM) combined with bit-interleaved coded modulation (BICM) and iterative decoding (ID) offers superior performance by providing unequal error protection, making data transmission more robust (Iyer et al., 2022).

Additionally, integrating maximum distance separable (MDS) codes with Orthogonal Frequency Division Multiplexing (OFDM) and RIS helps achieve reliable data transmission over complex wireless channels, improving BER performance. Sparse Code Multiple Access (SCMA) is another promising technique for 6G, addressing crucial issues of codebook design and decoding efficiency. These advancements not only enhance transmission efficiency but also improve error correction capabilities, ensuring robust and reliable communication for the diverse range of applications and services expected in 6G. This includes URLLC, mMTC, and more.

4.2.4 Quadrature Amplitude Modulation (QAM)

Quadrature Amplitude Modulation (QAM) stands as a foundational modulation technique in advanced wireless networks, and its significance in 6G is anticipated to be even more substantial. As 6G aims to deliver ultra-reliable, low-latency communication with immense data throughput, QAM enables an effective way to encode information by modulating both the amplitude and phase of the carrier signal. This combined modulation strategy enhances spectral efficiency, allowing for greater data transmission within the same bandwidth—essential for reaching the terabit-level data rates 6G aspires to achieve.

With 6G expanding into higher frequency ranges, such as millimeter-wave and terahertz bands, high-order QAM variants like 1024-QAM and beyond are under consideration to boost performance in these demanding spectral environments (Chack & Thool, 2023).

Additionally, coupling QAM with advanced error-correction techniques like LDPC and polar codes can significantly improve reliability, even in environments with substantial noise and interference. Alongside massive MIMO and RIS, QAM further strengthens spatial multiplexing and signal propagation, supporting robust, high-speed connectivity across large areas. Altogether, these advancements emphasize QAM's essential role in meeting the rising need for efficient modulation and coding solutions in 6G, optimizing both spectrum and power resources while ensuring high-integrity data transmission.

4.3 Network Protocol Innovations

4.3.1 Next-Generation IP Protocols

In the development of 6G networks, next-generation IP protocols are being designed to address the evolving needs of high-speed and high-capacity communications. These protocols aim to enhance network efficiency by supporting higher data rates and improved network flexibility. Advances include the integration of IPv6 with novel addressing schemes and hierarchical routing to manage the vastly increased number of connected devices and the diverse range of applications. Enhanced mechanisms for Quality of Service (QoS) and dynamic network management are also being incorporated to optimize performance and resource allocation in 6G environments (Adhikari et al., 2021).

4.3.2 Enhanced Transport Layer Protocols

Upgraded transport layer protocols are key to unlocking the advanced features of 6G networks. Innovations like the Low Latency Low Loss Scalable Throughput (L4S) architecture are being developed to cut down on delays and packet loss, ensuring data transmission is smooth and reliable. Efforts to improve transport protocols also aim at boosting throughput and reducing congestion with advanced congestion control algorithms and efficient packet scheduling methods. These upgrades are crucial for meeting the high-performance needs of applications such as real-time communications and high-definition streaming (Iyer et al., 2022).

4.3.3 Protocols for Ultra-Low Latency and High Reliability

To achieve ultra-low latency and high reliability in 6G networks, specialized protocols are being developed to meet the exacting needs of real-time applications. Time-Sensitive Networking (TSN) protocols are being used to ensure precise time synchronization and reliable packet delivery, which are crucial for applications like autonomous vehicles and augmented reality. Cross-Layer Design (CLD) approaches are also under exploration, integrating functions across different network layers to optimize performance for both low latency and high reliability (Adhikari et al., 2022). Hybrid Automatic Repeat Request (HARQ) techniques combine error correction with retransmission strategies to boost data reliability while keeping delays minimal. Furthermore, AI-driven protocols are being crafted to dynamically adjust network parameters and forecast network conditions, enhancing overall performance and reducing latency. These innovations are essential for meeting the demanding requirements of next-generation applications and ensuring the seamless, instantaneous communication that 6G networks aim to provide.

4. 4 Spectrum Management and Utilization

Effective spectrum management is crucial for optimizing 6G network performance, and cognitive radio and spectrum sensing technologies are leading the way. Cognitive radio systems are designed to intelligently manage spectrum usage by detecting and responding to the spectrum environment in real-time. They use advanced algorithms to find and utilize underused spectrum bands, making decisions about spectrum allocation and access on the fly. This approach helps reduce spectrum congestion and makes the most of available frequency bands.

Spectrum sensing techniques, such as energy detection, matched filtering, and cyclostationary feature detection, play a key role in this process. These methods monitor spectrum usage to identify open channels, which helps manage interference and optimize communication. In 6G networks, where high data rates and low latency are essential, the ability to allocate spectrum dynamically based on real-time data is critical for maintaining performance and user satisfaction. (Bhattacharya et al., 2022).

Moreover, cognitive radio networks offer features like spectrum aggregation and dynamic spectrum management, which boost overall network capacity and efficiency. By allowing simultaneous use of multiple spectrum bands and adapting to changing spectrum conditions, these technologies enhance network flexibility and adaptability.

Integrating cognitive radio and spectrum sensing into 6G networks addresses challenges related to spectrum scarcity and improves network resilience and efficiency. These technologies are set to be pivotal in the future of wireless networks, supporting a wide range of emerging applications and services.

4. 5. Multiple Access Techniques

4.5.1 Non-Orthogonal Multiple Access (NOMA)

Non-Orthogonal Multiple Access (NOMA) represents a significant advancement over traditional multiple access techniques by allowing multiple users to share the same time-frequency resources. Unlike orthogonal methods, which allocate distinct resources to each user, NOMA leverages power domain multiplexing to differentiate users. By assigning different power levels to different users, NOMA enhances spectral efficiency and increases the network's capacity (Mounir et al., 2022) This technique is particularly beneficial in scenarios with high user densities and varying channel conditions, as it allows for more effective utilization of available spectrum and supports a greater number of simultaneous connections.

4.5.2 Orthogonal Time Frequency Space (OTFS):

Orthogonal Time Frequency Space (OTFS) is an innovative multiple access technique designed to address the limitations of traditional orthogonal methods, particularly in dynamic and multipath-rich environments. OTFS combines orthogonal time and frequency resources to create a robust framework for managing time-varying channels and mitigating interference. This approach enhances the system's performance by providing better time-frequency resource management and improved resilience to channel distortions. OTFS is particularly well-suited for scenarios requiring high reliability and low latency, making it a promising candidate for future 6G networks. (Yuan et al., 2023).

4.5.3 Comparison with Traditional Access Methods

When comparing NOMA and OTFS with traditional access methods such as Time Division Multiple Access (TDMA), Frequency Division Multiple Access (FDMA), and Code Division Multiple Access (CDMA), several key differences emerge. Traditional methods partition the available spectrum into separate time slots, frequency bands, or codes, each with its own set of benefits and limitations in terms of capacity and interference management. In contrast, NOMA and OTFS offer enhanced capabilities by improving spectral efficiency and handling complex channel conditions more effectively. NOMA's ability to support a higher number of users within the same resource block and OTFS's advanced time-frequency resource management provide significant advantages in terms of capacity, latency, and network flexibility. These advanced techniques are well-suited to meet the demands of 6G networks, which require higher data rates, ultra-low latency, and increased user connectivity (Shah et al., 2022). Thus, the adoption of NOMA and OTFS represents a substantial leap forward in multiple access technologies, addressing the evolving needs of next-generation wireless communication systems.

4.6. AI-Driven Protocol Enhancements

4.6.1 AI for Protocol Optimization

Artificial Intelligence (AI) is vital for optimizing network protocols, enhancing efficiency, adaptability, and performance in 6G networks. Using AI algorithms, protocols can be dynamically adjusted based on real-time network conditions and traffic patterns, leading to better resource allocation, reduced latency, and higher throughput (Mao et al., 2020). AI-driven optimization also manages the complexity of heterogeneous networks, ensuring seamless communication across different network segments and technologies.

4.6.2 Machine Learning in Adaptive Protocol Design

Machine Learning (ML) facilitates the creation of adaptive protocols that learn from network data and adjust their behavior accordingly. These protocols can adapt to changes in network topology, user mobility, and traffic loads, ensuring optimal performance under diverse conditions. Techniques like reinforcement learning develop protocols that continuously improve their efficiency and effectiveness, making them ideal for the dynamic and complex environments of 6G networks (Tang et al, 2021).

4.6.3 AI for Predictive Network Management

AI is transforming predictive network management by enabling proactive and efficient handling of network resources. By using AI algorithms, particularly ML models, networks can anticipate states and user behaviors based on historical data and real-time analytics. This allows for dynamic adjustment of network parameters, enhancing performance and reliability. For instance, AI can predict traffic congestion or potential faults and initiate preemptive measures like load balancing or resource reallocation.

AI-driven predictive management also improves Quality of Service (QoS) by forecasting demand, optimizing energy consumption, and supporting the seamless integration of heterogeneous networks. This capability is crucial for maintaining the performance of latency-sensitive applications and improving overall network efficiency and security (Hossain et al., 2022).

4.7. Security Protocols in 6G

4.7.1 End-to-End Encryption Techniques

In 6G networks, end-to-end encryption (E2EE) is essential for ensuring data confidentiality and integrity throughout the entire communication process. Unlike traditional encryption methods that protect data only at specific points, E2EE secures data from the source to the destination, preventing access by intermediaries. Advanced encryption algorithms, including post-quantum cryptography, are being explored to defend against potential quantum computing threats. Additionally, 6G networks are expected to use lightweight encryption schemes to accommodate the limited resources of IoT devices and edge computing environments. Integrating blockchain technology can further strengthen E2EE by offering immutable and decentralized security frameworks, ensuring data security as it travels through complex and diverse network architectures. (Abdel Hakeem et al., 2022)

4.7.2 AI for Threat Detection and Mitigation

Artificial Intelligence (AI) is set to play a crucial role in detecting and mitigating threats in 6G networks. By leveraging machine learning (ML) algorithms, AI can analyze vast amounts of network traffic data in real time, identifying patterns that indicate malicious activities. Techniques such as anomaly detection, predictive analytics, and deep learning enable AI systems to detect and respond to threats more quickly and accurately than traditional methods. Additionally, AI-driven security systems can continuously learn and adapt to evolving threats, enhancing their effectiveness over time (Siriwardhana et al., 2021). Automating threat detection and response with AI reduces the time needed to counter cyber-attacks and minimizes the need for human intervention, thereby increasing the overall resilience and security of 6G networks.

4.7.3 Secure Transmission Protocols

Secure transmission protocols are fundamental to maintaining data integrity and confidentiality in 6G networks. These protocols ensure that data is securely transmitted across various network segments, including terrestrial and non-terrestrial networks (NTNs). Quantum key distribution (QKD) is being explored to provide unbreakable encryption keys, using quantum mechanics principles to create secure communication channels. Additionally, reconfigurable intelligent surfaces (RIS) can enhance physical layer security by dynamically altering the propagation environment to prevent eavesdropping. Protocols like Secure Sockets Layer (SSL) and Transport Layer Security (TLS) will evolve to support the higher data rates and lower latencies characteristic of 6G networks. Furthermore, multi-factor authentication and decentralized identity management systems will be crucial in ensuring that only authorized entities can access and transmit data, strengthening the overall security framework of 6G communications (Mitev et al., 2023).

4.8. Energy-Efficient Transmission Techniques

4.8.1 Green Communication Protocols

Green communication protocols are essential for minimizing the energy consumption of 6G networks, addressing the need for sustainable and environmentally friendly technologies. These protocols aim to optimize various aspects of communication, including power control, resource allocation, and data transmission methods, to reduce the overall energy footprint. Techniques such as energy-efficient routing, sleep mode operations for idle network components, and using renewable energy sources for network operations are integral to green communication strategies. Additionally, leveraging network function virtualization (NFV) and software-defined networking (SDN) enhances the adaptability and efficiency of network resources, contributing further to energy savings (Sujin et al., 2024). By implementing green communication protocols, 6G networks can balance high performance with a reduced environmental impact, supporting global sustainability goals. Energy-efficient transmission techniques are key to these protocols, with research suggesting novel approaches like joint data deepening-and-prefetching (JD2P) to reduce energy consumption in AI services by efficiently offloading data. Studies also focus on optimizing power allocation in NOMA-based systems to enhance energy efficiency and capacity, especially in IoT networks, highlighting the trade-off between energy efficiency and error rates.

4.8.2 AI for Energy Management in Transmission

Artificial Intelligence (AI) plays a crucial role in energy management for 6G transmissions by enabling intelligent and adaptive control of network resources. AI algorithms can analyze real-time data on network usage, traffic patterns, and environmental conditions to dynamically optimize energy consumption. Techniques such as machine learning and deep reinforcement learning predict network demand, adjust base station power levels, switch off unused network elements, and balance the load across different network segments. Furthermore, AI facilitates energy-efficient scheduling and routing of data packets, ensuring transmissions occur through the most energy-efficient paths. By integrating AI for energy management, 6G networks can significantly reduce their power consumption while maintaining high levels of performance and reliability. For example, the integration of AI and intelligent reflecting surfaces (IRS) in vehicular communication systems demonstrates improved energy efficiency and low-latency communication through deep reinforcement learning-based resource control and allocation schemes (Sujin et al., 2024). These advancements underscore the importance of leveraging AI and innovative transmission control strategies to achieve energy-efficient protocols in next-generation communication networks.

4.8.3 Low-Power Protocol Design

Low-power protocol design is crucial for developing energy-efficient 6G networks, particularly to support the proliferation of IoT devices and edge computing applications. These protocols are designed to minimize the power consumption of devices during data transmission and reception. Techniques such as duty cycling, where devices alternate between active and sleep modes, and the use of ultra-narrowband (UNB) communication for low-data-rate applications can significantly extend battery life. Additionally, protocols that support energy-harvesting technologies, allowing devices to draw power from ambient sources such as solar or kinetic energy, are increasingly important. Optimizing modulation and coding schemes for low power consumption without compromising data integrity and reliability is also a key focus. Advancing low-power protocol design ensures the long-lasting and sustainable operation of connected devices, even in resource-constrained environments. Research highlights novel approaches like joint data deepening-and-prefetching (JD2P) to reduce energy consumption in AI services by efficiently offloading data (Naser et al., 2023). Studies also explore optimizing power allocation in NOMA-based systems to enhance energy efficiency and capacity, particularly in IoT networks, illustrating the balance between energy efficiency and error rates.

4.9. Interoperability and Integration with Legacy Systems

4.9.1 Protocol Compatibility with 5G and Earlier Generations

Ensuring seamless interoperability and integration with legacy systems, 6G protocols need to be compatible with 5G and earlier generations. This backward compatibility is crucial for a smooth transition to 6G technologies without disrupting existing services (Xu et al., 2024) The Software-Defined Transitional Network (SDTN) architecture exemplifies this approach by integrating legacy network services into newly deployed 6G networks, gradually reducing reliance on physical legacy infrastructure. SDTN enables a phased transition, ensuring that 5G and earlier networks can coexist with 6G, allowing for a smoother upgrade path and minimizing potential disruptions.

4.9.2 Transition Strategies for Network Evolution

Effective transition strategies are essential for evolving networks from 5G to 6G. These strategies focus on incrementally deploying new technologies while maintaining the reliability and performance of existing networks. The Integrated Communication, Sensing, and Computation (ICSAC) framework for 6G aims to enhance the reliability, latency, accuracy, and security of communication, sensing, and computation functions, enabling them to work together for intelligent machine network applications. By gradually integrating new 6G capabilities with legacy systems, networks can evolve to support advanced applications without compromising service quality (Xu et al., 2024)

4.9.3 Hybrid Protocol Frameworks

Hybrid protocol frameworks are crucial for successfully integrating legacy systems with 6G networks. These frameworks combine elements of both traditional and new protocols to offer flexible and efficient communication solutions. For instance, using the Proxy Mobile IPv6 (PMIPv6) protocol in next-generation IoT networks ensures seamless mobility and resource efficiency, enhancing performance with demand-based, resource-efficient, location-aware extensions (Xu et al., 2024). As 6G evolves towards integrating terrestrial and non-terrestrial networks, including aeronautical platforms, emerging technologies like Mobile Edge Computing (MEC) and Software-Defined Networking (SDN) are proposed to integrate aeronautical systems into terrestrial networks, enabling high-performance services for 6G aeronautical applications. These hybrid frameworks facilitate the coexistence and interoperation of diverse network components, ensuring a cohesive and robust 6G ecosystem.

5.0 6G APPLICATIONS AND USE CASES

The applications of 6G networks are poised to revolutionize various sectors by providing unprecedented speed, low latency, and massive connectivity. 6G will make it feasible to perform remote surgery, real-time telemedicine, and ongoing patient monitoring using cutting-edge IoT devices in the healthcare industry. Improved Vehicle-To-Everything (V2X) connection will help the transportation industry by enabling intelligent traffic management and autonomous driving. 6G will make it easier for robotics and AI to work together effortlessly in industrial automation to optimize manufacturing processes. Furthermore, immersive AR and VR experiences will bring revolutionary changes in the entertainment and education sectors by facilitating high-fidelity streaming and virtual classrooms. 6G will be implemented in smart cities to provide smarter infrastructure, better public safety, and improved energy management (Chowdhury et al., 2020). All things considered, 6G networks will greatly impact daily life by fostering innovation, improving connection, and enhancing efficiency across a wide range of applications. Below is a detailed explanation of some of the main uses of 6G technology and also in **figure.3**

Figure 3. Applications of 6G

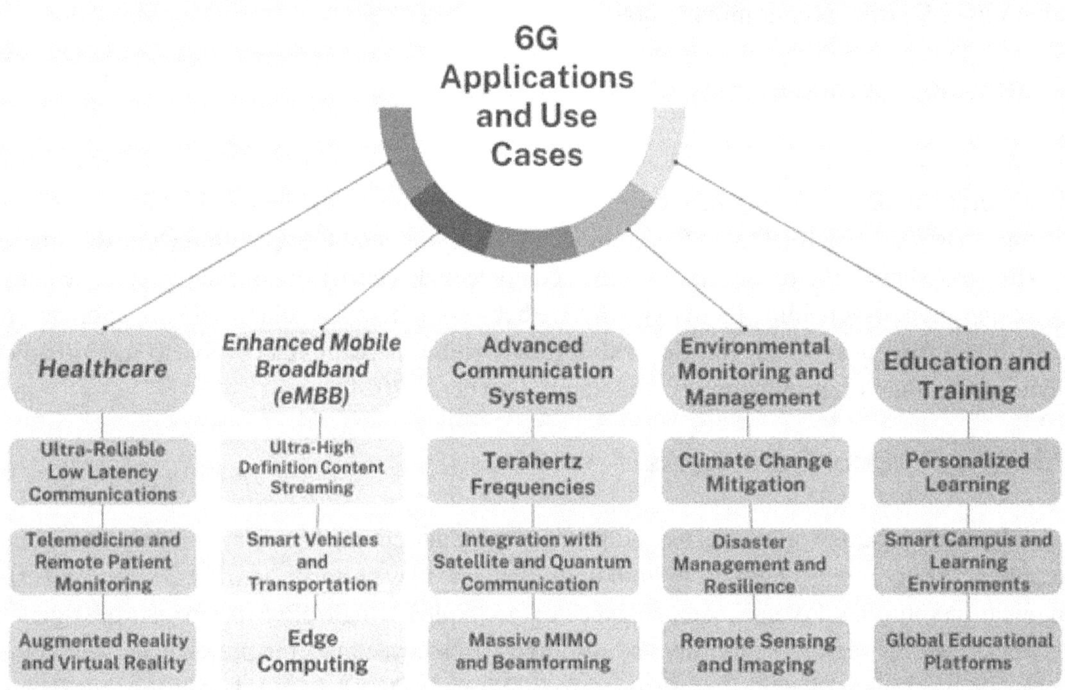

5.1 Healthcare

The role of 6G networks in healthcare is poised to be transformative, offering numerous benefits that can revolutionize the delivery of healthcare services. Here's an explanation of how 6G networks can impact healthcare:

5.1.1 Ultra-Reliable Low Latency Communications (URLLC)

URLLC promises to revolutionize healthcare by enabling real-time, highly reliable communication with minimal delay. In healthcare, URLLC capabilities are crucial for applications where split-second decisions and precise coordination are essential, such as remote surgery, emergency response, and patient monitoring.

Firstly, URLLC ensures that communication between medical devices, healthcare professionals, and patients occurs with ultra-low latency, virtually eliminating delays. For instance, in remote surgery scenarios, where a surgeon operates on a patient located in a different geographical location, any latency in communication could have severe consequences. With URLLC in 6G networks, the delay is minimized to imperceptible levels, allowing surgeons to perform complex procedures remotely with precision and confidence (Lin et al., 2018). Moreover, URLLC ensures the reliability of communication, minimizing the risk of data loss or interruption. This reliability is crucial for the continuous monitoring of a patient's vital signs in real time, especially for those in critical conditions. Healthcare providers can rely on the instantaneous transmission of data from wearable sensors or IoT devices to make timely clinical decisions and intervene promptly when necessary.

Furthermore, URLLC facilitates seamless communication between medical professionals during emergency response situations (Li et al., 2022). Whether it's coordinating with paramedics on-site or consulting specialists remotely, the ability to exchange information in real-time without delay ensures that patients receive the most effective and timely care possible.

5.1.2 Telemedicine and Remote Patient Monitoring

Telemedicine and Remote Patient Monitoring (RPM) stand as cornerstones of modern healthcare delivery, and their evolution with the advent of 6G networks promises revolutionary advancements. Telemedicine, facilitated by 6G's ultra-reliable low-latency communication, will enable seamless virtual consultations between healthcare providers and patients, transcending geographical barriers and enhancing access to medical expertise. Patients will have the convenience of accessing healthcare services from the comfort of their homes, reducing the need for physical visits to medical facilities and minimizing waiting times. Moreover, 6G's high data speeds and reliability will elevate Remote Patient Monitoring to new levels of effectiveness. Healthcare providers will be able to remotely monitor patients' vital signs, including heart rate, blood pressure, and glucose levels, in real-time with unprecedented accuracy and immediacy. IoT-enabled medical devices connected via 6G networks will facilitate continuous data collection and transmission, enabling early detection of abnormalities and proactive intervention to prevent adverse health events.

Additionally, the integration of advanced technologies such as augmented reality (AR) and virtual reality (VR) into Telemedicine and RPM, powered by 6G, will enhance the quality of care and patient engagement (Field et al., 2022). Healthcare professionals can use AR and VR to conduct virtual examinations, simulate medical procedures, and educate patients about their conditions and treatment plans in immersive and interactive environments. Furthermore, the security and privacy features inherent in 6G networks will ensure the confidentiality and integrity of patient data exchanged during Telemedicine and RPM sessions, fostering trust between patients and healthcare providers. With 6G's capabilities, Telemedicine and RPM will not only improve healthcare accessibility and convenience but also enable more personalized and proactive approaches to patient care, ultimately leading to better health outcomes and enhanced patient satisfaction.

5.1.3 Augmented Reality (AR) and Virtual Reality (VR)

In the context of healthcare, Augmented Reality (AR) and Virtual Reality (VR) technologies, powered by 6G networks, promise to revolutionize patient care, medical training, and therapeutic interventions. AR overlays digital information onto the user's real-world environment, enhancing their perception and interaction with the physical world. VR, on the other hand, immerses users in a completely virtual environment, often through the use of headsets or goggles, providing an immersive and interactive experience. With the ultra-high bandwidth and low latency capabilities of 6G networks, AR and VR applications in healthcare can reach unprecedented levels of realism, interactivity, and effectiveness (Kharche et al., 2023). In medical training, 6G-enabled AR and VR simulations can offer realistic scenarios for surgical procedures, allowing medical professionals to practice and refine their skills in a safe and controlled environment. Surgeons can rehearse complex surgeries, familiarize themselves with new medical equipment, and collaborate with colleagues in real time, regardless of their physical location. This immersive training approach can improve surgical outcomes, reduce medical errors, and enhance patient safety.

Moreover, 6G-powered AR and VR technologies hold immense potential in patient education and engagement. Healthcare providers can use AR overlays to explain medical conditions, treatment options, and surgical procedures to patients in a visually compelling and easily understandable manner. Patients can explore virtual models of their anatomy, visualize the progression of diseases, and gain a deeper understanding of their health status, empowering them to make informed decisions about their care. Therapeutic interventions can also benefit from 6G-enabled AR and VR applications in healthcare. Virtual reality environments can be used to manage pain, anxiety, and stress in patients undergoing medical procedures or recovering from injuries (Ahmad et al., 2023). By immersing patients in calming and immersive virtual experiences, healthcare providers can alleviate discomfort and enhance the overall patient experience. Additionally, 6G networks can facilitate remote AR and VR consultations between healthcare professionals and patients, enabling virtual examinations, remote diagnosis, and treatment planning without the need for in-person visits. This remote healthcare delivery model can improve access to specialized care, particularly for patients in rural or underserved areas, while reducing healthcare costs and minimizing travel-related burdens.

5.2 Enhanced Mobile Broadband (eMBB)

Enhanced Mobile Broadband (eMBB) refers to a pivotal aspect of wireless communication technology aimed at providing users with significantly faster data speeds, improved connectivity, and enhanced network capacity compared to previous generations. Enhanced Mobile Broadband (eMBB) is a key application of 6G networks, focusing on delivering significantly higher data rates, expanded network capacity, improved coverage, and lower latency. In 6G, eMBB plays a crucial role in providing ultra-high-speed internet access for bandwidth-intensive applications such as 8K/16K video streaming, virtual and augmented reality, cloud gaming, and holographic communications (Abdullah et al., 2021). With its ability to support massive device connectivity and advanced media delivery, eMBB enables immersive experiences, and seamless connectivity in dense urban areas, and extends broadband access to underserved regions. This technology plays a vital role in delivering a more immersive and interactive mobile experience while paving the way for emerging technologies and applications in the digital age.

5.2.1 Ultra-High Definition (UHD) Content Streaming

Ultra-High Definition (UHD) content streaming is poised to undergo a transformative evolution with the integration of Enhanced Mobile Broadband (eMBB) in 6G networks. As the successor to 5G, 6G promises to deliver significantly higher data rates, reduced latency, and improved reliability, thereby revolutionizing the way users consume and interact with multimedia content. In the context of eMBB within 6G, UHD content streaming refers to the seamless transmission of extremely high-resolution videos, such as 8K and potentially even 16K resolutions, over mobile networks. This capability will cater to the increasing demand for immersive multimedia experiences across various platforms, including video-on-demand services, live streaming applications, virtual reality (VR), and augmented reality (AR) environments.

One of the key challenges addressed by eMBB in 6G is the substantial increase in bandwidth capacity. Current 5G networks have already enabled UHD streaming to some extent, but 6G will further enhance this capability by supporting multi-gigabit per second data rates (Henrique et al., 2020). This advancement will ensure that users can access UHD content without experiencing buffering or delays, even in crowded network conditions or high-density urban areas. Moreover, eMBB in 6G will significantly reduce latency, which is critical for real-time interaction and responsiveness in UHD streaming applications. Lower latency will enhance the user experience by minimizing the delay between user actions and the corresponding feedback from streaming services, making interactions more natural and immersive.

Another crucial aspect of eMBB in 6G for UHD content streaming is its contribution to network reliability. By implementing advanced technologies such as network slicing and massive MIMO (Multiple Input Multiple Output), 6G networks can provide robust and consistent connectivity, ensuring uninterrupted streaming experiences across diverse scenarios and geographical locations. The deployment of eMBB in 6G networks represents a paradigm shift in UHD content streaming capabilities, offering unparalleled speed, reliability, and responsiveness (Rufino et al., 2021). This advancement not only meets current consumer demands for high-quality multimedia experiences but also sets the stage for innovative applications and services that leverage the full potential of ultra-high-definition content delivery over mobile networks.

5.2.2 Smart Vehicles and Transportation

In the realm of 6G networks, Enhanced Mobile Broadband (eMBB) promises to usher in a new era for smart vehicles and transportation systems. Building on the foundation laid by 5G, 6G's eMBB will provide unprecedented levels of connectivity characterized by ultra-high data rates, ultra-low latency, and enhanced reliability. These advancements are poised to revolutionize autonomous driving and vehicle-to-everything (V2X) communications, enabling vehicles to communicate seamlessly with each other, infrastructure, and pedestrians in real-time. This capability not only supports enhanced safety features and more efficient traffic management but also facilitates intelligent navigation systems that can adapt instantly to changing road conditions.

Moreover, eMBB in 6G will empower smart transportation ecosystems to integrate advanced technologies such as artificial intelligence (AI) and edge computing, optimizing decision-making processes and paving the way for more sustainable and interconnected mobility solutions (Serôdio et al., 2023). As a result, 6G's eMBB holds the potential to transform how we perceive and experience transportation, ushering in safer, more efficient, and environmentally friendly modes of travel for future generations.

5.2.3 Edge Computing

Edge computing integrated into Enhanced Mobile Broadband (eMBB) within 6G networks represents a transformative approach to data processing and service delivery, optimizing the performance of mobile networks and enabling new applications across diverse sectors. Unlike traditional cloud computing, which centralized data processing in remote data centers, edge computing brings computation and data storage closer to where it is needed—near the edge of the network, closer to users and devices. In the context of eMBB and 6G, edge computing plays a crucial role in reducing latency, which is the delay between sending data and receiving a response (Duan et al., 2023). By processing data locally at the edge of the network, rather than transmitting it back and forth to distant data centers, edge computing minimizes latency to imperceptible levels.

Moreover, edge computing enhances reliability by distributing computational tasks across multiple edge nodes. This decentralized approach improves fault tolerance and ensures continuous operation even in the event of network disruptions or failures in individual nodes. Additionally, edge computing in 6G enables real-time data analytics and decision-making. By analyzing data locally at the edge, organizations can extract valuable insights faster and respond promptly to changing conditions. This capability is beneficial across various industries, including healthcare (e.g., remote patient monitoring), manufacturing (e.g., predictive maintenance), and smart cities (e.g., traffic management), where timely insights drive efficiency and innovation (Syed et al., 2023). It also enhances privacy and security by processing sensitive data closer to its source, reducing the risk of data exposure during transmission to centralized cloud servers. This localized processing also facilitates compliance with data regulations and standards by keeping data within jurisdictional boundaries.

Edge computing integrated with eMBB in 6G networks represents a paradigm shift towards more responsive, reliable, and efficient mobile communication and computing. It enables innovative applications and services that leverage real-time data processing, improve user experiences, and drive advancements across industries, thereby shaping the future of connected ecosystems and intelligent infrastructure.

5.3 Advanced Communication Systems

Advanced communication systems in the context of 6G applications refer to the cutting-edge technologies and methodologies that will define the next generation of wireless communication networks. 6G is anticipated to surpass the capabilities of 5G by offering unprecedented speeds, ultra-low latency, massive connectivity, and reliability, enabling transformative applications across various sectors.advanced communication systems in applications of 6G represent a paradigm shift towards faster, more reliable, and more intelligent wireless networks. These systems will empower transformative applications across industries such as healthcare, transportation, entertainment, and manufacturing, ushering in a new era of connectivity and innovation (Banafaa et al., 2023). It represents a significant leap forward in wireless technology. These systems encompass a wide array of cutting-edge technologies and methodologies that will shape the next generation of wireless communication networks. 6G is expected to push the boundaries of what is possible with wireless communication, offering unprecedented speeds, ultra-low latency, massive connectivity, and reliability that will enable transformative applications across various sectors.

In essence, the advancement in communication systems for 6G applications heralds a paradigm shift towards faster, more reliable, and more intelligent wireless networks. These systems will not only empower transformative applications but also revolutionize industries such as healthcare, transportation, entertainment, and manufacturing. This will bring forth a new era of connectivity and innovation, creating possibilities that were previously unimaginable.

At its core, advanced communication systems in 6G applications encompass several key innovations:

5.3.1 Terahertz Frequencies

Terahertz frequencies are poised to revolutionize advanced communication systems within the context of 6G technology. Operating between 0.1 and 10 terahertz (THz), these frequencies offer unprecedented advantages for future networks. Their high bandwidth capabilities enable ultra-fast data transfer rates, supporting massive volumes of data exchanged between devices seamlessly. Moreover, the short wavelengths of THz waves facilitate the development of compact, high-resolution antenna arrays essential for dense network deployments and efficient spectrum usage. This advancement promises to reduce latency to imperceptible levels, critical for real-tiTerahertz frequencies, which fall within the range of 0.1 to 10 terahertz (THz), hold immense potential to revolutionize advanced communication systems within the realm of 6G technology (Alsharif et al., 2021). These frequencies boast exceptional properties that are expected to offer unprecedented advantages for future networks. Their high bandwidth capabilities pave the way for ultra-fast data transfer rates, facilitating seamless exchange of massive volumes of data between devices. Additionally, the short wavelengths of THz waves make it possible to develop compact, high-resolution antenna arrays, which are crucial for dense network deployments and efficient spectrum usage.

The utilization of terahertz frequencies is set to significantly reduce latency to imperceptible levels, a critical advancement that will benefit real-time applications such as autonomous vehicles and virtual reality. The implications of this progress extend beyond communications, as terahertz frequencies unlock new frontiers in imaging, sensing, and spectroscopy, thus positioning them as pivotal components in shaping the innovative landscape of 6G applications.me applications like autonomous vehicles and virtual reality (Jiang et al., 2024). Beyond communications, THz frequencies unlock new frontiers in imaging, sensing, and spectroscopy, making them pivotal in shaping the innovative landscape of 6G applications.

5.3.2 Integration with Satellite and Quantum Communication

In the realm of 6G applications, the integration of satellite and quantum communication technologies represents a transformative leap forward. Satellite communication plays a crucial role in extending coverage to remote and underserved regions, providing ubiquitous connectivity that complements terrestrial networks. By leveraging advanced satellite constellations, 6G aims to deliver seamless, high-speed internet access globally, supporting a wide range of applications from rural connectivity to disaster response and environmental monitoring.Furthermore, quantum communication stands at the forefront of secure and ultra-reliable communication systems envisioned for 6G . Quantum technologies harness the principles of quantum mechanics to encode, transmit, and decode information in ways that are fundamentally secure against eavesdropping, offering unparalleled levels of data protection. Integrating quantum communication into 6G networks promises to fortify cybersecurity measures, safeguarding sensitive information critical to industries such as finance, healthcare, and government (Akyildiz et al., 2020).

The synergy between satellite and quantum communication in 6G extends beyond traditional applications. Satellite-enabled quantum networks can establish secure links between ground stations and satellites, enabling global distribution of quantum keys for secure data transmission. Moreover, the low-latency capabilities of satellite systems complement the high-speed, low-latency requirements of emerging 6G applications such as autonomous vehicles, remote surgery, and augmented reality.

The integration of satellite and quantum communication technologies in 6G not only enhances global connectivity and data security but also paves the way for innovative applications that rely on robust, resilient, and ultra-fast communication infrastructures (Saeed et al., 2021). This convergence represents a cornerstone of next-generation networks, poised to redefine the boundaries of connectivity and technological advancement in the decades to come.

5.3.3 Massive MIMO and Beamforming

In the context of 6G technology, Massive Multiple Input Multiple Output (MIMO) and beamforming techniques are pivotal advancements driving the evolution of advanced communication systems. Massive MIMO involves deploying a large number of antennas at both the transmitter and receiver ends, allowing for simultaneous transmission of multiple data streams to multiple users. This capability significantly enhances spectral efficiency, data rates, and network capacity, making it ideal for accommodating the exponentially growing demand for high-speed data and the proliferation of connected devices.

Beamforming, on the other hand, complements Massive MIMO by focusing radio signals into narrow beams directed towards specific users or areas, rather than broadcasting in all directions. This technique improves signal strength, reduces interference, and enhances network coverage and capacity. By dynamically steering beams towards users, beamforming optimizes the use of available spectrum and boosts the overall efficiency of communication systems (Huo et al., 2023). Together, Massive MIMO and beamforming in 6G applications promise transformative benefits across various sectors. In urban environments, they can mitigate the challenges posed by dense populations and high data traffic, ensuring reliable connectivity and seamless user experiences. In industrial settings, they support the deployment of smart factories and Internet of Things (IoT) networks, facilitating real-time monitoring and control of machinery and processes. Moreover, in rural and remote areas, these technologies extend network coverage and improve service quality, bridging the digital divide and fostering economic development.

Furthermore, the integration of artificial intelligence (AI) and machine learning algorithms with Massive MIMO and beamforming enhances their capabilities further (Shi et al., 2024). AI-driven optimization enables adaptive beamforming, dynamic spectrum allocation, and predictive maintenance of network infrastructure, thereby optimizing performance and reducing operational costs. Massive MIMO and beamforming are cornerstone technologies in the advancement of 6G communication systems, offering unprecedented levels of performance, efficiency, and scalability. Their integration not only addresses current connectivity challenges but also opens doors to innovative applications that will define the next generation of wireless communication.

5.4 Environmental Monitoring and Management

Environmental Monitoring and Management (EMM) in the context of 6G applications encompasses the integration of advanced technologies for real-time data collection, analysis, and decision-making to address environmental challenges. 6G's enhanced connectivity, low latency, and high bandwidth facilitate the deployment of extensive sensor networks, drones, and IoT devices, enabling comprehensive monitoring of air, water, and soil quality, as well as wildlife habitats and climate conditions (Nayak et al., 2022). This robust data collection supports predictive analytics and AI-driven models to identify trends, detect anomalies, and inform proactive environmental management strategies. By leveraging 6G capabilities, EMM can achieve greater precision, scalability, and efficiency, ultimately contributing to sustainable development and environmental protection initiatives.

5.4.1 Climate Change Mitigation

Climate change mitigation in Environmental Monitoring and Management (EMM) through 6G applications represents a transformative approach to addressing one of the most pressing global challenges. The enhanced capabilities of 6G, such as ultra-high reliability, extremely low latency, and massive connectivity, enable a comprehensive and dynamic system for monitoring environmental parameters and implementing mitigation strategies. With 6G, vast networks of IoT sensors can be deployed in urban and rural areas, oceans, and remote regions, providing high-resolution data on various climate indicators like carbon dioxide levels, deforestation rates, glacier melt, sea level rise, and atmospheric conditions (Tomaszewski et al., 2023). This real-time data collection allows for immediate analysis and response, crucial for timely interventions. Advanced AI and machine learning algorithms process this influx of data, identifying patterns and predicting future trends with unprecedented accuracy.

For instance, 6G can facilitate the deployment of smart grids that optimize energy consumption and integrate renewable energy sources more effectively, reducing overall carbon emissions. Precision agriculture, enabled by 6G, can monitor soil health, water usage, and crop conditions, leading to more sustainable farming practices. Autonomous drones and vehicles equipped with 6G connectivity can continuously monitor forest health and detect illegal logging activities, enabling rapid response to prevent deforestation. Additionally, the high-speed, low-latency communication of 6G supports the development of global environmental monitoring networks, allowing researchers and policymakers worldwide to share real-time data and collaborate on climate solutions (Polymeni et al., 2024). This global network enhances the collective ability to track climate changes, implement coordinated mitigation efforts, and share successful strategies across borders.

In urban environments, smart city initiatives powered by 6G can manage and reduce urban heat islands through intelligent infrastructure and energy systems. 6G-enabled environmental monitoring systems in cities can provide real-time air quality data, enabling immediate public health responses and long-term urban planning to reduce emissions.

Overall, the application of 6G in EMM for climate change mitigation offers a multifaceted approach that combines real-time data collection, advanced analytics, global collaboration, and proactive management strategies. This integration not only enhances the precision and efficiency of climate change mitigation efforts but also supports the sustainable development goals by promoting more resilient and adaptive environmental management practices.

5.4.2 Disaster Management and Resilience

Disaster Management and Resilience within the framework of Environmental Monitoring and Management (EMM) using 6G technology represent a transformative approach to mitigating the impacts of natural and human-made disasters. 6G's promise of enhanced connectivity, low latency, and high data throughput allows for the seamless integration of diverse technologies such as Internet of Things (IoT) devices, AI-driven analytics, and autonomous systems. In the realm of disaster management, the deployment of dense networks of IoT sensors can provide granular, real-time data on environmental parameters like seismic activity, atmospheric conditions, water levels, and temperature variations (Khaloopour et al., 2024). These sensors, enabled by 6G's high reliability and low power requirements, can function continuously and transmit critical data with minimal delay. This facilitates the early detection of potential hazards, triggering early warning systems that can alert communities and authorities well in advance, thereby saving lives and reducing damage.

During disaster events, 6G's high-speed communication capabilities ensure that emergency response teams can operate with optimal coordination. Real-time data streams from affected areas can be analyzed using AI and machine learning algorithms to assess the situation quickly and determine the best course of action. For instance, in the event of a wildfire, real-time imagery and sensor data can help map the fire's spread and direct firefighting efforts more effectively. Autonomous drones and robots, guided by 6G networks, can be deployed for search and rescue operations, infrastructure inspections, and delivery of medical supplies or other necessities, often reaching areas inaccessible to human responders.

Post-disaster, 6G's capacity to support advanced technologies like augmented reality (AR) and virtual reality (VR) can enhance recovery efforts. For instance, AR can be used to overlay critical information onto the physical environment for field teams, while VR can facilitate virtual meetings and planning sessions among dispersed teams, enabling more efficient and effective recovery operations (Alhammadi et al., 2024). Furthermore, 6G-enabled networks can support the rapid restoration of communication infrastructures, which are often the first to be impacted during a disaster, ensuring that communities and responders remain connected.

Overall, the integration of 6G in EMM significantly bolsters disaster management and resilience by enhancing situational awareness, enabling real-time response, and supporting efficient recovery operations. The high-speed, reliable communication provided by 6G ensures that data flows seamlessly from the point of collection to the decision-makers, thereby transforming how disasters are managed and how communities recover and adapt in their aftermath.

5.4.3 Remote Sensing and Imaging

Remote Sensing and Imaging in Environmental Monitoring and Management, empowered by 6G technology, fundamentally enhance our capability to observe, analyze, and respond to environmental changes. Remote sensing involves the acquisition of information about Earth's surface without direct contact, typically through satellite or aerial imagery. With 6G's superior connectivity and bandwidth, data from an array of sensors—ranging from satellites to unmanned aerial vehicles (UAVs)—can be transmitted instantaneously, allowing for real-time monitoring of environmental parameters. High-resolution imaging techniques, such as hyperspectral and multispectral imaging, enable the detection of specific wavelengths of light reflected from the Earth's surface. This detailed spectral data allows for the identification and analysis of various materials and conditions, such as vegetation health, water quality, soil composition, and atmospheric pollutants. 6G's ability to support massive IoT ecosystems means that these imaging sensors can be deployed extensively, creating a dense network of observation points that provide comprehensive coverage and high granularity of data.

Artificial intelligence (AI) and machine learning (ML) play a pivotal role in processing and interpreting the vast amounts of data generated by these imaging systems. AI algorithms can quickly identify patterns and anomalies that may indicate environmental issues such as illegal logging, oil spills, or the spread of invasive species. Machine learning models can predict future environmental conditions based on historical data, enabling proactive management and mitigation strategies. The edge computing capabilities of 6G further enhance this process by allowing data to be processed close to the source of collection, significantly reducing latency and enabling near-instantaneous responses to environmental changes (Bourdoux et al., 2020). For instance, if a satellite detects a rapidly spreading wildfire, the information can be processed on the edge and immediately relayed to emergency services, facilitating a swift response to mitigate the disaster.

Additionally, the integration of 6G with other emerging technologies, such as blockchain, can ensure the security and integrity of environmental data. This is particularly important for regulatory compliance and for building trust in the data used for critical decision-making processes. The synergy of 6G technology with advanced remote sensing and imaging techniques revolutionizes environmental monitoring and management. It provides a powerful toolkit for understanding complex ecological systems, predicting environmental trends, and implementing timely interventions to protect and sustain our natural resources.

5.5 Education and Training

Education and Training in the context of 6G applications focus on utilizing the advanced capabilities of 6G networks to enhance learning experiences and skill development. With ultra-fast speeds, low latency, and reliable connectivity, 6G enables immersive and interactive educational tools such as virtual reality (VR) and augmented reality (AR) for more engaging and realistic simulations. It supports real-time collaboration between students and educators globally, providing access to remote labs, virtual classrooms, and instant feedback. AI and machine learning integrated with 6G can offer personalized learning experiences, adapting to individual needs and progress. Additionally, 6G's enhanced connectivity supports the use of large-scale data analytics to track educational outcomes and optimize training programs (Zhang et al., 2024) Overall, 6G technology promises to revolutionize education and training by making learning more accessible, interactive, and tailored to individual needs. Education and training focus on equipping individuals with the knowledge and skills necessary to leverage the next generation of wireless technology for various academic, professional, and personal development purposes. The following are detailed aspects of how 6G technology can transform education and training

5.5.1 Personalized Learning

In the realm of Education and Training, the advent of 6G technology offers transformative opportunities for Personalized Learning, fundamentally reshaping how educational content is delivered and experienced. With its ultra-low latency and high-speed connectivity, 6G facilitates the seamless integration of immersive technologies such as virtual reality (VR), augmented reality (AR), and mixed reality (MR) into learning environments. This allows for the creation of highly interactive and adaptive educational experiences that can be tailored to individual learner needs and preferences.

Advanced data analytics and artificial intelligence, supported by 6G's expansive bandwidth, enable the development of sophisticated learning platforms that can analyze student performance in real-time, adapt content dynamically, and provide personalized feedback. Moreover, 6G enables enhanced communication and collaboration tools, fostering interactive and remote learning opportunities that bridge geographical and socio-economic barriers (Liang et al., 2021). By supporting rich multimedia content, real-time data processing, and scalable applications, 6G empowers educators to deliver customized learning pathways, optimize instructional strategies, and ultimately enhance student engagement and outcomes in a more inclusive and effective manner.

5.5.2 Smart Campus and Learning Environments

6G technology revolutionizes Smart Campus and Learning Environments by enabling an unprecedented level of connectivity, interactivity, and personalization. With 6G's capabilities, educational institutions can deploy sophisticated digital tools such as virtual reality (VR) and augmented reality (AR) to create immersive learning experiences. VR can transport students to virtual laboratories, historical sites, or simulated environments, offering hands-on learning experiences that are both engaging and educational. AR can enhance traditional textbooks and classroom materials by overlaying interactive, multimedia content onto physical objects, thus making abstract concepts more tangible.

Furthermore, 6G supports the integration of extensive sensor networks and IoT devices within smart campuses, allowing for real-time monitoring and management of educational resources, campus safety, and environmental conditions. Intelligent learning platforms powered by AI can analyze student performance and engagement data to provide personalized learning paths and targeted support, addressing individual needs and optimizing educational outcomes (Pexyean et al., 2023). The high-speed, low-latency connectivity of 6G also facilitates seamless collaboration and communication among students and educators, breaking down geographical barriers and enabling more flexible and inclusive learning opportunities. Overall, 6G transforms educational environments into dynamic, adaptive spaces that cater to diverse learning styles and prepare students for the future.

5.5.3 Global Educational Platforms

Global educational platforms in the context of 6G applications represent a transformative leap in how education and training are delivered and experienced worldwide. With 6G's advanced capabilities, including ultra-low latency, ultra-high-speed connectivity, and massive device density, these platforms can facilitate immersive and interactive learning environments on an unprecedented scale. Virtual and augmented reality (VR/AR) can be seamlessly integrated into educational experiences, enabling students to engage in realistic simulations and interactive lessons regardless of their physical location. This capability supports global access to high-quality education, bridging gaps between different regions and providing equitable learning opportunities.

Moreover, 6G's high data transfer rates and reliability enable real-time collaboration and communication between students and educators across diverse geographical locations, enhancing remote learning experiences and enabling synchronous and asynchronous instruction with minimal delay. Artificial intelligence and machine learning algorithms, empowered by 6G's robust data handling capabilities, can further personalize learning paths, adapt educational content to individual needs, and provide real-time feedback (Agarwal et al., 2023). Consequently, global educational platforms can become more inclusive, flexible, and responsive, addressing diverse learning needs and promoting continuous professional development on a global scale.

REFERENCES

Abdel Hakeem, S. A., Hussein, H. H., & Kim, H. (2022). Security requirements and challenges of 6G technologies and applications. *Sensors (Basel)*, 22(5), 1969. DOI: 10.3390/s22051969 PMID: 35271113

Abdullah, D. M., & Ameen, S. Y. (2021). Enhanced mobile broadband (EMBB): A review. *Journal of Information Technology and Informatics*, 1(1), 13–19.

Abdulraqeb, A. (2024). Ibraheem, Shayea., Ayman, A., El-Saleh., Marwan, Hadri, Azmi., Zool, Hilmi, Ismail., Lida, Kouhalvandi., Sawsan, Ali, Saad. (2024). Artificial Intelligence in 6G Wireless Networks: Opportunities, Applications, and Challenges. *International Journal of Intelligent Systems*, 2024, 1–27. DOI: 10.1155/2024/8845070

Adhikari, M., & Hazra, A. (2022). 6G-enabled ultra-reliable low-latency communication in edge networks. *IEEE Communications Standards Magazine*, 6(1), 67–74. DOI: 10.1109/MCOMSTD.0001.2100098

Adhikari, M., Hazra, A., Menon, V. G., Chaurasia, B. K., & Mumtaz, S. (2021). A roadmap of next-generation wireless technology for 6G-enabled vehicular networks. *IEEE Internet of Things Magazine*, 4(4), 79–85. DOI: 10.1109/IOTM.001.2100075

Agarwal, A., Mishra, S. S., Panda, B. K., & Misra, G. (2023). A detailed review on 6G technology: Online learning system applications, challenges and research activities. Online Learning Systems, 147-176.

Ahammed, T. B., Patgiri, R., & Nayak, S. (2023). A vision on the artificial intelligence for 6G communication. *ICT Express*, 9(2), 197–210. DOI: 10.1016/j.icte.2022.05.005

Ahmad, H. F., Rafique, W., Rasool, R. U., Alhumam, A., Anwar, Z., & Qadir, J. (2023). Leveraging 6G, extended reality, and IoT big data analytics for healthcare: A review. *Computer Science Review*, 48, 100558. DOI: 10.1016/j.cosrev.2023.100558

Akyildiz, I. F., Kak, A., & Nie, S. (2020). 6G and beyond: The future of wireless communications systems. *IEEE Access : Practical Innovations, Open Solutions*, 8, 133995–134030. DOI: 10.1109/ACCESS.2020.3010896

Alhammadi, A., Abraham, A., Fakhreddine, A., Tian, Y., Du, J., & Bader, F. (2024). Envisioning the Future Role of 3D Wireless Networks in Preventing and Managing Disasters and Emergency Situations. arXiv preprint arXiv:2402.10600.

Alhammadi, A., Shayea, I., El-Saleh, A. A., Azmi, M. H., Ismail, Z. H., Kouhalvandi, L., & Saad, S. A. (2024). Artificial Intelligence in 6G Wireless Networks: Opportunities, Applications, and Challenges. *International Journal of Intelligent Systems*, 2024(1), 8845070. DOI: 10.1155/2024/8845070

Alsharif, M. H., Albreem, M. A., Solyman, A. A. A., & Kim, S. (2021). Toward 6G communication networks: Terahertz frequency challenges and open research issues. *Computers, Materials & Continua*. Advance online publication. DOI: 10.32604/cmc.2021.013176

Banafaa, M., Shayea, I., Din, J., Azmi, M. H., Alashbi, A., Daradkeh, Y. I., & Alhammadi, A. (2023). 6G mobile communication technology: Requirements, targets, applications, challenges, advantages, and opportunities. *Alexandria Engineering Journal*, 64, 245–274. DOI: 10.1016/j.aej.2022.08.017

Bhattacharya, P., Patel, F., Alabdulatif, A., Gupta, R., Tanwar, S., Kumar, N., & Sharma, R. (2022). A deep-Q learning scheme for secure spectrum allocation and resource management in 6G environment. *IEEE Transactions on Network and Service Management*, 19(4), 4989–5005. DOI: 10.1109/TNSM.2022.3186725

Bourdoux, A., Barreto, A. N., van Liempd, B., de Lima, C., Dardari, D., Belot, D., . . . Miao, Y. (2020). 6G white paper on localization and sensing. arXiv preprint arXiv:2006.01779.

Chack, D., & Thool, S. N. (2023). High Capacity 64-Quadrature Amplitude Modulation Based Optical Coherent Transceiver for 60 GHz Radio over Fiber System. *Wireless Personal Communications*, 132(1), 183–204. DOI: 10.1007/s11277-023-10606-x

Chataut, R., Nankya, M., & Akl, R. (2024). 6G networks and the AI revolution—Exploring technologies, applications, and emerging challenges. *Sensors (Basel)*, 24(6), 1888. DOI: 10.3390/s24061888 PMID: 38544151

Chataut, R., Nankya, M., & Akl, R. (2024). 6G networks and the AI revolution—Exploring technologies, applications, and emerging challenges. *Sensors (Basel)*, 24(6), 1888. DOI: 10.3390/s24061888 PMID: 38544151

Chergui, H., Ksentini, A., Blanco, L., & Verikoukis, C. (2022). Toward zero-touch management and orchestration of massive deployment of network slices in 6G. *IEEE Wireless Communications*, 29(1), 86–93. DOI: 10.1109/MWC.009.00366

Chowdhury, M. Z., Shahjalal, M., Ahmed, S., & Jang, Y. M. (2020). 6G wireless communication systems: Applications, requirements, technologies, challenges, and research directions. *IEEE Open Journal of the Communications Society*, 1, 957–975. DOI: 10.1109/OJCOMS.2020.3010270

R, Deepa., A., Sivasamy., S., Selvam. (2024). Intelligent networking unleashed: exploring ai and ml for enhanced 6g wireless communication systems. 228-242. DOI: 10.58532/V3BINC3P8CH3

Dib, O., & Toumi, K. (2020). Decentralized identity systems: Architecture, challenges, solutions and future directions. Annals of Emerging Technologies in Computing (AETiC), Print ISSN, 2516-0281.

Dogra, A., Jha, R. K., & Jain, S. (2020). A survey on beyond 5G network with the advent of 6G: Architecture and emerging technologies. *IEEE Access : Practical Innovations, Open Solutions*, 9, 67512–67547. DOI: 10.1109/ACCESS.2020.3031234

Dogra, A., Jha, R. K., & Jain, S. (2020). A survey on beyond 5G network with the advent of 6G: Architecture and emerging technologies. *IEEE Access : Practical Innovations, Open Solutions*, 9, 67512–67547. DOI: 10.1109/ACCESS.2020.3031234

Duan, Q., Huang, J., Hu, S., Deng, R., Lu, Z., & Yu, S. (2023). Combining federated learning and edge computing toward ubiquitous intelligence in 6G network: Challenges, recent advances, and future directions. *IEEE Communications Surveys and Tutorials*, 25(4), 2892–2950. DOI: 10.1109/COMST.2023.3316615

Farzaneh, H., Malehmirchegini, L., Bejan, A., Afolabi, T., Mulumba, A., & Daka, P. P. (2021). Artificial intelligence evolution in smart buildings for energy efficiency. *Applied Sciences (Basel, Switzerland)*, 11(2), 763. DOI: 10.3390/app11020763

Field, M. J., & Grigsby, J. (2002). Telemedicine and remote patient monitoring. *Journal of the American Medical Association*, 288(4), 423–425. DOI: 10.1001/jama.288.4.423 PMID: 12132953

Hamdan, S., Ayyash, M., & Almajali, S. (2020). Edge-computing architectures for internet of things applications: A survey. *Sensors (Basel)*, 20(22), 6441. DOI: 10.3390/s20226441 PMID: 33187267

Henrique, P. S. R., & Prasad, R. (2020, October). The road for 6G multimedia applications. In 2020 23rd International Symposium on Wireless Personal Multimedia Communications (WPMC) (pp. 1-6). IEEE.

Hossain, M. A., Hossain, A. R., & Ansari, N. (2022). AI in 6G: Energy-efficient distributed machine learning for multilayer heterogeneous networks. *IEEE Network*, 36(6), 84–91. DOI: 10.1109/MNET.104.2100422

Huo, Y., Lin, X., Di, B., Zhang, H., Hernando, F. J. L., Tan, A. S., Mumtaz, S., Demir, Ö. T., & Chen-Hu, K. (2023). Technology trends for massive MIMO towards 6G. *Sensors (Basel)*, 23(13), 6062. DOI: 10.3390/s23136062 PMID: 37447911

Iyer, S., Patil, A., Bhairanatti, S., Halagatti, S., & Pandya, R. J. (2022). A survey on technological trends to enhance spectrum-efficiency in 6g communications. *Transactions of the Indian National Academy of Engineering : an International Journal of Engineering and Technology*, 7(4), 1093–1120. DOI: 10.1007/s41403-022-00372-w

Jiang, W., Zhou, Q., He, J., Habibi, M. A., Melnyk, S., El-Absi, M., Han, B., Renzo, M. D., Schotten, H. D., Luo, F.-L., El-Bawab, T. S., Juntti, M., Debbah, M., & Leung, V. C. (2024). Terahertz communications and sensing for 6G and beyond: A comprehensive review. *IEEE Communications Surveys and Tutorials*, 26(4), 2326–2381. DOI: 10.1109/COMST.2024.3385908

Kaur, K., Mangat, V., & Kumar, K. (2020). A comprehensive survey of service function chain provisioning approaches in SDN and NFV architecture. *Computer Science Review*, 38, 100298. DOI: 10.1016/j.cosrev.2020.100298

Kenner, A., & Oladele, O. (2024). Software-Defined Networking (SDN) and Network Function Virtualization (NFV) in Telecommunication: A Computer Science Perspective.

Khaloopour, L., Su, Y., Raskob, F., Meuser, T., Bless, R., Würsching, L., . . . Jamali, V. (2024). Resilience-by-Design Concepts for 6G Communication Networks. arXiv preprint arXiv:2405.17480.

Khan, L. U., Yaqoob, I., Imran, M., Han, Z., & Hong, C. S. (2020). 6G wireless systems: A vision, architectural elements, and future directions. *IEEE Access : Practical Innovations, Open Solutions*, 8, 147029–147044. DOI: 10.1109/ACCESS.2020.3015289

Kharche, S., & Kharche, J. (2023). 6G Intelligent Healthcare Framework: A Review on Role of Technologies, Challenges and Future Directions. J. *Mobile Multimedia*, 19(3), 603–644. DOI: 10.13052/jmm1550-4646.1931

Li, C., Li, J., & Chen, W. (2022). U.S. Patent No. 11,284,416. Washington, DC: U.S. Patent and Trademark Office.

Li, J., Dang, S., Wen, M., Li, Q., Chen, Y., Huang, Y., & Shang, W. (2023). Index modulation multiple access for 6G communications: Principles, applications, and challenges. *IEEE Network*, 37(1), 52–60. DOI: 10.1109/MNET.002.2200433

Li, J., Della Penda, D., Sahlin, H., Schliwa-Bertling, P., Folke, M., & Stattin, M. (2021). An overview of 5g system accessibility differentiation and control. *IEEE Communications Standards Magazine*, 5(4), 104–111. DOI: 10.1109/MCOMSTD.0001.2000067

Li, Y., Huang, J., Sun, Q., Sun, T., & Wang, S. (2021). Cognitive service architecture for 6G core network. *IEEE Transactions on Industrial Informatics*, 17(10), 7193–7203. DOI: 10.1109/TII.2021.3063697

Liang, Y., & Zeng, J. (2021, October). Ai application in english vocational education through 6g revolution. In International Conference on 5G for Future Wireless Networks (pp. 532-545). Cham: Springer International Publishing.

Lin, H. L., Liao, P. K., & Wu, W. D. (2018). U.S. Patent Application No. 15/835,768.

Lu, L., Liu, C., Zhang, C., Hu, Z., Lin, S., Liu, Z., Zhang, M., Liu, X., & Chen, J. (2023). Architecture for self-evolution of 6g core network based on intelligent decision making. *Electronics (Basel)*, 12(15), 3255. DOI: 10.3390/electronics12153255

2023. Machine Learning Techniques and Routing Protocols in 5G and 6G Mobile Network Communication System - An Overview. DOI: 10.1109/ICOEI56765.2023.10125697

Mansouri, Y., & Babar, M. A. (2021). A review of edge computing: Features and resource virtualization. *Journal of Parallel and Distributed Computing*, 150, 155–183. DOI: 10.1016/j.jpdc.2020.12.015

Mao, B., Kawamoto, Y., & Kato, N. (2020). AI-based joint optimization of QoS and security for 6G energy harvesting Internet of Things. *IEEE Internet of Things Journal*, 7(8), 7032–7042. DOI: 10.1109/JIOT.2020.2982417

Mitev, M., Chorti, A., Poor, H. V., & Fettweis, G. P. (2023). What physical layer security can do for 6G security. *IEEE Open Journal of Vehicular Technology*, 4, 375–388. DOI: 10.1109/OJVT.2023.3245071

Moro-Visconti, R., Cruz Rambaud, S., & López Pascual, J. (2023). Artificial intelligence-driven scalability and its impact on the sustainability and valuation of traditional firms. *Humanities & Social Sciences Communications*, 10(1), 1–14. DOI: 10.1057/s41599-023-02214-8

Mounir, M., El_Mashade, M. B., & Mohamed Aboshosha, A. (2022). On the selection of power allocation strategy in power domain non-orthogonal multiple access (PD-NOMA) for 6G and beyond. *Transactions on Emerging Telecommunications Technologies*, 33(6), e4289.

Naser, S., Bariah, L., Muhaidat, S., & Basar, E. (2023). Zero-energy devices empowered 6G networks: Opportunities, key technologies, and challenges. *IEEE Internet of Things Magazine*, 6(3), 44–50. DOI: 10.1109/IOTM.001.2200235

Nayak, S., & Patgiri, R. (2022, April). 6G communication: A vision on the potential applications. In Edge Analytics: Select Proceedings of 26th International Conference—ADCOM 2020 (pp. 203-218). Singapore: Springer Singapore.

Nayak, S., & Patgiri, R. (2022). 6G Communication: A Vision on the Potential Applications. In Patgiri, R., Bandyopadhyay, S., Borah, M. D., & Emilia Balas, V. (Eds.), *Edge Analytics. Lecture Notes in Electrical Engineering* (Vol. 869). Springer. DOI: 10.1007/978-981-19-0019-8_16

Nguyen, V. L., Lin, P. C., Cheng, B. C., Hwang, R. H., & Lin, Y. D. (2021). Security and privacy for 6G: A survey on prospective technologies and challenges. *IEEE Communications Surveys and Tutorials*, 23(4), 2384–2428. DOI: 10.1109/COMST.2021.3108618

Pana, V. S., Babalola, O. P., & Balyan, V. (2022). 5G radio access networks: A survey. *Array (New York, N.Y.)*, 14, 100170. DOI: 10.1016/j.array.2022.100170

Paolini, E., Valcarenghi, L., Maggiani, L., & Andriolli, N. (2023). Real-time clustering based on deep embeddings for threat detection in 6G networks. *IEEE Access : Practical Innovations, Open Solutions*, 11, 115827–115835. DOI: 10.1109/ACCESS.2023.3325721

Pexyean, T., Saraubon, K., & Nilsook, P. (2023, August). IoT, 6G and Digital Twin For Smart Campus. In 2023 Research, Invention, and Innovation Congress: Innovative Electricals and Electronics (RI2C) (pp. 46-50). IEEE.

Polymeni, S., Skoutas, D. N., Sarigiannidis, P., Kormentzas, G., & Skianis, C. (2024). Smart Agriculture and Greenhouse Gas Emission Mitigation: A 6G-IoT Perspective. *Electronics (Basel)*, 13(8), 1480. DOI: 10.3390/electronics13081480

Ramesh, B., Changala, R., Kalangi, P. K., & Bhavsingh, M. (2024). Optimizing 6G Network Slicing with the EvoNetSlice Model for Dynamic Resource Allocation and Real-Time QoS Management. *International Research Journal of Multidisciplinary Technovation*, 6(3), 325–340.

Rowshan, M., Qiu, M., Xie, Y., Gu, X., & Yuan, J. (2024). *Channel Coding Towards 6G: Technical Overview and Outlook.* IEEE Open Journal of the Communications Society.

Rufino Henrique, P. S., & Prasad, R. (2021). 6G networks for the next generation of digital TV beyond 2030. *Wireless Personal Communications*, 121(2), 1363–1378. DOI: 10.1007/s11277-021-09070-2 PMID: 34566262

Saeed, N., Almorad, H., Dahrouj, H., Al-Naffouri, T. Y., Shamma, J. S., & Alouini, M. S. (2021). Point-to-point communication in integrated satellite-aerial 6G networks: State-of-the-art and future challenges. *IEEE Open Journal of the Communications Society*, 2, 1505–1525. DOI: 10.1109/OJCOMS.2021.3093110

Serôdio, C., Cunha, J., Candela, G., Rodriguez, S., Sousa, X. R., & Branco, F. (2023). The 6G ecosystem as support for IoE and private networks: Vision, requirements, and challenges. *Future Internet*, 15(11), 348. DOI: 10.3390/fi15110348

Shah, A. S. (2022, January). A survey from 1G to 5G including the advent of 6G: Architectures, multiple access techniques, and emerging technologies. In 2022 IEEE 12th Annual Computing and Communication Workshop and Conference (CCWC) (pp. 1117-1123). IEEE.

Shen, X., Liao, W., & Yin, Q. (2022). A novel wireless resource management for the 6G-enabled high-density Internet of Things. *IEEE Wireless Communications*, 29(1), 32–39. DOI: 10.1109/MWC.003.00311

Shi, E., Zhang, J., Du, H., Ai, B., Yuen, C., Niyato, D., Letaief, K. B., & Shen, X. (2024). RIS-aided cell-free massive MIMO systems for 6G: Fundamentals, system design, and applications. *Proceedings of the IEEE*, 112(4), 331–364. DOI: 10.1109/JPROC.2024.3404491

Shi, Y., Lian, L., Shi, Y., Wang, Z., Zhou, Y., Fu, L., Bai, L., Zhang, J., & Zhang, W. (2023). Machine learning for large-scale optimization in 6g wireless networks. *IEEE Communications Surveys and Tutorials*, 25(4), 2088–2132. DOI: 10.1109/COMST.2023.3300664

Singh, R., Kaushik, A., Shin, W., Di Renzo, M., Sciancalepore, V., Lee, D., . . . Dobre, O. A. (2024). Towards 6G Evolution: Three Enhancements, Three Innovations, and Three Major Challenges. arXiv preprint arXiv:2402.10781.

Siriwardhana, Y., Porambage, P., Liyanage, M., & Ylianttila, M. (2021, June). AI and 6G security: Opportunities and challenges. In 2021 Joint European Conference on Networks and Communications & 6G Summit (EuCNC/6G Summit) (pp. 616-621). IEEE.

Sirohi, D., Kumar, N., Rana, P. S., Tanwar, S., Iqbal, R., & Hijjii, M. (2023). Federated learning for 6G-enabled secure communication systems: A comprehensive survey. *Artificial Intelligence Review*, 56(10), 11297–11389. DOI: 10.1007/s10462-023-10417-3 PMID: 37362891

Sodhro, A. H., Pirbhulal, S., Luo, Z., Muhammad, K., & Zahid, N. Z. (2020). Toward 6G architecture for energy-efficient communication in IoT-enabled smart automation systems. *IEEE Internet of Things Journal*, 8(7), 5141–5148. DOI: 10.1109/JIOT.2020.3024715

Sujin, K. (2024, August). Won-Yong, Shin., Seong-Lyun, Kim., Seung-Woo, Ko. (2024). Energy-Efficient Edge Learning via Joint Data Deepening-and-Prefetching. *IEEE Transactions on Wireless Communications*, 23(8), 9927–9942. Advance online publication. DOI: 10.1109/TWC.2024.3367352

Syed, J. N., Sharma, S. K., Patwary, M. N., & Asaduzzaman, M. (2023). Enhanced URLLC-Enabled Edge Computing Framework for Device-Level Innovation in 6G. *Authorea Preprints*.

Tang, F., Mao, B., Kawamoto, Y., & Kato, N. (2021). Survey on machine learning for intelligent end-to-end communication toward 6G: From network access, routing to traffic control and streaming adaption. *IEEE Communications Surveys and Tutorials*, 23(3), 1578–1598. DOI: 10.1109/COMST.2021.3073009

Tomaszewski, L., & Kołakowski, R. (2023, January). Mobile services for smart agriculture and forestry, biodiversity monitoring, and water management: Challenges for 5G/6G networks. In Telecom (Vol. 4, No. 1, pp. 67-99). MDPI.

Umoga, U. J., Sodiya, E. O., Ugwuanyi, E. D., Jacks, B. S., Lottu, O. A., Daraojimba, O. D., & Obaigbena, A. (2024). Exploring the potential of AI-driven optimization in enhancing network performance and efficiency. *Magna Scientia Advanced Research and Reviews*, 10(1), 368–378. DOI: 10.30574/msarr.2024.10.1.0028

Wang, C. X., You, X., Gao, X., Zhu, X., Li, Z., Zhang, C., Wang, H., Huang, Y., Chen, Y., Haas, H., Thompson, J. S., Larsson, E. G., Renzo, M. D., Tong, W., Zhu, P., Shen, X., Poor, H. V., & Hanzo, L. (2023). On the road to 6G: Visions, requirements, key technologies, and testbeds. *IEEE Communications Surveys and Tutorials*, 25(2), 905–974. DOI: 10.1109/COMST.2023.3249835

Xu, C. (2024, May 07). Zhiyong, Feng., Andrew, Zhang., Zhaohui, Yang., Xin, Yuan., Xuhua, He., Ping, Zhang. (2024). Integrated Communication, Sensing, and Computation Framework for 6G Networks. *Sensors (Basel)*, 24(10), 2968. Advance online publication. DOI: 10.3390/s24102968

Yajun, Zhao., Guanghui, Yu., Hanqing, Xu. (2019). 6G mobile communication networks: vision, challenges, and key technologies. DOI: 10.1360/N112019-00033

Yang, H., Alphones, A., Xiong, Z., Niyato, D., Zhao, J., & Wu, K. (2020). Artificial-intelligence-enabled intelligent 6G networks. *IEEE Network*, 34(6), 272–280. DOI: 10.1109/MNET.011.2000195

Yang, Y., Ma, M., Wu, H., Yu, Q., You, X., Wu, J., Peng, C., Yum, T.-S. P., Aghvami, A. H., Li, G. Y., Wang, J., Liu, G., Gao, P., Tang, X., Cao, C., Thompson, J., Wong, K.-K., Chen, S., Wang, Z., & Shu, H. (2022). 6G network AI architecture for everyone-centric customized services. *IEEE Network*, 37(5), 71–80. DOI: 10.1109/MNET.124.2200241

Yuan, W., Li, S., Wei, Z., Cui, Y., Jiang, J., Zhang, H., & Fan, P. (2023). New delay Doppler communication paradigm in 6G era: A survey of orthogonal time frequency space (OTFS). *China Communications*, 20(6), 1–25. DOI: 10.23919/JCC.fa.2022-0578.202306

Zhang, Y., Azizui, M. F. A. B., & Yan, C. (2024). The Application of 6G and Augmented Reality Technology in Education and Training. *Wireless Personal Communications*, •••, 1–19. DOI: 10.1007/s11277-024-11038-x

Ziegler, V., Viswanathan, H., Flinck, H., Hoffmann, M., Räisänen, V., & Hätönen, K. (2020). 6G architecture to connect the worlds. *IEEE Access : Practical Innovations, Open Solutions*, 8, 173508–173520. DOI: 10.1109/ACCESS.2020.3025032

Ziegler, V., Viswanathan, H., Flinck, H., Hoffmann, M., Räisänen, V., & Hätönen, K. (2020). 6G architecture to connect the worlds. *IEEE Access : Practical Innovations, Open Solutions*, 8, 173508–173520. DOI: 10.1109/ACCESS.2020.3025032

Chapter 6
Enhancing Public Safety in Developing Countries Using a Mobile Application Solution:
Nigeria as a Case Study

Moyinoluwalogo Mayowa
Nottingham Trent University, UK

Richard I. Otuka
https://orcid.org/0009-0006-0198-8999
Nottingham Trent University, UK

Nemitari Ajienka
Nottingham Trent University, UK

Augustine O. Nwajana
https://orcid.org/0000-0001-6591-5269
University of Greenwich, UK

ABSTRACT

In today's world, the escalating global insecurity poses significant challenges for both developed and undeveloped countries. The high crime rates and shortcomings in reporting crimes are pressing issues, emphasizing just how crucial it is to ensure the safety of citizens. While law enforcement agencies play a vital role, relying solely on them is insufficient. Embracing modern technology and fostering active citizen participation are essential to address these concerns effectively. This research adopts an agile methodology to tackle the above concerns. The result of this approach is the development of a user-friendly mobile application that leverages cutting-edge technology. The primary objective of this application is to effectively map incidents of crime and provide real-time alerts, thereby significantly enhancing citizen security. With this, users are empowered to report criminal incidents directly from their smartphones, eliminating the need to physically approach law enforcement authorities.

DOI: 10.4018/979-8-3693-8799-3.ch006

1. INTRODUCTION

According to a report by the United Nations Office on Drugs and Crime (UNODC), there has been a rise in crime rates in developing countries, with some regions experiencing an increase of up to 3.3% per year (Heikkilä et al., 2021) . Crimes such as theft, assault, and robbery are the most reported, but there has also been an increase in more violent crimes like rape and murder. Due to the threats they have on economic growth and public safety, these crimes have increased to be a major concern for most governments (O'malley, 1992; Kure & Nwajana, 2022) . Particularly, the governments of developing countries have been facing challenges in ensuring the safety of their citizens, and the situation has become more concerning. While the government has implemented various measures to combat these crimes, the problem persists. Therefore, it is essential to explore alternative solutions that could potentially reduce these incidents and improve public safety (Achumba et al., 2013) .

Using Nigeria as a case study, this book chapter focuses on the development of a mobile app solution for enhancing public safety. The purpose of the app is to provide citizens with an easy-to-use platform for reporting incidents of crime and emergencies in real-time, as well as to access important safety information and tips. This will provide the authorities with vital information needed to respond quickly and efficiently to reported crime incidents (Desouza & Bhagwatwar, 2012). This study will explore the effectiveness of the app on improving public safety and reducing crime in Nigeria. Through this, we hope to explore the potential of mobile technology as a tool for enhancing public safety.

The research was conducted by adopting a quantitative research approach. This involved conducting online survey/questionnaires and receiving user insights with a purposive sample of 200 people living across the six geopolitical zones of Nigeria. The users were selected based on their willingness to participate in the study and their experience with crime in their community. The users were recruited using google forms, an online survey platform. The sample size was determined based on the principles of statistical power and precision (Das, Mitra, & Mandal, 2016) . The sample selection was random to ensure that the respondents are representative of users across Nigeria. The survey aimed to gather data on the respondents' perceptions of the proposed mobile app solution, their reactions to public safety mobile apps, and their willingness to adopt it in the future.

The outcome of this research is a comprehensive mobile app solution that addresses the security concerns of citizens in developing countries, with a focus on Nigeria. The mobile app will serve as a platform for reporting incidents of crimes, providing real-time information to the authorities and enhancing collaboration between citizens and the law. Based on the results of the survey, the app was designed to provide relevant safety features, such as emergency contacts, safety tips, upload of media – audio, photos, and videos, SOS alert, and real time location tracker. The app was designed to be user-friendly and accessible, with a simple interface that requires minimal technical skills to use. The research findings was used as a guide in the development of the mobile app solution, ensuring that it meets the needs and expectations of the target audience. By providing an innovative and practical solution to the security challenges in developing countries, this research aims to contribute to the improvement of public safety and enhance the quality of life of citizens.

1.1. Aim and objectives

1.1.1. Aim

The primary aim of this research is to develop a mobile app solution that will improve public safety in developing countries by providing a platform for reporting real-time crime incidents. In addition to the main goal, this research also seeks to achieve the following objectives in the next section.

1.1.1. Research Objectives

For the research to be successful, the following objectives must be met:

1. Investigate the current state of public safety in developing countries, using Nigeria as a focus
2. Examine the existing measures taken by the government to address security issues in Nigeria.
3. Identify the key features of a mobile app that can improve public safety.
4. Investigate the challenges that may hinder the adoption of the proposed app solution.
5. Develop a user-friendly mobile app solution that promotes real-time reporting of security incidents and provides safety tips and measures to citizens.
6. Evaluate the effectiveness of the mobile app solution in enhancing public safety in developing countries.

The research findings will contribute to the existing body of knowledge on public safety aided by mobile technology. This would be achieved by improving the reporting and promoting fast response to crime incidents. The mobile app solution developed through this research will enhance the safety and security of citizens in developing countries, ultimately leading to a better quality of life for all.

2. BACKGROUND

The use of mobile apps to enhance public safety in developing countries has become a trending topic in recent years. Developing countries like Nigeria face significant challenges in maintaining public safety due to the high rate of crime, inadequate infra-structure, and limited resources. The development of mobile apps that can help improve public safety has, therefore, become a priority for governments in these countries.

Nigeria has been facing several public safety challenges, including terrorism, kidnapping, armed robbery, and cultism (Omede, 2011). The Boko Haram insurgency, which started in 2009, has resulted in the loss of thousands of lives and displacement of millions of people in Nigeria (Awortu, 2015), Crime, generally, has become a lucrative business for criminals in Nigeria, with several high-profile cases of kidnapping reported in recent years. The rise of armed robbery and cultism in urban areas has further exacerbated the public safety situation in Nigeria.

Several studies have been conducted on the use of mobile apps to enhance public safety in developing countries (Mathijssen & Tanner, 2022). Timothy (2020) developed a mobile app that uses GPS technology to track criminal activities in Nigeria. The app provides users with real-time up-dates on criminal incidents in their vicinity and alerts security agencies. Bhole et al. (2021) developed a mobile app that enables users to report crime incidents directly to the police, with features such as real-time video streaming and location-based tracking. Also, Lips, Amola, & Draheim (2021) developed a mobile app that enables users to request emergency medical services in Nigeria. The app provides users with information on nearby hospitals and ambulance services and allows them to track the progress of their requests.

The proposed mobile app solution for enhancing public safety in Nigeria is a step in the right direction. It addresses a critical need in the country and has the potential to significantly improve public safety. The app's features, such as the ability to report incidents in real-time, provide emergency security response teams with accurate information to respond effectively to emergencies. Additionally, the app's ability to provide safety tips and information on emergency services would promote public safety awareness.

Despite the potential benefits of existing public safety solutions, some limitations must be addressed. One of the primary concerns is the need for a stable and reliable internet connection for the app to function optimally. Given the erratic nature of internet connectivity in some parts of Nigeria, this could limit the proposed app's effectiveness in the areas without internet connectivity. Additionally, the proposed solution may face the challenge of adoption, as some members of the public may be hesitant to embrace modern technology or may not have access to smartphones.

To improve the effectiveness of the proposed mobile app, the app would explore alternative ways of transmitting data that do not rely solely on internet connectivity. This would include SMS and USSD technology, which are more accessible and reliable in areas with poor internet connectivity.

This literature review answers the below questions.:

1. What are the key features and functionalities that are most important for a mobile app solution to enhance public safety?
2. How can the mobile app solution be tailored to address the specific needs and preferences of different user groups?
3. What are the best practices and lessons learned from existing mobile app solutions for public safety in other countries?
4. What is the anticipated impact and effectiveness of the mobile app solution in enhancing public safety, and how can it be measured and evaluated?
5. What are the current challenges and gaps in public safety measures in the Nigeria?

2.1. Review of existing mobile app soultions for public safety

The bSafe personal safety application for enhancing the safety and well-being of individuals, was launched in 2012 and founded by a father-daughter team from Norway, after Charlen, the daughter, was raped. She struggled with guilt, shame, anxiety and suicidal thoughts for many years, until she reflected with her father on how the incident could have been avoided. That birthed the bSafe safety app, available for both iOS and Android devices, employing the programming languages and frameworks specific to these platforms, such as Swift or Objective-C for iOS and Java for An-droid. It offers various features designed to improve personal safety, such as live GPS tracking, a panic button, and the ability to create a social safety network. Users can add trusted contacts who can monitor their location and receive alerts in case of emergencies (Azman et al., 2018).

The launch of the app received traction and the acceptance rate across the country was high. Some of the key features found on the app include its user-friendly interface, allowing individuals to easily navigate and access its features. It also possesses an intuitive design that enables quick activation of safety features during emergencies. Another key strength of the bSafe app is its live GPS tracking feature. which allows users to share their location in real-time with their selected contacts, enhancing their safety and enabling a faster response in case of emergencies. Another key feature is the panic button. The app incorporates a panic button that, when activated, sends an emergency alert to designated contacts. This feature can be a crucial lifeline in urgent situations, ensuring that help is notified promptly. Also, the app enables users to create a social safety network, adding trusted contacts who can be alerted in case of emergencies. This feature fosters a sense of community safety and provides an additional layer of protection.

The bSafe app offers several other strengths that contribute to personal safety, including its user-friendly interface, live GPS tracking, panic button, and social safety network. However, it is important to recognize the limitations of the app, such as de-pendency on smartphones and network connectivity, reliance on user activation, and privacy concerns.

Despite its strengths, the bSafe app is not without its limitations. Some of which include the dependency on smartphone and network connectivity because the effectiveness of the app relies on the availability of a functioning smartphone and a stable network connection. In situations where these resources are limited or unavailable, the app's functionalities may be compromised. Another limitation is the reliance on user activation as the app requires users to proactively activate safety features during emergencies. There may be high-stress situations where individuals may forget or maybe unable to activate the panic button, which could hinder the app's effectiveness.

Another app solution built for public safety is the Eyewatch app. Eyewatch app is a personal safety application designed to enhance security and aid during emergency situations. The Eyewatch app was developed in India in 2012, with the aim of lever-aging technology to improve personal safety and enable swift response during emergencies. The app offers various features and functionalities to enhance user security and peace of mind.

The strengths of the Eyewatch app include its intuitive interface and user-friendly design. The app allows users to quickly access emergency contacts and send distress signals with just a few taps. It offers multiple options for alerting contacts, such as SMS, email, and social media notifications, ensuring that help can be sought even in situations where making a voice call is not feasible (Leema et al., 2021).

The Eyewatch app utilizes GPS technology to provide accurate location tracking, allowing emergency responders or designated contacts to locate the user in need of assistance. Additionally, the Eyewatch app offers real-time audio and video recording capabilities, which can be vital for capturing evidence or providing context during an emergency.

Another notable strength of the Eyewatch app is its community safety feature. Users can join local or interest-based safety groups, enabling them to share safety tips, incidents, and updates within their community. This fosters a sense of collective security and promotes a collaborative approach to personal safety.

However, the Eyewatch app also has some limitations. One limitation is its dependence on network connectivity and smartphone functionality. In areas with weak or no network coverage, the effectiveness of the app may be compromised. Moreover, reliance on smartphones implies that users must always carry their devices and keep them charged for the app to be fully functional.

Privacy concerns may also arise with the Eyewatch app, particularly regarding the access and storage of personal data. Users must carefully consider the app's privacy policy and ensure that their information is protected and used responsibly. In terms of technologies used, the Eyewatch app incorporates several key elements. It utilizes GPS technology for accurate location tracking, allowing for quick and efficient response during emergencies. The app also leverages networking capabilities to send distress signals and notifications to emergency contacts and authorities. Furthermore, the Eyewatch app integrates with audio and video recording functionalities on smartphones to capture critical evidence or record incidents. A tabular representation of the review of other existing applications in other countries is given in Table 1.

Table 1. Review of existing apps (in other countries).

Existing app	Country	Strength	Limitation
iPolice App	South Africa	Location Tracking Sends report to police force	Cannot record Situation
CitizenCop App	India	SOS Alert Location Tracking, News Update, Vehicle Search Send Help to Police, Emergency Calls.	No Live Streaming Service
Namola App	South Africa	Location Sharing, Get Help used to connect citizens to emergency services, resulting in faster response times and increased safety	Does not connect to Emergency Authority Services, such as the Police Force
Kitestring safety app	United States	Automatic Recording Includes motion sensor feature. Panic Button	Does not allow audio/video recordings. No real-time location
Walk Safe App	United States	User can view their live location	Does not support Video Live Stream

The success rate of public safety apps in Nigeria is difficult to quantify, as there is no comprehensive data on the number of crimes prevented or solved based on their usage. However, there are several reports and case studies that suggest that these apps can be effective in improving public safety (Desouza & Bhagwatwar, 2012). Several crime reporting systems, in the form of mobile apps and web applications have been created by different privately held companies as well as by the Nigerian Government in a bid to improve the security situation of the country (Imasuen & Nwajana, 2022). Each of these solutions has unique features, all of which will be covered below.

In 2017, the Lagos State Police Command launched a public safety app called "Safer Lagos". The app allows users to report crimes, suspicious activity, and traffic incidents. In the first year of its operation, the app received over 10,000 reports, and the police were able to successfully investigate and resolve several crimes.

Another existing mobile app is the "767" emergency hotline, which was launched by the Nigerian Police Force in 2015 (Ogunyombo, Azeez, & Odunlami, 2018). The hotline allows citizens to report crimes and emergencies, and it has been credited with helping to reduce crime rates in some parts of the country.

In 2019, the Federal Government of Nigeria launched the Nigerian Internal Security and Public Safety Alert System (NIPSAS) mobile app called "N-Alerts" for sending alerts to security agencies. This is a technological innovation that can be installed on mobile smart phones running on either the Android or IOS OS, to send alerts to all security agencies in real time. The mobile application utilizes a combination of audio-visual and geospatial location identification of the caller and scenes of incidents on the ministry inter-agency situation room (MISR). The mobile application enables registered users within the country to report crime and the data collected from the app is made available to the command-and-control centers which are the Nigerian Immigration service, correctional service, federal fire service, civil defense as well as police headquarters for joint inter-agency operations, this is to allow for quick response time to crime incidents reported via the app.

Looking to enhance public safety while giving a user's family and friends a real-time location of the user, Sety Inc (a privately held company that develops solutions with a focus on safety) developed a mobile app named Sety, it was released in 2020. Sety is a sub-scription-based mobile application that works by sending notifications to an individual's circle members and keeps a help request open for 1 hour, while communication occurs in real-time between all notified parties. The Sety mobile app's major features is to broadcast to both saved circle as well as to people around the user. It also keeps the user updated when they are approaching a known danger zone. This works in the form of frequent push notifications sent to the user's mobile device.

Another popular existing mobile app is the NPF rescue me mobile app, proposed by the Nigerian police force. The app has an emergency button which, once triggered automatically sends an alert to the Nigerian police force command and control team where trained operators will track the user's location with GPS and dispatch necessary help swiftly. The initiative was launched in 2019. The app is valuable to effectively report emergencies, criminal activities, and professional misconducts of police officers for rapid and immediate reaction by operatives of the Force.

Another popular mobile app solution for public safety in Nigeria is the Emergency Response App. This app was developed by a private security company and is de-signed to provide users with access to emergency services such as ambulance, fire service, and police. The app allows users to send distress signals and provides real-time updates on the status of their requests.

Despite the existence of these mobile app solutions, there are still several challenges in enhancing public safety in Nigeria. One of these challenges is the limited access to the internet and smartphones. According to the World Bank, only about 40% of the Nigerian population has access to the internet, and even fewer have access to smartphones. This limits the reach of mobile app solutions for public safety in the country (Stork, Calandro, & Gillwald, 2013).

Another challenge is the lack of trust in government and authorities, including the police force. Many citizens are reluctant to report crimes and emergencies to the police due to their perceived inefficiency and corruption. This affects the effectiveness of mobile app solutions that are developed by the police force.

Below is a tabular representation of existing safety mobile app solutions in Nigeria, their strengths, and their limitations.

Table 2. Review of existing apps (in Nigeria).

Existing app	Country	Strength	Limitation
iReport App	Nigeria	Allows citizens to alert local authorities on human trafficking incidents	No video or audio recording feature Can't capture photos. No real-time location sharing feature
NPF Rescue Me App	Nigeria	Designed to allow citizens request for quick help during emergency. Report Crime Report Police Officers Request an ambulance	No video or audio recording feature Can't capture photos. No real-time location sharing feature
Lagos State Emergency Management Agency (LASEMA) app	Nigeria	Report road accident occurrences	No video or audio recording feature Can't capture photos. No real-time location sharing feature

These apps are designed to help citizens access emergency services and provide them with necessary information on how to stay safe in different situations. However, there are still several challenges that need to be addressed to improve the effectiveness of these mobile app solutions in the country. The limitations of these existing apps will be considered during the development of the proposed solution.

2.3. The proposed solution

The proposed public safety app solution will build on existing safety apps in Nigeria, while also implementing additional features which have been implemented in more developed countries. The public safety app will leverage on existing systems available in Nigeria as well as introduce additional key features which the existing apps are lacking. The proposed solution will be designed to store information regarding any type of crime incidents, its description, location, and media – image, video, or audio, which would be entered by the user at the point of the reporting.

Then, the reports submitted by the users are viewed by the responsible authorities or officials with access to do so. In the case of an emergency, the civilian's location is immediately sent to their emergency contacts. A push notification alert is sent to other people in the area to inform them about the possible ongoing reported (active) crime. The updates of the crime reported, and the action taken on them will be updated to the person that reported the incident. Also, for emergency cases, there will be an SOS feature on the app. This will be an easy to navigate option for a user who feels threatened to be able to share their location conveniently and easily with their saved contacts (family and friends), as well as the authorities to seek help swiftly. This part of the proposed app will leverage on existing in-built features of mobile phones which includes GPS tracking and location sharing and monitoring.

The proposed mobile application would aim at enhancing public safety in developing countries such as Nigeria. Some features would include rating the occurrences of crimes in certain areas to alert other citizens who might be new to the area. Other features will include being able to make rapid calls to security authorities and requests for ambulance services, sharing real-time location tracking with friends and family.

The proposed app will be developed using Flutter. Flutter is an open-source mobile application development framework developed by Google. It is based on the Dart programming language and allows the creation of high performance, cross-platform mobile apps for both iOS and Android operating systems. The choice of Integrated development environment (IDE) is Visual Studio Code (VS Code). Visual Studio Code provides a rich set of tools for developing and debugging Flutter applications. The app will be integrated with Firebase which will be used as a Backend-as-a-Service. Firebase Realtime Database and Firebase Cloud Firestore are both cloud-based NoSQL databases offered by Google as part of the Firebase suite of tools. The database provides real-time data synchronization across the app, allowing real-time activity changes on the app.

2.4. Required Resources for the Proposed Application.

To develop the proposed application for this research, we will be building it on an integrated development environment (IDE). For this research, the choice of IDE is Visual Studio Code (VS Code). VS Code is a source-code editor made by Microsoft for writing, debugging, syntax highlighting, and refactoring of software code. Some of the other resources required for this research are listed below.

- Visual Studio Code
- Flutter Framework and Dart Programming Language
- Android Mobile phone or an iOS device.
- Internet Connectivity
- Database (Firebase)

2.5. Key Features and Functionalities.

The key features and functionalities of the proposed public safety mobile app include the following:

1. Camera Handling: The Camera capture feature will be handled on the report action screen of the app where a user can attach a photo while inputting the information for the reporting of an active crime to be submitted to the authorities. The photo attachment will be optional. A user can also share a photo with their friends and family (their circle). Once the Camera option has been selected, the photo is captured, and the app sends the captured photo to the intended recipient or recipients.
2. Data Storage: The proposed mobile app will store app data on Firebase Realtime Database and Firebase Cloud Firestore. The real time database is used for storing user and app data, and the Firebase cloud Firestore keeps a record of all the inter-actions across the app. Basic CRUD operations such as Creating, Retrieving, Up-dating, and deleting of records are performed on the databases.
3. Real-time Location Tracking: The proposed mobile app will make use of GPS and other location-based technologies to track the user's location in real-time and provide them with assistance during an emergency. On a user's first log in to an application, there is the request for permission to user's current location, this receives the user's consent to use their location on the app. This feature makes use of the google location services and users can view their exact location on the map. This also includes allowing a user to share their current location with friends and family.
4. Panic (SOS) Feature: A panic button which allows users to alert emergency services, friends, or family in case of an emergency. This feature would be easily accessible so that in a situation where a user feels threatened for any reason, they would be able to easily access the app to notify the authorities as well as their close friends and family of their current situation.
5. Emergency Contacts: The app will allow users to add emergency contacts to their profile, contacts who can be notified in case of an emergency. On the proposed mobile app solution, users will be able to save their close friends and family as emergency contacts so that in a situation where they need to send out a communication of their well-being or distress in a situation, they can easily notify their saved emergency contacts, who would be notified of the user's most recent location.
6. In-App Messaging & Push Notifications: The app will have a messaging feature, which can be used to communicate with emergency services or designated contacts. This is a valuable addition to the app because it helps to keep the civilians informed and engaged with the app. When a user enters an area regarded as a danger zone, they are immediately notified. Also push notifications can be sent to saved family and friends who have the app installed to alert them of the location of their loved one. This feature also includes sending alerts and notifications to users in the event of an ongoing crime, accident, or any other emergency.

7. Safety Tips: The proposed app will provide users with safety tips and advice on how to stay safe in various situations. These safety tips will be able to advise a user on what to do when they find themselves in certain situations and how they can get out of that situation safely. It will also be able to provide them with general safety awareness tips and info.
8. GPS Location tracking: GPS (Global Positioning System) location tracking is a technology that enables the determination of an object or person's precise geographical location using signals from satellites. GPS tracking is a valuable tool in enhancing public safety by enabling accurate and efficient location monitoring and assistance during emergencies. It uses a network of satellites orbiting the Earth to provide positioning information with high accuracy.
9. GPS location tracking is widely used in various applications and devices, including smartphones, navigation systems, fitness trackers, and vehicle tracking systems. These devices receive signals from multiple GPS satellites to calculate their exact position based on the time it takes for the signals to reach them.
10. In terms of public safety, GPS location tracking can play a crucial role. Public safety agencies and emergency services can utilize GPS tracking to locate individuals in distress, monitor the movement of emergency response vehicles, or coordinate search and rescue operations. Additionally, mobile apps and wearable de-vices with GPS capabilities can help users send their precise location during emergencies, ensuring prompt assistance.

For this research work, GPS location tracking will be used to determine the location of a user who requires assistance from a situation.

2.6. Functional and Non-Functional Requirements.

2.6.1. Functional Requirements

This is the detailed description of the functionalities of a System, how it will react to certain input, and how the System should behave in certain conditions.

Information-Oriented

- Users can view forums and get the latest information from the news feed section.
- Administrators can view all reported incidents.
- Administrators can generate reports based on reported incidents.
- The system can retain users' activities.
- The system can track and provide real-time information on citizen safety.
- Users can view messages with their emergency contacts (Friends and Family).

Process-Oriented

- Users can login to system to access their profile.
- Users can sign up on system for an account via the proposed mobile app.
- The system must allow registered users to only view their own related information such as emergency contact details.
- Users can report an incident.

- Users can request emergency services.
- The system alerts the administrators to report crime incidents.
- Users can update their profile information.
- Users can add, edit, or delete an emergency contact.
- Emergency contacts are notified when a user is in SOS mode.

2.6.2. Non-Functional Requirements

These are the constraints put on the functionalities of a System. A system is basically useless if the non-functional requirements are not met. The following the non-functional requirements of the system:

Operational

- The system should be platform independent, that is, it must run on android and iOS mobile devices.
- System should notify users including emergency contacts when there is a system process update.
- Administrators are notified of a newly reported incident.

Performance

- Any interaction between the user and the system should not exceed 3 seconds.

Security

- Only administrators can access all user records.
- Only administrators can view reported crime incidents.
- All user's personal details are protected in compliance with the Data Protection Act

Key System User Types

The system user types are as follows:

- The Citizens (Users/Reporters)
- The Administrators (Police officers, People in authority)

3. RESEARCH METHODOLOGY

In this chapter, we will focus on the methodology used in the design and development of the proposed application. Given the project scope, the most suitable system development methodology is the Agile methodology. The Agile approach involves breaking the project into phases allowing for flexibility in adapting to changes in plans or features during application development (Flora & Chande, 2014).

The mobile application will consist of two main components: the user or citizen interface (UI) and the administrative backend for the authorities. The user interface enables users to send information to the database server, while the administrative backend provides access to the information submitted by the users or victims.

The process begins with designing the application, drawing out the mockups of the system design and subsequently employing software, such as Visual Studio Code, for its development. The choice of programming language is Dart, using the Flutter Framework, following the material design.

Figure 1. Agile Methodology.

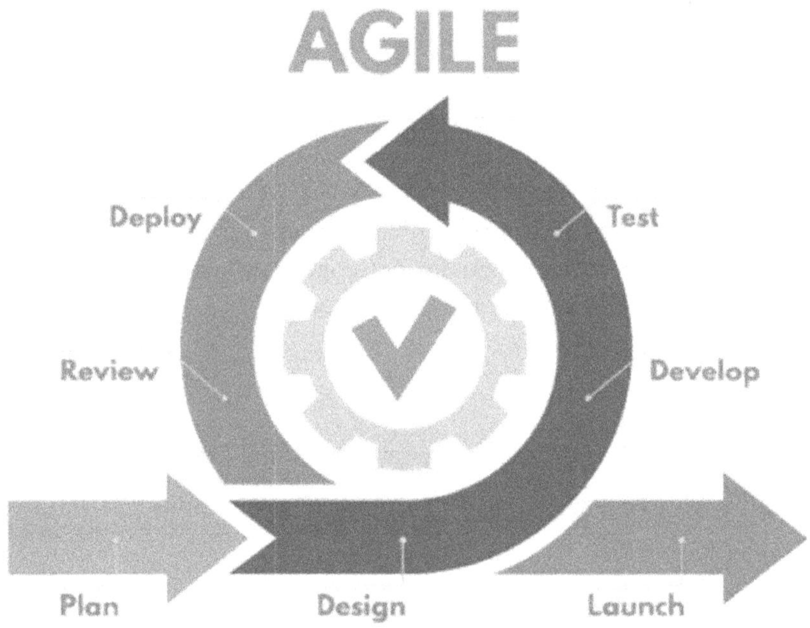

3.1. Justification of Model

Flexibility and Adaptability: One of the advantages of using Agile methodology is its flexibility and adaptability to changing requirements and priorities. This approach embraces changes in the project scope without major disruptions.

Iterative Development: Agile follows an iterative approach, breaking the project into smaller increments. This allows for the early delivery of working features, promoting continuous feedback from users. It ensures that the project is continually refined based on real-world feedback and changing needs.

Early Detection of Issues: Agile' methodology's short development cycles facilitate the early detection and resolution of issues, minimizing the risk of major problems arising later in the development of the mobile app. Regular testing and continuous integration of code leads to higher software quality.

Continuous Improvement: Agile encourages regular retrospectives, where the perfor-mance of the product is reflected upon and areas for improvement are identified. This focus on continuous improvement and refinement enhances overall project perfor-mance over time.

3.3. System design and Architectire

Design Technique

The system design was achieved through a systematic process that involved gathering requirements and developing various architectural components. This includes creating a user interaction design architecture, a use case diagram, an entity-relationship diagram, user activity flow and activity diagrams. The primary focus is on designing a mobile-based crime reporting system, and UML (Unified Modeling Language) was used as the diagrammatic representation to specify the Object-Oriented Programming design.

UML is a powerful tool that allows the visualization of systems enabling the creation of models at different levels of detail. In this research, we used UML diagrams to provide a clear and comprehensive understanding of the system.

Object Oriented Design (OOD)

In this section, we will be reviewing the platform user interaction, use case model, activity diagram, user flow diagram and databased design of the proposed system.

Advantages of using object-oriented design (OOD) are as follows:

- Structured Planning and Design: OOD facilitates a well-structured approach to planning and designing software systems, leading to better code maintenance.
- Object Reusability: Objects can be reused across different programs, promoting code efficiency, and reducing development time for similar functionalities.
- Ease of Maintenance: The modular nature of objects allows for separate maintenance, making it easier to locate and address issues or bugs within the system.
- Modularity: Each object functions as a distinct entity with its own internal workings isolated from the rest of the system.
- Extensibility: Introducing new features or adapting to changing environments can be accomplished by introducing a few new objects and modifying existing ones, rather than reworking the entire codebase.

Platform User Interaction

The result of this research project will be a mobile app which will be used by the public (citizens). A user installs the mobile app on their phones their preferred app store, given the device meets the minimum requirement for installing the software. Once installed, the user registers their information, all of which is stored on the database (Firebase). The user then has access to the public safety app features. They can proceed to add emergency contacts, update their profile information, or give permission to some inbuilt services on the app such as access to Google Maps location.

Figure 2. User interaction design architecture

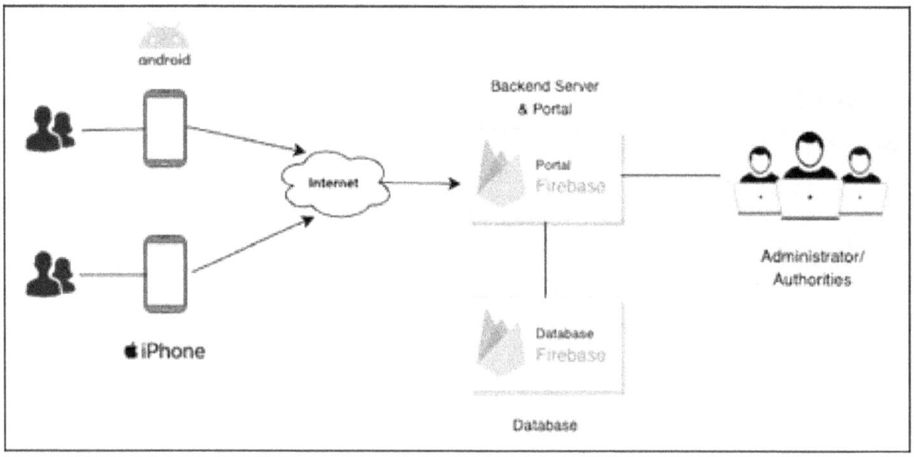

Use Case Diagram

A Use case describes the activities that a user can perform within a system (Bittner, Spence 2003). This is usually represented by horizontal eclipses. In a use case diagram, a user, described as an actor is a person, organization or an external entity that plays a role in one or more interactions with the system. Below is the system's use case diagram showing the proposed system's actors and the actions they can perform on the system.

Figure 3. System Use Case Diagram

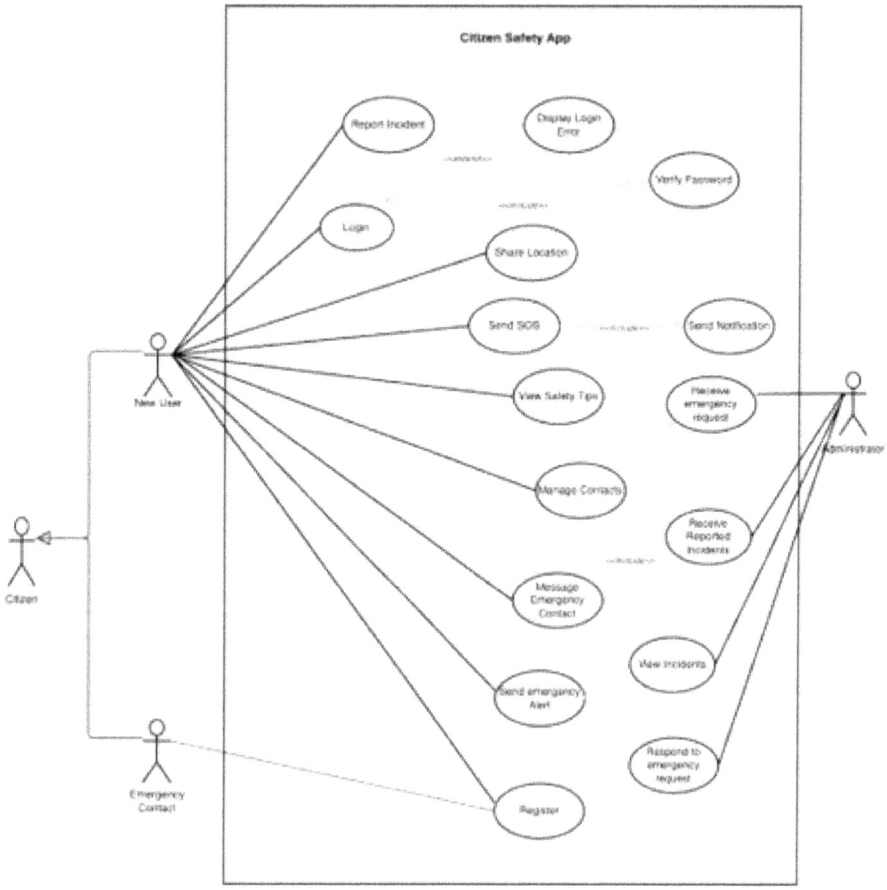

Activity Flow Diagram.

Figure 4. Activity Flow Diagram

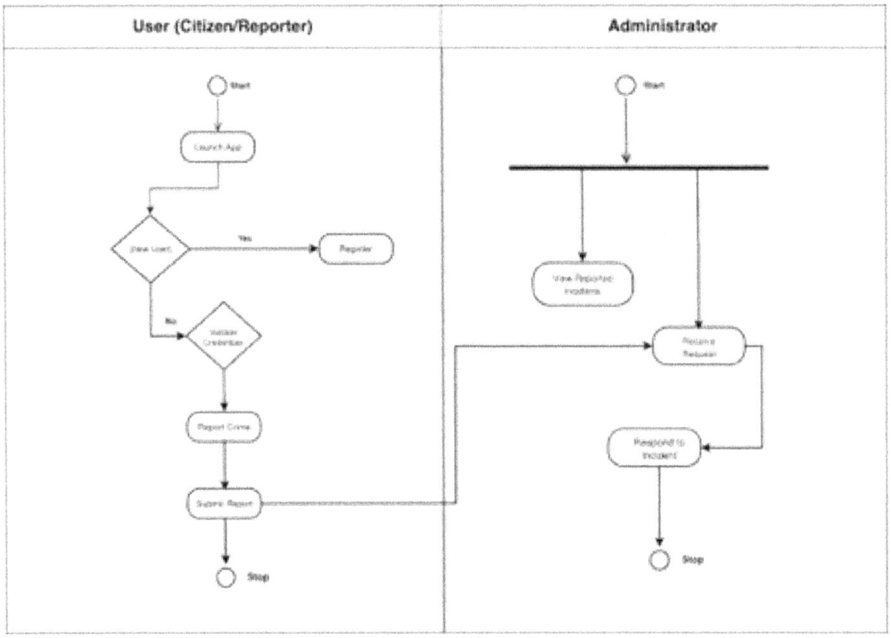

Figure 5. Activity Diagram

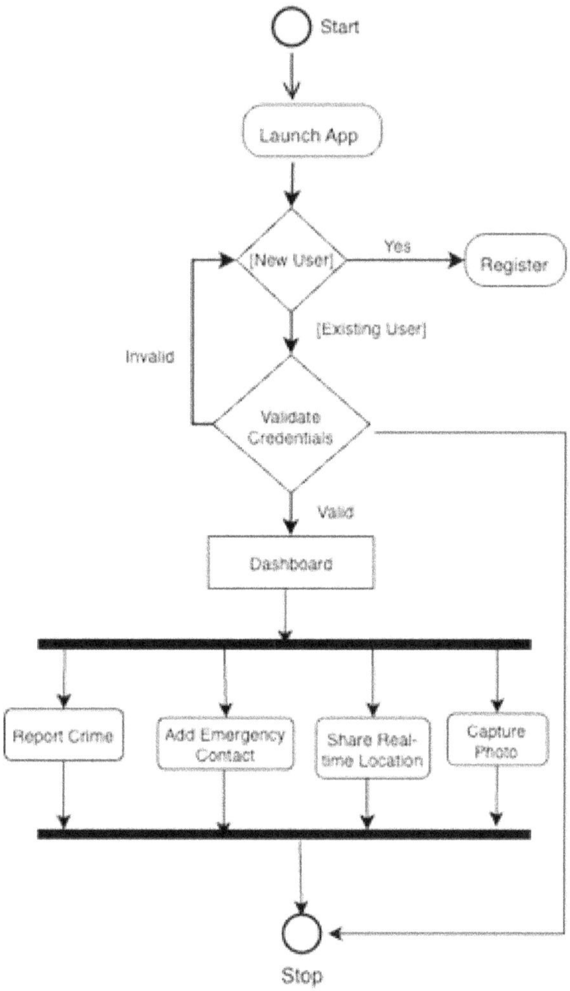

Figure 6. Database Design

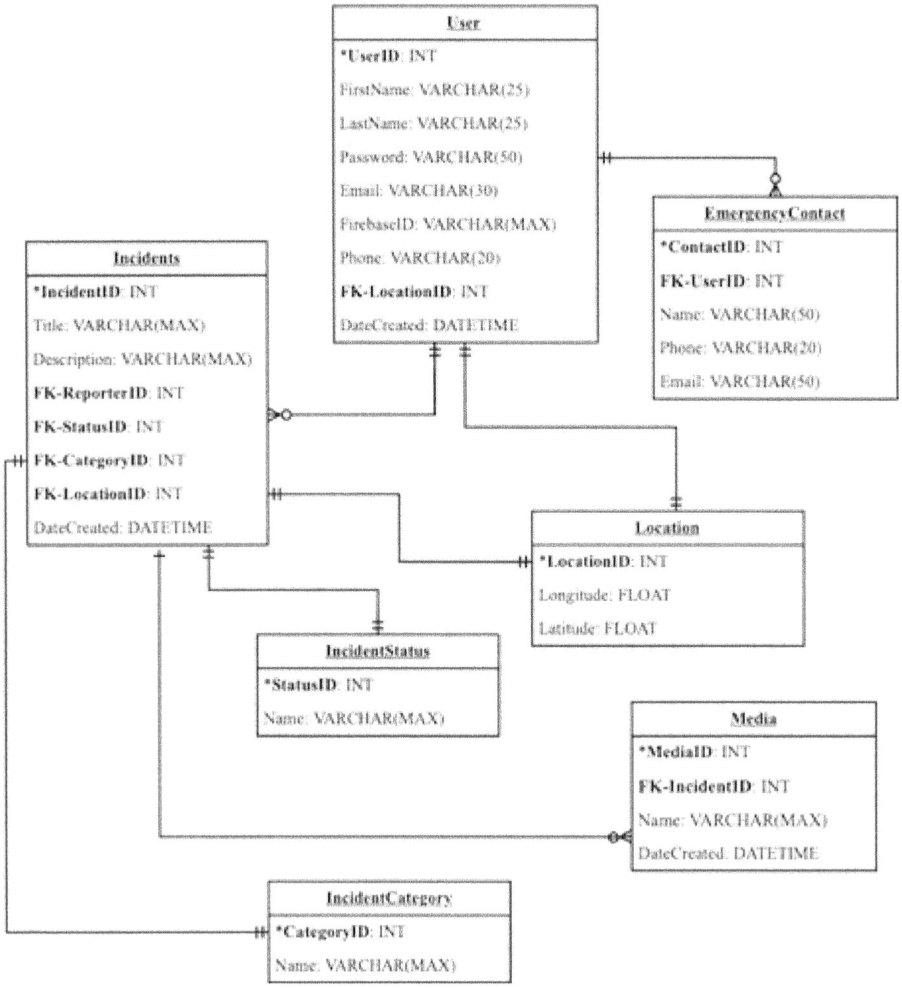

User Interface (UI) Design.

The system users interact with the application interface. Therefore, the UI design is an important thing to succeed in the system. Important interfaces have been shown in Chapter 5 of this research. Please refer to Chapter 5.

3.4. Application Implementation

In this section, we will focus on the tools and technologies used in the development of the mobile application and the admin dashboard for the proposed mobile and web system.

Development Technologies and Framework used.

There is a wide range of development technologies and frameworks available for creating mobile-based applications. In this section, the various technologies and frameworks are outlined, along with their advantages and disadvantages. Also, the primary technologies and frameworks that are well-suited for the proposed system are identified.

Flutter is an open-source mobile application development framework developed by Google. It is a modern framework that allows the creation of high-performance, cross-platform mobile apps for both iOS and Android and it uses a reactive programming model. This means that the UI is constantly being rebuilt in response to changes in the underlying data. This makes Flutter apps feel very responsive and smooth. Flutter also has a rich set of widgets that can be used to create beautiful and com-plex UIs. These widgets are all written in Dart, which is a modern programming language that is designed to be concise and expressive. It is based on the Dart programming language and allows the creation of high-performance, cross-platform mobile apps for both iOS and Android.

Dart is a client-optimized programming language for developing fast apps on any platform. It is de-signed to be concise, expressive, and scalable. Dart is used to build web, mobile, and desktop applications.

Firebase is the designated backend technology used for this research. Firebase offers two types of non-relational databases - Firebase Realtime Database and Cloud Fire-store. Both databases allow to store and sync data between users in real-time using listeners that get invoked every time a change is observed. The Firebase Realtime Database is a cloud-hosted database. Data is stored as JSON and syn-chronized in real-time to every connected client. When integrated with our proposed system, the app automatically receives updates with the newest data.

Cloud Firestore is a flexible, scalable database for mobile, web, and server development from Firebase and Google Cloud. Like Firebase Realtime Database, it keeps data in sync across client apps through real-time listeners and offers offline support for mobile and web for building responsive apps that work regardless of network latency or Internet connectivity.

3.5. Data Collection Method

The data collection method is a critical point in the research method as this helps to collect user insights into the proposed system. The primary form of collecting user feedback in this research work is the use of Google Forms. Google Forms is a survey administration software included as part of the free, web-based Google Docs Editors suite offered by Google (Bhalerao,2015). Survey will be sent out in the form of questions asking users for their opinion on existing safety measures as well as their reaction towards a citizen safety app. Questions are asked in the form of multiple-choice questions, dropdown, linear scale etc.

4. RESEARCH SURVEY RESULT AND DATA ANALYSIS

Survey analysis involves the process of turning user survey data into meaningful conclusions from data obtained. With the use of illustrative tools like bar charts and pie charts, the insights gathered from the survey are represented in this chapter (Bearfield et al, 2023). .

The primary objective of this survey was to gain valuable insights from users regarding their preferred courses of action when faced with potentially dangerous situations. 11 survey questions were asked and sent to users from all the six geopolitical zones of Nigeria. In total, a dataset of 164 responses was collected.

4.1. Online Survey Questions

Demographic Information:

- What is your gender?
- What Geopolitical Zone do you live in?

Mobile Device Usage:

- Do you own a smartphone?

Area Safety Issues:

- Do you encounter safety issues in your area?
- If yes, what kind of safety issues do you encounter?

App Features and Functionality:

- What features would you expect in a mobile app designed to enhance public safety?
- How important are features such as emergency alerts, real-time location tracking, and reporting incidents for public safety?

Reporting Incidents:

- How would you handle an ongoing crime if you witnessed one?
- How likely are you to use a mobile app to report incidents related to public safety?
- If there was a mobile app with the ability to capture/record an incident, contact the authorities and send an SOS to your friends and family, how likely are you to use this app?
- Would you report an incident and share the information or media captured with the police?

3.2. Responses to the Survey

1. What is your gender?

The objective of this question is to collect information about the gender distribution among users and determine which gender is more likely to utilize the application frequently.

Figure 7. Pie chart representation of gender

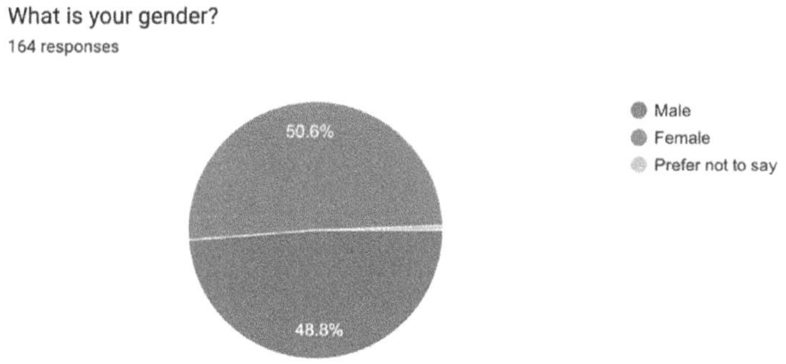

2. What Geopolitical Zone do you live in?

Collecting geopolitical survey responses from across Nigeria is essential for a well-informed and inclusive process, ensuring that the user data used in the development of the application will address the diverse needs and provide insights into the public safety issues faced by the users in the different parts of the Country.

Figure 8. Pie chart representation of the user geopolitical zones

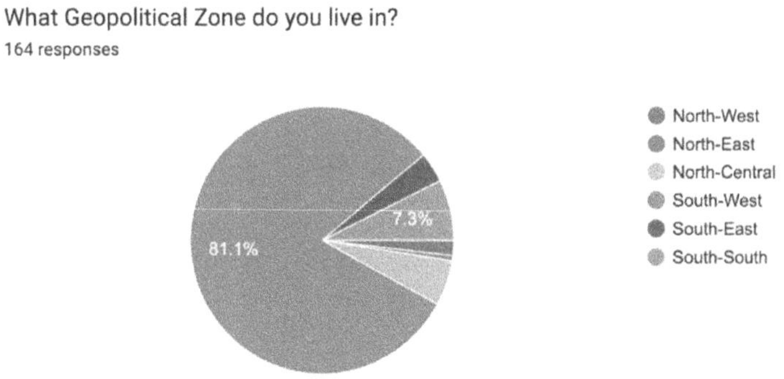

3. Do you own a smartphone?

The proposed mobile application will only be available on smartphones which have some specific built-in features that support user safety data collection; therefore, it is important to know the number of users who have a smartphone device.

Figure 9. Pie chart representation of users who own a smartphone.

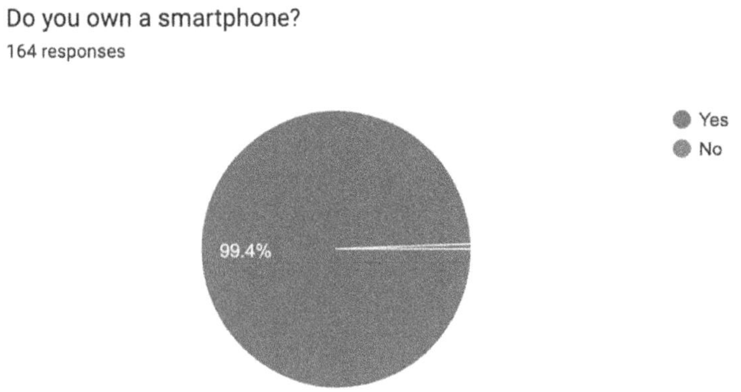

From the representation in Figure 9 showing user feedback, 99.4% of the users own a mobile phone, bringing this survey question to a conclusion that the majority of the potential app users have the minimum required device.

4. Do you encounter safety issues in your area?

Below is the survey response to whether users face safety issues in their area. 73.4% of the respondents face personal safety issues in the areas they live in, while 26.2% have stated that they do not face safety issues in their area.

The conclusion from the chart is that most people in the different parts of the country face some safety issues.

Figure 10. Pie chart representation on encountering safety issues.

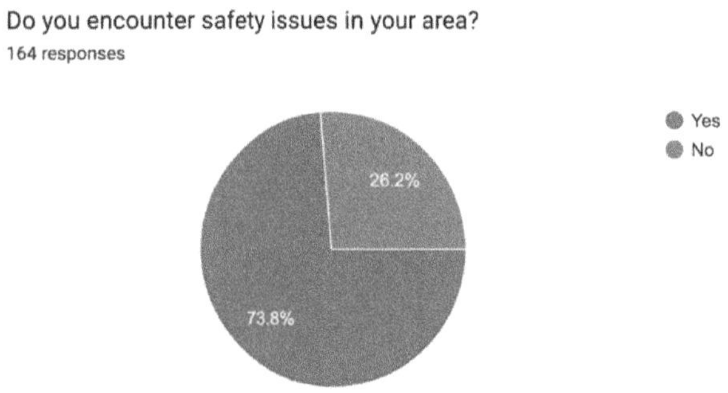

5. If yes, what kind of safety issues do you encounter?

To collect more details from the survey question above "Do you encounter safety issues in your area?", users who faced safety issues in their areas were asked to give more details as to the specific or dominant safety issues they face in their area. Below bar chart shows the respondents' feedback.

The dominant crime faced in the different parts of the country is armed robbery with 58.3%. Followed by general violence which is 46.7%. Next is the safety issue of kidnapping which totals at 32.5%, and then cases of rape, others, theft, cultism, and house burglary which are 9.2%, 10%, 0.8%, 0.8% and 0.8% respectively.

Figure 11. Bar chart representation of safety issues faced.

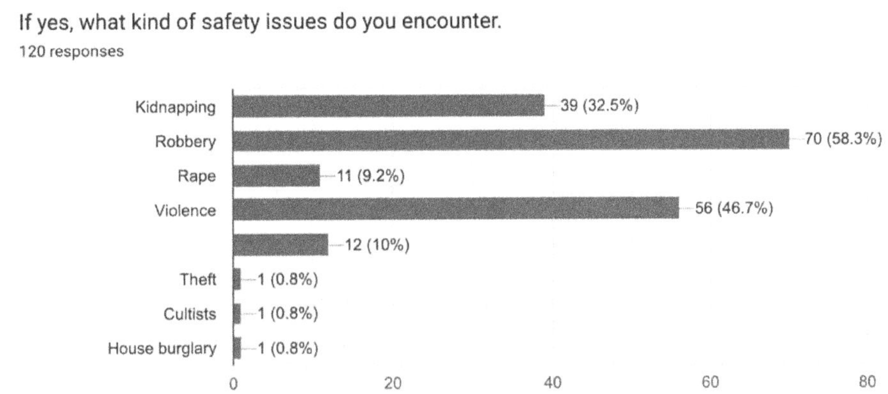

6. What features would you expect in a mobile app designed to enhance public safety?

In this question, users were asked to select mobile app features they would expect to find in a public safety mobile application. Most of the users thought Location Tracking is a key feature in any safety app with a percentage rating of 70.7. The other response shows that other features they would like to see in a safety mobile app include Push Notifications 20.7%, direct connect with the police and authority 58.5%, ability to report crimes and send media 51.8%, inclusion of a SOS panic button 50.6%, and having safety tips on the app 45.7%.

Figure 12. Bar chart representation of expected app safety features.

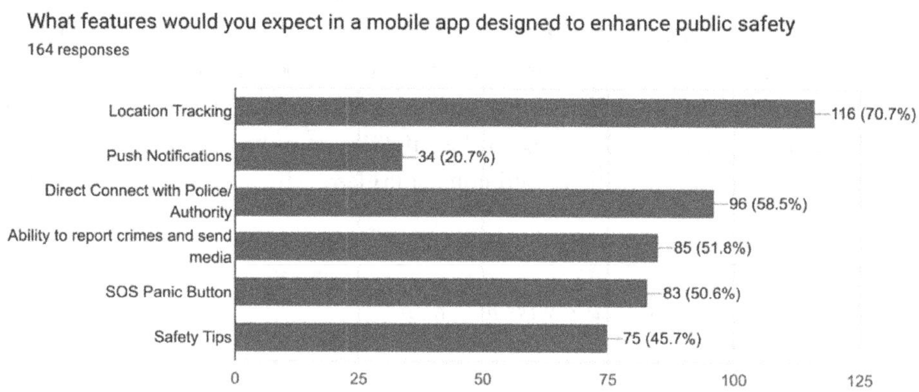

7. How important are features such as emergency alerts, real-time location tracking, and reporting incidents for public safety?

For this question, 100% of the respondents think that emergency alerts, real-time location tracking, and reporting incidents for public safety are very important features to have on a public safety app. Below is the diagrammatic representation of the response.

Figure 13. Pie chart representation of importance of safety app features

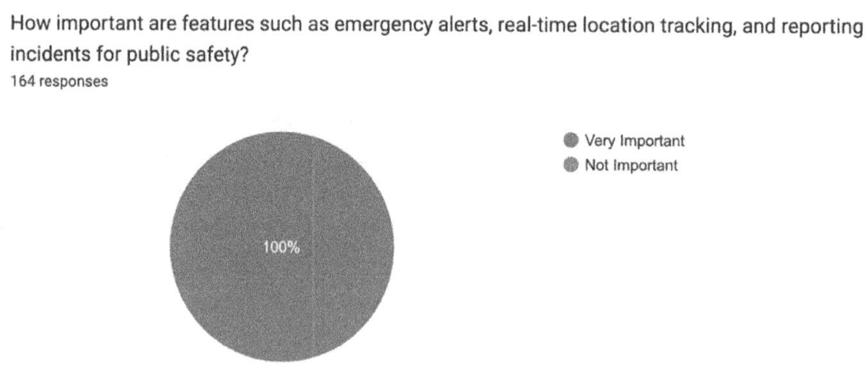

8. How would you handle an ongoing crime if you witnessed one?

In this question, the users were asked how they would respond to a real time incident if they witnessed one. Below bar chart shows that coming in second place as the next preferred option besides calling the police or authorities – 69.5% would be to record the incident using a public safety mobile app. This has a respondents' rating of 60.4%, with other options being going to the nearest police station – 11.6% and calling a friend or family at 18.3%.

Figure 14. Bar chart representation of user response to ongoing crime.

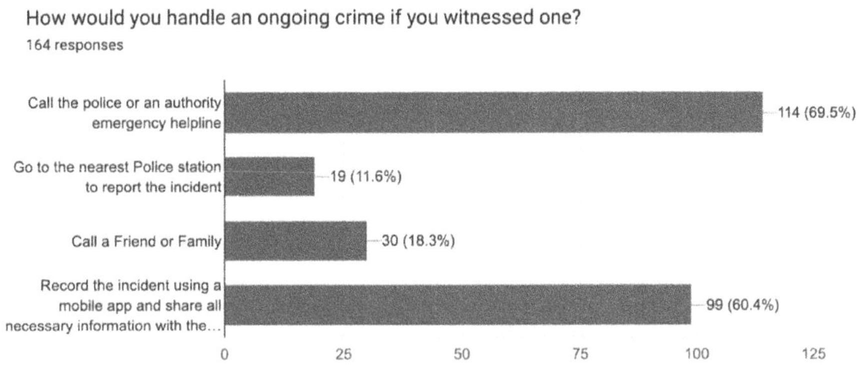

9. How likely are you to use a mobile app to report incidents related to public safety?

This question was asked to understand if users would likely use a public safety application to report crime incidents. The below chart shows that 89.6% of the respondents are very likely to use a public safety mobile app while 10.4% stated that they are not likely to use the app.

Figure 15. Pie chart representation of User likeliness to report incidents.

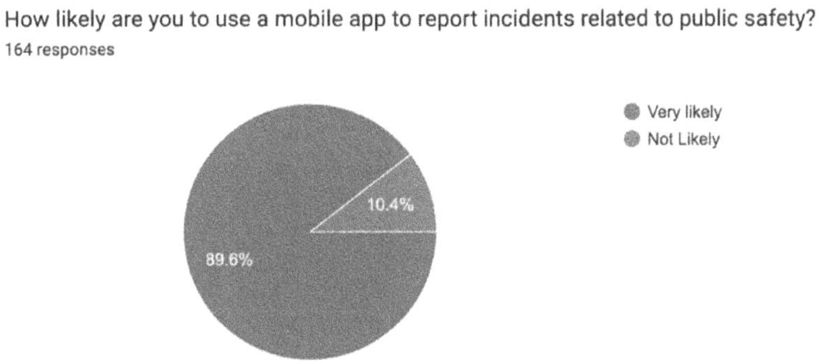

10. If there was a mobile app with the ability to capture/record an incident, contact the authorities and send an SOS to your friends and family, how likely are you to use this app?

This question asks if users report any crime incidents via a public safety app if such an app were to be developed. 1.8% of the respondents' state that they will not use the app while most of the users, with a percentage rating of 98.2, state that they will use the app to report crimes and incidents.

Figure 16. Pie chart representation of Application Usage.

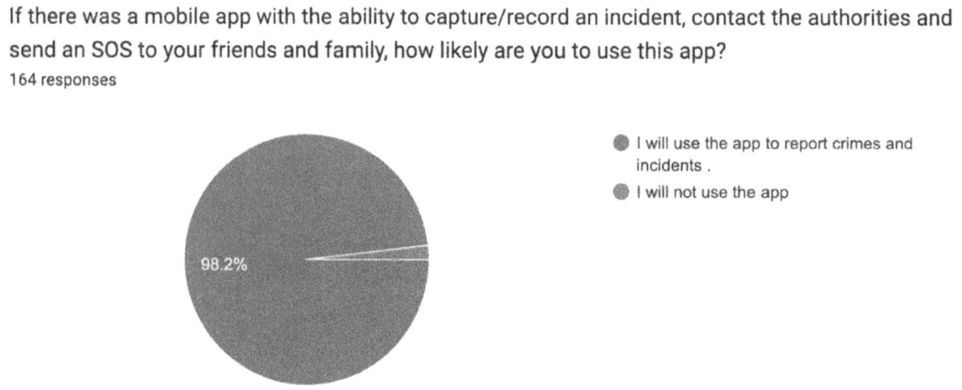

11. Would you report an incident and share the information or media captured with the police?

Finally, users were asked if they would report an incident and share supporting information for a crime with the police to assist with enhancing public safety. 12.9% of the users reported they will not report or share information with the police while most of the users with a percentage rating of 87.1 stated that they will report an incident and share media and information with the police.

Figure 17. Pie chart representation of reporting a crime.

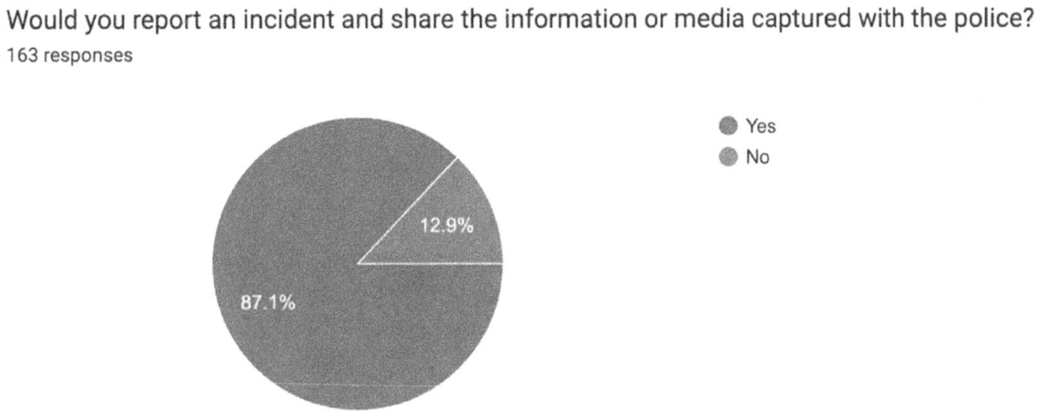

5. DESIGN AND IMPLEMENTATION

System implementation is the process of defining how the system should be built. It involves taking the design and development work from the earlier stages of the software development life cycle and making it operational (Bueno, Pereira, & Fabricio, 2018).

Based on the survey feedback from users in Chapter 4, the proposed system was de-signed and implemented. Material Design, which is an adaptable system of guidelines, components, and tools that support the best practices of user interface design was used in designing the system. The app was implemented using Flutter's rich set of pre-designed widgets and customizable components which provide the canvas from which the user interface, navigation, and visual appeal are built upon.

With design principles established, the focus shifts to the implementation phase, where the flexibility and performance of Dart programming with Flutter's framework (using the material design) come to the fore. In this chapter, we will go over the coding, from integrating real-time location tracking and emergency alerts to facilitating two-way communication channels between users and authorities. Firebase's backend services allow data management, authentication, and real-time updates that underpin the app's functionality. Through a step-by-step breakdown of code snippets and widget (UI) screenshots, the following subsections show insights into the technical orchestration used to transform the designs into tangible app features.

5.1 Splash Screen and Application Walkthrough

The splash screen is the initial view a user has with the application. This is displayed for 3 seconds and then if the system identifies a user as a first-time user, they are redirected to the walkthrough screens, otherwise they are redirected to the login screen. A new user may wish to skip the walkthrough and navigate directly to the login screen.

Figure 18. The splash and walkthrough screens

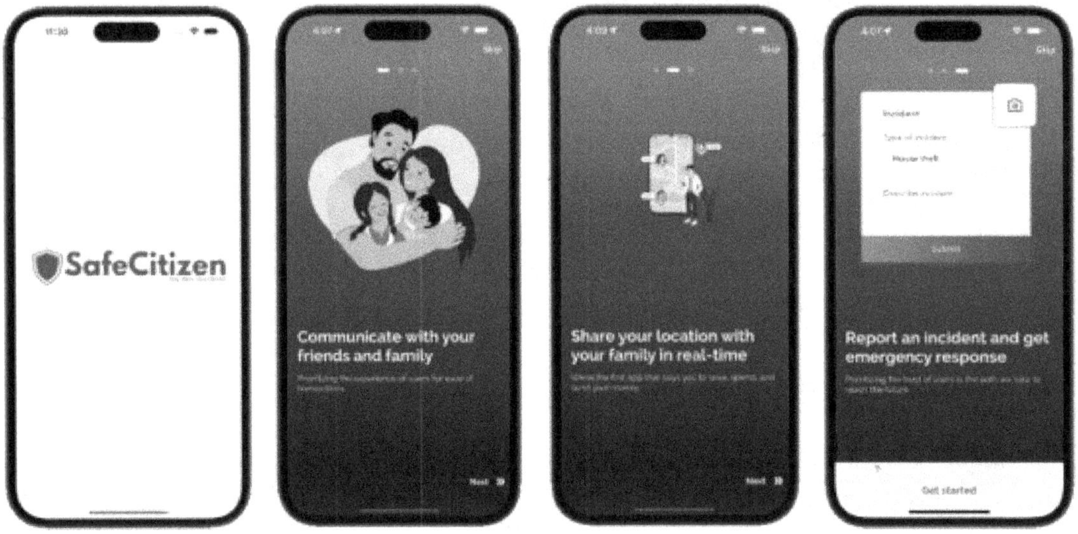

5.2. Login Screen and Registration Screen

The login screen is the interface that grants access into the mobile app. An existing user is required to input their login credentials to be able to access the app. For first time users, a permission request notification for allowing in app notifications is dis-played to the user. All user authentication is done against the firebase real time data-base. Also, on the login screen are other options such as resetting a forgotten pass-word or navigating to the user registration screen.

The registration screen displays a form for accepting user details. All fields are re-quired for a suc-cessfully registration, and simple user validations such as email and phone number format checks are carried out. The user information collected via the screen is stored on firebase.

Figure 19. The login and signup screens.

5.3. The Administrative portal – User Records

When a user registers on the app, their bio data is stored on Firebase and the associat-ed email address and password are saved on the authentication table on Firebase. Dur-ing login, the inputted credentials are validated against the details on Firebase. Below table shows user records on the authentication table.

Figure 20. The administrative portal

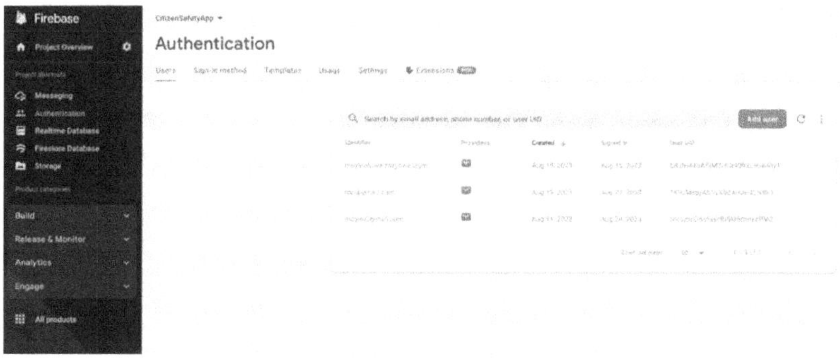

5.4. User home screen

The system navigates a user to the home screen when they have successfully logged on the application. On the screen, the user can see the map with an indication pin of their exact real time location and route movements.

Figure 21. The home screen

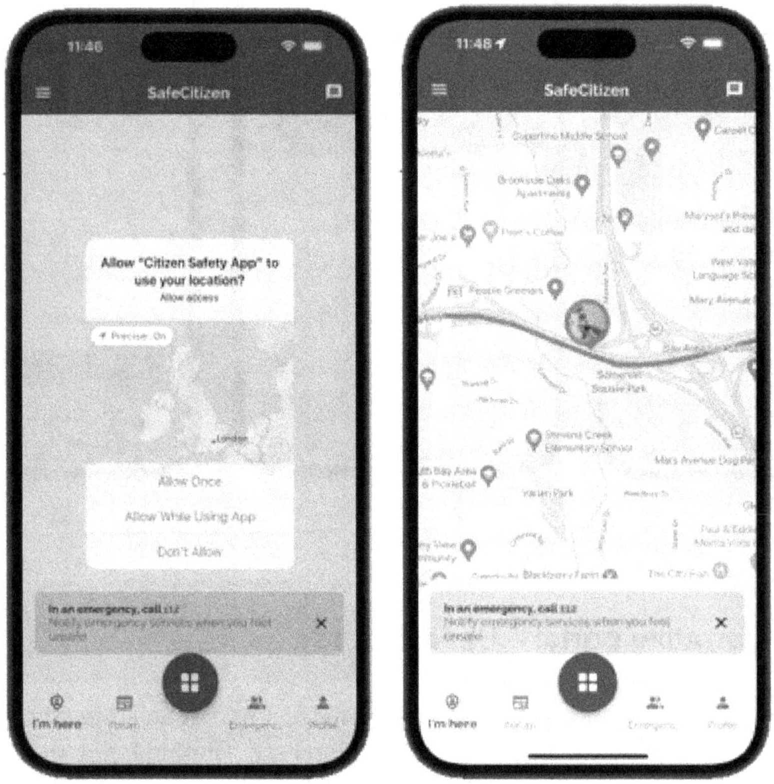

5.5. . Menu screens

The user has two choices of accessing the menu options that the app provides. Each of the menu screens can be accessed via the home screen. One is on the top left corner of the home screen, and the other on the center floating bottom navigation bar.

Figure 22. The menu screen

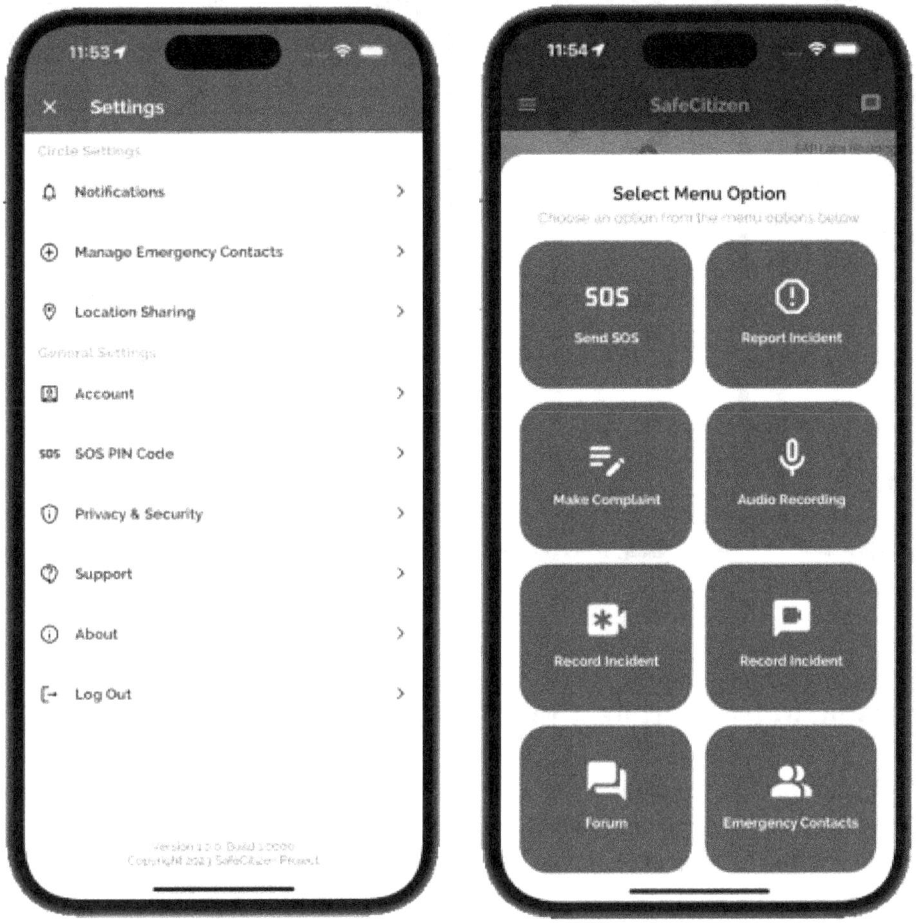

5.6. Emergency Contacts Management

The emergency contacts screens, shown below, allow all management of family and friends that the system contacts should the user be in danger. The screens comprise of the initial view for messaging where a user can have conversations with their added contacts. In the other view, a user can view existing contacts as well as add a new emergency contact.

Figure 23. The emergency management screen

5.7. The Administrative portal – Emergency Contact Records

On the Firebase real-time database, emergency contacts saved on the mobile app for each user can be found. The details saved on the emergency contact table include the contact's e-mail, their name, and phone number. In the case of an emergency or when a user sends out an SOS alerts each emergency contact is alerted of the situation along with the user's last known location.

Figure 24. Emergency Contacts - The Administrative Portal

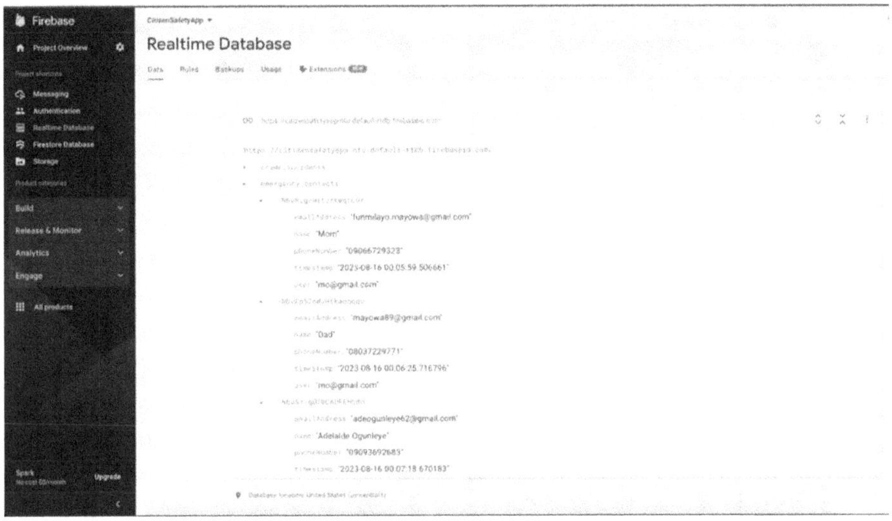

5.8. Profile screen

On the profile screen, the user can update their profile image and personal information. The details that can be updated are the phone number or email after inputting the valid password.

Figure 25. The profile screen

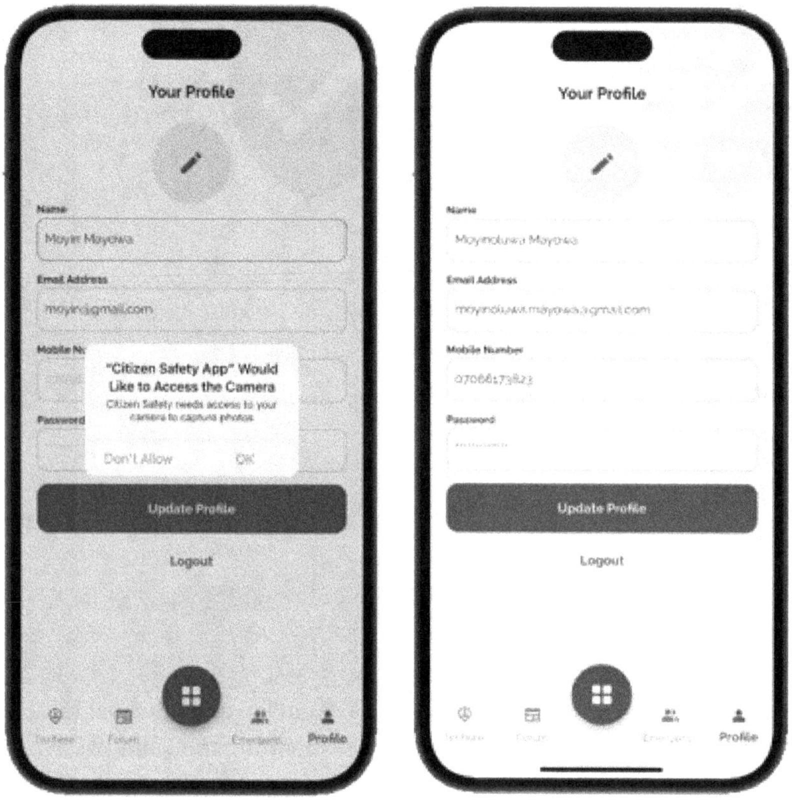

5.9. SOS Screen

The SOS screen sends the user's current location to their saved emergency contacts as well as to the authorities and calls out to them for immediate help.

Figure 26. The profile screen

5.10. The Administrative portal -SOS

The mobile app has the SOS alert feature. When a user is in a situation where they require help, they send out an SOS alert which notifies the authorities as well as their emergency contacts. The SOS alerts are sent along with the user's last known location.

Figure 27. The Administrative portal

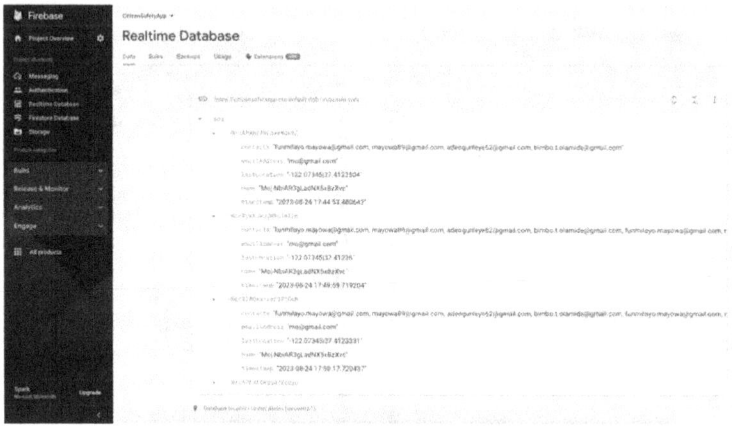

5.11. Report a crime screen

On this screen, a user can submit an incident report. The user selects the type of inci-dent they are reporting and allows them to be able to attach a supporting media.

Figure 28. The incident report screen

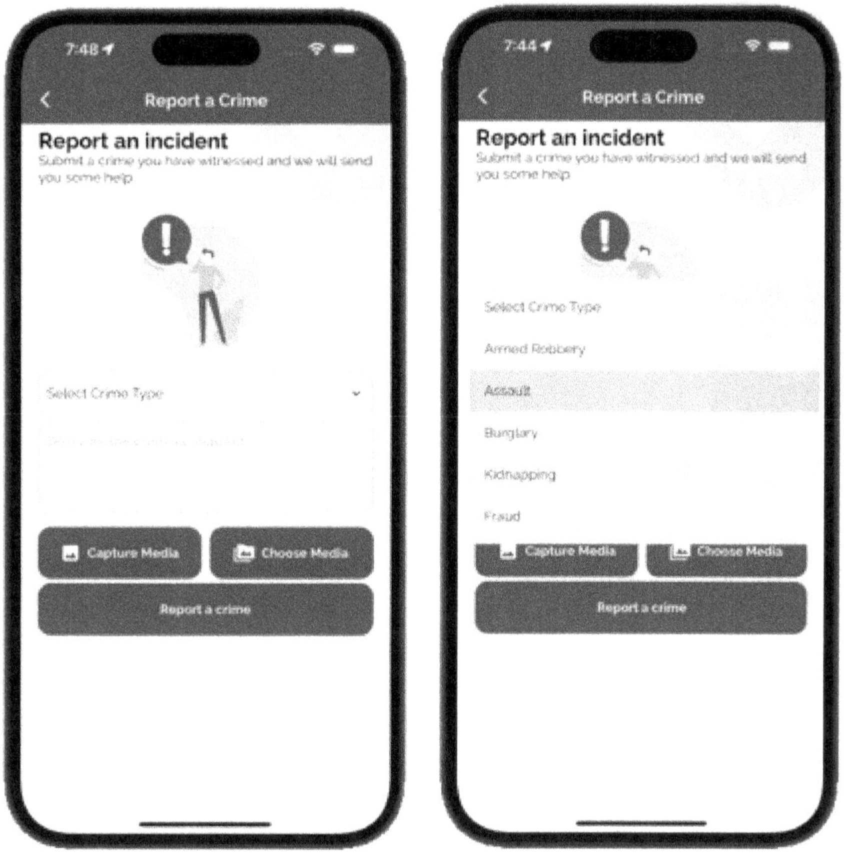

5.12. The administrative portal – incident report

The Realtime database has a collection for incident reports. Users can submit inci-dents which they are a witness to. The details submitted in an incident report are the type of crime, the description, image, or video – if available, the location of the incident and the timestamp.

Figure 29. Incident reports on administrative portal.

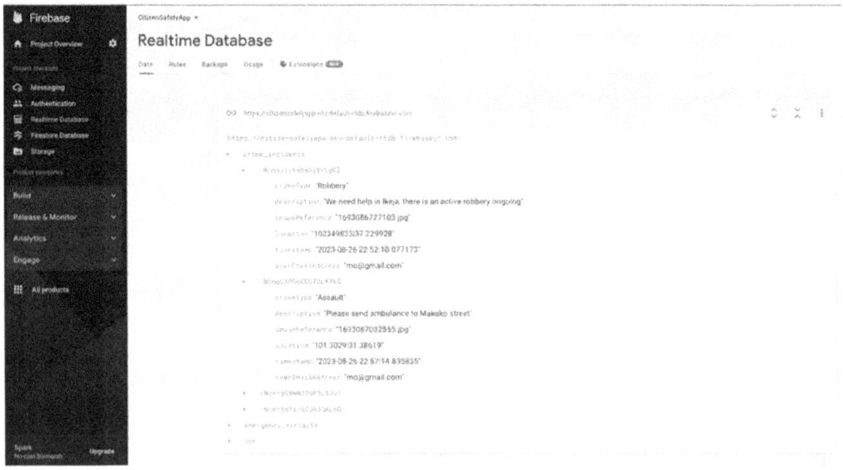

5.13. The Administrative Dashboard – Firebase (Other collections)

The admin backend, via Firebase is the platform where the admins can view all re-quests and users registered on the application. All data including incident reports, SOS, and supporting media are saved on the real-time and Firestore database. Via the portal, in app push notifications can also be sent to users to inform them of updates and other important information.

5.14. Cloud Firestore – Emergency Contacts

The Firebase Cloud Firestore is a flexible NoSQL database for easily storing, syncing, and querying data for mobile and web apps. The messaging activities on the app are managed and stored on the cloud Firestore.

Figure 30. Administrative cloud messaging for emergency contacts

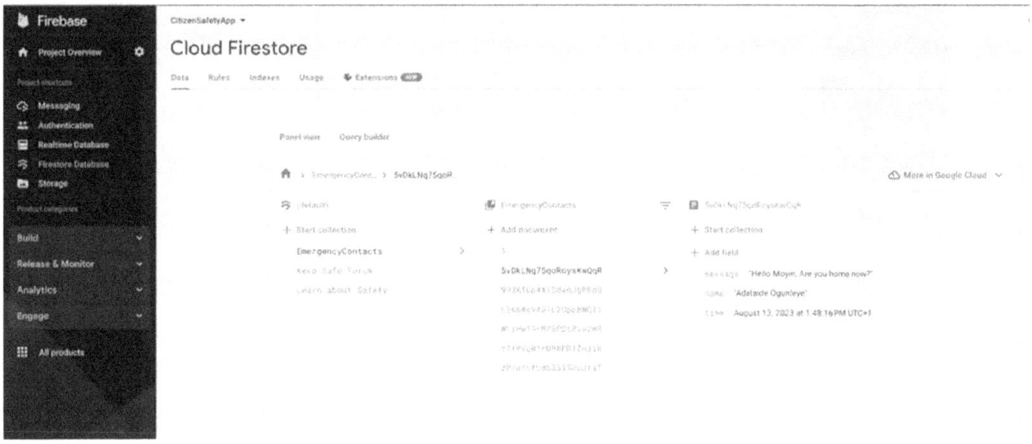

5.15 Firebase Storage

Cloud Storage for Firebase is a powerful and simple object storage service used for uploading and downloading files for Firebase apps, regardless of network quality. The storage folder contains three folders - images, recordings, and video, all of which are media uploading as supporting files for incident reports, audio recording or video recordings.

Figure 31. Administrative storage for media folder

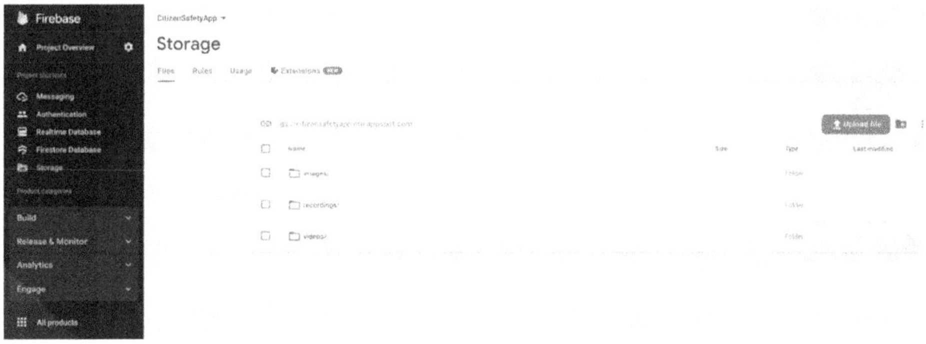

5.16 Firebase Storage – Images Folder

The images folder contains photos, in .png or .jpg format of supporting attachments for incident reports from users via the app.

Figure 32. Administrative storage for images

5.17 Firebase Storage – Audio Recordings Folder

The audio recordings folder contains files in .mp3 format of submitted audio from users via the app.

Figure 33. Administrative storage for audio recordings

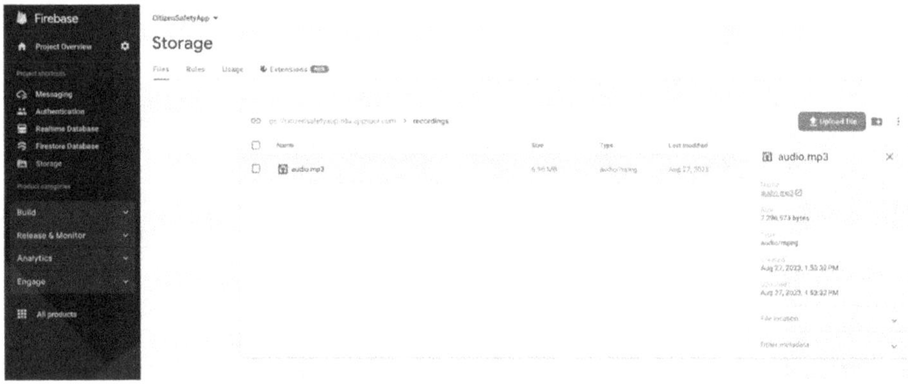

5.18 Firebase Storage – Video Recordings Folder

The video recordings folder contains files in .mp4 or .mov format of submitted video from users via the app.

Figure 34. Administrative storage for video recordings

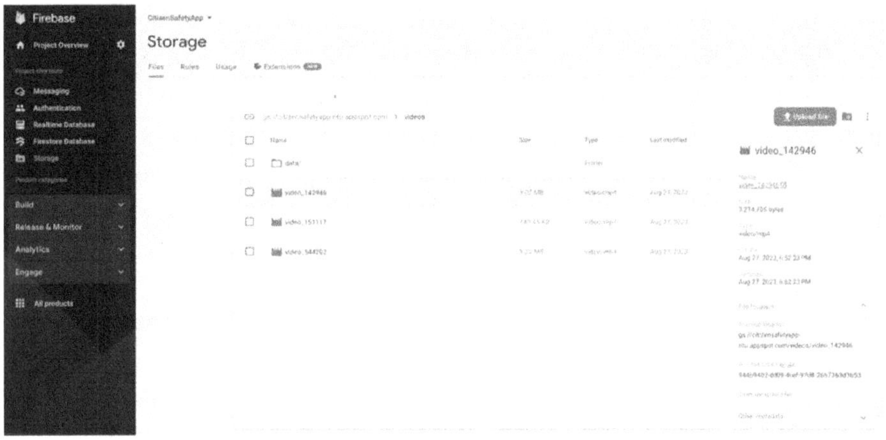

5.19. Other mobile app screens

Figure 35. Support and privacy screens

6. SYSTEM TESTING

App testing is the process of assessing and confirming a software application's usabil-ity, performance, and security (Zein, Salleh, & Grundy, 2016). Before making an app available to the public, it is crucial to find and correct bugs, errors, and other problems. This is accomplished during the software development life cycle.

There are several tests that can be performed on a mobile app, some of these include:

1. Functional Testing: This ensures that an app works as intended and satisfies user needs. It comprises testing the user interface, user experience, and functionalities.
2. Performance Testing: This type of test examines how quickly, steadily, and effi-ciently an application runs under a variety of scenarios, such as high traffic, limited network bandwidth, or a significant data load.
3. Usability Testing: This type of testing measures how user-friendly the app is by testing its ease of use, learnability, and efficiency.
4. User Acceptance Testing: In this type of test, actual users evaluate an app's usabil-ity, functionality, and user experience in a real-world setting.
5. Compatibility testing: This verifies that the program works with a range of hard-ware, software, browsers, and screen sizes.

6.1. Functional Testing

Functional tests are used to check that a system meets its functional requirements. By verifying the system's inputs, outputs, and behavior, functional testing makes sure the system behaves as planned (Kemmerer, 1985).

Below is a table showing the test cases, expected result and actual result carried out on the Citizen safety mobile app.

6.1.1. Test Plan

Table 3. Test plan to check for bug free code.

Test plan	Test plan 1
Test name	Check the programming code to confirm it is bug free
Test Purpose	To find out any errors in my code and to successfully run the application
Test Condition	Run the app on physical and virtual devices
Test Data	Running the app on a physical or virtual device
Expected Result	Application launches successfully and is bug free.

Table 4. Test plan for login screen.

Test case	Test description	Outcome	Result
Successful login	User enters valid credentials	Login is successful	PASS
Failed login	Login is successful	Login fails	PASS
Blank Fields Login	User leaves username/password blank	Login fails	PASS
Register screen navigation	User clicks on "Sign up" button	System navigates user to registration screen	PASS

Table 5. Test plan for registration screen.

Test case	Test description	Outcome	Result
Successful Sign-up	User signs up with valid details	Sign-up is successful	PASS
Existing User Sign-up	User signs up with existing user data such as phone number or email address.	Sign-up fails	PASS
Blank Fields Sign-up	User leaves username/password blank	Sign-up fails	PASS
Login screen navigation	User clicks on "Login" button	System navigates user to login screen	PASS

Table 6. Test plan for home screen.

Test case	Test description	Outcome	Result
Clicking on bottom navigation bar item	User is logged in to app or clicks on "I'm here" option	User is navigated to the home screen.	PASS
Click on side menu hamburger icon	The side drawer displays the menu screen	The side menu screen is shown to user	PASS
Click on center menu options icon	User clicks on the bottom navigation bar menu sheet	The menu bottom sheet is displayed	PASS
Click on message icon	User clicks on the message icon on the home screen	User is navigated to message their emergency contacts	PASS
View user real time location	User is shown their location	User can view their live location	PASS

Table 7. Test plan for forum screen.

Test case	Test description	Outcome	Result
Click on the Forum option on the bottom navigation bar	User is navigated to the forum/ news screen	User is navigated to the forum/news screen	PASS
Slide between the Forum and Safety feed (news) tabs	System slides between the forum or the news screen	User can scroll through information on the forum or news screens	PASS

Table 8. Test plan for SOS screen.

Test case	Test description	Outcome	Result
Click on SOS on the menu options	System navigates the user to the SOS screen	User can view the SOS screen	PASS
Click on send SOS button	System sends user location and message to emergency contacts	SOS is sent successfully	PASS

Table 9. Test plan for emergency screen.

Test case	Test description	Outcome	Result
Click on the Emergency contacts option on the bottom navigation bar	System navigates user to the emergency contacts screen	User is navigated to screen	PASS
Click to add a new contact	User enters emergency contacts details	New emergency contact is created successfully	PASS
Send a message to emergency contacts on the group	User inputs a message in the text field	Message is sent successfully	PASS

Table 10. Test plan for profile screen.

Test case	Test description	Outcome	Result
Click on the Profile option on the bottom navigation bar	User clicks on the profile option in the navigation bar	System navigates user to the profile screen	PASS
Input mobile number or email address	User enters the phone number or email address to be updated	User's profile details are updated successfully	PASS

Table 11. Test plan for logout screen.

Test case	Test description	Outcome	Result
App Logout	Click on the Logout button on profile screen	User is logged out of the app successfully	PASS

Table 12. Test plan for incident reporting screen.

Test case	Test description	Outcome	Result
Report an incident	User clicks on the Report an incident option on the menu options	System navigates user to the report an incident screen	PASS
Empty data on submission	The side drawer displays the menu screen	The side menu screen is shown to user	PASS
Missing information report	User submits a report with missing fields	Report submission fails	PASS
Report submission retry	User submits a report, connection fails, then retries	Report submission is successful on retry	PASS

Table 13. Test plan for home screen.

Test case	Test description	Outcome	Result
Navigate to Audio recording option on the menu options	User clicks on the Audio recording option on the menu options	System navigates to Audio recording screen	PASS
Click on start audio recording button	User clicks on start audio button	Audio recording is successfully started	PASS
Click on message icon	User clicks on the message icon on the home screen	User is navigated to message their emergency contacts	PASS
Send audio recording and location by clicking on submit button	User clicks on submit button	Audio and user location are successfully sent to the Firebase backend	PASS

Table 14. Test plan for home screen.

Test case	Test description	Outcome	Result
Clicking on bottom navigation bar item	User is logged in to app or clicks on "I'm here" option	User is navigated to the home screen.	PASS
Click on side menu hamburger icon	The side drawer displays the menu screen	The side menu screen is shown to user	PASS
Click on center menu options icon	User clicks on the bottom navigation bar menu sheet	The menu bottom sheet is displayed	PASS
Click on message icon	User clicks on the message icon on the home screen	User is navigated to message their emergency contacts	PASS
View user real time location	User is shown their location	User can view their live location	PASS

6.2. User Feedback on App

At the completion of the app's development, an Android APK file was shared with 20 users to evaluate their feedback on the completed mobile app solution. Below is the response from the 20 users.

Based on the bar chart figure below, it is gathered that most of the users found the app to be very user friendly.

Figure 36. Bar chart representation of app's user-friendliness

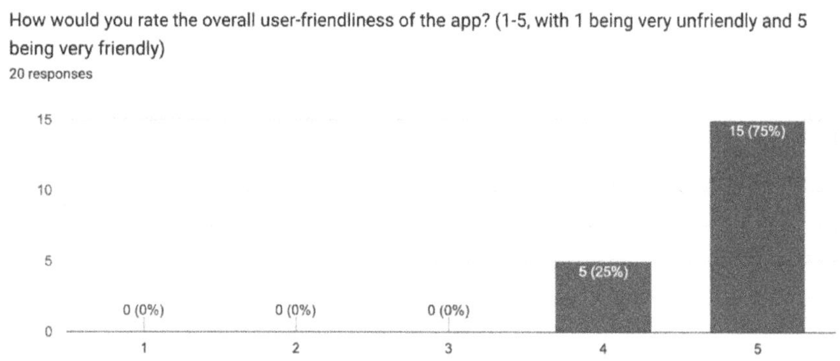

Based on the pie chart below, all users will recommend the use of the mobile app to their friends and family.

Figure 37. Pie chart representation of app recommendation

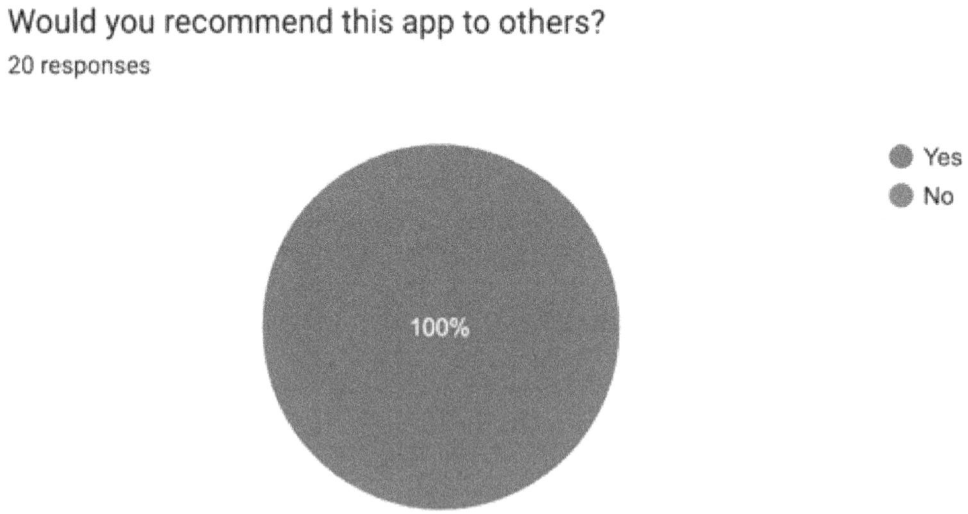

The feedback from the users, based on the pie chart below shows that all the users found it easy to navigate across the app for the features they needed.

Figure 38. Pie chart representation of ease of navigation

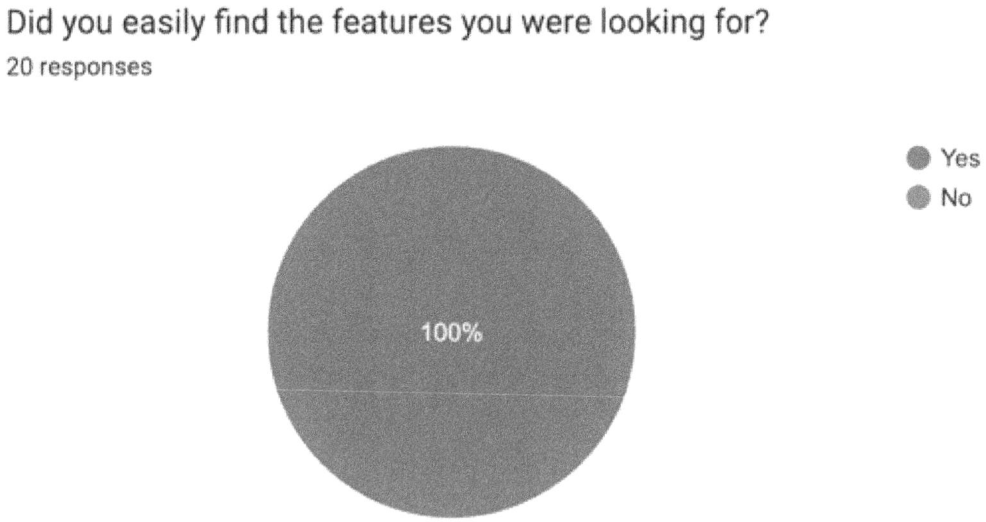

95% of the users answered that the app met its intended purpose of enhancing citizen safety.

Figure 39. Pie chart representation app's requirement satisafaction

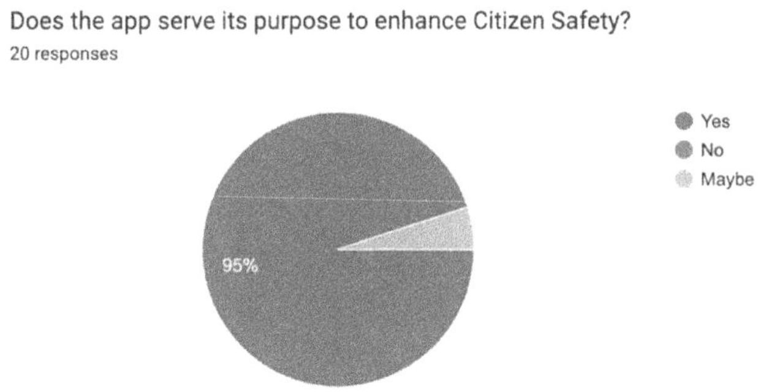

Below are other user suggestions of features that can be factored into future versions of the citizen safety app.

Figure 40. User app suggestions

Do you have any other suggestions to improve the app's functionality and experience?

13 responses

No

None

Add live chat feature with the police

Direct call with local police and hospital

Add live news to the news page so that we can see security news around the world

In-app calls with family

Add a way for users to learn self-defense techniques.

Can you make the app to work without internet connection?

No suggestions, the app is nice

Add a way for users to learn self-defense techniques.

Can you make the app to work without internet connection?

No suggestions, the app is nice

I like the app, it is okay, I do not have any other suggestions

Allow people to be able to call the police directly

7. CONCLUSION AND FUTURE WORK

In Chapter 1, a detailed analysis of the project requirements, associated project risks, essential resources, and considerations based on PSEL criteria were described. Following this is a comprehensive literature review of existing systems. The literature review highlighted the potential of mobile app solutions for enhancing public safety, with a focus on Nigeria. By carrying out a detailed review and comparison of existing world class mobile apps for enhancing public safety, critical functionalities lacking in existing solutions were identified and integrated into the proposed mobile application.

A successful implementation of the app's designs was carried out using standard mo-bile development practices and applying well-known modern technologies such as Flutter, Dart, Firebase and Google Maps API. The final application contains an in-app SOS feature for alerting emergency contacts when a user is in danger as well as functionality to report crime incidents. The app's map functionality also plays a crucial role, offering real-time location tracking and preserving a user's last known location. Both functional and non-functional requisites were successfully addressed.

The next step was to perform usability testing. This was carried out by sending out a survey to people spanning Nigeria's six geopolitical to gather a diverse range of user safety data. In total, 164 user responses were collected. The outcome of the usability testing has shown features to be included in future improvements, particularly inte-grating real-time communication with authorities and worldwide security news inte-gration (for providing real time global news) in future app versions.

In conclusion, the central objective of this research - to design a mobile application fostering crime incident reporting and public safety enhancement - has been successfully achieved, effectively bridging the gap between the initial project goals and the desired outcomes. Through careful planning and implementation, every part of the project has been successfully completed. The user-friendly app, leveraging on modern technologies like Flutter and Firebase empowers citizens to actively engage in prompt and precise reporting of crime incidents. The potential for improving crime incident reporting in developing countries, specifically in Nigeria remains largely untapped, particularly given the increased use of smartphone devices. Therefore, the development of a mobile application is well-suited to address this gap and offer users an efficient and effective solution for reporting crime incidents.

By leveraging the widespread use of Android and IOS smartphones, the research has successfully provided a solution to streamline the reporting process and therefore can be used as a tool to enhance the overall safety and security measures.

REFERENCES

Achumba, I. C., Ighomereho, O. S., & Akpor-Robaro, M. O. M. (2013). Security challenges in Nigeria and the implications for business activities and sustainable development. *Journal of economics and sustainable development, 4*(2).

Awortu, B. E. (2015). Boko Haram insurgency and the underdevelopment of Nigeria. *Research on Humanities and Social Sciences*, 5(6), 213–220.

Azman, F., Suraya, Q., Rahim, F. A., Mohd, M. S., & Ariffin, N. A. M. (2018, November). My guardian: A personal safety mobile application. In *2018 IEEE Conference on Open Systems (ICOS)* (pp. 37-41). IEEE. DOI: 10.1109/ICOS.2018.8632808

Bearfield, C. X., Stokes, C., Lovett, A., & Franconeri, S. (2023). What does the chart say? grouping cues guide viewer comparisons and conclusions in bar charts. *IEEE Transactions on Visualization and Computer Graphics*. PMID: 37792647

Bhalerao, A.K., 2015. Application and performance of Google Forms for online data collection and analysis: A case of North Eastern Region of India. *Indian Journal of Extension Education, 51*(3and4), pp.49-53.

Bhole, C., Malviya, V., Nadar, S., & Patel, J. (2021, May). CitiSafe–Interactive Crime Reporting. In *Proceedings of the 4th International Conference on Advances in Science & Technology (ICAST2021)*.

Bueno, C., Pereira, L. M., & Fabricio, M. M. (2018). Life cycle assessment and environmental-based choices at the early design stages: An application using building information modelling. *Architectural Engineering and Design Management*, 14(5), 332–346. DOI: 10.1080/17452007.2018.1458593

Das, S., Mitra, K., & Mandal, M. (2016). Sample size calculation: Basic principles. *Indian Journal of Anaesthesia*, 60(9), 652–656. DOI: 10.4103/0019-5049.190621 PMID: 27729692

Desouza, K. C., & Bhagwatwar, A. (2012). Citizen apps to solve complex urban problems. *Journal of Urban Technology*, 19(3), 107–136. DOI: 10.1080/10630732.2012.673056

Flora, H. K., & Chande, S. V. (2014). A systematic study on agile software development methodologies and practices. *International Journal of Computer Science and Information Technologies*, 5(3), 3626–3637.

Heikkilä, H., Maalouf, W., & Campello, G. (2021). The United Nations Office on Drugs and Crime's efforts to strengthen a culture of prevention in low-and middle-income countries. *Prevention Science*, 22(1), 18–28. DOI: 10.1007/s11121-020-01088-5 PMID: 31953793

Imasuen, I. I., & Nwajana, A. O. (2022). Balancing WTO-TRIPS standard against Nigeria counterfeit regulatory efforts: A critique. [IJPPPHCE]. *International Journal of Public and Private Perspectives on Healthcare, Culture, and the Environment*, 6(1), 1–19. DOI: 10.4018/IJPPPHCE.301575

Kemmerer, R. A. (1985). Testing formal specifications to detect design errors. *IEEE Transactions on Software Engineering*, SE-11(1), 32–43. DOI: 10.1109/TSE.1985.231535

Kure, H. I., & Nwajana, A. O. (2022). *Protection of critical infrastructure using an Integrated Cyber-security Risk Management (i-CSRM) framework. In 5G Internet of Things and Changing Standards for Computing and Electronic Systems*. IGI Global.

Leema, R. G., Rajesh, R., Rajeswari, M., Akshaya, V., Saravanan, D., & Sangeetha, N. (2021, July). Women Safety Android Application with Hardware Device. In *2021 International Conference on System, Computation, Automation and Networking (ICSCAN)* (pp. 1-5). IEEE. DOI: 10.1109/ICSCAN53069.2021.9526474

Lips, S., Amola, O., & Draheim, D. (2021, November). Designing a Crisis Management Mobile Application Solution in Nigeria. In *The 23rd International Conference on Information Integration and Web Intelligence* (pp. 571-579). DOI: 10.1145/3487664.3487807

Mathijssen, M., & Tanner, M. (2022, May). The use of a user-centric smart mobile application prototype for supporting safety and security in a city: a design science method. In *International Conference on Social Implications of Computers in Developing Countries* (pp. 270-294). Cham: Springer International Publishing. DOI: 10.1007/978-3-031-19429-0_17

O'malley, P. (1992). Risk, power and crime prevention. *Economy and Society*, 21(3), 252–275. DOI: 10.1080/03085149200000013

Ogunyombo, O. E., Azeez, O. Y., & Odunlami, D. (2018). Exploring the awareness of emergency helplines among taxi drivers in Ikeja administrative division of Lagos state.

Omede, A. J. (2011). Nigeria: Analysing the security challenges of the Goodluck Jonathan administration. *Canadian Social Science*, 7(5), 90–102.

Stork, C., Calandro, E., & Gillwald, A. (2013). Internet going mobile: Internet access and use in 11 African countries. *info, 15*(5), 34-51.

Timothy, M. (2020). An android location-based crime reporting system using the Google Map API. *University of Pitesti Scientific Bulletin: Electronics and Computers Science, 20*(1).

Zein, S., Salleh, N., & Grundy, J. (2016). A systematic mapping study of mobile application testing techniques. *Journal of Systems and Software*, 117, 334–356. DOI: 10.1016/j.jss.2016.03.065

Chapter 7
Millimetre-Wave Communication for 5G/B5G Applications

Rahul Koshti
NMIMS Hyderabad, India

ABSTRACT

Emerging millimeter wave (mm Wave) wireless communication systems reflect the culmination of more than a century of advancement in modern communications. Millimeter wave (mm Wave) communications have recently piqued the interest of researchers due to the enormous available bandwidth, which has the potential to result in rates of multiple Gbps (gigabit per second) per user. Though mm Wave can be easily used in stationary scenarios such as indoor hotspots or backhaul, it can be challenging to use in mobile networks because the transmitting/receiving nodes may be travelling, the channels may have a complex framework, and coordination among multiple nodes is difficult. In spite of this, mm wave is significantly important in implementation of 5G/B5G communication networks. The novelty in the chapter is to suggesting a modern antenna design for mm wave application and link budget analysis so as to achieve flaw less communication with reduced call drop rate. In this chapter we are focusing on introduction and background. The chapter further will focus on mm wave in cellular mobile communication along with the applications in 5G/B5G wireless communication networks. The chapter will also deal with system architecture of 5G communication employing mm wave and antenna in mm wave communication will be discussed. Also various challenges associated with deployment of mm wave technology in 5G networks are discussed. At the last advantages and disadvantages of mm wave will be discussed in brief.

1. INTRODUCTION

With the continued usage of wireless technology, wireless data traffic has been growing at a rate of over 50% annually per user, and this trend is anticipated to intensify over the next ten years. IoT (Internet of Things) growth and video (Gubbi, J., Buyya, R., Marusic, S., & Palaniswami, M. et al (2013) and Rappaport, T. S. et al (2016) The wireless sector is adapting to meet this demand by Fifth-generation (5G) cellular technology will be utilized. frequencies at millimeter wave (mm Wave) to provide unprec-

DOI: 10.4018/979-8-3693-8799-3.ch007

edented multi-Gigabit-per-second (Gbps) data rates and spectrum to a portable gadget (Rappaport et al., 2013). Cell phones and other portable electronics are commonly known as user equipment (UE). a quick evaluation demonstrated the potential for 1 GHz wide channels at 28 or 73 GHz. with a small phased array, provide UE with many Gbps of data rate. antennas on a mobile device(Rangan et al., 2014). Wireless channel bandwidths for 5G mm Wave will be greater than ten times more than the 20 MHz cellular channels used by 4G Long-Term Evolution (LTE) today. Due of the wavelengths' amplification, order of magnitude at mm Wave against current 4G microwave frequencies, material penetration, and diffraction will experience more attenuation, increasing the significance of propagation, reflection, and scattering of line-of-sight (LOS) signals. When designing novel systems, accurate propagation models are essential. protocols for mm Wave signaling, such as air interfaces. more than In recent years, data and models for a wide range of numerous businesses and have presented by many research groups (Rappaport et al., 2017)

The most recent paradigm-shifting advancement in wireless networks is communication at mm Wave operating frequencies. Consumers will continue to demand faster data rates for the consumption of media while demanding lower latency and consistent connectivity on wireless devices as mmWave gains traction. When compared to cellular and WLAN microwave systems that operate at frequencies below 10 GHz, the accessible spectrum at mmWaves is unmatched. The unlicensed spectrum at 60 GHz in particular provides 10 to 100 times more spectrum than is typically available for traditional unlicensed wireless local area networks in the Industrial, Scientific, and Medical (ISM) bands (such as at 900 MHz, 2.4 GHz, or 5 GHz) or for users of WiFi and 4G (or older) cellular systems that are not 5 GHz-capable. (Rappaport et al., 2014) In order to share network resources between local D2D communications and worldwide device-to-device (D2D) communications, it is possible to take advantage of the properties of mm Wave propagation and the usage of directional antennas.

In this chapter in the subsequent sections, we are summarizing different generations of wireless communication, mm wave propagation for device to device (D2D) communication, mm wave 5G system architecture, spectrum allocations in mm wave application and also the applications of mm wave are discussed along with the advantages disadvantages of mm wave.

Table 1. Wireless Generations

Wireless Generations	Features	Applications
1G	Analog Technology Deployed in 1980	Only Voice communication
2G	GSM, CDMA, deployed in 1990	SMS. Voice and low-rate data
3G	WIMAX & CDMA	Trans reception of video signals.
4G	Long term evolution (LTE)	Voice, video and data on high data rates (MBPS)

2. MM WAVE PROPAGATION FOR D2D COMMUNICATION

There are numerous basic principles in Mm Wave communication (with wavelength on the order of millimeters), in the aspects of propagation (Geng et al., 2009). First off, the propagation loss at 60 GHz is 28 dB more than it is at 2.4 GHz, making it significantly worse than the microwave bandh) because the propagation loss in empty space is proportional to the carrier frequency squared. It is preferred to use a high-gain directional antenna. make up for the massive propagation loss as well as lessen the shadowing effect. Next, the mm Wave bands' short wavelengths lead to obstructions present complications for diffracting. Transmissions made via line-of-sight (LOS) are quite simple. obstructed by the barriers. Considering non-LOS (NLOS) Due to the considerable attenuation and lack of multipath in mm Wave channels, LOS links may fail and cause link outages. blocked. Third, mm Wave signals encounter challenges. passing through dense solids (for instance, at 40 GHz, attenuation of 178 dB for brick walls and higher a painted board will attenuate sound by 20 dB). The restricted penetration potential might restrict outdoor mm Wave signals to exterior areas like streets structures, while a small amount of signal power reach the interior of the buildings via glass windows wood doors, too. These propagation characteristics make it difficult to establish reliable coverage that is seamless (Zheng et al., 2013). Local D2D communications and global D2D communications are two types of D2D communications that can be enabled in mm Wave 5G cellular networks. When the LOS link between two wireless devices is blocked, local D2D communications provide a conduit between them, either directly or through relays.

They make it easier to find nearby devices and cut down on the cost of communication between them. Global D2D communications hop between the backbone networks to link two wireless devices connected to different base stations. They consist of base station to base station (B2B) and device to base station (D2B) communications. For B2B communications in mm Wave 5G cellular networks, mm Wave communication with a highly directed antenna offers wireless connections with high data rates, in contrast to 4G cellular networks where communications between base stations are carried out via fibre lines.

D2D communications, as previously said, are anticipated to be a crucial component of mm Wave 5G cellular networks in order to increase network capacity and create connections between two wireless devices. Due to the directional antenna's high propagation loss and low multi-user interference (MUI), mm Wave communication is able to enable simultaneous communications. The network bandwidth can be increased even more by enabling many concurrent D2D links. D2D communications in mm Wave 5G cellular networks may encounter two types of potential interference within each cell: interference between various local D2D communications (if there are several local D2D communications) and interference between local D2D communications and D2B/B2B communications. The majority of the current research on D2D communications is on creating resource-sharing algorithms that are optimized while controlling interferences [9, 10]. Using dynamic data arrival parameters, the performance of frequency reuse among D2D lines is examined in (Lei et al., 2013) to determine the average queue length, mean throughput, average packet latency, and packet dropping probability. According to (Yu et al., 2011), the system seeks to maximize throughput while meeting prioritized cellular service constraints. The D2D underlay system's performance is assessed in both a single-cell scenario and the Manhattan grid setting. One local D2D connection and one cellular connection are taken into account when discussing resource sharing. In order to further increase network capacity and boost spectrum efficiency, more D2D links can be enabled in each cell of mm Wave 5G networks by taking use of high propagation loss and directional antennas.

In order to support simultaneous D2D communications in mm Wave 5G cellular networks, a new resource sharing mechanism that takes into account directional interference is required.

3. MM WAVE BASED CELLULAR NETWORK ARCHITECTURE

Figure 1. 5G mm wave Network Architecture

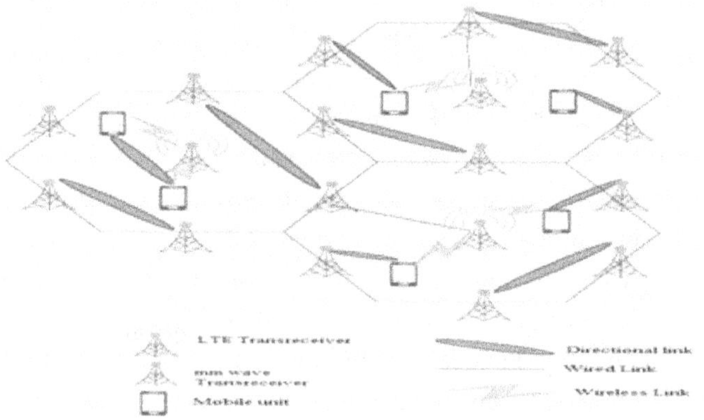

Because of the lower frequency range, it is anticipated that the present 4G cellular networks can offer dependable communications and seamless coverage. To ensure a seamless and economical upgrade from 4G to 5G, 5G cellular networks employ the 4G+mmWave hybrid system architecture Fig. 1 in order to accomplish seamless coverage. very high rate in the majority of the regions covered. Low-cost apps and management information (such as text, phone, and web browsing) are sent by 4G networks, whereas the mm Wave Bands for high-speed multimedia applications are available. Mobile devices, mm Wave Transceivers, and LTE transceivers make up the 5G cellular networks. The entire region is divided into cells in 4G networks, and one or more 4G base stations are stationed in each cell. High directional antennas, the foundation of mm Wave transmission and reception, can significantly lessen mutual interference between mm Wave base stations. It has been shown and demonstrated (Mudumbai et al., 2009) that directional mm Wave communication links can be thought of as pseudo-wired in an outside setting, and the interference among mm Wave concurrent links is minimal. Consequently, it is not necessary to deploy mm Wave base stations in cells. In order to achieve high rates and aggregate capacity, dense mesh networks with grid topology deployment are used for the mm Wave backbone in this article.

Figure 2. Operation of 5G mm wave wireless network

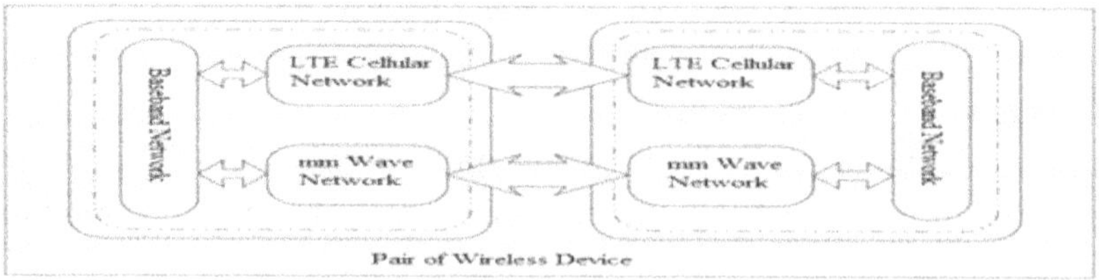

Every wireless device, as depicted in Fig. 2, is capable of operating in both mm Wave and LTE modes, and can switch between them quickly. It is possible for two devices to converse in the same mode.

Following Table summarizes the features of 5G mm wave wireless network.

Table 2. features of 5G mm wave wireless network

Features	Description
Data Rate	10 Gbps or Higher
Distance Coverage	300 Meters (Outdoor) 2 meters (in door)
Frequency Band	40-100 Ghz
Frame Type	Time division duplex
Modulation	OFDMA
Antenna	Massive MIMO
Latency	About 1ms

Figure 3. 5G mm wave frame structure

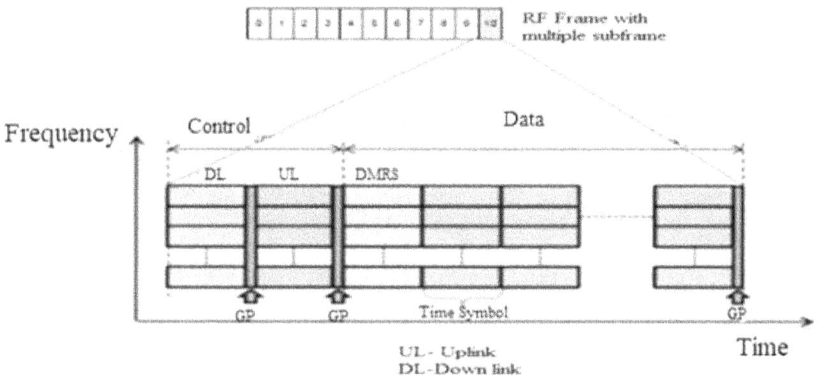

A 5 G mm wave frame consist of radio frame with multiple sub frames. One sub frame consists of multiple OFDM sub frames. Here, DL refers to down link transmissions from Nodes to user equipment and UL refers to uplink transmission in reverse direction. As shown in the figure control and data planes are separated to achieve lesser latency as the processing of control and data signals can be executed in parallel.

4. SPECTRUM ALLOCATION IN MM WAVE APPLICATIONS

Communication over both mm Wave bands (also known as Frequency Range 2) and sub-6 GHz bands (also known as Frequency Range 1) is being planned for Fifth Generation (5G) cellular systems. Ample capacity is available in mm Wave spectrum, allowing for multi-Gbps transmission speeds per user. Historically, local multipoint distribution services (LMDS) and fixed/mobile satellite services (FSS/MSS) have been provided using mm Wave frequencies. The Federal Communications Commission (FCC) recently made almost 11 GHz of the millimeter-wave spectrum available for mobile broadband in order to facilitate wireless LANs (such as Wi-Gig) and 5G cellular systems. As shown in fig 4 The newly allotted bands consist of a new unlicensed band at 64-71 GHz as well as licenced bands at 24 GHz (24.25-24.45 GHz and 24.75-25.25 GHz), 28 GHz (27.5-28.35 GHz), 37 GHz (37-38.6 GHz), and 39 GHz (38.6-40 GHz) – . When coupled with the unlicensed 57–64 GHz band that was previously introduced, this adds about 18 GHz of fresh spectrum for next-generation wireless networks. Along with spectrum above 95 GHz, the FCC also plans to add 15.8 GHz extra spectrum at the 31.8-33.4 GHz, 42-42.5 GHz, 47.2-50.2 GHz, 71-76 GHz, and 81-86 GHz bands.

Figure 4. Frequency bands in mm wave spectrum (Arizona Wireless Communication Laboratory, n.d.)

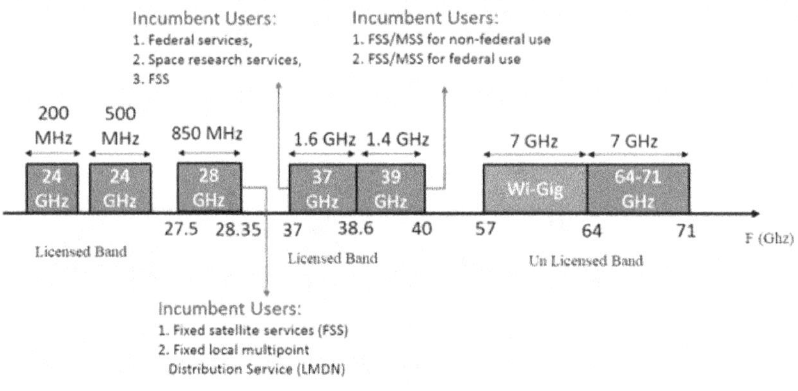

In spite of this ample capacity, severe propagation losses, low penetration, obstruction, rain sensitivity, and other factors can make RF communications at mm Wave frequencies extremely difficult. Simultaneously, a single device (mobile phone, tiny base station, or access point) may accommodate tens or even hundreds of antenna components thanks to the shorter wavelengths. It is possible to conduct electronically steerable, highly directed transmissions by properly processing the signals input into these antennas. Thus, the resulting beamforming and spatial reuse benefits can compensate for the significant signal attenuation. (Arizona Wireless Communication Laboratory, n.d.)

Beamforming is applied in multiple ways at mmWave frequency. The capacity to use fully digital beamforming requires multiple RF links to be available. An RF chain's modulated baseband signal is multiplied by a certain precoding vector. On the other side, one RF chain is sufficient for analogue beamforming. Through the use of phase shifters, it applies the beamforming weights in the RF domain. Last but not least, hybrid (analog/digital) beamforming achieves similar performance to digital beamforming with fewer RF chains by dividing the signal processing between the analogue and digital domains. Analogue beamforming is a desired architecture because of its low power consumption, especially for user equipment (UE).

4.1 Sharing and dynamic allocation of spectrum in mm wave.

Figure 5. Illustration of Dynamic Frequency spectrum allocation

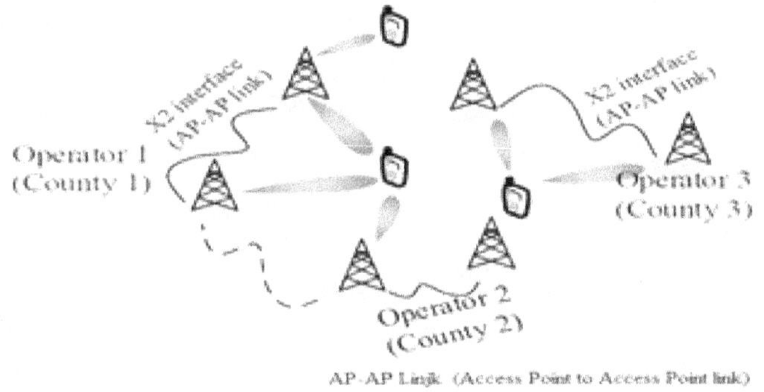

The FCC made the decision to implement various spectrum regulations for several millimeter-wave bands. Geographic area licencing was specifically developed to give licensees the ability to deliver localised services, permit tailored deployments based on consumer demand and market forces, and make access easier for both small and major wireless operators. A 600 MHz band (37-37.6 GHz) was made available for "coordinated coequal shared access" between federal and non-government UEs in order to safeguard currently operating federal military operations in the 37 GHz range. In order to create a total of 14 GHz ultra-wide band for unlicensed devices, such as WLAN systems (like WiGig), short-range devices for interactive motion sensing, and others, the same technical specifications of the 57-64 GHz band were applied to the 64-71 GHz band.

The geographical area licencing policy assigns a single licence to a designated area within the licenced bands (see Figure 5). This could lead to low efficiency of spectrum utilisation in locations, like many urban areas, where several operators have similar interests. For instance, even if other nearby operators are unlikely to cause detrimental interference, a cautious operator might not be ready to share its licenced bands if it does not receive specific information about their traffic, direction, and location. It is also unclear how to encourage operators to cooperate in sharing or leasing their spectrum without revealing their personal data. The FCC's spectrum policy tends to encourage large and small operators to share the same spectrum with each other in the unlicensed bands(Arizona Wireless Communication Laboratory, n.d.)

Mobile carriers are moving towards the high-frequency millimetre-wave spectrum bands (Qamar et al., 2021) and making efficient use of the available spectrum (Qamar et al., 2020) due to the ever-increasing demand for huge capacity and the limited amount of available radio spectrum. As of right now, spectrum management policies are considered to be a useful strategic tool for meeting user service demand, based on a radio spectrum availability (Foster, 2008). Over the past few decades, mobile network operators' (MNOs) needs for high data rates, capacity, and an increase in available radio spectrum have not increased proportionately (Saha, 2020a). As each MNO is permitted to operate only in its assigned licenced spectrum, usually for an extended period of time, the previous static licenced spectrum allocation paradigm is no longer prevalent (Foster, 2008). Each MNO in a nation with such an exclusive licence to use spectrum underutilizes the whole amount of spectrum allotted to that nation in terms of time, frequency, and space. Thus, this calls for the adoption of a more progressive spectrum management policy with elements like flexibility and liberalization (Foster, A. M. et al (2008) and Saha, R. K. et al (2020).

Liberalization, which primarily addresses problems from competitive assignments (e.g., from the spectrum auction to the secondary spectrum trading), sets the upper limit to which spectrum usage rights can be governed by methods (Saha, 2020b). Flexibility, on the other hand, focuses on reducing restrictions on the use of spectrum (Foster, 2008). Examples of these include managed shared-use models, in which unlicensed users can dynamically access licenced users' spectrum as long as they don't interfere with the licenced users (R. K. et al (2020) and Hassan, M. R et al 2017) and spectrum commons, in which everyone has equal access to the spectrum.

The coordination between nodes in the system to make the spectrum sharing more efficient is covered in this section. The two primary methods of coordination are dispersed and centralised (Tehrani et al., 2016). Nodes under centralised coordination are unable to speak with one another directly. Instead, they interact through a central body known as a spectrum controller (such as the NRA), which manages the distribution of spectrum across all of the linked nodes. Decentralised coordination is typically helpful for static spectrum sharing, in which nodes' spectrum modifications don't require as much attention. Long delays and increased protocol/infrastructure complexity are drawbacks of centralised allocation. Furthermore, single-point failure is a possibility in centralised coordination. Distributed coordination, on the other hand, involves direct, dispersed communication between nodes. Locally and often updating the status of nodes is accomplished by distributed coordination. However, because of the information exchanged among nodes, which increases with the quantity of coordinated nodes, distributed spectrum sharing suffers from substantial control signalling overhead. In this technique, during the primary level, the National Regulatory agency (NRA) centrally allots nationwide spectrum to each mobile network operator (MNO) in accordance with their individual user demand during each renewal period, for a minimum of three to six months. At the secondary level, each MNO's allotted spectrum is subsequently distributively divided with the others. In level 1, the centralised spectrum allocation is preferred because the spectrum allotted to each MNO changes infrequently. However, in level 2, the spectrum allotted to each MNO is further shared by all MNOs. Owing to the rapid evolution of radio access networks, each MNO's level 2 spectrum sharing status must be modified often in order to obtain best quality signal. Since distributed coordination is delay-sensitive and cellular standards transmit at the millisecond level, it is better suited for implementation on the radio access side of spectrum sharing. In this sense, tiny cells of each MNO continuously detect the existence of other MNOs' UEs inside every area. A small cell can share its allotted spectrum with other active small cells serving customers in the same area if no UE of that small cell is found within that level. Figure 6, depicts two-level dynamic spectrum allocation scheme (DSAS) technique while taking into account the absence of UEs for MNO 1 and MNO 3 small cells, which means that their licenced spectrum is shared with an MNO 2 small cell located in the same area.

Figure 6. Implementation of Dynamic Spectrum Allocation Scheme

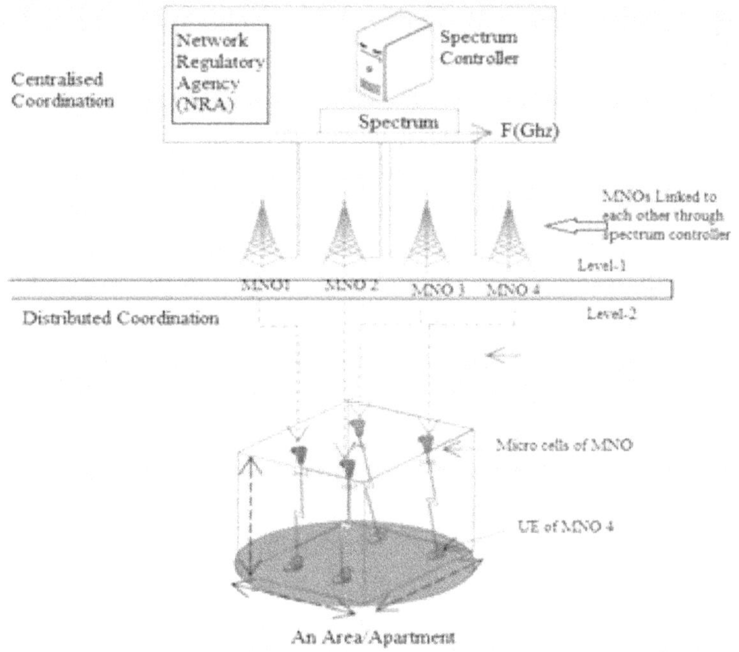

5. ANTENNA FOR 5G MM WAVE APPLICATIONS AND BEAM PATTERN OF ANTENNA ARRAYS

An antenna is a vital element in any wireless communication system. The transmitting and receiving ability of an antenna is determined by it's gain. The antenna's ability to give off radiation more or less in any direction in comparison to a theoretical antenna is known as antenna gain(Khattak et al., 2017). The newest cellular technology, 5G, will make sure billions of gadgets are connected, which is why it's necessary to Attend to all the essential needs for this technology that is being implemented right now(Rahman, 2020). Since mmWave frequencies are extremely high and have a limited range, attenuation problems need to be addressed. Beamforming, massive MIMO, and taking care of topological issues throughout the rollout are some of the strategies used to mitigate signal loss and guarantee optimal performance of 5G systems. Even though the 5G roll out is underway, work is still being done in this area of antenna gain optimisation to make sure that any shortcomings in the anticipated issues are addressed. High path loss, which requires a high gain for an antenna operating at these frequencies, is the primary problem with antennas at these frequencies. A high gain by itself requires a short beam width, which lowers coverage. Therefore, antenna designers must take into account the trade-off between gain and beam width (Muzaffar et al., 2019). It is well known that in order to meet user mobility requirements in access links during severe path loss situations in outdoors scenarios, an appropriate technologies is the beamforming scheme for the use of cellular communications at the mmw ranges of frequencies (Roh et al., 2014). In this article, we review the radiation patterns of various antenna array architectures in order to maximize the performance of millimeter-wave communication systems through the application of appropriate antenna array prototypes. As seen in Fig. below(Zhang et al., 2017) four common antenna array architecture types are taken into consideration: 8 8 rectangle array, 64 circular array, 61 hexagonal array, and 16 crisscross shape array. The antenna elements in the 8 X8 rectangle array are arranged in a square grid with λ2 equal inter-element space in both the x and y axes.

Two linear arrays of eight elements each, arranged vertically against one another, are multiplied to create the 16-crisscross shape array's three-dimensional (3D) beam pattern. The hexagonal array's characteristic is that any two neighbouring antenna elements have the same amount of inter-element space (λ /2). Four concentric 16-element circular arrays of diameters of λ /2, 3 λ /2, 5 λ /2, and 7 λ /2, respectively, also constitute the circular array. Below fig shows, the normalized two-dimensional (2D) beam patterns that were derived from the three-dimensional (3D) beam patterns at various angles. It is clear from different angles that the primary lobes in the patterns of the rectangular and crisscross shape arrays fluctuate in gain between 3 and 5 dB. But in the case of the circular array, it is found that because of the axial symmetry of the antenna array layout, the primary lobe remains unaffected by changes in angle. Regarding the side lobe of the array patterns, it is noted that the hexagonal and circular arrays exhibit greater gain fluctuation than the rectangular and crisscross shape arrays. Nonetheless, by utilizing the theory of pattern multiplication, the element antenna design can efficiently compress the side lobe(Zhang et al., 2017) This study also examines the radiation patterns of rectangular arrays with varying number of antenna elements. Although not displayed here, it is observed that, The major lobe of the radiation pattern exhibits a gain of 3 dB ~ 5 dB . As the number of antenna elements increases, the width of main lobe reduces. However, the distance between nearby elements varies with angle. Large rectangular antenna arrays with narrow main lobes exhibit more dramatic gain fluctuations with different angles than small antenna arrays with wide main lobes. Using huge antenna arrays may not always be optimal.(Zhang et al., 2017)

Figure 7. The radiation patterns of: a) 8 X 8 rectangular array; b) 16 cross shape array; c) 64 circular array; d) 61 hexagonal array in different angles.(Zhang et al., 2017)

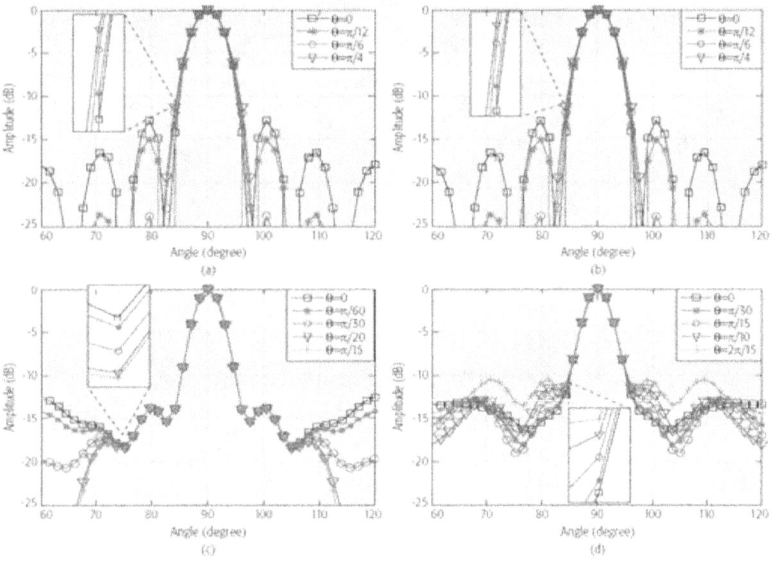

5.1 5G mm wave beam forming:

mm-wave beamforming networks use two different kinds of array feeding mechanisms. These beam formers are circuit- and lens-based. The capacity to steer beams with radio frequency (RF) and a large beam width are crucial for producing multiple beamforming with an array. This makes it possible for 5G wireless networks to regulate the amplitude and phase at every array element (Rahimian et al., 2016). The researchers evaluated the viability of realising beamforming using inkjet technology for application in 5G mm-Wave systems by proposing a unique and highly effective Rotman lens beam former. The authors of (Rahimian et al., 2016) use RF beam formers in their research. A highly directional antenna system can be form by using Quadrature baseband beam forming. By using a modified Bessel beam former, the adaptive beamforming algorithm, as seen in Figure 8, is able to optimise the SNR of the antenna and update the weights dynamically. The Bessel beam former effectively calculates adaptive weights and has automatic gain control (Heath, 2013)

Figure 8. Bessel Beam former

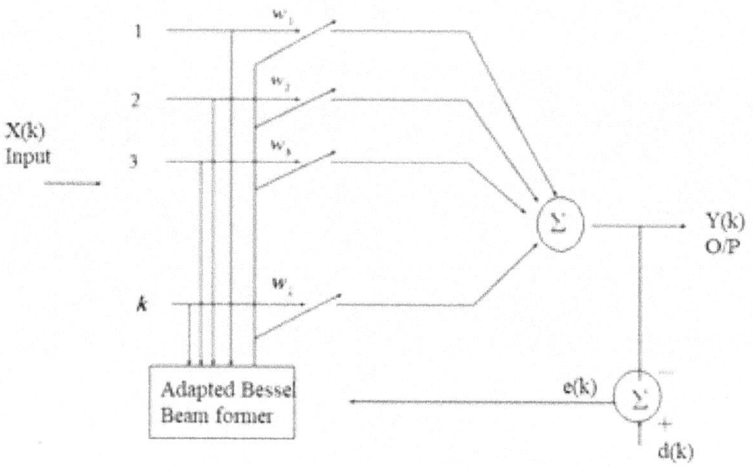

5.2 Antenna and it's coverage for efficient 5G mm wave cellular Communication networks.

The backhaul links and access links make up the two parts of the cellular system that make up the 5G wireless network's transmission portion. To accommodate user mobility and high data rate transmission needs, the access links often use a highly directed and flexible beam. The effective coverage inside the small cells and the high directivity of mmw beams should ensure that the interference level is reduced. The mmw antenna array shows more benefits in a backhaul link that is flexibly deployed for backhaul links where base station (BS) and access point (AP) locations are typically fixed. This is because it allows high data rate transmissions between a BS and an AP that are relatively far apart. The received power from the mmw antenna array is determined by the antenna array gain and the polarisation loss between the transmit and receive antennas, as stated by the Friis transmission formula.

$$P_R = \frac{P_t G_t A_{er}}{4\pi r^2} Cos^2 \alpha$$

where r indicates the distance between the transmit and receive antennas, Aer indicates the receiving antenna's effective aperture, and PT and PR indicate the transmitted and received power, respectively. The term indicates the loss factor resulting from the polarisation differences between the transmit and receive antennas. From the above equation, Three types of advantages can be used for mmw communications' benefit. The shorter mmw antenna array can achieve a greater transmission gain with the same antenna array area.

wavelength, increasing the rate of transmission. Additionally, at shorter wavelengths, the interference power and transmit power leaks can be efficiently lessened to other users by the significant path loss and attenuation were encountered by microwave transmissions. However, in contrast to discover additional crucial methods in upcoming 5G wireless communication networks, such as Large MIMO, CoMP, and SU MIMO, the millimeter-wave communication technology exhibits a notable capacity advantage due to the substantially greater accessible resources in the spectrum (Heath, 2013) High antenna gains are possible with highly directed antennas, but there are drawbacks as well. First, numerous impediments may often hinder mmw communications in both access and backhaul networks (Rappaport et al., 2013). The radio waves' diffraction and penetration properties determine whether electromagnetic energy can go through the barriers. Furthermore, for communications to be effective in millimeter-wave bands, both the transmitter and the associated receiver must be within line of sight (LOS). In such cases, the nearby structures and trees may have a major impact on mmw cellular networks' performance, particularly for mobile users. For millimeter-wave communication systems, scattering—which depends on particle diameter—is an additional critical aspect that must be taken into account in addition to ground barriers. In general, particles in the transmission route scatter more power at shorter wavelengths. Extra Mie scattering, which happens in the mmw frequency bands when raindrop diameters range from 3 micrometres to 3 mm, which is on the order of the mmw wavelength, differs from normal Rayleigh scattering, which always happens in the microwave frequency bands.

The mmw channels have properties. Highly directional antennas enable LOS channels with little shadowing and multi-path components. This ensures maximum capacity for mmw communication systems while minimizing interference to other users. However, the high frequency of mmw communications might lead to frequent blockage due to uneven terrain, obstructions, and user mobility. Non-line-of-sight (NLOS) channels significantly reduce the performance of mmw communication systems compared to typical microwave systems that use multi-path components to combat fading. To ensure good coverage and seamless network connectivity in mmw cellular networks, link blockage must be addressed to transfer signals over LOS pathways to the targeted users. Existing techniques, such as distributed antenna systems (DAS) and cooperative multi-hop relaying, can be used to address the peculiarities of mmw communication systems (Zhang et al., 2017)

5.3 Distributed Antenna System (DAS) for mm wave applications:

There are little barriers in the way of mmw communication systems' propagation in flat suburban areas, ensuring LOS transmissions. Therefore, to cover a specific service area, it would be practical and cost-effective to install a centralised antenna array at the mmw BS/AP, such as a 16 X 16 antenna array.

However, in a densely populated region, the received signal may become too weak to be effective over even a short distance, like 200 metres and beyond, due to route loss caused by difficult terrain, surrounding barriers, and/or the highly reflecting nature of modern structures (Rappaport et al., 2013) Then, it would be more advantageous to install several smaller antenna arrays, such as 16 4 X 4 antenna arrays, which are geographically spread to cover a specific area, as opposed to a single, centralised antenna array. Each distributed antenna array is linked by optical fibres to a single central controller, as shown in Fig. 9

Figure 9. mm wave communication with distributed antenna system

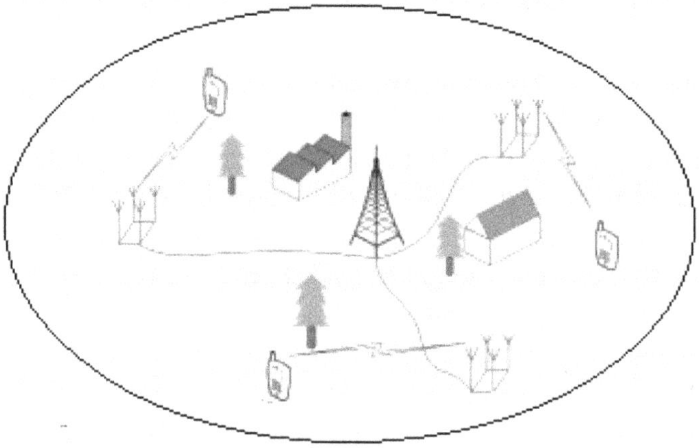

The actual application of DAS varies depending on the circumstances, taking into account the diverse applications of DAS and the features of mmw communication systems and they are worth investigating, the ideal antenna array sites and accompanying signal processing methods that ensure good coverage while enabling high data throughput. The purpose of dispersed antenna arrays is to bring the mmw Access points (APs) closer to their intended users by placing them at predetermined positions based on building locations and terrain. On the one hand, less energy is lost in compensating for the blockage/shadowing, reflection, and penetration losses. However, there is less fading loss and a shorter delay spread with a LOS channel since it is available more often than with a centralised antenna array. Stated differently, system energy consumption can be efficiently lowered while greatly boosting the effective coverage, dependability, and possible network throughput by dividing the broadcast power among numerous smaller antenna arrays that are separated in space (Wang et al., 2014).

5.4 Cooperative Multi hop relaying for mm wave communication

DAS can be used to address a blockage that is caused by relatively static obstacles, such as rough terrain, buildings, and so on. However, cooperative relays can be used as an additional solution to deal with a blockage caused by user mobility or surrounding obstacles, such as vehicles, walking people, and so on. This allows multi-hop NLOS transmissions to be used to reroute the transmission from the blocked path (Sakarellos, 2012), as shown in Fig. 9. When a targeted user combines numerous versions of the same signal received through distinct pathways, cooperative diversity can be utilised. This is widely acknowledged as a beneficial technique for improving network performance, such as IEEE 802.16, 3GPP, LTE, and furthermore. Densely deployed dedicated relays in the network could be one way to boost the likelihood of locating an appropriate path when a blockage arises (Boccardi et al., 2014). Then, as shown in Fig. 10, anytime a blockage happens, the Base station (BS) may direct the beam direction to a relay that has a LOS link to the intended user.

Figure 10. Cooperative multi hop relaying

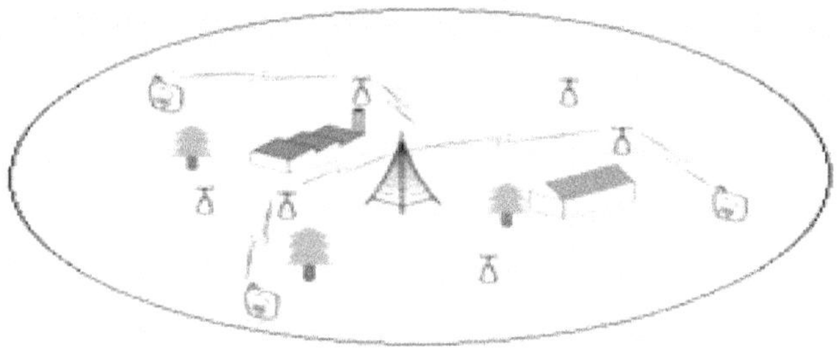

Relay selection and routing systems can be implemented in a variety of ways depending on different performance objectives.

An intermediate relay terminal is simply chosen that is close to the direct channel between the source and the destination while having LOS connections to both of them, for example, when a blockage arises between a source and its corresponding destination. More complex relay selection/routing algorithms will be needed if the system is subject to limitations, such as the number of hops (or latency), co-channel interference, possible throughput, and so on.

6 DESIGN CONSIDERATION OF MM WAVE COMMUNICATION SYSTEM

This section covers the main needs for creating a mmWave channel model and the architecture of mmWave systems. Also the link budget analysis for a network is covered.

6.1 Consideration for channel model.

Understanding the propagation characteristics of channel effects can lead to effective treatment and a high-integrity communication system, similar to disease symptoms (Lee, 2014). At mmWaves, understanding channel characteristics and their differences from lower frequency bands is crucial for successful system design.

Sections II and III covered the propagation characteristics of mmWaves, highlighting the sparse CIR and significant attenuation. Future systems should account for these constraints.

Figure 11. Design consideration of mm wave system(Hemadeh et al., 2017)

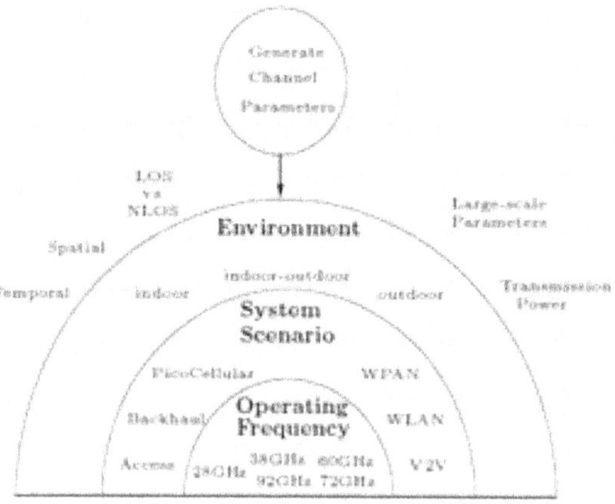

To simulate a mmWave system in a certain frequency range, the channel model must include accurate measurements of channel properties. Figure 11 illustrates how to construct the mmWave simulation channel requires three critical points (Hemadeh et al., 2017)

1) Identification of system environment: To design a mmWave system, it's important to select its environment. Indoor systems (WPAN, WLAN), outdoor systems (cellular, D2D), and indoor-outdoor systems (WLAN-cellular interface) are all options.
2) Distinguish the channel scenario: Indoor scenarios may include WLAN, WPAN, or indoor festivals, while outdoor systems may use UMi, UMa, or backhaul.
3) Frequency of operation: The most essential parameter is propagation properties, which vary across frequency bands within the mm Wave spectrum.

Obtaining channel characteristics from extended measurement campaigns is crucial for accurately characterizing a system. The channel model can be developed based on system specifications and scenario selection. The spatial and temporal features of large-scale and small-scale fading in LOS and NLOS situations are widely known for many frequency bands, including 28 GHz, 38 GHz, 60 GHz, and 72 GHz.

Figure 12 illustrates the process of creating the mm Wave channel model for based on the 3GPP channel. (Hemadeh et al., 2017)

Figure 12. various techniques in MM wave Technology (Hemadeh et al., 2017)

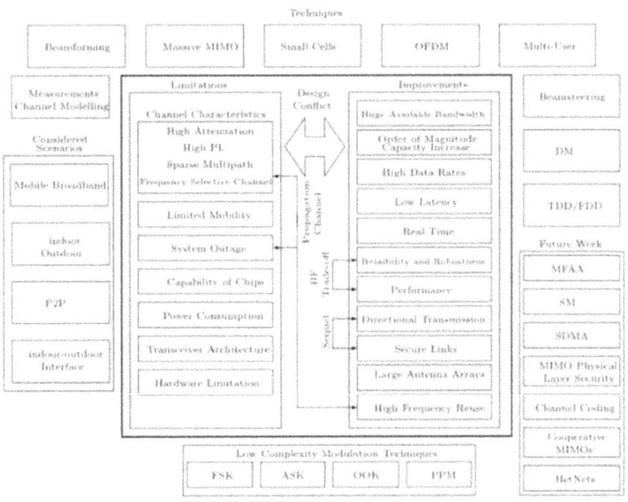

6.2 Link Budget Analysis

To establish a communication link between a transmitter and a receiver, the transmit power and received power must be carefully adjusted to maintain the desired data rate for a particular channel while accounting for thermal noise at the receiver. This power analysis is necessary for building a link based on the link budget.

A communication system's link budget is mostly determined by the specifications of the base and mobile stations, including transmit power, antenna gain, signal-to-noise ratio (SNR), and link throughput requirements.

Table 3. Link Budget Parameters

Link Budget Parameters	**Specification of up link**	**Specification of downlink**
Distance (km)	.5	.5
Gain of Transmitting antenna (dbi)	**17-19**	**25-27**
Transmit power (dBm)	**20-22**	**28-41**
Gain of receiving antenna(dbi)	**25-28**	**12-15**
Power of receiving antenna	-58.32	-73.12
Carrier frequency (GHz)	**25-28**	**25-28**
Bandwidth (GHz)	0.5	0.5
Noise figure	**7**	**7**

continued on following page

Table 3. Continued

Link Budget Parameters	Specification of up link	Specification of downlink
Thermal noise (dBm)	-82.01	-82.01
Spectral efficiency	1.25	6.25
Data rate (Gbps)	3.6	0.8
Free space propagation loss (dB)	115	115
Signal to Noise ratio (dB)	6.69	21.72

Table 3 (Hemadeh et al., 2017) shows the link budget for a typical uplink and downlink in mmWave communications. downlink scenario uses a 28 GHz carrier with a base station transmit power of 40 dBm . The path loss model in (best fits the 28 GHz communication system) In a LOS environment (np = 2) with an average shadowing loss of 20 dB, the propagation loss over 0.5 km is 115.32 dB. Our goal is to attain a data rate of 3.1 Gbps inside a 0.5GHz bandwidth. System-level simulations show that data rates of up to 7.5 Gbps are possible. For successful construction of the link at a 3.1 Gbps throughput, using Shannon's channel capacity (Cover & Thomas, 2006), it requires 15.69 dB of SNR. Currently, the gain-to-bandwidth product of RF components at high frequencies remains low (Rangan et al., 2014). Due to other system losses, such as power dissipation from circuitry. Hence considering all the losses, a SNR of 21.72 dB is required to establish the link. We assume a noise figure of 7dB (Agilent Technologies, n.d.) and thermal noise of -174 dBm/Hz. We assume 12 dBi of receive antenna gain. Achieving 12 dBi gain at the mobile station is feasible, as it can be accomplished with a four-element array with 6dBi gain per element via patch antennas.

Using the values in Table and (Rappaport et al., 2014) in Pt = Gt + Gr + PL, the base station requires an antenna gain of around 25 dBi. We assume a base station antenna element with a gain of 10dBi (Rajagopal et al., 2011). A low gain antenna allows for better steering and less power leakage to side lobes (Huang & Edwards, 2008). To reach 3.1 Gbps, an additional increase of around 15 dBi is required.

The antenna array with d =λ/2 spacing has a total gain of Gantenna = (μGelementN), where μ represents the array's capacity to steer the beam in the desired direction. On a dBi scale, Gantenna gain is the sum of the array and element gains.

On dbi scale the gain of the antenna can be written as- $G_{Antenna} = 10log$μ(dbi) + $G_{element}(dbi) + G_a$ Where, = gain of single antenna element and is gain of antenna array.

The array provides an additional 15dBi gain. To get 15dBi gain with μ = 1, the array requires 1015 antenna elements (10 x 32). Nevertheless, in Practice to account for non-negligible power (radiation) in other directions, especially while steering. Having a narrow pencil beam is not realistic for maximizing gain from antenna arrays. A high-integrity array may require more than 32 antenna elements, depending on the array's directivity. Figure 36 depicts array gain at three different efficiencies. To get a 15dBi array gain at μ = 0.75, around 40 antenna elements are required, but at μ = 0.5, up to 60 antenna elements may be needed.

Table 2 shows that the mobile station's receive antenna gain of 12 dBi may be calculated using a four-antenna array with a 6dB element gain.

Increasing the frequency from microwave to millimeter wave enhances bandwidth availability for mobile users. However, this comes with a large propagation loss, as illustrated in Figure 13. Doubling the frequency causes a 6dB loss in link budget for a given antenna gain. The decrease at 60GHz is due to oxygen absorption. Therefore, the 60 GHz frequency is not appropriate for outdoor situations.

Figure 13. power of receiving antenna at different frequencies at a transmit power =40dbm (Hemadeh et al., 2017)

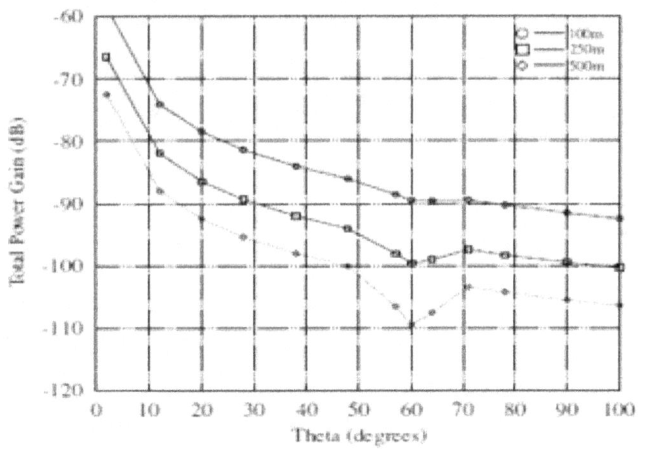

Beamforming with antenna arrays can help overcome propagation loss, as explained above. At millimeter wave frequencies, the distance between antennas is very small allowing for the use of several antenna components at the base station to obtain significant gains using beamforming

7 CHALLENGES IN 5G MM WAVE TECHNOLOGY

7.1 Penetration

The fact that mmw signals are easily spied on is one of the main barriers to the implementation of mmw technology in some outdoor environments, like forests. Utilising huge MIMO and extremely dense relays are two potential solutions to this issue.

Lower frequency signals can be used by both methods to provide reflection and scattering in order to get around the obstruction. The latter shortens the distance to minimise path loss, while the former increases the number of transmitting antennas to improve antenna gain. Lower frequency communications have a bandwidth constraint that must be made up for with other resources like power, hardware costs, scheduling, and so forth.

7.2 Energy Efficiency

Shorter wavelength microwave communications suffer more severe attenuation within the same communication distance, even if the mmw antenna array can achieve better gains than a microwave communication system with the same array area. During signal transmissions, free-space loss uses up the majority of energy. Therefore, the mmw APs should be placed with high density to shorten the distance between the AP and the user terminal in order to increase the energy efficiency of mmw communication systems.

7.3 Communication Security

The transmitter in millimetre wave communications can achieve significant antenna gains by using highly directed beams. But this could be vulnerable to listening in, especially if someone is blocking the beam.

Consequently, more access points or relays need to be installed in the cells for the mmw cellular network. In order to ensure communication security, the receivers should select the proper APs and relays and compile channel state information.

7.4 Hardware Development

While millimeter-wave communication has great promise for next-generation 5G systems, its hardware implementation presents significant challenges. The power consumption of analogue to digital converters, for instance, often increases exponentially with sample resolution and scales linearly with sampling rate. This is a problem for communication systems operating in the millimetre wave range.

Therefore, the creation of mmw communication circuits is essential to the eventual deployment of mmw cellular networks.

8 ADVANTAGES AND DISADVANTAGES OF 5G MM WAVE TECHNOLOGY

8.1 Advantages:

- offers additional bandwidth, making it possible to support more subscribers.
- Since first order dispersion pathways (Scattering) are feasible, coverage is not restricted to line of sight.
- The Channel Sounding feature is utilised to address various loss scenarios at millimetre wave frequencies, ensuring optimal 5G network performance.
- As antennas are physically small, many antennas are crammed into a limited space. As a result, massive MIMO is used in access points to increase capacity.
- Since dynamic beamforming is used, higher path loss at millimetre wave frequencies is reduced.
- Multi-gigabit backhaul up to 400 metres and cellular access up to 200–300 metres are supported by 5G millimetre wave networks.

5G millimetre wave is suited for mobile communication over sub-6GHz wireless technologies because of these advantages.

8.2 Disadvantages

- Millimetre waves experience a variety of losses, including attenuation from rain and penetration. This reduces the mm wave's required distance coverage for 5G-based cellular mobile deployment. In addition, path loss at millimetres is a function of square of frequency . Depending on the channel and the height of the AP and eNB above the ground, it can support up to two metres indoors and between 200 and 300 metres outside.

- solely allows for Line of Sight (LOS) propagation. Coverage is therefore restricted to LOS.
- Significant foliage loss occurs at these millimetre wave frequencies.
- At millimetre wave, power consumption is higher since there are more RF modules and antennas. Hybrid architecture, which uses fewer RF chains than antennas at the reception, is required to get around this problem. Moreover, mm wave hardware is developed with low power analogue processing circuits.

When calculating the budget for 5G millimetre wave links, these drawbacks must be taken into account. This is important for the effective deployment of 5G millimetre wave technology.

CONCLUSION

In this chapter the basics of 5G mm wave technology along with the network structure are discussed. The mm Wave band's usefulness for 5G cellular networks has been discussed. Additionally, for concurrent D2D communications in mm Wave 5G cellular networks, a resource sharing mechanism is discussed that can greatly increase network capacity while maintaining good network connectivity. Also the frequency spectrum sharing and dynamic allocation of the spectrum in discussed in brief. The effect of antenna array designs on mmw communication systems in real-world outdoor settings was covered in this chapter. MMw systems that are prone to frequent blockages need to maintain effective coverage and connectivity. To this end, viable solutions including DAS and multi-hop relaying are studied in this chapter. The link budget analysis is aslo discussed in this chapter to identify the no. of antenna elements and antenna array so as to achive the desired gain. A significant gain can be achived with the help of beam forming with reduced propagation loss.

In addition, the deployment of millimeter-wave communications presents a plethora of intriguing and demanding novel research issues that manufacturers, operators, and investigators will need to confront in the times ahead. those issues/challenges are also discussed.

REFERENCES

3GPP. (2010). Further advancements for E-UTRA physical layer aspects. 3rd Generation Partnership Project (3GPP), TR 36.814.

Agilent Technologies. (2000). Fundamentals of RF and microwave noise figure measurements.

Akdeniz, M. R., Liu, Y., Samimi, M. K., Sun, S., Rangan, S., Rappaport, T. S., & Erkip, E. (2014). Millimeter wave channel modeling and cellular capacity evaluation. *IEEE Journal on Selected Areas in Communications*, 32(6), 1164–1179.

Boccardi, F., Heath, R. W., Lozano, A., Marzetta, T. L., & Popovski, P. (2014). Five disruptive technology directions for 5G. *IEEE Communications Magazine*, 52(2), 74–80.

Geng, S., Kivinen, J., Zhao, X., & Vainikainen, P. (2008). Millimeter-wave propagation channel characterization for short-range wireless communications. *IEEE Transactions on Vehicular Technology*, 58(1), 3–13.

Gubbi, J., Buyya, R., Marusic, S., & Palaniswami, M. (2013). Internet of Things (IoT): A vision, architectural elements, and future directions. *Future Generation Computer Systems*, 29(7), 1645–1660.

Hassan, M. R., Karmakar, G. C., Kamruzzaman, J., & Srinivasan, B. (2017). Exclusive use spectrum access trading models in cognitive radio networks: A survey. *IEEE Communications Surveys and Tutorials*, 19(4), 2192–2231.

Heath Jr, R. W. (2013). What is the role of MIMO in future cellular networks: Massive? coordinated? mmwave?. ICC Wksp. Plenary: Beyond LTE-A.

Hemadeh, I. A., Satyanarayana, K., El-Hajjar, M., & Hanzo, L. (2017). Millimeter-wave communications: Physical channel models, design considerations, antenna constructions, and link-budget. *IEEE Communications Surveys and Tutorials*, 20(2), 870–913.

Huang, K. C., & Edwards, D. J. (2008). *Millimetre wave antennas for gigabit wireless communications: a practical guide to design and analysis in a system context*. John Wiley & Sons.

Iffat Naqvi, S., Hussain, N., Iqbal, A., Rahman, M., Forsat, M., Mirjavadi, S. S., & Amin, Y. (2020). Integrated LTE and millimeter-wave 5G MIMO antenna system for 4G/5G wireless terminals. *Sensors (Basel)*, 20(14), 3926.

Khattak, M. K., Kahng, S., Khattak, M. S., Rehman, A., Lee, C., & Han, D. (2017, July). A low profile, wideband and high gain beam-steering antenna for 5G mobile communication. In 2017 IEEE international symposium on antennas and propagation & USNC/URSI national radio science meeting (pp. 2575-2576). IEEE.

Lee, W. C. (2015). *Integrated wireless propagation models*. McGraw-Hill Education.

Lei, L., Zhang, Y., Shen, X. S., Lin, C., & Zhong, Z. (2013). Performance analysis of device-to-device communications with dynamic interference using stochastic Petri nets. *IEEE Transactions on Wireless Communications*, 12(12), 6121–6141.

Mudumbai, R., Singh, S. K., & Madhow, U. (2009, April). Medium access control for 60 GHz outdoor mesh networks with highly directional links. In *IEEE INFOCOM 2009* (pp. 2871–2875). IEEE.

Muzaffar, K., Magray, M. I., Karthikeya, G. S., & Koul, S. K. (2019, September). High gain broadband Vivaldi antenna for 5G applications. In *2019 International Conference on Electromagnetics in Advanced Applications (ICEAA)* (pp. 496-497). IEEE.

Qamar, F., Hindia, M. N., Rahman, T. A., Hassan, R., Dimyati, K., & Nguyen, Q. N. (2021). Propagation characterization and analysis for 5G mmWave through field experiments. *Computers, Materials & Continua*, 68(2), 2249–2264.

Qamar, F., Siddiqui, M. U. A., Hindia, M. N., Hassan, R., & Nguyen, Q. N. (2020). Issues, challenges, and research trends in spectrum management: A comprehensive overview and new vision for designing 6G networks. *Electronics (Basel)*, 9(9), 1416.

Rahimian, A., Alomainy, A., & Alfadhl, Y. (2016, November). A flexible printed millimetre-wave beamforming network for WiGig and 5G wireless subsystems. In 2016 Loughborough Antennas & Propagation Conference (LAPC) (pp. 1-5). IEEE.

Rajagopal, S., Abu-Surra, S., Pi, Z., & Khan, F. (2011, December). Antenna array design for multi-gbps mmwave mobile broadband communication. In 2011 IEEE Global Telecommunications Conference-GLOBECOM 2011 (pp. 1-6). IEEE.

Rangan, S., Rappaport, T. S., & Erkip, E. (2014). Millimeter-wave cellular wireless networks: Potentials and challenges. *Proceedings of the IEEE*, 102(3), 366–385.

Rangan, S., Rappaport, T. S., & Erkip, E. (2014). Millimeter-wave cellular wireless networks: Potentials and challenges. *Proceedings of the IEEE*, 102(3), 366–385.

Rappaport, T. S., Xing, Y., MacCartney, G. R., Molisch, A. F., Mellios, E., & Zhang, J. (2017, December). Overview of millimeter wave communications for fifth-generation (5G) wireless networks—With a focus on propagation models. *IEEE Transactions on Antennas and Propagation*, 65(12), 6213–6230. Advance online publication. DOI: 10.1109/TAP.2017.2734243

Rappaport, T., Heath, R., Daniels, R., & Murdock, J. (2014). Millimeter wave wireless communications (Prentice Hall Communications Engineering and Emerging Technologies Series). Pearson Education. Retrieved from https://books.google.co.uk/books?id=0Jh6BAAAQBAJ

Rappaport, T. S. (2016, March 10). Spectrum Frontiers: The new world of millimeter-wave mobile communication. Keynote presentation at the Federal Communications Commission (FCC) Headquarters.

Rappaport, T. S., Heath, R. W., Daniels, R. C., & Murdock, J. N. (2013). Broadband millimeter-wave propagation measurements and models using adaptive-beam antennas for outdoor urban cellular communications. IEEE Transactions on Antennas and Propagation, 61(4), 1850–1859. DOI: 10.1109/TAP.2013.2245062

Rappaport, T. S., Heath, R. W., Daniels, R. C., & Murdock, J. N. (2013). Millimeter wave mobile communications for 5G cellular: It will work! IEEE Access, 1, 335–349. DOI: 10.1109/ACCESS.2013.2260813

Rappaport, T. S., Heath, R. W., Daniels, R. C., & Murdock, J. N. (2014). Millimeter wave wireless communication. Prentice Hall.

Roh, W. (2014). Millimeter-wave beamforming as an enabling technology for 5G cellular communications: Theoretical feasibility and prototype results. IEEE Communications Magazine, 52(2), 106–113. DOI: 10.1109/MCOM.2014.6736752

Saha, R. K. (2020a). Interweave shared-use model for dynamic spectrum access in millimeter-wave mobile systems for 6G. Proceedings of the 2020 IEEE 92nd Vehicular Technology Conference (VTC2020-Fall), 1–6.

Saha, R. K. (2020b). A flexible licensed spectrum allocation technique for millimeter-wave mobile systems toward 6G. Proceedings of the 2020 23rd International Symposium on Wireless Personal Multimedia Communications (WPMC), 1–6.

Sakarellos, V. K. (2012). Cooperative diversity performance in millimeter wave radio systems. IEEE Transactions on Communications, 60(12), 3592–3603. DOI: 10.1109/TCOMM.2012.092612.110885

Samimi, M. K., & Rappaport, T. S. (2015). 3-D statistical channel model for millimeter-wave outdoor mobile broadband communications. In Proceedings of the IEEE International Conference on Communications (ICC) (pp. 2430–2436). DOI: 10.1109/ICC.2015.7248615

Tehrani, R. H., Vahid, S., Triantafyllopoulou, D., Lee, H., & Moessner, K. (2016). Licensed spectrum sharing schemes for mobile operators: A survey and outlook. IEEE Communications Surveys & Tutorials, 18(4), 2591–2623. DOI: 10.1109/COMST.2016.2596343

Thomas, M. T. C. A. J., & Joy, A. T. (2006). *Elements of information theory*. Wiley-Interscience.

Wang, C. X., Gao, X., & Kadoch, M. (2014). Cellular architecture and key technologies for 5G wireless communication networks. IEEE Communications Magazine, 52(2), 122–130. DOI: 10.1109/MCOM.2014.6736751

Yu, C. H., Huang, Y. C., & Chen, H. (2011). Resource sharing optimization for device-to-device communication underlying cellular networks. IEEE Transactions on Wireless Communications, 10(8), 2752–2763. DOI: 10.1109/TWC.2011.060911.101290

Zhang, J., Ge, X., Li, Q., Guizani, M., & Zhang, Y. (2017). 5G millimeter-wave antenna array: Design and challenges. IEEE Wireless Communications, 24(2), 106–112. DOI: 10.1109/MWC.2017.1600236

Zheng, K., Jiao, Y., & Chen, L. (2013). Stochastic performance analysis of a wireless finite-state Markov channel. IEEE Transactions on Wireless Communications, 12(2), 782–793. DOI: 10.1109/TWC.2012.110212.121883

Chapter 8
The Role of 6G in Empowering Smart Cities Enabling Ubiquitous Connectivity and Intelligent Infrastructure

Hitesh Mohapatra
https://orcid.org/0000-0001-8100-4860
Kalinga Institute of Industrial Technology, India

ABSTRACT

Mobile technology communication is evolving towards the seventh generation, the '6G' network. These networks will revolutionize society to the extent of connectedness and intelligence that we have not seen before. 6G will be important to implement the new generation smart cities by enhancing the digital communication technology with terahertz communication, ultra-low latency and artificial intelligence to operate the network system. This will result in smart cities that are fully interconnected to real time data as required for instance in self-drive cars, energy consumption, or security. 6G is found to be promising for URLLC, eXR and massive sensing at the edge necessary for Smart City instantiation. But there are issues such as technicality, data privacy, and regulations to consider. The present chapter contends using a 6G empowered smart city based on IoT, artificial intelligence, and edge computing for the increased efficiency of urban processes. These include dynamic network slicing to fulfill the latency and bandwidth requirement that is supported by Intelligent Resource management as well as Edge AI processing. The chapter therefore concludes by highlighting the need for joint efforts and forward planning to overcome hindrances in the 6G's delivery to support sustainable cities.

1. INTRODUCTION

1.1. Brief Overview of 6G Technology and Its Potential Impact

The appearance of 6G technology can be considered as a sophisticated step forward in the develop of the wireless communication services, provided that the reports which claim for the improvements compared to the previous generation of the 5G networks seem to be quite realistic. Hence it is predicted

DOI: 10.4018/979-8-3693-8799-3.ch008

that 6G is destined to be in the middle of the development of connected environments as societies rely on digital structures to solve multifaceted urban issues (Trichias et al., 2024). This next-generation network technology will not only be about the speed and the capacity but will be defined by the integration of artificial intelligence (AI), edge computing and the Internet of Things (IoT) to create smarter, efficient systems. At the center of these innovations is the desire to achieve maximum potential of the smart cities concept, where connectivity and intelligent decisions are expected to seamlessly create sustainable and resilient city (Mahmood et al., 2020).

Over the last few years, the concept of smart cities has become the central area of focus in the debates about urbanization and sustainability. Today's cities face several problems that include population density, pollution, resources, and security among others. Current approaches and technologies, despite having their utility, can be lack scalability, flexibility and real-time performance (Yrjölä et al., 2023). The emerging 6G is set to overcome these barriers and offer the base on which cities are smart, connected and can learn and adapt in actual time. An exciting development to note regarding the 6G is the forecast of the network to work at terahertz band, which promises data transfer speed beyond imagination, among other benefits. This level of integration is needed for new use cases like autonomous transport, telemedicine and extended reality beyond the vehicle, respectively. Furthermore, 6G is expected to welcome AI at the heart of the network that also support the malleable offering of resources, fall preventive workmanship, and upgraded safety measures (Inomata et al., 2021). The integration of these technologies in 6G will therefore result in cities that are not only connected but also self-sufficient.

In addition to these, 6G creates opportunities for equity and inclusion in the delivery of services digitally as well. However, there is an opportunity of reducing the 'digital divide' where knowledge and use of information and services are not restricted to the well-off citizens. Some of the Intelligent urban systems that can be supported by 6G could include, for instance, public health monitoring that occurs in real-time, adaptive energy grids as well as smart governance systems that are in direct response to population needs (Antón-Haro et al., 2024). However, the processes of moving to the 6G smart cities are not without various challenges. Those questions, which are connected with data protection, security, and ethical aspects of the AI utilization, will arise as these technologies become the organic part of the public utilities. Furthermore, the large-scale application of 6G entails the need to incorporate massive physical and digital investments at operational and national levels and well-developed policies for the further evolution of 6G.

The goal of this chapter is to discuss how 6G would help modernized the urban socio-sphere, regarding how it would upgrade the connectivity, intelligence systems, and eventually, smart cities (Cordova Morales et al., 2023). In this discussion, an attempt has been made to understand the concepts of 6G Wiki news by exploring 6G's characteristics and opportunities and proposed new model of 6G-based urban systems. Over the course of the development of the 6G, more and more people are thinking about its potential and potential consequences for the life of megacities.

1.2. The Significance of 6G in the Context of Smart Cities

As for smart cities, the 6G technology will be the infrastructure enabling an uninterrupted interconnected intelligent environment of the city. While cities are becoming ever sprawling and complex, sensible allocation of resources, better supply of public utilities and a higher quality of life are paramount concerns. It is expected that 6G will deliver the high reliability supporting the efficient connection density for Billions of devices and sensors in urban environment to exchange relevant data and to make intelligent decisions promptly (Zong et al., 2022). Such a level of integration will be vital in the provision of ITS, AVs, and SEG, because they need real-time interaction and information transfer rate.

Furthermore, the integration of AI, edge computing and IoT by 6G will open new opportunities for improvement of the urban planning, government and citizens' interaction. The application of digital twins as virtual models of physical assets will enable city planners and administrators to plan and improve infrastructures and related services (Liu et al., 2022). Moreover, novel networks based on the 6G will help to progress more inclusive and sustainable cities that would provide digital services for all the inhabitants and those in the weak signal area. Indeed, 6G will be the primary enabler of connected cities, that are, more importantly, smart, green, and resilient (Liu et al., 2022).

1.3. Outline of the Chapter's Objectives and Focus Areas

Specifically, as will be discussed in this chapter, the development of the 6G technology will fundamentally alter how people live in cities. The discussion is structured around three core objectives: The discussion is structured around three core objectives:

- **Understanding the Fundamentals of 6G Technology:** To begin with, the chapter presents an analysis of the prospects and possibilities of the 6G, which encompasses terahertz communication, AI propelled network, and ultra-low latency. This section offers the background needed for understanding of 6G as the part of future urban systems (Siriwardhana et al., 2021).
- **Analyzing the Applications of 6G in Smart Cities:** Instead, the notions of 6G are introduced in the context of their utilization in smart city sectors like transportation systems, infrastructures management, safety and security provision, and environment monitoring. By these examples, the chapter demonstrates the way, how 6G will help improving smarter and more efficient urban services (Suyama et al., 2022).
- **Proposing a 6G-Enabled Smart City Model and Addressing Challenges:** This last chapter is a summary and provides a conceptual model, that is a smart city ecosystem enabled by 6G. This model focus on technology enablers such as artificial intelligence, internet of things, and edge computing, and also gives brief about some concern of implementation, data privacy, and regulation (Norp et al., 2024).

2. EVOLUTION OF COMMUNICATION TECHNOLOGIES AND THE PATH TO 6G

2.1. Brief History: From 2G to 5G and the Advancements in Each Generation

The development of wireless communication has been a process that has experienced awesome shifts in performance, connections and functionalities. Every generation of cellular networks brought in new set of advancements that have revolutionized the manner in which society interact with technologies.

- **2G (1990s):** The second generation or the 2G brought the concept of using digital instead of analog to send messages. It launched the Global System for Mobile Communications (GSM) allows voice communication, that text messaging, (SMS) and fundamental over data services similar to multimedia messaging (MMS). 2G networks have been important in the development of mobile communication since they were responsible for enhancing the quality of calls as well as providing secure data transfer (Inomata et al., 2024).

- **3G (2000s):** With the Bar 3G, the mobile communication went beyond the voice call, test message to internet browsing and video telephony. With this generation the commercial data rates of up to 2 Mbps were made and supported web browsing, email and low-quality multimedia streaming services. The WCDMA (Wideband Code Division Multiple Access) and HSPA (High-Speed Packet Access) technologies became crucial for 3G networks and positively impacted the Mobile internet usage (Ma et al., 2023).
- **4G (2010s):** The fourth-generation (4G) networks, mainly LTE (Long-Term Evolution), changed the mobile communication through providing broadband-speed connection with comparatively low delay time. 4G was a great enabler of high-definition video streaming, mobile gaming, and 'real-time' communication applications such as video conferencing. The advancement in technology showed a great deal of improvement with the introduction of 4G that offered internet speed of up to 100 Mbps making smartphones a necessity in people's daily lives (Wang et al., 2023).
- **5G (2020s):** The change to 5G technology resulted in the data speed jump up to 10 Gbps, minimum latency of 1 ms, and devises connection up to 1M. Best known as the fifth generation of networks, they incorporated further ideas such as network slicing and edge computing to offer specific services and to process information in real time. This development led to IoT, self-driving cars and smart city solutions, which in turn makes the 5G technology as an enabling technology of upcoming technologies (Akgul et al., 2024).

2.2. Key Limitations of 5G That 6G Aims to Overcome

Although 5G enhanced the existing communication network through enhanced speed, capacity, and connectivity, it has its drawbacks, which 6G wants to solve. These limitations include:

- **Coverage and Penetration:** It has been a challenge for 5G networks especially those operating in the millimeter-wave (mm/Wave) to provide broad coverage as is the case with 4G networks. This limitation makes the 5G network to be ineffective in the provision of its services to the rural and remote areas. To overcome these barriers, it is expected that 6G will get enhanced spectrum efficiency and will function at higher range of frequencies (Pires et al., 2024).
- **Energy Efficiency:** And given the fact that current 5G deployments plan to handle tens of billions of connected devices, power is a major issue. Applying the energy-efficient solutions like AI-controlled resource management, and intelligent power distribution 6G's goal is to minimize the CO2 emissions of communication networks (Toumi & Dimitrovski, 2024).
- **Latency and Real-Time Responsiveness:** While 5G has low latency some applications such as holographic communication, brain computer interfaces and real-time digital twins require lower latency than 5G. 6G aims to offer as low latency as possible starting from a few microseconds which will enable instant communication and retaliation (Wang et al., 2023).
- **Network Intelligence:** Enhanced by 5G, the AI engineering inserted some fundamental operations; however, these are rather restricted. 6G will introduce AI into the core of the network so that it will create self-optimizing networks, predictive as well as more intelligent decision-making at the edge.

2.3. Defining 6G: Features, Technologies, and Anticipated Deployment Timeline

6G is another generation which is defined as the enhancement in wireless communication with superior speed, connectivity as well as intelligence compared to the previous generations. Stll in the research and development stage 6G is expected to be commercially deployed within the early 2030s. There is supposed to work in the range of terahertz (THz) and is designed to provide data transmission rates is up to one terabit per second which is hundred time faster than 5G (Huynh & Huh, 2021). Such an increase in speed and bandwidth will enable content of ultra-high-definition, next-generation XR, and IoT demands of almost unimaginable scale. Key Features and Technologies of 6G.

- **Terahertz Communication:** Located in the range of 0. 1-10 THz, the 6G will be capable of delivering incredible bandwidth with low latency for applications such as high-density traffic in the urban area and real time augmented reality and holography.
- **AI-Driven Networks:** 6G will involve the use of AI throughout the network structure with regards to self-organizing, self-healing and self-optimizing network. AI will make networks more intelligent and adaptive by improving the data routing, power usage, and resources utilizing in real time (Faye et al., 2024).
- **Edge and Quantum Computing:** 6G will adopt edge computing that will help in handling data near the source to enhance the speed of analysis. Also, the utilization of quantum computing can be implemented in 6G network giving data security along with higher computational facilities for solving various problems (Ahokangas et al., 2024).
- **Holographic and Extended Reality (XR):** AR, VR, and MR will further become mainstream in the 6G era owing to the increased density provided for them by 6G. Such applications will revolutionize communication, education, health care and entertainment as the simulated augmented environments will be real time and life-like (Borgianni et al., 2023).
- **Ubiquitous Sensing and Connectivity:** In 6G networks, it will be possible to integrate sensing systems which will include the sensing of the environment, people's health, and infrastructure. This ever-entwining network will be the basis for the emergence of smart cities, automobile mobility on own.

3. KEY ENABLING TECHNOLOGIES OF 6G FOR SMART CITIES

Some key technologies which have to be effectively harnessed to ensure that 6G fulfills the promise of making urban environments into smart cities include the following; There is also highlights that these enabling technologies not only extend the functionality of the urban infrastructure but also provide the foundation to the more coordinated and resilient urban system. In the following, we analyze how each of these technologies will contribute to the development of smart cities based on 6G (Strinati et al., 2024).

3.1. Terahertz Communication: Speed, Bandwidth, and Latency Improvements

Another unique characteristic of 6G is the use of the terahertz (THz) frequency bands which has data transfer rates of up to 1 Tbps. Such a drastic rise in speed proves critical to developing the next generation of applications that involve the transfer of huge chunks of data at the blink of an eye. The ability for smart cities will, therefore, be revolutionary as it will support URLLC and high throughput for highly complex tasks such as real-time DT, XR, and massive numbers of AVs (Trichias et al., 2024). There are also drastic decreases in latency with the transition to terahertz communication; it is estimated that 6G networks will achieve latency that can reach at most a few microseconds. In a smart city setting, many decisions may be required within minutes if not seconds for instance in traffic control or in responding to an incident. This speed and reliability will make it possible for different integrated systems to operate optimally and the smart city structure will continue to be effective in its operations regardless of prevailing conditions (Mahmood et al., 2020).

3.2. AI-Driven Networks: How AI and Machine Learning Will Optimize Network Management and Smart City Applications

Intelligent automation and optimization will be carried by AI and embedded in all the components of the communication system in 6G networks. In contrast with previous generations that implemented AI in particular use cases, 6G is going to incorporate the AI into the very core of the network providing features of self-optimization and self-repair. Such integration of AI and machine learning will help to design dynamic network management, where the resources can be adjusted in real-time hence can solve the problem of densely populated urban areas (Yrjölä et al., 2023). Automated functions in smart cities will eventually improve traffic flow, the maintenance of infrastructure, and a host of other factors that define a smart city. Artificial intelligence will involve in range identification out of billions of sensors and connected devices, thereby providing city administrators with a range of data to be used in decision-making concerning urban planning, energy use, and security. Besides, AI will be one of the main drivers to automate most routine operations in the cities, minimizing the need for human interferences, and managing the city's resources more efficiently (Inomata et al., 2021). Due to the AI adaptive approach, new 6G networks will contribute to the development of Intelligent Smart self-sufficient Cities.

3.3. Edge Computing: The Role of Edge Devices in Processing Data Closer to the Source for Real-Time Decision-Making

Another key technology that will influence the 6G networks is known as Edge computing especially concerning Smart cities. Classical models of cloud service virtualized services for computing in which users submit data to centralized processing centers which may take time and so hinder real time reactions. While edge computing is a way of handling data near the source, which might be a sensor on an illuminating streetlamp, a camera at an intersection, or a smart meter in a constructing. Edge computing decreases the distance that data has to travel for analysis, and by doing so, decreases the latency meaning the time it takes for the data to be processed in real time which is essential in application that needs quick decision making (Antón-Haro et al., 2024).

In smart cities, Edge computing will become almost mandatory to support, for example, fully self-driving cars, which require processing of huge amounts of information as soon as possible in order to avoid hazards. It will also enable real time monitoring of the environment in terms of air quality, noise level or the use of water among others. Furthermore, edge devices can assist in the decentralization of some of the workloads from centralized cloud systems making urban networks more scalable and robust. With 6G envisaging AI and Machine learning at the edge, this will enable cities to function with improved local intelligence for decision making in order to improve efficiency as well as service delivery (Cordova Morales et al., 2023).

3.4. IoT and Sensor Networks: Expanding Connectivity and Intelligence Across Urban Infrastructure

The integration of IoT with the features of 6G facilitates an increased and improved connective layer of objects including billions of devices and sensors embedded in cities. These combined devices will be the 'neural network' of smart cities, constantly creating information that underpins choices and underpins numerous operations. Self-sensing traffic lights that change with real-time traffic conditions, self-regulating energy networks that continually adjust to supply and demand, and many others will be IoT-based and will be supported and driven by 6G-enabled smart cities. 6G networks will improve IoT by increasing device density; more devices by square kilometer hence improving the quality of the network. It will also create a smarter sensor network by the integration of AI with edge computing that will help in processing the data locally on the edge rather than continuously communicating with central systems (Zong et al., 2022). These advancements will result in gains in automation and optimization like intelligent garbage collection where collection routes are determined by collecting information or the intelligent lighting systems where brightness is dependent on activity or weather. However, there will be higher level IoT applications that will be supported by 6G, some of which include ubiquitous sensing in which sensors will be integrated into objects and environments.

Moreover, these sensors will monitor the physical state of a city and interact instantaneously with citizens and devices, providing services both intimate (personal) and contextual. This means that cities will be reactive and flexible according to each inhabitant's requirements, leading us into the intelligent smart connected city experience (Liu et al., 2022). It is expected that the major enabling technologies of 6G will co-work in constructing a resilient and intelligent infrastructure for smart cities. Terahertz communication will provide the high speed and bandwidth needed to support demanding applications, with AI-powered networks and edge computing minimizing delays and enabling real-time decision-making. At the same time, a broader number of IoT and sensor networks will enable cities to be more continuously tracked and managed with higher precision. Collectively, these technologies are going to form the building blocks of the smart cities of tomorrow, including connectivity, intelligence, and sustainability.

4. APPLICATIONS OF 6G IN SMART CITIES

With the increasing complexity and connectivity of cities, 6G networking is going to be crucial for transforming urban services. 6G will usher in a new era of smart cities, built to leverage the power of real-time data and improved communication on everything from intelligent transportation systems to more sophisticated environmental monitoring. The following are the major applications of 6G in smart cities:

4.1. Intelligent Transportation Systems: Autonomous Vehicles, Smart Traffic Management, and Vehicle-to-Everything (V2X) Communication

In fact, it will revolutionize how 6G technology is seamlessly integrated into auto and traffic systems that make up a large share of total energy consumption within smart cities, where autonomous vehicles and traffic management will be at the forefront in making solutions effective. 6G technology will be able to deliver ultra-low latency and high-speed communication that will allow vehicles to partake in vehicle-to-everything (V2X) interactions, via which cars can communicate with other cars, infrastructure, pedestrians, as well as other networks in real-time. The ability to share data on the fly, such as traffic information and road conditions, will create an even smarter network of autonomous cars that will, in theory, make driving a lot less dangerous.

6G capacity will extend to smart traffic management as well. With AI and real-time data analytics playing a significant role in 6G networks, adaptive traffic systems in future highways will be able to optimize signal timings, reduce congestion, and improve overall traffic flow. These systems will be able to monitor and react to real-time circumstances, changing routes and signals on the fly in response to traffic incidents or emergencies. Some of these advancements in intelligence can result in reduced drive time — not to mention reduced pollution and energy consumption by reducing idling or stop-start driving (Porambage et al., 2021).

4.2. Urban Planning and Smart Infrastructure: Real-Time Monitoring, Digital Twins, and Adaptive Infrastructure

6G technology will bring about massive changes in urban planning and infrastructure management by merging real-time data with digital twins. These digital twins—essentially virtual replicas of real-world urban environments—will run on 6G networks and help city planners track infrastructure performance and visualize the outcome of proposed changes for better land-use planning. These digital twins will be updated in real-time through data being shared from sensors, providing reliable information that can then be used for construction, maintenance, or disaster management (Suyama et al., 2022). In addition, 6G will give us adaptive infrastructure that can change its basic characteristics in response to changing conditions. Smart buildings will, for instance, automatically adjust lighting, heat, and ventilation in real-time based on occupancy and weather data to reduce energy consumption and operational costs. Smart utilities, such as water and energy distribution systems, will be able to regulate supply and demand effectively, reduce wastage, and provide uninterrupted service using predictive analytics. Additionally, detailed information on population migration, primary use, and environmental impact would enable better urban planning with the help of 6G. This data will support planners in creating cities with a higher quality of life and more sustainability, allowing better use of space, public transportation systems, and green areas (Norp et al., 2024).

4.3. Public Safety and Emergency Response: Enhanced Communication Networks for First Responders and Smart Surveillance Systems

For higher-order emergency situations and public safety, ultra-high-speed and real-time networks suitable for quick and coordinated implementation will be possible through 6G. In emergencies, the local response to critical information needs instant access, like live location tracking, medical data, or situation awareness. With 6G, ultra-reliable low-latency communication (URLLC) will make it possible to securely pass this information between different branches of government and devices. Using AI-driven tools, smart wearables, and augmented reality (AR) interfaces for real-time guidance will facilitate quick action by first responders when dealing with high-pressure situations. 6G will also introduce major improvements in smart surveillance systems to enhance urban security using high-definition video analytics and artificial intelligence. The software will include real-time facial recognition, crowd monitoring, and anomaly detection, providing instant alerts to potential threats or suspicious behavior. Additionally, 6G networks will be critical for enabling large-scale deployment of drones and robots, particularly for surveillance, search and rescue missions, and hazard detection, ensuring faster and safer operations (Inomata et al., 2021).

4.4. Environmental Monitoring and Sustainability: Smart Grids, Pollution Control, and Efficient Resource Management

This shows sustainability is also the main priority for today's cities, and 6G technology will help in keeping a sustainable environment. 6G, powered by expansive IoT (internet of things) networks and advanced sensing functionalities, will allow real-time environmental monitoring. Smart grids will balance energy production and consumption in real-time, allowing renewable forms of energy to be integrated more effectively and reducing reliance on fossil fuels. This will result in a robust, sustainable energy grid. The power capabilities of 6G are also well-suited to pollution control. Real-time sensors, deployed throughout cities to track air and water quality, will take this data to the next step by offering actionable insights about pollution. Smart waste management systems will optimize collection routes with real-time data, minimizing fuel usage and therefore emissions. Further, 6G-driven analytics will project environmental adjustments that a city needs to plan for, such as optimized water distribution in drought conditions or smart cooling systems to reduce excessive heat during prolonged heat waves. It will also enhance efficient resource management in sectors like agriculture, water, and energy (Mohapatra et al., 2023). For example, more accurate monitoring and control provided by 6G networks can help cities become much better at resource management, cutting wastage and building on sustainability initiatives. In effect, 6G-enabled sensors across urban farms will enable precision agriculture, including soil health monitoring, enhanced irrigation practices, and better crop yields. This implies that the scope of 6G applications and services across various urban sectors contribute significantly to transforming cities into smarter, livable, and sustainable communities. 6G will empower real-time communication and data-driven insights through advanced automation to create new dimensions for urban living, benefiting both the quality of life for residents and the operational efficiency needed in city management.

4.5. Transition from 5G to 6G-Based Smart City

Moving from the 5G to the 6G in smart city applications is a unique and challenging process. Some of the world's cities have already begun trials of 6G technologies to support better connection of individuals and organizations, optimize the delivery of public services, and manage data in real time. For instance, Helsinki has initiated a smart city programme in which 6G testbeds are employed to facilitate high-speed links between sensors, vehicles and the environment. For instance, the city of Shanghai has adopted 5G to manage traffic with plans to switch to 6G for better control of autonomous vehicles and traffic flow. These projects show how the next generation of network – 6G – could help to enhance the workings of smart cities. Nonetheless, there are several challenges, including the demand for significant capital expenditure to develop infrastructure and the issue of data protection. When cities grow to have 6G or even 6G, there are new conditions that will require new approaches to regulation.

Current rules on spectrum management and privacy are a crucial factor that needs to change for 6G to be implemented effectively. One of the key themes being explored by many governments is spectrum management, as 6G will require even higher frequencies than 5G, which could cause problems with other technologies. New rules also consider cybersecurity where data flow would be enhanced in 6G connected cities to ensure that it doesn't pose a threat to public safety. South Korea and the United States have started working on their 6G policies that assume a world where connectivity is always seamless, but such policies require a delicate touch lest they hamper innovation. Also, governments and private organizations must work together to outline the guidelines that will govern 6G infrastructure. This is because, to meet the international interconnectivity of the 6G system, these regulations must be consistent across various regions. Despite this, there are still aspects which offer great opportunities, yet present challenges in the development of 6G for smart cities.

5. PROPOSED 6G-ENABLED SMART CITY MODEL

6G technology, when implemented into smart cities, calls for an all-encompassing modular and adaptive design that integrates advanced communication technologies with the urban infrastructure. The next section explains a 6G-based smart city model, proposing and elaborating on the system architecture, data flow and processing, a conceptual use case, as well as discussing issues related to scalability and adaptability (Mohapatra et al., 2023).

5.1. System Architecture: Overview of the Components

A 6G-enabled smart city requires its system architecture to have three main parts: network layers, devices, and control systems. They combine to offer a continuum of networked data transmission, immediate information analysis, and informed decision-making. Smart City of 6G is illustrated in Fig and Table 1.

$$\text{Bandwidth (B)=Frequency Bandwidth (F)}\times\text{Spectral Efficiency } (\eta) \tag{1}$$

$$\text{Latency (L)=Propagation Delay + Processing Delay + Queueing Delay} \tag{2}$$

$$\text{Throughput (T)=Bandwidth} \times \log_2 \left(1+\frac{\text{Signal Power}}{\text{Noise Power}}\right) \tag{3}$$

Table 1. Components of the 6G-Enabled Smart City Architecture

Component	Description
Network Layers	Includes the Terahertz (THz) communication layer, AI-driven network management, and edge computing infrastructure.
Edge Devices	Devices such as sensors, cameras, and IoT nodes that collect and process data at the edge of the network.
Data Aggregators	Centralized and decentralized systems that gather, analyze, and store data from various sources across the city.
Control Systems	AI-based systems that make real-time decisions and manage urban services, including traffic management, public safety, and infrastructure maintenance.

Figure 1. System Architecture of a 6G-Enabled Smart City

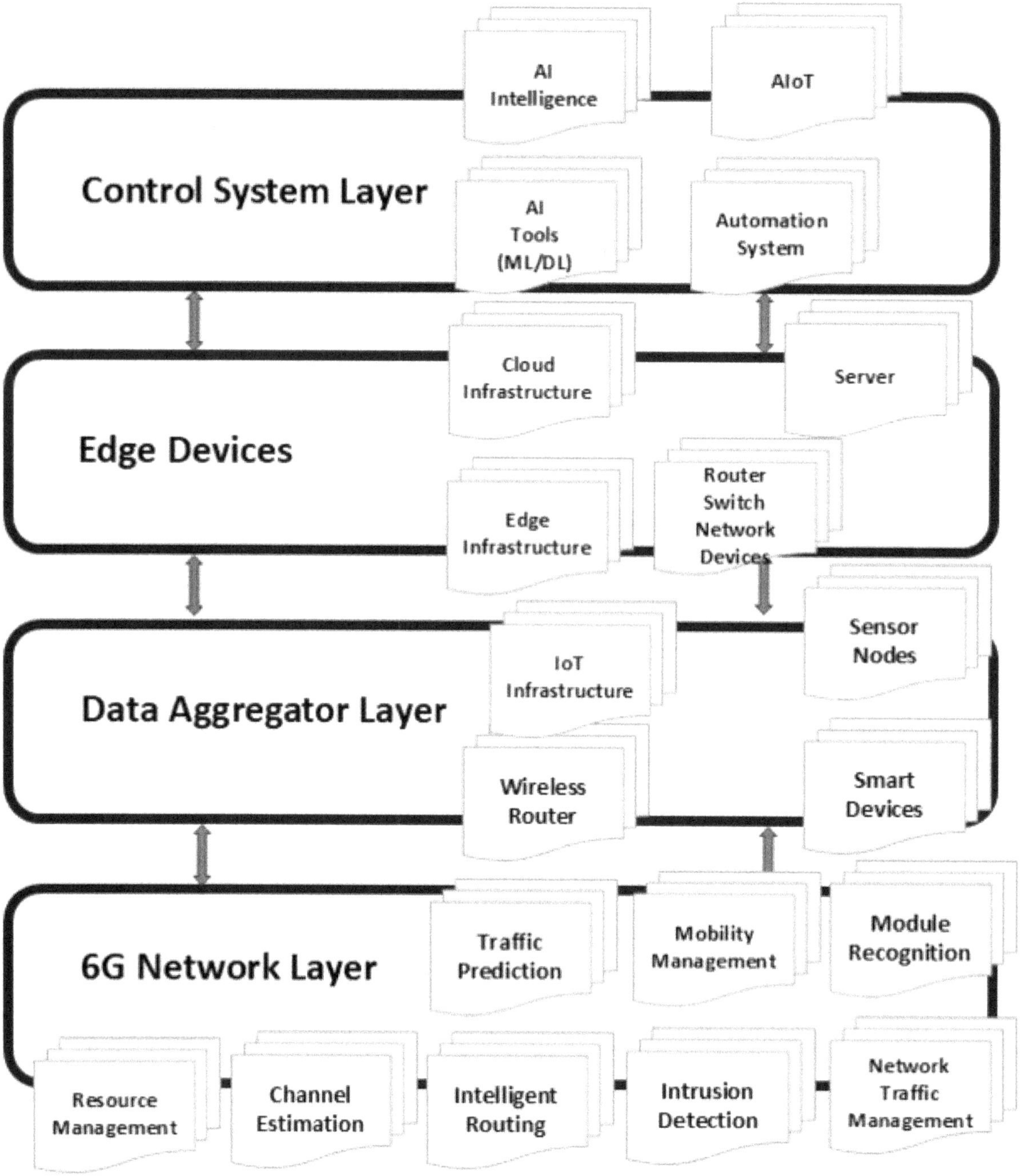

5.2. Data Flow and Processing: How Data Is Collected, Processed, and Utilized in Real-Time

Like running an efficient city management protocol on data in a 6G-capable smart city. The process involves multiple stages (Table 2).

1. **Data Collection:** Collected from diverse sources such as IoT sensors, surveillance cameras, traffic management systems, and environmental monitors. The data these devices generate is wirelessly transmitted to the edge nodes and central servers via 6G networks.
2. **Real-Time Processing:** Edge computing for real-time processing reduces latency so that you can take real-time action on the given data. The data is processed, and AI algorithms extract actionable insights (Stojanovska et al., 2018).

Table 2. Data Processing Stages and Technologies

Stage	Description	Technologies
Data Collection	Gathering data from sensors and devices.	IoT Sensors, Cameras, Wearables
Real-Time Processing	Immediate analysis of data at the edge.	Edge Computing, AI Algorithms
Data Aggregation	Combining and storing data from multiple sources.	Centralized Data Hubs, Cloud Storage
Utilization	Using processed data for managing city services and operations.	AI Control Systems, Analytics

3. **Data Aggregation:** Processed data from edge devices is aggregated at centralized data hubs for further analysis. This aggregation helps in maintaining a holistic view of city operations and supports long-term planning.
4. **Utilization:** The data analyzed by the system can be used to drive urban services, including traffic control, public safety, and incident response. These data-driven control systems can be automated by AI, which makes decisions from the data, optimizing city operations and improving the quality of life for citizens.

5.3. Case Study Example: A Conceptual Smart City Ecosystem Illustrating the Role of 6G

For example, Smart City 2030 demonstrates how 6G technology would reinvent an urban environment of the future (Uusitalo et al., 2021). Showing how the technology would be applied in practice, one of its subsections is dedicated to the city that hosts different smart services, all powered by 6G (Ref: Fig. 2).

Figure 2. Conceptual Layout of Smart City 2030

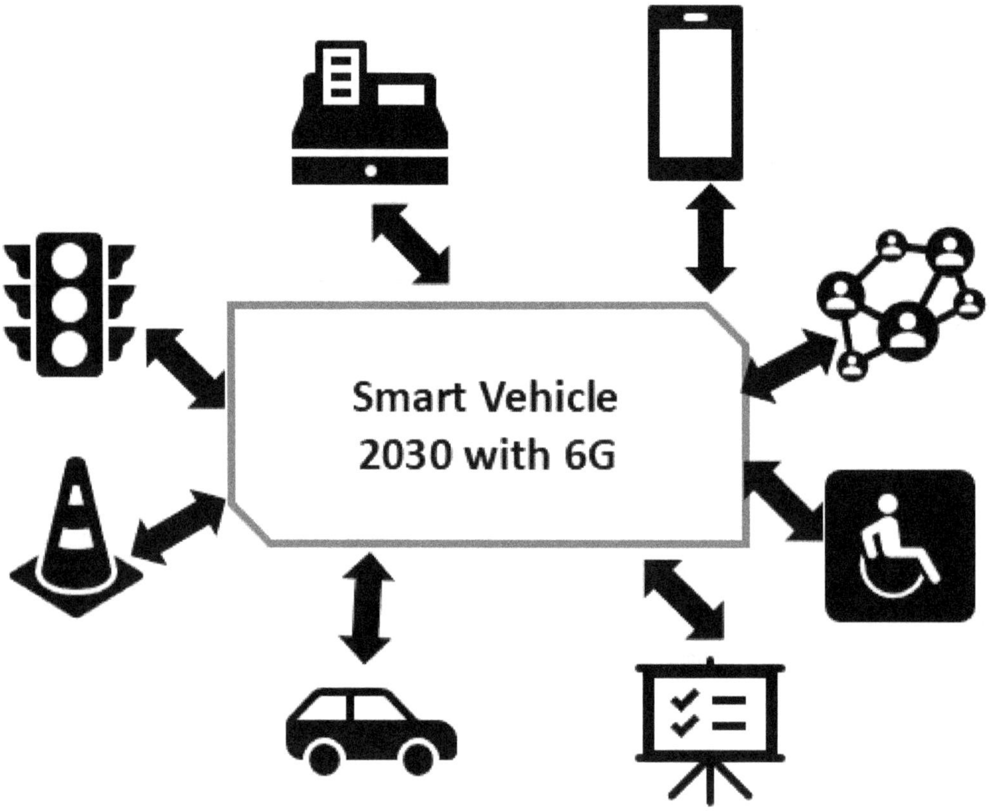

Table 3. Key features of smart city 2030

Feature	Description
Autonomous Vehicles	Self-driving cars with real-time V2X communication, improving safety and efficiency.
Smart Traffic Management	Dynamic traffic signals and adaptive routing based on real-time data analysis.
Smart Infrastructure	Real-time monitoring of buildings and utilities for optimized management and maintenance.
Public Safety Systems	AI-driven surveillance and emergency response systems for enhanced security and faster incident resolution.

Components:

- **Autonomous Vehicles:** 6G enables mixed-traffic (manned, self-driving, or tethered caravan) with vehicle-to-vehicle and vehicle-to-infrastructure communication for safer and smoother traffic.
- **Smart Traffic Management:** V2I communication and AI enable traffic lights to adjust themselves on the fly, reducing congestion and increasing the efficiency of public transportation.
- **Smart Infrastructure:** Buildings and infrastructure are embedded with sensors using digital twins that monitor and manage energy use, structural health, and maintenance.

5.4. Scalability and Flexibility: Designing a System That Adapts to Different City Sizes and Needs

The potential of a 6G-enabled smart city model needs to be designed and implemented with flexibility in terms of scaling to fit various city sizes and requirements [37]. Key considerations include:

1. **Modular Architecture:** The solution can be designed with modular components that are customizable to fit cities of all sizes and needs. This allows for a phased rollout and adaptation.
2. **Adaptive Network Management:** 6G networks must support dynamic resource allocation and automatic optimization to handle different load conditions. AI-driven management systems can adjust network parameters in real-time based on usage patterns.
3. **Customizable Solutions:** Cities face different challenges and priorities. These customizable solutions should adapt to specific needs, such as traffic management in busy areas or environmental monitoring in industrial zones.
4. **Integration with Legacy Systems:** The technology must integrate with existing infrastructure, ensuring smooth communication and migration for higher efficiency, while coexisting with legacy systems.

Table 4. Design considerations for scalability and flexibility

Aspect	Consideration
Modular Architecture	Enables incremental deployment and adaptation to different city sizes.
Adaptive Management	Dynamic resource allocation and real-time optimization based on network conditions.
Customizable Solutions	Tailored applications to meet specific city needs and priorities.
Legacy Integration	Compatibility with existing systems to facilitate a smooth transition and integration.

The smart city model in which 6G will be integrated appears to comprise communication solutions and urban infrastructure as an integrated system architecture. Centered around real-time data processing, adaptive management, and scalability, this model illustrates how 6G could leverage future smart city applications, empowering rapid urban developments in living and operations.

6. CHALLENGES AND CONSIDERATIONS

This lays the groundwork for 6G technology to bring about game-changing evolutions in smart cities, but there are many challenges and considerations to tackle in order to successfully realize these changes. These consist of technological challenges, data privacy and security issues, regulatory landscapes as well as wider economic and social implications. All these components have a fundamental part in the building and shaping of future cities (Garg et al., 2023). Some challenges that may hinder the deployment of 6G are within the realms of regulation, digital divide, and security. Regulatory constraints on the part of the governments might pose a threat to the approval and launching of the technology which may take time. It will also be crucial for the future since not every person will have the means to get an access to 6G technology. Cybersecurity threats are another key challenge, as more complex networks are at greater risk of being compromised. To overcome these problems, the government can reduce complexities of policies and encourage the rapid implementation. In this regard, the collaboration between public and private sectors can contribute to the elimination of the digital gap and make 6G available to everyone. Furthermore, the implementation of standard cybersecurity protocols and periodic checks will secure 6G systems in smart cities.

6.1. Technical Challenges: Network Deployment, Infrastructure Readiness, and Interoperability with Legacy Systems

- **Network Deployment:** Building a 6G network presents significant technical challenges, especially concerning the creation and installation of base stations. The use of terahertz (THz) frequencies introduces the need for novel antenna and base station designs to operate at these higher frequencies. To maintain high-speed, low-latency connections, a dense deployment of these components is required across urban areas. Additionally, ensuring consistent network coverage in both densely populated and less developed regions pose logistical and engineering difficulties [39].
- **Infrastructure Readiness:** Many cities currently lack the infrastructure needed to support the advanced requirements of 6G technology. Upgrading or replacing outdated infrastructure to accommodate new technologies, such as advanced fiber optics and high-frequency antennas, requires significant investment and planning. This includes both physical upgrades to existing infrastructure and the integration of new technologies.
- **Interoperability with Legacy Systems:** A major challenge is ensuring that new 6G systems can effectively interact with existing legacy systems. Many cities rely on older communication technologies and infrastructures that may not be immediately compatible with 6G. Developing solutions that ensure seamless interoperability between new and existing systems is crucial to avoid disruptions and guarantee a smooth transition (Nguyen et al., 2022).

6.2. Data Privacy and Security: Managing Data Ethically While Ensuring Robust Cybersecurity

The importance of data protection and cyber security is paramount to the implementation of the 6G in smart cities. For the sixth generation of networks, the number of connected devices is expected to grow by several orders of magnitude, including traffic lights and household appliances. This leads to more data and more potential targets for cyber criminals to attack. This is because of the increased cases of hacking and soring of personal information. If not protected, then the confidential data can be mishandled. This is because encryption and security networks are crucial to avoid breaches. Thus, governments and companies have to set up strong security measures and policies. This will foster trust in smart cities as they expand with the 6G technology for data privacy.

- **Managing Data Ethically:** With sensors and data recording tools sprouting all over a smart city, running on 6G, managing information ethically becomes crucial. Ensuring that data is collected, stored, and processed in a way that respects individuals' privacy rights is essential. Building public trust requires transparency in data practices, clear consent processes, and user access controls, in line with emerging privacy regulations (Abdel Hakeem et al., 2022).

- **Robust Cybersecurity:** The improvement in networks like 6G will also bring strengthened connectivity and increased data flow, which could attract cyberattacks. Robust cybersecurity measures are critical to protecting personal data and ensuring smart city operations. This includes advanced encryption methods, real-time threat detection systems, and ongoing security checks to prevent breaches and unauthorized access.

6.3. Regulatory and Policy Frameworks: The Role of Governments and International Bodies in Setting Standards and Guiding Development

The role of 6G in smart cities is very promising for sustainability and socioeconomic change since 6G networks can contribute to the implementation of green technologies that save energy and increase the effectiveness of city management. Advanced utility networks such as smart grids, enhanced transportation systems, and IoT-based waste management have the potential to reduce wastage greatly. Nonetheless, the socioeconomic effects may be positive or negative. On the job front, 6G can be seen to open up new avenues for employment in the tech and service sectors but it can also deepen the digital divide between those who are well connected and those who are not. There may also be worries that with connectivity, privacy and data protection may be at risk. For the sustainability, it is crucial that every citizen gains from these developments in the same manner.

- **Government Role:** Governments are central to creating regulatory and policy frameworks that ensure the safe and fair deployment of 6G technologies. This includes setting standards for network performance, data protection, and environmental impact. Governments should also engage resources through research and development programs or financial incentives to accelerate the implementation of 6G technologies.
- **International Standards:** Developing 6G technology is a global undertaking, requiring coordination between international bodies and industry stakeholders. Organizations like the International Telecommunication Union (ITU) and various standards organizations are responsible for creating global standards to ensure compatibility and interoperability across regions. Collaborative efforts are necessary to develop comprehensive guidelines that address the complexities of 6G networks (Antón-Haro et al., 2024).

6.4. Economic and Social Impact: The Implications of 6G on Employment, Digital Equity, and Citizen Well-Being

- **Employment:** The transition to 6G technology will greatly impact employment. While it will create new opportunities in fields such as network engineering, data analytics, and cybersecurity, it may also eliminate roles tied to older technologies. To minimize the negative effects, education and re-training programs will be crucial in helping workers adapt and succeed in this evolving job market.
- **Digital Equity:** Given 6G's high speeds, it can help bridge digital divides in underserved areas. However, digital equity is essential to ensure that all layers of society can access the benefits of 6G. This involves addressing disparities in technology access, affordability, and digital literacy. Policymakers must prioritize inclusive strategies to provide fair access to 6G services and infrastructure (Cordova Morales et al., 2023).

- **Citizen Well-Being:** Enhanced connectivity and smart services from 6G will improve healthcare, transportation, and public services, contributing to better citizen well-being. However, concerns around privacy, surveillance, and the digital divide must be carefully managed. Ensuring that technology is used to improve quality of life while protecting individual rights is key.

While 6G technology promises revolutionary benefits for smart cities, addressing technical challenges, data privacy, security, regulatory frameworks, and social and economic impacts is crucial for successful implementation. By tackling these challenges, cities can leverage 6G to build connected, efficient, and equitable urban environments.

7. FUTURE TRENDS AND INNOVATIONS

When we consider the broader changes 6G technology will bring, it is evident that future cities and societies will be fundamentally informed by an array of novel trends and breakthroughs. These developments have the potential to deeply change the way we interact with technology and one another. We explore some of the most promising trends on the horizon, such as quantum communication, brain-computer interfaces, and AI-powered governance, and consider how 6G may continue to redefine urban life and modern societies at large.

7.1. Emerging Trends

- **Quantum Communication:** Quantum communication is a breakthrough in secure communications. By leveraging principles like entanglement and superposition from quantum mechanics, it offers unparalleled security. Unlike classical methods, quantum systems can detect eavesdroppers, ensuring that data remains secure. As 6G networks evolve, integrating quantum communication could boost data protection and privacy, adding a new layer of security for both personal and institutional communications (Ravikumar et al., 2024).
- **Brain-Computer Interfaces (BCIs):** Brain-computer interfaces work by translating brain activity into commands, allowing users to control devices or machines. These interfaces could enable users to operate smart devices, navigate virtual environments, or even communicate telepathically. For smart cities, BCIs could revolutionize accessibility, enabling individuals with disabilities to interact more seamlessly with their surroundings. The integration of BCIs with 6G networks would allow real-time, high-bandwidth communication between the brain and digital systems, opening new possibilities for both personal and professional applications (Inomata et al., 2024).
- **AI-Driven Governance:** The future of urban governance will be built around Artificial Intelligence (AI). AI-driven systems can analyze large datasets to make data-backed decisions, optimize resource management, and predict trends. In smart cities, AI could help manage traffic, reduce energy consumption, and support emergency responses and public health. By harnessing AI's data processing capabilities, city administrators can make more precise and efficient decisions, improving urban management and quality of life (Liu et al., 2022).

7.2. Continuous Advancements in 6G

- **Redefining Urban Living:** As 6G technology evolves, it will make cities smarter and more interactive. Enhanced connectivity will support advanced applications like real-time augmented reality (AR) and virtual reality (VR) experiences, increasing the interactivity of urban spaces. For example, AR could help residents by projecting navigation information onto their surroundings or enabling virtual community events. 6G's faster data speeds and reduced latency will also allow seamless integration of smart devices into real-time urban response systems (Ma et al., 2023).
- **Transforming Societal Structures:** As 6G advances, societal structures will shift. Remote work and digital collaboration will become more common, potentially reshaping traditional workspaces and social interactions. Telepresence and real-time data analysis will enhance personalized education and healthcare. Additionally, 6G innovations could drive sustainable urban planning and climate resilience, contributing to more livable and environmentally friendly cities (Singh et al., 2023).
- **Innovation Ecosystems:** The convergence of 6G with emerging technologies will create new innovation ecosystems. Collaboration between tech companies, universities, and government bodies will be essential to address integration challenges and ensure these technologies benefit society in a sustainable and equitable way (Norp et al., 2024).

Urban living and societal structures will be shaped by a range of groundbreaking technologies. As 6G continues to develop, it will evolve beyond current smart city applications, enabling new forms of interaction and governance. Cities embracing these trends will create connected, efficient, and inclusive environments, improving the quality of life for all residents (Kumar Singh et al., 2021).

8. CONCLUSION AND FUTURE SCOPE

This conclusion ties back to the topic "6G Technology: A Game-Changer for Smart Cities" We hope that we were able to create an image in your mind about how much this next-generation of a network holds potential. The incorporation of 6G in urban areas signifies a great stride towards better connectivity, efficiency, and innovation. 6G will enable ultra-fast data transmission, low latency communication and next generation networks for new era of technology usage and city management. One of the next areas that 6G technology plans to tackle are smart cities, enabling a whole new level of connectivity and data processing. It will provide the infrastructure for real-time interaction and smart decision-making as technology continues to progress, e.g., terahertz communication, AI-driven networks. These will create the way for smarter infrastructure, renewable energy grid and overall enhance life quality of its residents. 6G, affecting everything from autonomous vehicles and smart traffic systems to advanced healthcare and environmental monitoring.

Thoughts on Cross-Industry Collaboration and Forward-Looking Policies:

It is important to mandate new rules default settings, that push us forward and no back. 6G networks pose many a challenge, which must be met by the concerted efforts of technology providers, city planners, regulatory bodies and academic institutions. By working together, we can ensure that technology is deployed responsibly to maximize the benefits and minimize the potential harms in terms of privacy, security, and digital equity. In addition, we need to develop forward-looking policies and standards to ensure that the emerging technologies and architectures respond effectively to future needs. Such policies should include issues such as privacy-by-design principles, which are particularly relevant in 6G when data assimilation will come from a large variety of sources. Policies should also be developed on protectionism vs. global integration, innovation promotion measures with regards to economic behavior (such as patents), access and divide issues (as technology advances between generations are always uneven) or flexible spectrum licensing. Finally, and very important, regulatory agencies must be able to keep pace the technological evolution so that regulations can be adopted before technology arrives not after.

Preparing for the 6G Era

With the advent of 6G, it is important that strategic planning and continued research efforts are made to fully capitalize on its potential. Stakeholders should allocate funds for 6G infrastructure, R&D, and education in preparation for these technological revolutions, including:

- **Strategic Planning:** Develop overall plans to incorporate CoG technologies within current and future urban infrastructures, where broad domains to address will be recognized as well as potential challenges.
- **Research and Development:** This shall encompass ongoing research on 6G technologies and their applications along with stimulating AI innovations in the areas of network design, security, and user experience. This also includes promoting such developments in applications based on AI systems.
- **Workforce Training:** Ensure that the workforce is ready and able to effectively manage and operate using new 6G technologies. Develop training programs, and educational assistance programs to support professionals in adapting to their changing technology environment.

REFERENCES

Abdel Hakeem, S. A., Hussein, H. H., & Kim, H. (2022). Security requirements and challenges of 6G technologies and applications. *Sensors (Basel)*, 22(5), 1969. DOI: 10.3390/s22051969 PMID: 35271113

Ahokangas, P., Matinmikko-Blue, M., Basaure, A., & Yrjölä, S. (2024). Use cases for local 6G networks. *2024 Joint European Conference on Networks and Communications & 6G Summit (EuCNC/6G Summit)*, 1127-1132. https://doi.org/DOI: 10.1109/EuCNC/6GSummit60053.2024.10597019

Akgul, O. U., Ericson, M., Burguera, P. G., Arar, B., Khorsandi, B. M., Querio, R., Varvara, A., Harkous, H., Wänstedt, S., Groshev, M., De la Oliva, A., Pieczerak, J., Kukliński, S., Nande, S. S., Bassoli, R., & Fitzek, F. H. (2024). Discussion on 6G architecture evolution: Challenges and emerging technology trends. *2024 Joint European Conference on Networks and Communications & 6G Summit (EuCNC/6G Summit)*, 664-669. https://doi.org/DOI: 10.1109/EuCNC/6GSummit60053.2024.10597056

Antón-Haro, C., Trichias, K., De Majo, C., Kaloxylos, A., & Beriere, J. (2024). 6G smart networks and services: Global strategies, main work directions & Future outlook. *2024 Joint European Conference on Networks and Communications & 6G Summit (EuCNC/6G Summit)*, 1115-1120. https://doi.org/DOI: 10.1109/EuCNC/6GSummit60053.2024.10597046

Borgianni, L., Adami, D., & Giordano, S. (2023). Optimizing network performance and reliability with an integrated SD-WAN and satellite 6G architecture. *2023 2nd International Conference on 6G Networking (6GNet), 21*, 1-4. https://doi.org/DOI: 10.1109/6GNet58894.2023.10317704

Cordova Morales, D. A., Nguyen, T., & Pujolle, G. (2023). Towards a blockgraph-based trustless authentication scheme for future 6G technology. *2023 2nd International Conference on 6G Networking (6GNet)*. https://doi.org/DOI: 10.1109/6GNet58894.2023.10317665

Faye, S., Camelo, M., Sottet, J., Sommer, C., Franke, M., Baudouin, J., Castellanos, G., Decorme, R., Fanti, M. P., Fuladi, R., Kesik, G., Mendes, B., Murphy, C., Parker, S., Pryor, S., Senouci, S. M., & Turcanu, I. (2024). Integrating network digital twinning into future AI-based 6G systems: The 6G-TWIN vision. *2024 Joint European Conference on Networks and Communications & 6G Summit (EuCNC/6G Summit)*, 883-888. https://doi.org/DOI: 10.1109/EuCNC/6GSummit60053.2024.10597058

Garg, S., Kaur, K., Aujla, G. S., Kaddoum, G., Garigipati, P., & Guizani, M. (2023). Trusted explainable AI for 6G-Enabled edge cloud ecosystem. *IEEE Wireless Communications*, 30(3), 163–170. DOI: 10.1109/MWC.016.220047

Huynh, L. N., & Huh, E. (2021). Envisioning edge computing in future 6G wireless networks. *2021 Fifth World Conference on Smart Trends in Systems Security and Sustainability (WorldS4)*, 307-311. https://doi.org/DOI: 10.1109/WorldS451998.2021.9514026

Inomata, M., Yamada, W., Kuno, N., Sasaki, M., Kitao, K., Nakamura, M., Ishikawa, H., & Oda, Y. (2021). Terahertz propagation characteristics for 6G mobile communication systems. *2021 15th European Conference on Antennas and Propagation (EuCAP)*. https://doi.org/DOI: 10.23919/EuCAP51087.2021.9411143

Kumar Singh, S., Pan, Y., & Hyuk Park, J. (2021). OTS scheme based secure architecture for energy-efficient IoT in edge infrastructure. *Computers, Materials & Continua*, 66(3), 2905–2922. DOI: 10.32604/cmc.2021.014151

Liu, H., Zong, J., Wang, Q., Liu, Y., & Yang, F. (2022). Cloud native based intelligent RAN architecture towards 6G programmable networking. *2022 7th International Conference on Computer and Communication Systems (ICCCS)*. https://doi.org/DOI: 10.1109/ICCCS55155.2022.9846266

Ma, G., Khalili, M., Parthiban, R., & Katz, M. (2023). A low-complexity handover scheme using unsupervised learning techniques for 6G multi-networking. *2023 2nd International Conference on 6G Networking (6GNet), 101*, 1-5. https://doi.org/DOI: 10.1109/6GNet58894.2023.10317655

Mahmood, N. H., Alves, H., Lopez, O. A., Shehab, M., Osorio, D. P., & Latva-Aho, M. (2020). Six key features of machine type communication in 6G. *2020 2nd 6G Wireless Summit (6G SUMMIT)*. https://doi.org/DOI: 10.1109/JIOT.2022.3216402

Mohapatra, H., & Dalai, A. K. (2022). IoT based V2I framework for accident prevention. *2022 2nd International Conference on Artificial Intelligence and Signal Processing (AISP), 0*, 1-4. https://doi.org/DOI: 10.1109/AISP53593.2022.9760623

Mohapatra, H., Mohanta, B. K., Nikoo, M. R., Daneshmand, M., & Gandomi, A. H. (2023). MCDM-based routing for IoT-enabled smart water distribution network. *IEEE Internet of Things Journal*, 10(5), 4271–4280. DOI: 10.1109/JIOT.2022.3216402

Nguyen, D. C., Ding, M., Pathirana, P. N., Seneviratne, A., Li, J., Niyato, D., Dobre, O., & Poor, H. V. (2022). 6G Internet of things: A comprehensive survey. *IEEE Internet of Things Journal*, 9(1), 359–383. DOI: 10.1109/JIOT.2021.3103320

Norp, T., Sachdeva, P., Anton, C., Trichias, K., & Mohnani, P. (2024). 6G research & Innovation activities in Europe: An overview of EU & Nationally funded programmes. *2024 Joint European Conference on Networks and Communications & 6G Summit (EuCNC/6G Summit)*, 1109-1114. https://doi.org/DOI: 10.1109/EuCNC/6GSummit60053.2024.10597076

Pires, R., Blue, H., Malinen, J., De Angelis, M., Giardina, P. G., Landi, G., Laukkanen, M., Ahola, K., & Porambage, P. (2024). Closed-loop automation in 6G for minimum downtime task continuity in surveillance Cobots. *2024 Joint European Conference on Networks and Communications & 6G Summit (EuCNC/6G Summit)*, 860-865. https://doi.org/DOI: 10.1109/EuCNC/6GSummit60053.2024.10597048

Porambage, P., Gur, G., Moya Osorio, D. P., Livanage, M., & Ylianttila, M. (2021). 6G security challenges and potential solutions. *2021 Joint European Conference on Networks and Communications & 6G Summit (EuCNC/6G Summit)*. https://doi.org/DOI: 10.1109/EuCNC/6GSummit51104.2021.9482609

Ravikumar, S. S., Betgeri, S., & Singh, S. K. (2024). Strategies for mitigating security concerns in IoT-enabled smart cities. *Advances in Computational Intelligence and Robotics*, 239-273. https://doi.org/DOI: 10.4018/979-8-3693-2373-1.ch012

Singh, S. K., Yang, L. T., & Park, J. H. (2023). FusionFedBlock: Fusion of blockchain and federated learning to preserve privacy in industry 5.0. *Information Fusion*, 90, 233–240. DOI: 10.1016/j.inffus.2022.09.027

Siriwardhana, Y., Porambage, P., Liyanage, M., & Ylianttila, M. (2021). AI and 6G security: Opportunities and challenges. *2021 Joint European Conference on Networks and Communications & 6G Summit (EuCNC/6G Summit)*. https://doi.org/DOI: 10.1109/EuCNC/6GSummit51104.2021.9482503

Stojanovska, S., Madzova, V., & Matlievska, M. (2018). The impact of human capital on sustainable growth - The case of Republic of Macedonia. *Knowledge based sustainable economic development*, 884. https://doi.org/DOI: 10.1016/j.inffus.2022.09.027

Strinati, E. C., Di Lorenzo, P., Sciancalepore, V., Aijaz, A., Kountouris, M., Gündüz, D., Popovski, P., Sana, M., Stavrou, P. A., Soret, B., Cordeschi, N., Scardapane, S., Merluzzi, M., Zanzi, L., Renato, M. B., Quek, T., Pietro, N. D., Forceville, O., Costanzo, F., & Li, P. (2024). Goal-oriented and semantic communication in 6G AI-native networks: The 6G-GOALS approach. *2024 Joint European Conference on Networks and Communications & 6G Summit (EuCNC/6G Summit), 5*, 1-6. https://doi.org/DOI: 10.1109/EuCNC/6GSummit60053.2024.10597087

Suyama, S., Okuyama, T., Nonaka, N., & Asai, T. (2022). Recent studies on massive MIMO technologies for 5G evolution and 6G. *2022 IEEE Radio and Wireless Symposium (RWS)*. https://doi.org/DOI: 10.1109/RWS53089.2022.9719949

Toumi, N., & Dimitrovski, T. (2024). AI-native architecture for 6G networks and services with model dependencies. *2024 Joint European Conference on Networks and Communications & 6G Summit (EuCNC/6G Summit)*, 901-906. https://doi.org/DOI: 10.1109/EuCNC/6GSummit60053.2024.10597080

Trichias, K., Kaloxylos, A., & Willcock, C. (2024). 6G global landscape: A comparative analysis of 6G targets and technological trends. *2024 Joint European Conference on Networks and Communications & 6G Summit (EuCNC/6G Summit)*, 1-6. https://doi.org/DOI: 10.1109/EuCNC/6GSummit60053.2024.10597064

Uusitalo, M. A., Rugeland, P., Boldi, M. R., Strinati, E. C., Demestichas, P., Ericson, M., Fettweis, G. P., Filippou, M. C., Gati, A., Hamon, M., Hoffmann, M., Latva-Aho, M., Parssinen, A., Richerzhagen, B., Schotten, H., Svensson, T., Wikstrom, G., Wymeersch, H., Ziegler, V., & Zou, Y. (2021). 6G vision, value, use cases and technologies from European 6G flagship project Hexa-X. *IEEE Access : Practical Innovations, Open Solutions*, 9, 160004–160020. DOI: 10.1109/ACCESS.2021.3130030

Wang, C., You, X., Gao, X., Zhu, X., Li, Z., Zhang, C., Wang, H., Huang, Y., Chen, Y., Haas, H., Thompson, J. S., Larsson, E. G., Renzo, M. D., Tong, W., Zhu, P., Shen, X., Poor, H. V., & Hanzo, L. (2023). On the road to 6G: Visions, requirements, key technologies, and Testbeds. *IEEE Communications Surveys and Tutorials*, 25(2), 905–974. DOI: 10.1109/COMST.2023.3249835

Wang, Q., Liu, Y., Wang, Y., Xing, Y., Zong, J., & Liu, H. (2023). Deep edge node enables AI native toward 6G*. *2023 IEEE International Symposium on Broadband Multimedia Systems and Broadcasting (BMSB)*, 1-6. https://doi.org/DOI: 10.1109/BMSB58369.2023.10211613

Yrjölä, S., Matinmikko-Blue, M., & Ahokangas, P. (2023). Developing 6G visions with stakeholder analysis of 6G ecosystem. *2023 Joint European Conference on Networks and Communications & 6G Summit (EuCNC/6G Summit)*. https://doi.org/DOI: 10.1109/EuCNC/6GSummit58263.2023.10188379

Zong, J., Liu, Y., Liu, H., Wang, Q., & Chen, P. (2022). 6G cell-free network architecture. *2022 IEEE 2nd International Conference on Electronic Technology, Communication and Information (ICETCI).* https://doi.org/DOI: 10.1109/ICETCI55101.2022.9832308

Chapter 9
Multisensor Data Fusion and Its Application

Gregory Omozuhiomwen Egbe
https://orcid.org/0009-0005-5813-9691
University of Benin, Nigeria

Stella Ndidi Arinze
https://orcid.org/0000-0003-3762-0433
Enugu State University of Science and Technology, Nigeria

Emenike Raymond Obi
https://orcid.org/0000-0003-4976-6181
RaySoft AssetAnalytics, Canada

Augustine O. Nwajana
https://orcid.org/0000-0001-6591-5269
University of Greenwich, UK

ABSTRACT

Smart industrial processes are impossible without the use of machinery. Majority of these machines comprise rotating machines. As supporting tools for developing maintenance strategies, a Data-driven approach for multifault diagnosis of Rotating Machines is currently used in smart industry. This Chapter presents Multisensor Data Fusion and its application in Predictive maintenance methodology for multifault diagnosis of Industrial Rotating Machines. The main objective of Multisensor Data Fusion, rather than the different sensors being applied independently, are optimally fused to take advantage of their respective strengths, by combining data from multiple sensors and related information to achieve specific inferences than could be achieved by a single sensor. This chapter presents a methodology using data extraction features domain parameters for multifault diagnosis of Rotating Machines. The analysis shows that Root Mean Square Amplitude increases as fault develops, Crest Factor is more for healthy bearing and less for faulty bearing, these can provide early warning of faults

DOI: 10.4018/979-8-3693-8799-3.ch009

1. INTRODUCTION

Smart industrial processes are impossible without the use of machinery. Majority of these machines comprise rotating machines. As supporting tools for developing maintenance strategies, a Data-driven approach for multifault diagnosis of Rotating Machines is currently used in smart industry. The concept of multisensor data fusion is hardly new. As humans and animals have evolved, they have developed the ability to use multiple senses to help them survive. Data fusion is carried out by the human brain when, for example, associating images and sound while watching television. Some interesting examples of human and other animal data fusion processes are discussed by Luo and Kay (1989). The fusion of multiple information by humans occurs every time the senses are stimulated by appropriate signals. Our sensors, for example eyes, ears, nose, tongue and skin, are fusing sight, hearing, smell, taste and tactile information in our brain. The sound of a voice combined with visual information helps in identifying a person; Figure 1 illustrates this example.

Multisensory data fusion is naturally performed by animals and humans to assess more accurately the surrounding environment and to identify threats, thereby improving their chances of survival. While the concept of data fusion is not new, the emergence of new sensors, advanced processing techniques, and improved processing hardware have made real-time fusion of data increasingly viable.

In recent years, significant attention has focused on multisensor data fusion for both military and nonmilitary applications. Data fusion techniques combine data from multiple sensors and related information to achieve more specific inferences than could be achieved by using a single, independent sensor.

Figure 1. Illustration of the human data fusion system

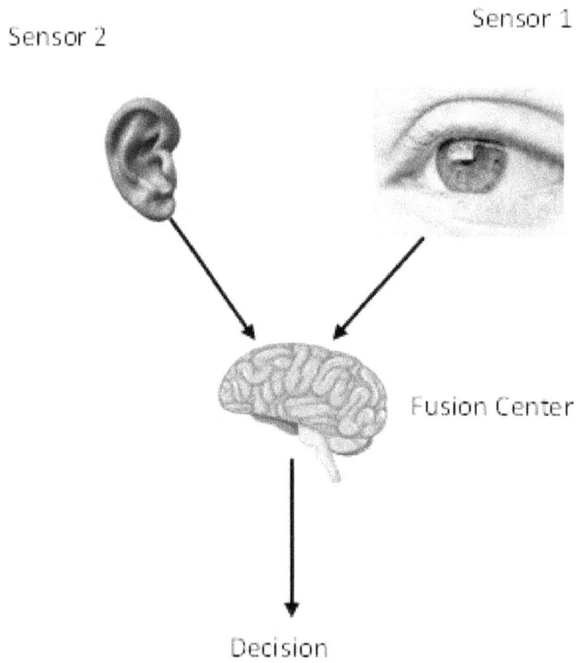

Hall and Llinas (1997) Defined data fusion as a multi-level process dealing with the association, correlation, combination of data and information from single and multiple sources to achieve refined position, identify estimates and complete and timely assessments of situations, threats and their significance. Hence Multi-sensor data fusion is the process of combining sensory data or data derived from disparate sources such that the resulting information has less uncertainty than would be possible when these sources used individually. A multisensor data fusion is an emerging technology applied to areas such as automated target recognition (ATR), identification-friend-foe-neutral (IFFN) recognition systems, battlefield surveillance, and guidance and control of autonomous vehicles, monitoring of complex machinery such as industrial rotating machines, environmental surveillance and monitoring systems, medical diagnosis, and smart buildings.

Multiple heuristic and analytical techniques for data fusion have appeared in the literature during the last 20 years. Table 1 presents a survey of the most common of these. From the literature, no optimal combination technique has been proposed, all varying from one application to another. However, the most commonly used theories appear to be Bayesian probabilistic reasoning and Dempster-Shafer theory of evidence.

Table 1. Survey of most common data fusion and integration methods

Fusion method	Applications	References
Pixel level fusion	Image processing, image segmentation	Buede (1998)
Bayesian theory	Decision making between multiple hypotheses	Durrant-Whyte (1991)
Dempster-Shafer theory of evidence	Decision making, beliefs intervals	Parra-Loera (1991); Blackman (1988)
Neural network	Signal interpretation	Liu et.al., (2018)
Neyman-Pearson criteria	Decision making	Viswanathan (1998)
Fuzzy logic	Handle vagueness	Xi-yang (2018)
Knowledge based system	Pattern recognition	Okoh (2017)
Markov random field	Image processing	Lee (1989)

Federico Castanedo (2013) enumerated and explained different classification schemes for data fusion, and reviewed the most common algorithm. His methods and algorithms were presented using three different categories: (i) data association, (ii) state estimation, and (iii) decision fusion. He observed that the selection of the most appropriate technique depends on the type of the problem and the established assumptions of each technique. Lionei et al (2022) discussed on the task of data association via uniquely defined similarity functions and reviewed current literature about various techniques which have been used to advance the performance in Multiple Object Tracking (MOT) for vehicle and pedestrian scenarios. These various techniques are show in figure 2. This classification of techniques has been categorized based on the association structure and mechanism, algorithms used and its underlying theoretical structure.

Figure 2. Data association techniques

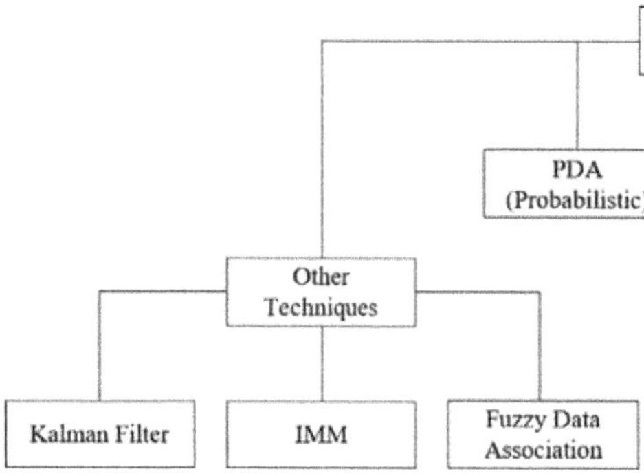

The classification of Probabilistic Data Association is show in figure 3 and the literature review summary advantages and disadvantages are in table 2.

Figure 3. A classification of probabilistic data association methods

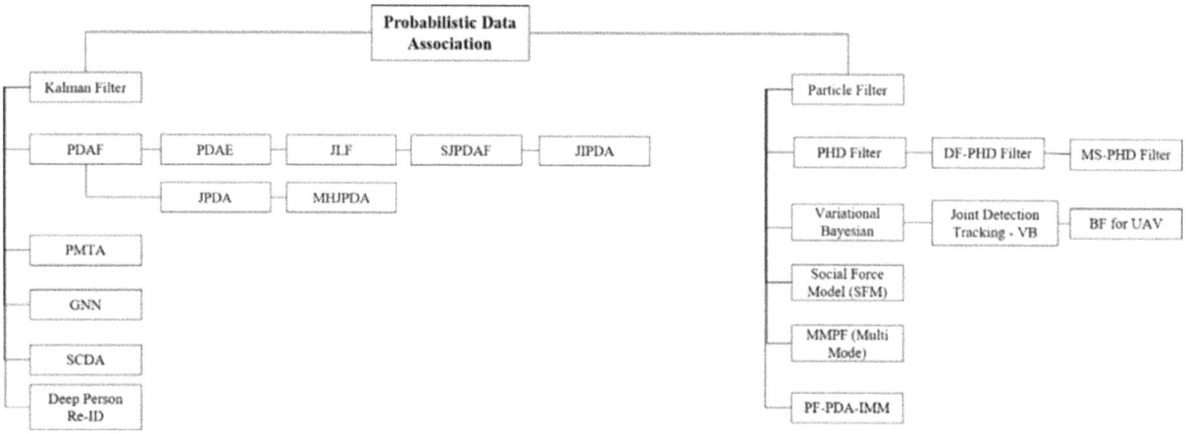

Table 2. Probabilistic methods with a summary of the advantages and disadvantages for each method

Technique	Advantages	Disadvantages
Permutation Matrix Track Association (PMTA) (Lee, Kanazawa et al., 2018)	Stability, Robustness	Customized to a specific scenario
Multi Model Smooth Variable Structure Filter (MMSVSF) (Luo et al., 2019)	Accuracy, robustness	Need more dynamic movements of surrounding objects
Global Nearest Neighbour (GNN) for Field-of-View Sensors (Cabrera, 2018)	Low number of lost tracks	Performance reduces over more objects entering the experiment

continued on following page

Table 2. Continued

Technique	Advantages	Disadvantages
Social Force Model (SFM) within Markov Chain Mote Carlo (MCMC) (Feng et al., 2016)	Improved localization and cardinality	Computational complexity
Structural Constraints Data Association (Yoon et al., 2019)	Validated under unexpected camera motion, reduction in mis-detections and false positives	Less focus on temporal constraints like limited video lengths
Multi-Mode Particle Filter (PF) Data Association (Ritter et al., 2018	Improved accuracy in detection and tracking	No method to estimate bias, enhancement in tracking accuracy is small
Joint Detection and Tracking - Variational Bayesian (JDT-VB) (Lan et al., 2016)	Improved state estimation	No simulation verification
Joint Likelihood Filter	Improvement in handling occlusion	Confusion in classification of noise for some scenarios
Bayesian Filtering for Unmanned Aerial Vehicles (UAV) (Barkley & Paley, 2017)	Maintains tracker accuracy	Average accuracy for motion model of targets
Sample Based JPDAF (SJPDAF) (Zhang & Tang, 2018)	Accuracy improvement with inter-object occlusion	Computational complexity
Markov Chain Data Association with Joint Integrated PDA (JIPDA) (Lee et al., 2017b)	Reduced computational time, improved accuracy in dense clutter	Average results with false track discrimination and target retention
Poisson Spatial Measurement Model for JPDA (Yang et al., 2018)	Requires no clustering and partitioning	Number of target tracks is fixed
Likelihood Based Data Association with Sampling Methods (Granström, Svensson et al., 2017)	Improved tracking performance	Complexity with implementation efficiency
JPDA for extended target tracking (ETT) (Vivone et al., 2015	High tracking accuracy, limited false alarm rate, support real time requirements	Results consider the complete MOT framework with marine radar data
Improved JPDA (Hamid Rezatofighi et al., 2015)	Reduced computational complexity, good performance with noisy detections and occlusion	Average speed and not generalized
Multiple Hypothesis JPDA (MH-JPDA) with fixed interval (Stauch et al., 2017)	Tracking precision and accuracy with close spaced objects, real-time functionality	Initial ambiguity
Tracklet level association in MHT(Sheng et al., 2018	Computational efficiency in solving MWIS problems with MHT	No specific improvement in tracking performance compared to existing algorithms
JPDA for automated multi target tracking (Mahemuti et al., 2016)	High confidence measurement data association can improve tracking performance	Reduced performance in more complex traffic scenarios
PF-PDA-IMM scheme (Chalvatzaki et al., 2018)	Improved accuracy, robustness to occlusion	Requires development for more complex motion models
JPDA in trajectory optimization (He et al., 2020b)	Single scan approach that reduces computational burden	Accuracy reduces as the count of modality increases
MHT in Graph neural networks (Rangesh et al., 2021)	Capable of working in real time	A difficult estimation problem with lower particles
JPDA with reinforcement learning (Qu et al., 2020)	Shorter execution time and effective in dense clutter	Temporal range is set to a limited time space

The classification of Hierarchical Data Association is show in figure 4 and the literature review summary advantages and disadvantages are in table 3.

Figure 4. Breakdown of Hierarchical Data Association Methods

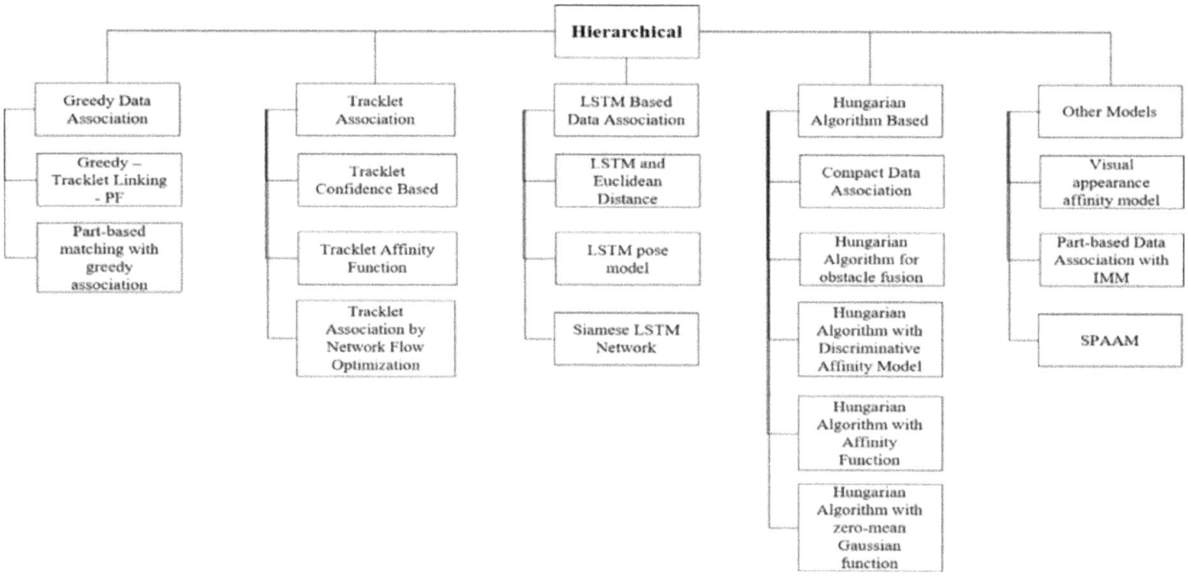

Table 3. Hierarchical Methods with a summary of the advantages and disadvantages for each method.

Technique	Advantages	Disadvantages
Greedy data association with tracklet linking and particle filters (Singh et al., 2017)	Focus on high-speed performance	Slight decrease in qualitative performance
Hierarchical data association tracking based on tracklet confidence(Chen et al., 2018)	Better single target tracking with appearance change and object re-identification	High false alarm rate of detections and poor re-identification after long term occlusion
LSTM and Euclidean distance (Tan et al., 2018)	Good adaptation to appearance changes of objects, fast performance	Average computational efficiency
Visual appearance affinity model (Bewley, Ott et al., 2016)	Model can learn without explicit labelling	Offline based only, some track fragmentation
Tracklet confidence-based data association (Bae & Yoon, 2017)	Accuracy, robustness	Only applies to linear motions
Part-based matching, linear programming and greedy data association (Zhang et al., 2018)	High track initiation rate	Poor results on occlusion and low-resolution video
Greedy data association with tracklet linking and particle filters (Jiang & Huynh, 2017)	Reliable tracklet generation, occlusion handling, linear computational complexity	Average robustness, requires appearance and motion model evaluations
LSTM pose model with region-based appearance model model (Xu & Zhou, 2018)	Speed with deep feature extraction, occlusion handling	Requires more tracking accuracy by adding more association cues
Hungarian algorithm for obstacle fusion and tracking association (Allodi et al., 2016)	Robust, fast processing time	Need improvement in maintaining track
Siamese LSTM network (Wan et al., 2018)	Accuracy	Computational complexity

continued on following page

Table 3. Continued

Technique	Advantages	Disadvantages
Discriminative affinity model with Hungarian algorithm (Li et al., 2017)	Tracks through occlusion and long periods of time	Requires more generalization, accuracy scores are comparable but not significantly improved
Tracklet affinity function association by generalized linear assignment (Yang et al., 2018)	Improved tracking consistency	Low computational efficiency
Confidence score-based appearance model (Liu et al., 2018)	Accuracy, fast processing time, robustness	Results are comparable on MOT datasets
Part-based data association with IMM tracking (Jiang et al., 2017)	Robustness	Does not capture temporal information effectively
Hungarian algorithm with zero-mean Gaussian function(Wu & Li, 2016)	Speed promotion	Missing detections lead to false tracking alarms
Stochastic Progressive Association Across Multiple Frames (SPAAM) (Elliethy & Sharma, 2018)	Accuracy, computational efficiency	Challenge with limited spatial details
LSTM-based aggregated mode (Chang et al., 2020)	Improved prediction over LSTM	some environmental factors make it difficult to verify testing values
Hungarian algorithm in R-CNN (Daniłowicz, 2020	Used in real time situations	Limited information about dealing with temporal information
Hungarian algorithm for the assignment problem (Meneses et al., 2020)	improved tracking accuracy and efficient computation	IOU and SORT trackers compare a faster identity switch rate
LSTM in a Graph Neural Network (Weng et al., 2020	improves discriminative feature learning	limited information on speed
Mahalanobis distance and Hungarian algorithm (Salscheider, 2021)	reduced miss rate and localization error	requires some computational time
Adaptive fusion model based on kalman filtering and LSTM LSTM (Wang et al., 2021)	accuracy and speed improvement	limited specific information on robustness
Hungarian algorithm applied in an LSTM framework (Yu et al., 2020)	Shows stability in both long-term and short-term	Some noise such as fog creates lesser performance
Bilinear LSTM and greedy association (Kim et al., 2021	Achieves real time tracking performance	Limited information on comparison of computational cost
LSTM with tracking association (Farhodov et al., 2020)	Supports long term tracking in real time and can work in online mode	Limited information on comparison of computational cost

The classification of Other Data Association is show in figure 5 and the literature review summary advantages and disadvantages are in table 4.

Figure 5. Breakdown of other techniques; IMM, Kalman Filter, Fuzzy Association.

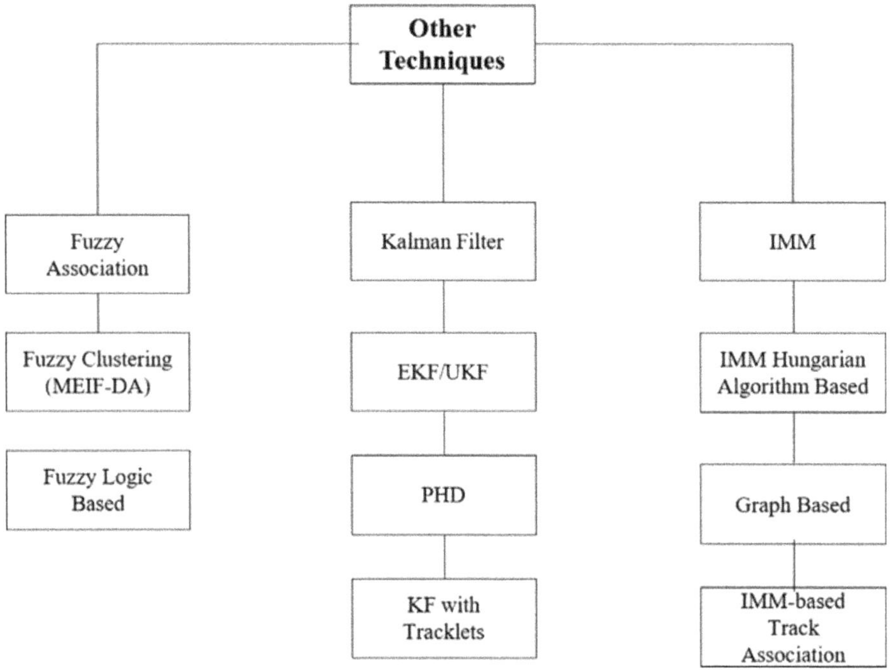

Table 4. Other Methods with a summary of the advantages and disadvantages for each method.

Technique	Advantages	Disadvantages
Kalman MOANA: An Online Learned Adaptive Appearance Model for Robust Multiple Object Tracking in 3D (Xi-yang et al., 2018)	More robust and efficient	Only works on static cameras, future work on moving cameras
Kalman filter for multi view object tracking by data association (MCMC based approach) (Tang et al., 2018	Efficiency in multi view tracking	Less information on computational efficiency
IMM with The Munkres Algorithm (Yuan et al., 2017a)	Robust	Cannot handle more complex or crowded scenes
IMM-SMHT (IMM with sequential multiple hypothesis test) (Li et al., 2017a)	Robustness and high accuracy	Computational cost
Fuzzy logic data association (Li et al., 2017a	Better performance	Little information on stability
Kalman filter with maximum entropy intuitionistic data association (Xi-yang et al., 2018)	Robustness and accuracy	No information on computational time
Kullback–leibler differential entropy equation-based measurement data association (Hu et al., 2020)	The model is more adaptable with increased noise	Computational cost increase
Self-supervised approach to associating objects using Kalman filter and Mahalanobis distance (Wang et al., 2020)	Useful suppression of incorrect associations during self-supervised training	Unclear if human efficiency correlates to the computational efficiency of the model
Data association between perception and V2V communication sensors (Cantas et al., 2021)	Method can be effective for curved roads and intersections	More tests need to be performed in complex scenarios and crowded scenes

continued on following page

Table 4. Continued

Technique	Advantages	Disadvantages
Extended Kalman filter using IP (innovation projections) (Joerger & Hassani, 2020)	Lower level of wrong associations	Complex input parameters with Lidar
A priority data association policy for multitarget tracking (Zeng et al., 2020)	Improve performance for risk assessment	Performance diminishes with multi target tracking
Two stage fuzzy logic association integrated with Kalman filtering (Liu et al., 2020)	Improve performance with multiple targets and complex movements	Sometimes false sizes of targets are generated due to reflections or disturbance
GSM: graph similarity model for multi-object tracking (Liu et al., 2020)	Focuses on improving robustness and accuracy in association	Comparisons are only made with online settings for simplicity
Neighbour graph with GCN (graph convolutional networks) (Liang et al., 2020)	Applies features of the spatio-temporal domains based on neighbour selection	Requires more research to improve object detection

The classification of New techniques Data Association literature review summary advantages and disadvantages are show in table 5.

Table 5. New Methods with a summary of the advantages and disadvantages for each method.

Technique	Advantages	Disadvantages
Hybrid data association as a min-cost multi commodity flow problem (Yang et al., 2017)	Improved performance for online tracking, accuracy	Need more definition on motion and shape cues
Modified Frank Wolfe algorithm with SWAP steps(Dehghan & Shah, 2017)	Reduce computational complexity	Run time increases as people increases
Particle Labelling Association for grid-based object tracking tracking (Steyer et al., 2018)	Robust	Less information on speed and computation cost
Multi-frame data association with sparse representation (Fagot-Bouquet et al., 2016)	Robust, less association errors	Computational cost
Hybrid data association with affinity and detection-detection association (Dai et al., 2018)	Accuracy in complex sequences	Need richer interaction model and shape information model
CNN based data association (Leal-Taixé et al., 2016	Efficiency in speed	Need more accuracy in complex situations
Gain functions for data association (Jaiswal et al., 2018)	Accuracy	No generalization yet - applies only to a specific trained task
GLMB Joint Object Clutter model (Punchihewa et al., 2018)	Improved performance rate	More complexity as the scene becomes crowded
On detection data association and segmentation (Tian et al., 2018	Accurate segmentation results	Poor performance with overall comparison
Enhanced identity association (Gan et al., 2018)	High accuracy and precision with low ID switches	Confusion by occlusion and interaction with other targets, not generalized.
Stochastic Optimization (Granström, Renter et al., 2017	Accuracy, computationally efficient	Less information on robustness and generalization
Tracklets (long term and short term (Yang et al., 2020)	Learns better appearance features for more effective association	Cannot be used on real time situations

continued on following page

Table 5. Continued

Technique	Advantages	Disadvantages
Attention with transformer network data association (Hung et al., 2020)	Improves results with occlusion	Requires improvement with spatio-temporal dependencies
TransTrack, Box association with Kuhn–Munkres (KM) algorithm (Sun et al., 2020)	Improves accuracy	Limited information about long term tracking
Attention association with Track Former (Meinhardt et al., 2021	Attention operations improve appearance modelling and improves segmentation	Limited information about computational cost
DASOT: a unified framework integrating data association (Chu et al., 2020)	Focuses on improving computational cost through learning discriminative features	Limited information on adaptability to long term tracking
Data association using geometry priors (Chen et al., 2020)	Generates a combined cost matric to improve accuracy	Requires further study for the inclusion of more sensor information
Embedding association using attention (Guo et al., 2021	Improves robustness and tracking performance	Requires further study for computational efficiency

In the present era of Industry 4.0, engineers came up with another logical approach: the philosophy of predictive maintenance Okon et.al., (2016). Predictive Maintenance is based on Condition Monitoring, abnormality detection, and AI algorithms. It integrates predictive models that can estimate the remaining machine runtime left or diagnose the type of fault in machinery according to detected abnormalities. This approach uses a wide range of tools, such as statistical analyses and Machine Learning to predict the state of the equipment. The evolution of Maintenance strategies is as shown in figure 6 Sayyad et.al., (2021)

Figure 6. Evolution of Maintenance Strategies.

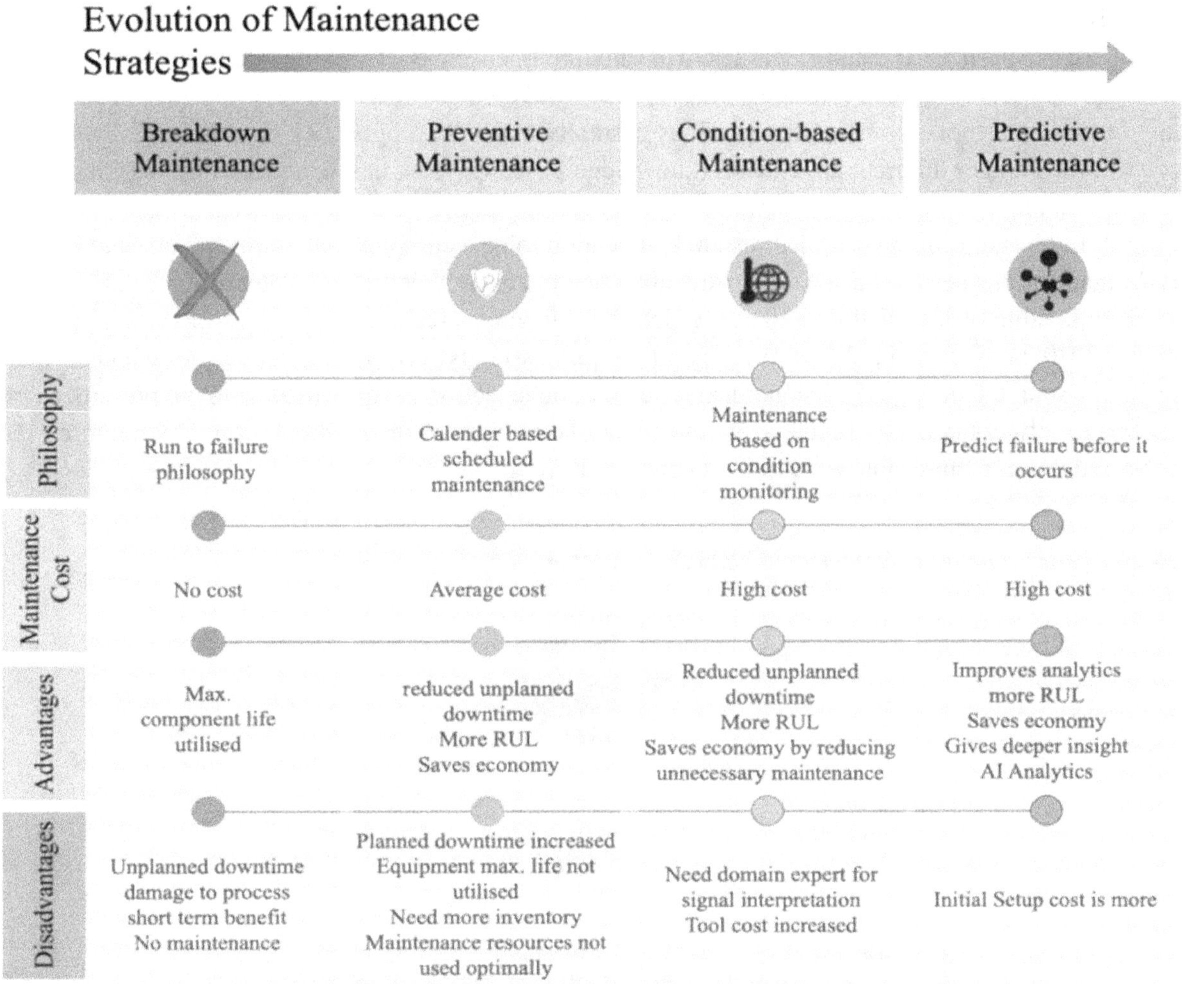

Predictive maintenance (PdM) is a type of condition-based maintenance that monitors the condition of assets using sensor devices. These sensor devices supply data in real-time, which is used to predict faults using AI, and intimate when the asset will require maintenance and prevent equipment failure. Specific widely used parameters and methods Mohanty, (2014) to detect the fault and predict the condition of Industrial rotating machines are Vibration analysis Khadersab (2018), Motor Current Signature Analysis Lee (2015), Temperature analysis, Magnetic chip detectors Zhou (2007), Infrared Thermography, Acoustic Emission Zhou (2007) Airborne Ultrasound, Lubrication analysis, etc. A combination of these methods improves the accuracy of analysis to predict the faults. The PdM approach is classified into three major categories: Physics-based Approach, Knowledge-based Approach, and Data-Driven Approach Liao (2014), Okoh(2017) and Shao (2017). The knowledge-based approach combines field experience and computational knowledge from domain experts and sets rules to interpret the faults. Physics-based models are methods that use knowledge of a system's failure mechanisms to create a mathematical equation model for the system's degradation process. These physics-based models include the Finite Element Model (FEM), Kalman filter (KF), and Particle filter (PF), which all rely heavily on mathematical models. Data-driven models, the most widely used approach, depend highly on big data and analysis. The Analysis allows us to predict the system's condition or match comparable examples in the set of past experiences. It mainly includes Artificial Intelligence models. Using different AI Algorithms, viz., Machine learning, deep learning is employed to build these models. Another effective and emerging approach that anchors the benefits of various currently available prognostics models is the hybrid approach. It combines different approaches (knowledge-based, physics-based, and data-driven) that result in a hybrid model with better predicting ability. These approaches are summarized in figure 7. Sayyad (2021). Egbe et.al (2024) described resource allocation to be a complex task in maintenance activities, and organizations need to have a systematic approach to ensure that they allocate resources effectively.

The pipeline of a typical data-driven based fault diagnosis techniques consists of three steps. Lei et.al., (2020) and Lei et.al., (2015). first, signal pre-processing is performed to reduce noise and normalize multi-source fault signals; second, feature extraction\selection is performed from the pre-processed signals to reflect the health status of the machines. Third, pattern recognition is used by inputting the extracted defect features into a classifier to predict defect types.

Figure 7. Approaches to Implement Predictive Maintenance

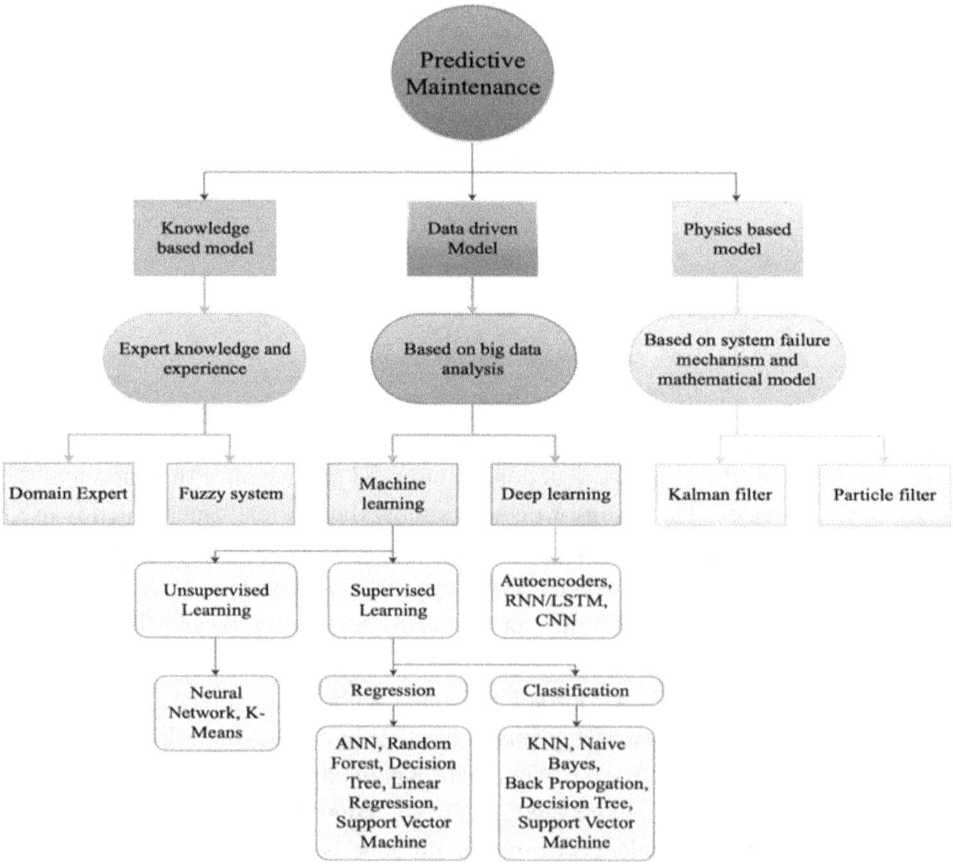

1.1 State-of-the-Art Techniques in Industry 4.0

In the era of the fourth industrial revolution, several concepts have arisen in parallel with this new revolution, such as predictive maintenance, which today plays a key role in sustainable oil and gas, manufacturing and production systems by introducing a digital version of machine maintenance. The data extracted from production processes have increased exponentially due to the proliferation of sensing technologies. Even if Maintenance 4.0 faces organizational, financial, or even data source and machine repair challenges, it allows for minimizing machine downtime and associated costs, maximizing the life cycle of the machine, and improving the quality and cadence of production. There are various methods and applied tools for intelligent predictive maintenance models in Industry 4.0, these models are: condition-based maintenance (CBM), prognostics and health management (PHM), and remaining useful life (RUL). The technologies of the Internet of Things (IoT) and Big Data, and the integration of artificial intelligence (AI) methods and cyber-physical systems (CPS) play an important role in this context by introducing cognitive automation and, consequently, implementing the concept of intelligent production, leading to intelligent products and services. Predictive maintenance has become a promising approach, providing solutions for the remaining life of equipment through the prediction of data collected by various sensors on equipment Tiddens et al (2020). The combination of cyber-physical systems, the Internet of Things (IoT), and the Internet of Systems make Industry 4.0 possible and the smart factory a reality. It has radically transformed the industrial and manufacturing world by enabling intercommunication between equipment through the Internet of Things, Big Data, computer intelligence, and decision-making systems Li et al (2017). Industry 4.0 devices can automatically communicate with each other, allowing them to coordinate with each other and with other remote systems over the Internet. Figure 8 provides an overview of the components of Industry 4.0.

Figure 8. Components of Industry 4.0 technologies in the context of intelligent manufacturing.

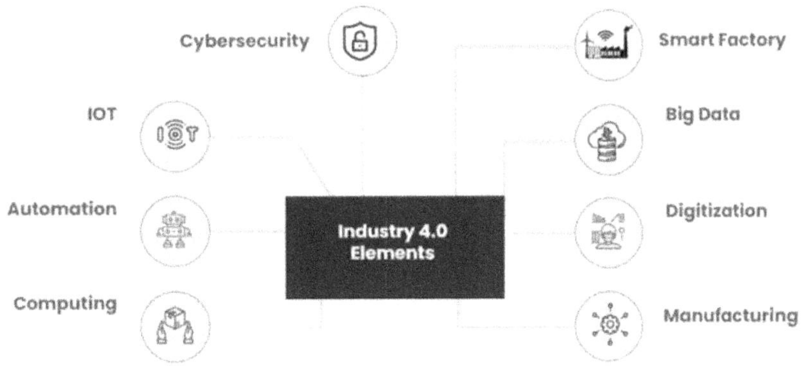

1.2 Predictive Maintenance Tools and Materials

In the era of the fourth industrial revolution, computational and advanced visualization tools based on the latest technologies have become the main and necessary components of digital transformation in Industry 4.0. They have been adopted in multiple industrial applications and, more specifically, in predictive maintenance, named maintenance 4.0 et al. (2012). The main technological tools applied in maintenance 4.0 are.

1.2.1 Cyber-Physical Systems

The most recent advancement in intelligently connected production devices and sensors, and autonomous intelligent monitoring, control, sensing, and diagnostic systems are known as "cyber-physical systems (CPS)", which are outfitted with a variety of computational and physical methods and tools that are perfectly suited to meeting human needs Okano et al (2017).

1.2.2 Industrial Internet of Things

Connecting machines, devices, and systems is the most fundamental step in implementing Internet of Things (IoT) applications. The IoT concept provides convenience in industrial environments, where it is called Industrial Internet of Things (IIoT) Pech et al (2021). It is a network that connects cyber-physical systems and provides interconnection and interaction between physical devices in the process, enabling the automatic collection and retrieval of a huge flow of data [8]. This constitutes the notion of Big Data.

1.2.3 Big Data

Collecting, processing, and analyzing real-time Big Data from cyber-physical systems is a strategic phase for the intelligent transformation of the maintenance function, especially with respect to failure prediction, planning, and risk management Wang et al (2021). This is done using artificial intelligence tools and techniques such as machine learning and deep learning or through statistical models and approaches based on the data and information collected by the different sensors Silvestri et al (2021).

1.2.4 Digital Twin

A Digital Twin (DT) is a digital or virtual copy of physical assets or products. DTs connect the real and virtual world by collecting real-time data from the installed sensors. The collected data is either locally decentralized or centrally stored in a cloud. The data is then evaluated and simulated in a virtual copy of assets Hosamo et al (2022). After receiving the information from the simulation, the parameters are applied to real assets. The integration of data in real and virtual representation helps in optimizing the performance of real assets

1.2.5 Augmented Reality

Augmented reality (AR) is an emerging technology developed on the basis of virtual reality (VR), which generates three-dimensional virtual information through a computer system, including virtual scenes, virtual objects, etc. Wang et al (2022), and then superimposes this information into the real scene to realize the function of real-world enhancement and improve the user's perception of the real world.

1.2.6 Artificial Intelligence

Artificial intelligence (AI) is the keyword in the transition to Industry 4.0; it is a powerful technology that compensates for the deficiencies and ineptitude of traditional techniques and approaches practiced in industry Daniyan et al (2020). It has emerged as a powerful tool for developing intelligent predictive algorithms in many applications

1.2.7. Machine Learning and Deep Learning

Machine Learning (ML) is a subfield of AI that has evolved from pattern recognition to analyzing data structures and integrating them into models that users can understand and regenerate. Figure 9 identifies all ML categories, methods, and models applicable to maintenance-related projects Nacchia, et al (2021). Furthermore, ML is classified into four categories, namely supervised learning, unsupervised learning, reinforcement learning, and deep learning.

Figure 9. Classification of data-driven techniques and machine learning categories.

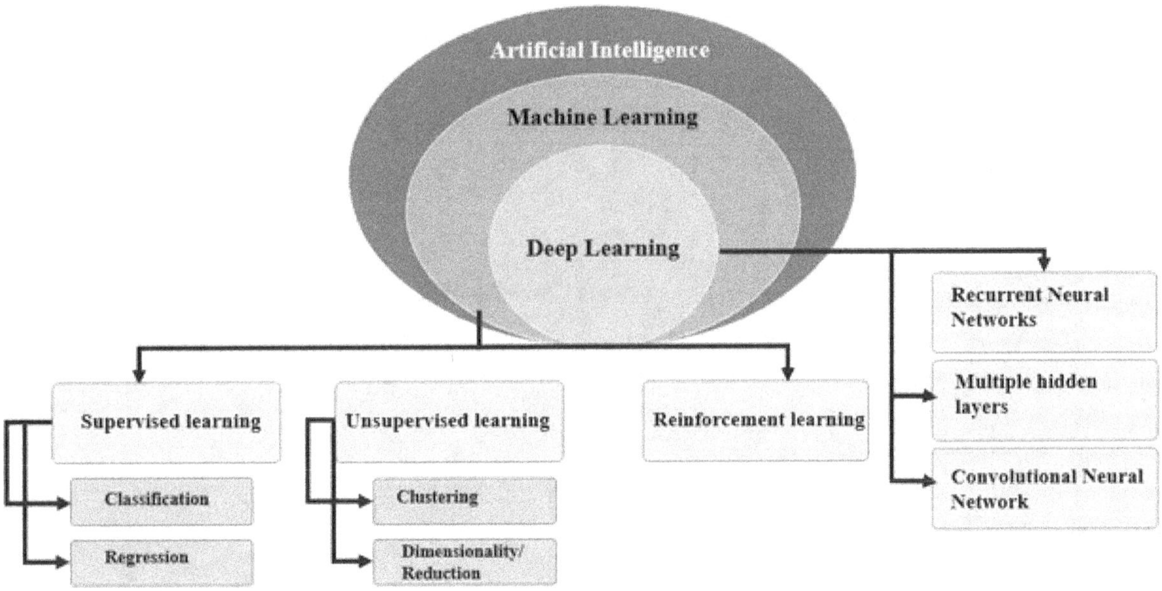

1.3 Predictive Maintenance Models

Depending on the technical and digital maturity level, different types of PdM models can be offered. Among others, three main PdM models are presented in the following subsections.

1.3.1 Condition-Based Maintenance

It is defined as a maintenance program that recommends maintenance solutions based on information gathered from condition monitoring and consists of three main steps: data collection, data processing, and maintenance decision-making Fernandes et al (2021). CBM can take many forms, from simple to complex. However, the most common form adopted by most industries is the one shown in Figure 10 where the focus is on data collection, prediction, and correction with varying degrees of automation.

Figure 10. The seven modules of a conditional maintenance system

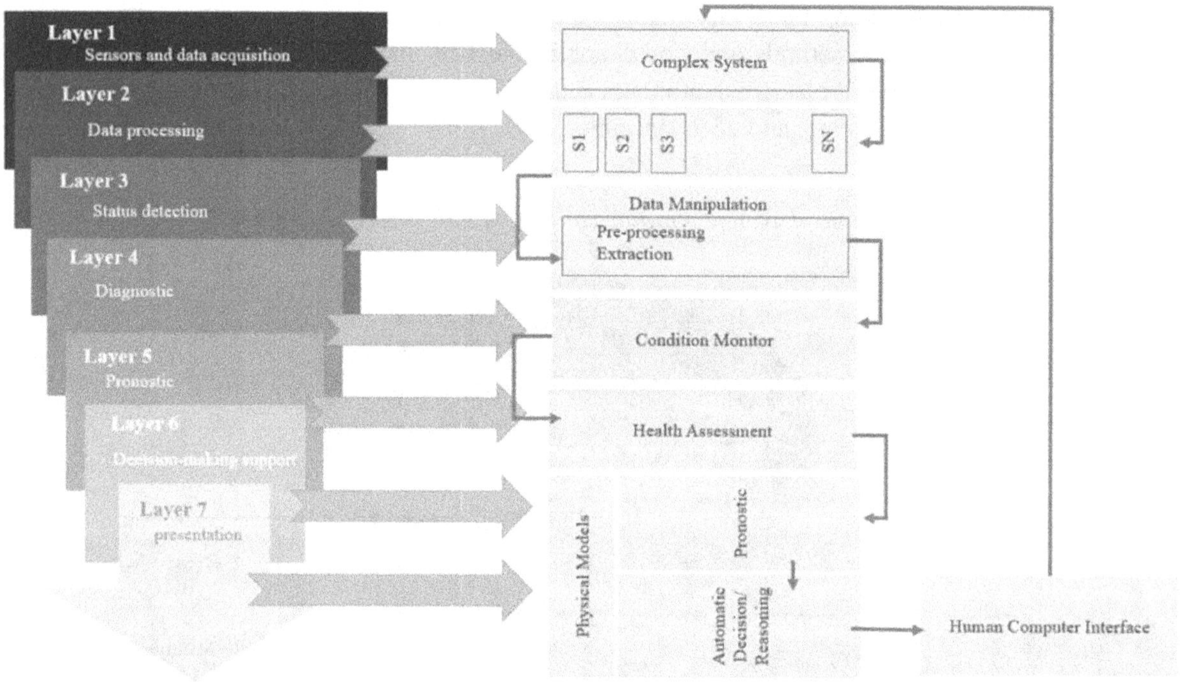

1.3.2 Prognostics and Health Management

The objective behind the implementation of a PHM architecture includes the dynamic monitoring of the system state for the purpose of a predictable type of maintenance Omri et al (2021). In this respect, the prognostic activity is considered a key process; unlike diagnostics, which concerns the ability to detect and isolate the cause of an existing error, PHM has the ability to assess the development of future degradation or errors of a system. it guarantees the operation of the system and is an important step before being able to describe the different maintenance scenarios used for the prediction and prevention of malfunctions Zheng et al (2020). PHM methods can be classified into three categories as shown in Figure 11. These categories are model-based approaches, data-based approaches and hybrid approaches that combine the advantages of the two previous approaches.

Figure 11. Prognostics and Health Management methods

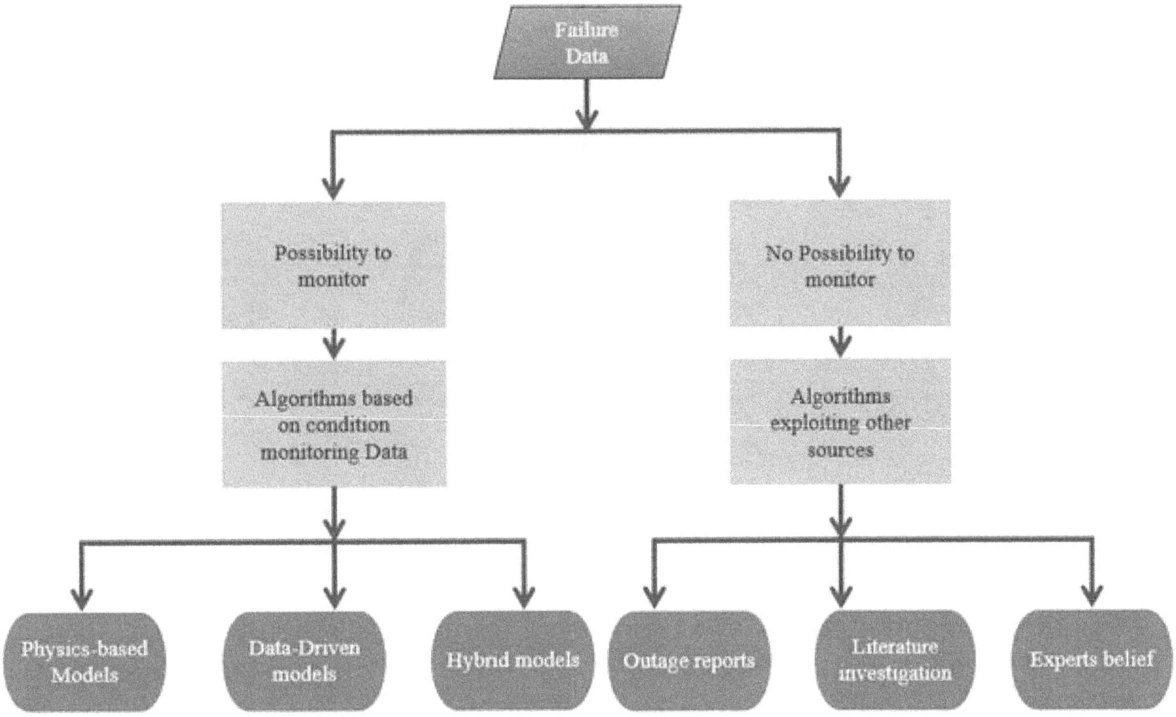

1.3.3 Remaining Useful Life

The remaining useful life (RUL) is the period of time that a piece of equipment is likely to operate before it needs to be repaired or replaced. Service life prediction provides early warnings of failure and has become a key component of prognostication and systems health management. In the last decade, many tools and methods have been proposed as shown in Figure 12. Generally, two types of methods are used to predict equipment life: physics-based methods and data-based methods de Pater et al (2022). The physics-based approach uses mathematical principles such as statistics and probability, which attempt to create accurate mathematical models from prior knowledge to predict equipment life. In order to improve the prediction of RUL, Kitai et al. (2020) used convolutional neural networks (CNN) and hierarchical Bayesian regression.

Figure 12. A brief explication of RUL models.

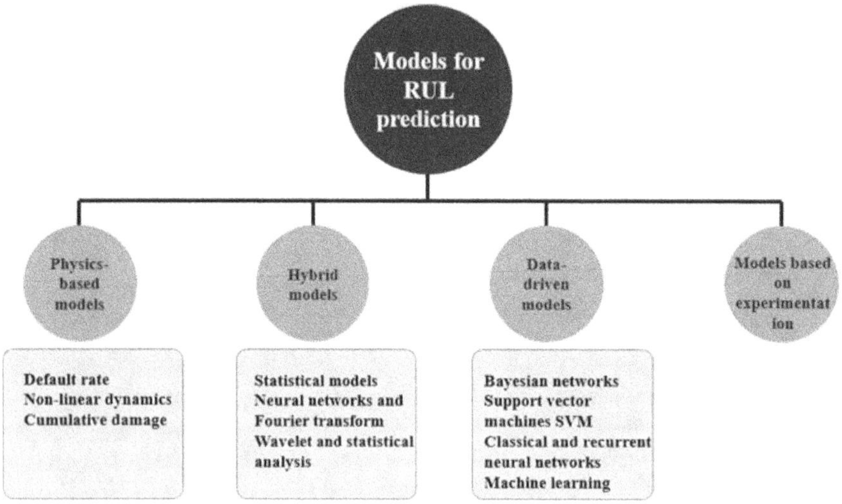

1.4 Predictive Maintenance Case Study on a Centrifugal Compressor

In this section, a practical case study based on a PdM approach is presented to estimate the remaining life of a centrifugal compressor. Compressors are used throughout industry to supply air to shops or instruments, power air tools, paint sprayers and sandblasting equipment, propel gas in pipelines, etc. Zhang et al (2020). Therefore, they play a key role in the production line. This requires a study on the remaining life of this equipment in order to predict future failures and avoid them while applying machine learning methods. Figure 13 represents the components of the CBM/PHM approach for an industrial compressor.

Figure 13. Components of the CBM/PHM approach for a compressor.

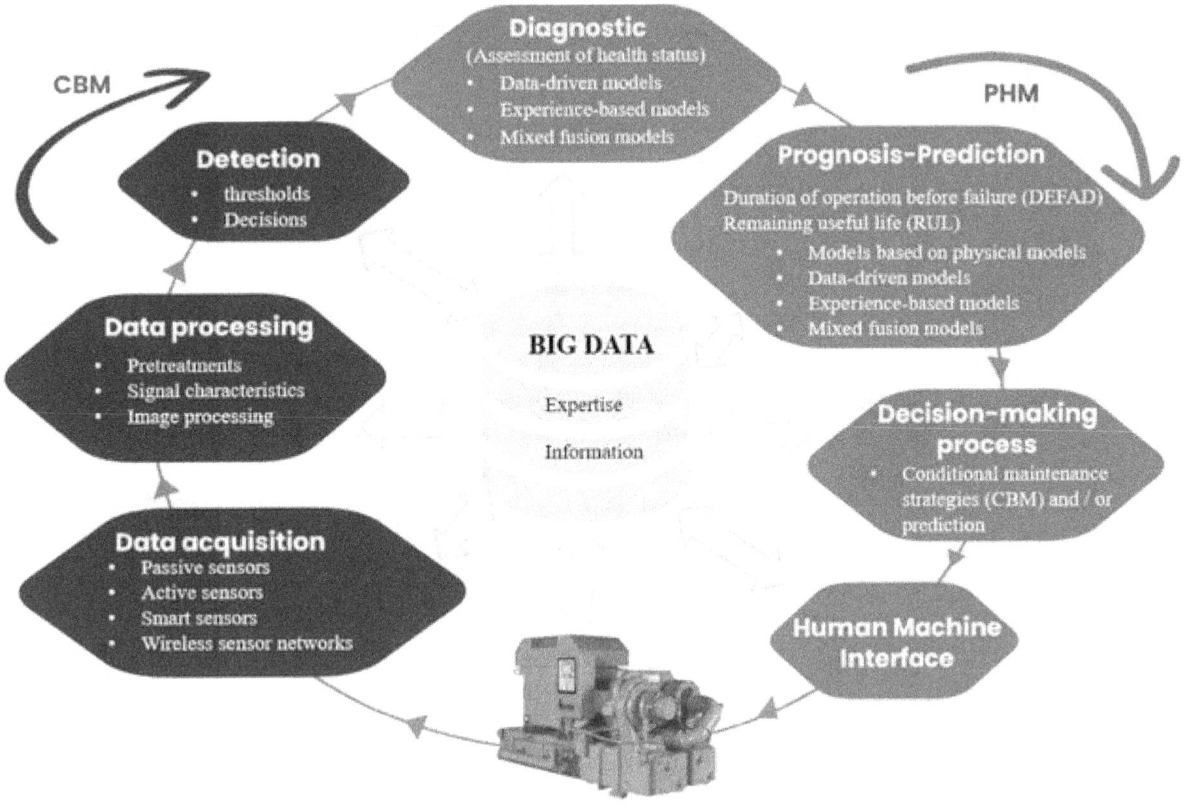

It can be noticed that the application of maintenance 4.0 is mostly focused on a specific machine or industrial equipment. The idea of creating a user-friendly and multifunctional platform, as depicted in Figure 14, that integrates different types of sensors and enables in-situ PdM on different types of equipment in real-time can be considered revolutionary in the field [87]. This platform highlights support in maintenance management and decision-making based on acquired data and developed models of PdM. It allows users and the maintenance team, depending on their predefined access level to the platform, to manipulate and modify proposed decisions anytime and at any step of the maintenance planning process. The platform gives remote access via user authentication protected by a VPN (private virtual network). Thus, a cloud infrastructure is provided that allows mobile and web connections in order to receive notifications received by the alarms. Apart from remote connectivity, this platform will provide real-time on-site data monitoring collected by machines and equipment in the form of a dashboard that calculates key performance indicators and supports API (application programming interface) protocols for data customization and alarm notifications. This provides remote plant monitoring from any browser and also ensures high security against attacks by preserving the integrity and confidentiality of information.

Figure 14. The proposed multifunctional platform of predictive maintenance based on industrial asset management

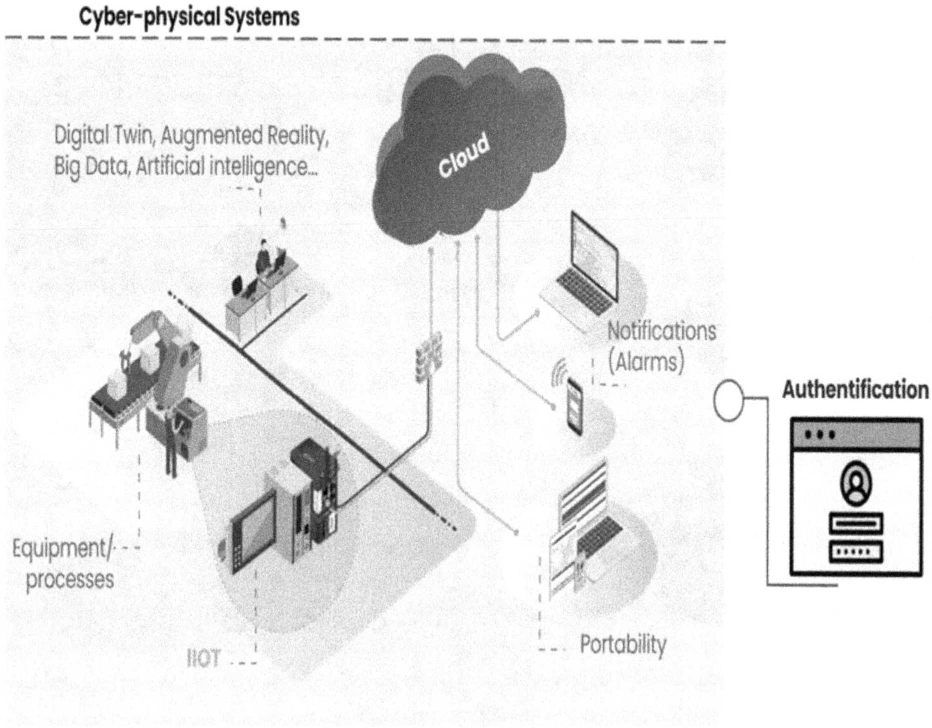

Shreyas et.al (2022) established that faults are broadly classified at three levels: Component-level faults, System-level faults, and Interrelated faults. Figure.15 summarizes different types of component-level faults that can arise the rotating machinery.

Figure 15. General component-level faults in rotating machines

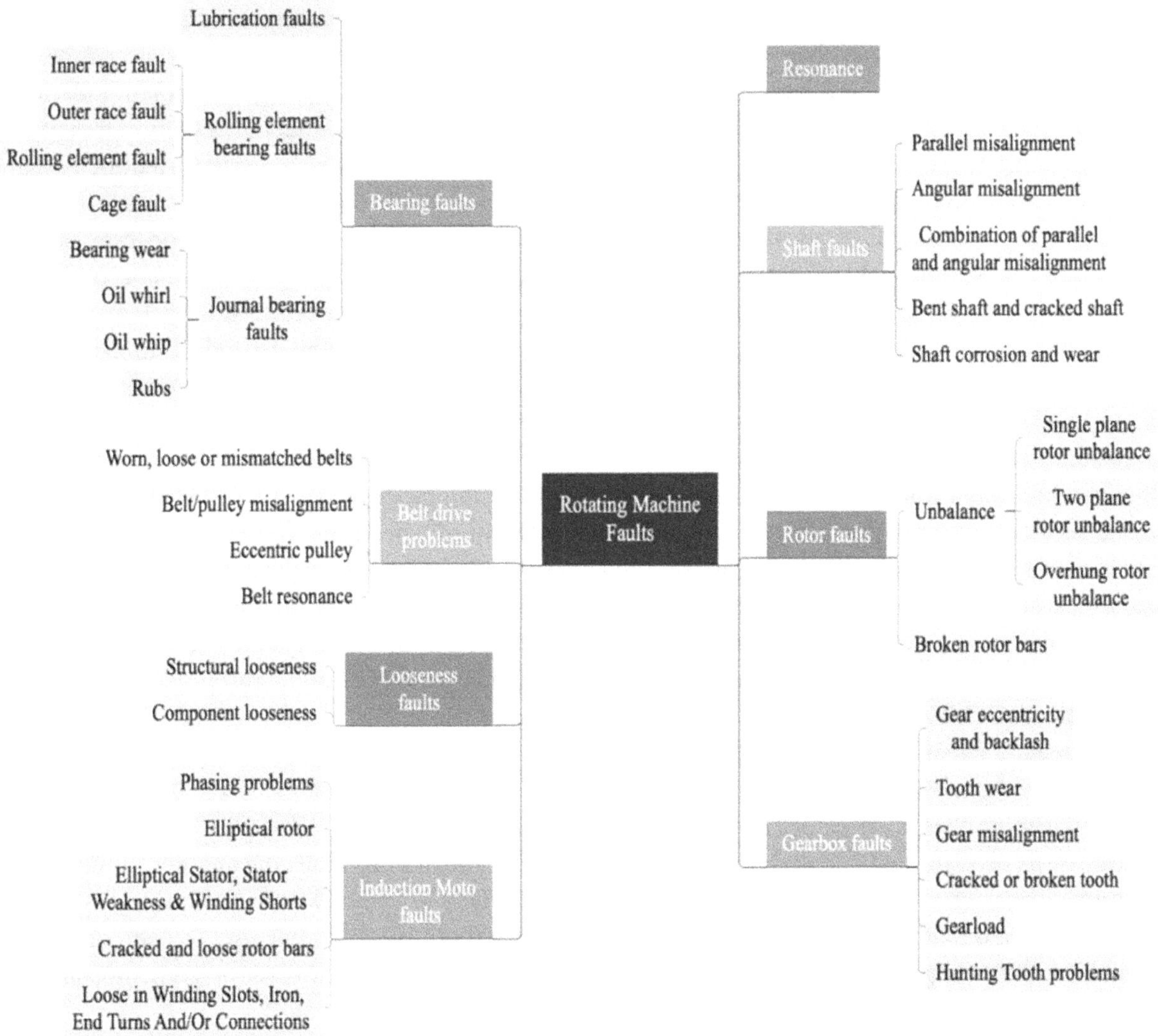

1.5 Economic Impact of Multisensor Data Fusion

The economic implications of AI-driven maintenance are significant. By reducing downtime and preventing catastrophic failures, AI maintenance systems can save millions of dollars in lost production and repair costs. For stakeholders, particularly investors and company executives, this translates to improved profitability and a stronger competitive edge in the market. Moreover, as Mukherjee (2022) highlighted, the application of AI in the industry is not without its limitations, particularly in terms of the initial investment required for implementation.

1.6 Data Privacy and Security

Data privacy and security are essential concerns in the implementation of sensor fusion systems. As multiple sensors collect and share a significant amount of data, the risk of unauthorized access or data breaches increases. Such breaches can result in the loss of sensitive information, violation of individual privacy. One challenge in securing sensor fusion systems is the need to protect data both in transit and at rest. Ensuring the integrity of data exchanged between sensors and fusion systems requires secure communication protocols and encryption mechanisms. For example, Transport Layer Security (TLS) can be employed to protect data transmitted over networks, while symmetric encryption algorithms, such as Advanced Encryption Standard (AES), can secure stored data. Furthermore, asymmetric encryption such as Rivest-Shamir-Adleman (RSA), provides robust security for key exchange processes. Another challenge is the potential for malicious actors to tamper with or spoof sensor data, which can lead to incorrect or misleading fusion results. Counter measures against such attacks include sensor data authentication and integrity checks, such as digital signatures or cryptographic hashes. Additionally, robust sensor fusion algorithms can be designed to detect and mitigate the impact of compromised sensor data by considering the credibility and trustworthiness of each sensor in the fusion process. Abrera, (2024) suggested that cloud service providers must adopt and rigorously implement advanced encryption techniques, not only for data at rest but also for data in transit and in use, ensuring end-to-end protection. Additionally, the deployment of multi-factor authentication and continuous monitoring systems is critical to identify and mitigate potential threats in real-time. Firewalls and Intrusion Detection Systems (IDS) cloud-based firewalls need to be more dynamic and scalable to match the elasticity of cloud resources. Multi-Factor Authentication (MFA) is consistently acknowledged as a critical security measure for cloud computing environments. The use of hardware tokens, biometric data, and time-based one-time passwords (TOTP) were among the most popular methods implemented.

Furthermore, based on extent review literature of signal processing techniques in data extraction features was used for multi-fault diagnosis of industrial processing machines the and not capturing the features of the signals such as the crest factor (CF), clearance Factor and kurtosis. Hence, it is against this background that the chapter is presented to investigate these areas of concern. To overcome these limitations, this chapter integrated crest factor (CF), clearance Factor and kurtosis and other techniques, to determine the health state of industrial rotating machine

2. METHODOLOGY

Vibration signals recorded using vibration sensors from a rotating machine are frequently in the time domain. Depending on the type of sensor used to acquire the data, they are a collection of time-indexed data points gathered across historical time, signifying acceleration, velocity, or proximity. Signal processing is frequently required to clean data and put it into a format from which condition indicators may be extracted, called features. Feature extraction is a part of signal processing that gives the essence of the data. These features are used to distinguish between two different vibration signals. Hence in this chapter, multi-fault diagnosis of Industrial Rotating Machines was done using data extraction features.

The analysis utilizes three signal processing tools; Vibration time domain feature, Vibration frequency domain feature and Vibration time-frequency domain feature.

Multisensory data fusion technology was divided into three fusion methods: data-level fusion, feature-level fusion, and decision-level fusion. At the data-level fusion approach, data from commensurate sensors are fused directly. The fused data is then subjected to feature extraction, followed by a pattern recognition method for classification. In the feature-level fusion, each sensor is used to collect a kind of signal. Feature extraction is subsequently applied to obtain a feature vector. All features are combined to identify the best subset of features, then input into a classifier or decision-level fusion. At decision-level fusion, features calculation and pattern recognition processes are applied in sequence for single-source data obtained from each sensor. Decision-level fusion methods are then utilized to fuse the decision vectors. The integration of the decision layer mainly includes Bayesian reasoning, D-S Reasoning. An overview of feature fusion and decision level fusion is shown in figure 16

Figure 16. Information Fusion at Feature-Level and Decision-Level.

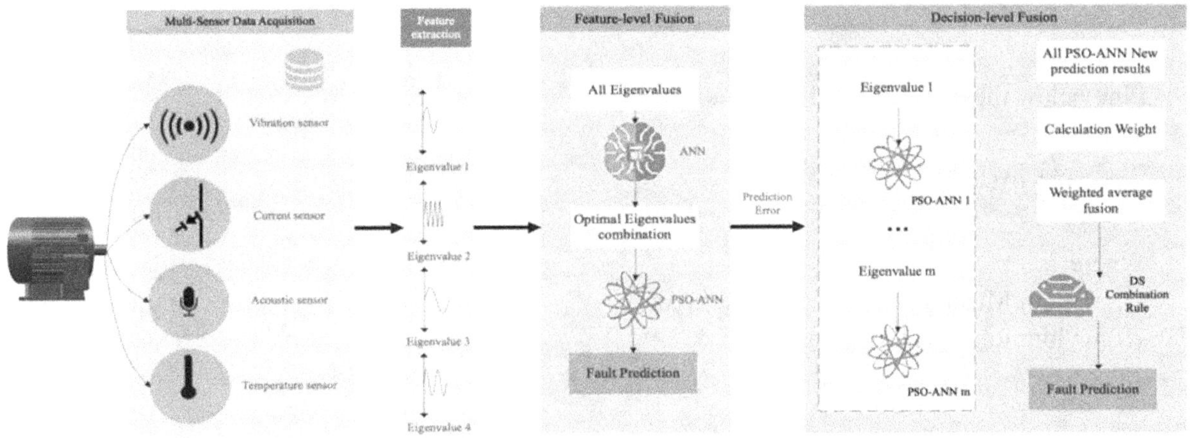

Table 6. Time Domain Features along with their Definition and Formula

Feature	Definition	Formula	Significance / Note
Root Mean Square (RMS) Amplitude	Variance of the vibration signal magnitude. It is the measure of power contained in the vibration signal	$x_{RMS} = \sqrt{\frac{1}{T}\int \lvert x(t)\rvert^2 dt}$	It increases as fault develops
Crest Factor	Ratio of the vibration signal's peak amplitude to its RMS amplitude. Faults often first manifest themselves in changes in the peakiness of signal before they manifest in energy represented by RMS.	$x_{CF} = \frac{x_p}{x_{RMS}}$	Healthy bearing = more CF. Calculates how much impact occurs during the rolling element and raceway contact. CF can provide early warning for faults when they first develop.
Clearance Factor	Ratio of the maximum value of the input vibration signal to the mean square root of the absolute value of the input vibration signal	$x_{CLF} = \frac{x_{max}}{\left(\frac{1}{N}\sum_{i=1}^{N}\sqrt{\lvert x_i\rvert}\right)^2}$	This feature is max. for healthy bearings and decreases for the defective ball, outer race, and inner race, respectively. CF has the highest separation ability for defective inner race faults.
Kurtosis	The measure of the peak value of the input vibration signal through its PDF.	$x_{KURT} = \frac{\sum_{i=1}^{N}(x_i - \overline{x})^4}{N\sigma_x^4}$	Good bearing: Kurtosis value ≤ 3. defective bearing: Kurtosis value > 3

Find below the equation used in this chapter

$$x_{RMS} = \sqrt{\frac{1}{T}\int \lvert x(t)\rvert^2 dt} \quad 1$$

Where:
x_{RMS}= Root Mean Square (RMS) Amplitude
$x(t)$= vibration signal magnitude

$$x_{CF} = \frac{x_p}{x_{RMS}} \quad 2$$

Where:
x_{CF}= Crest Factor
x_p= vibration signal's peak amplitude
x_{RMS}= Root Mean Square (RMS) Amplitude

$$x_{CLF} = \frac{x_{max}}{\left(\frac{1}{N}\sum_{i=1}^{N}\sqrt{\lvert x_i\rvert}\right)^2} \quad 3$$

Where:
x_{CF}= Crest Factor
x_{max}= maximum value of the input vibration signal

$$x_{KURT} = \frac{\sum_{i=1}^{N}(x_i - \overline{x})^4}{N\sigma_x^4} \quad 4$$

Where:

x_{KURT}= Kurtosis

x_i= inputed signal

$\bar{x}$ = mean of the inputed signal

σ_x= standard deviation of the inputed signal

Table 7. Contribution of Data-driven Approach for Fault Diagnosis of Rotating Machine

Algorithm	Algorithm type	Diagnosis Application
Deep Neural Network (DNN	Deep Learning	Fault in Bearings and gearbox
Convolutional Neural Network (CNN)	Deep Learning	Fault in bearings, unbalance, motor faults, pump faults
Artificial Neural Network (ANN)	Deep Learning	Fault in bearing, misalignment, rotor bar faults
Deep belief network (DBN)	Deep Learning	Fault in Induction motor, Bearing faults
Extreme Learning Machine (ELM)	Machine Learning	Fault in bearings, gearbox

Table 8. Contribution of Hybrid Approach for Fault Diagnosis of Rotating Machine

Algorithm	Hybrid PdM Approach	Definition	Diagnosis Application
Fuzzy logic +NN	Knowledge-based + Data-Driven	Fuzzy Logic is an approach to variable processing that allows for multiple values to be processed through the same variable	Gearbox faults, RUL Prediction
Fuzzy logic + Wavelet Neural Network	Knowledge-based + Data-Driven	Wavelet networks are a new class of networks that combine the classic sigmoid neural networks (NNs) and the wavelet analysis (WA)	RUL prediction of bearing
GPR+KF/PF	Data-driven + physics-based approach	Gaussian process regression is nonparametric, Bayesian approach to regression that is making waves in machine learning. Particle filters or Sequential Monte Carlo (SMC) methods are a set of Monte Carlo algorithms used to solve filtering problems arising in signal processing and Bayesian statistical interference	RUL Prediction
SK+ELM	Data-driven + physics-based approach	The spectral kurtosis (SK) is a statistical tool that can indicate the presence of series of transients and their locations in the frequency domain	Bearing faults

The purpose of this chapter is to use multisensory data fusion in predictive maintenance of industrial rotating machines. To achieve the objective, we used a combination of Physics- based Approach, Data-Driven and Hybrid-based Approach for the multi-fault diagnosis of the industrial rotating machine.

2.1. Abbreviations

AI: Artificial Intelligence

ANNs: Artificial Neural Networks

ATR: Automated Target Recognition

BN: Belief Networks

Bpa: Basic Probability Assignment

BPNN: Back Propagation Artificial Neural Network

CBM: Condition-Base Monitoring
CF: Crest Factor
CLF: Clearance Factor
CM: Condition Monitoring
CMFH: Competitive Major Feature Histogram
CNN: Convolutional Neural Network
DBN: Deep Belief Network
DFT: Discrete Fourier Transform
DNN: Deep Neural Network
DPMM: Dirichlet Process Mixture Models
ELM: Extreme Learning Machine
FC: Frequency Centre
FFT: Fast Fourier Transform
FIS: Fuzzy Inference System
FLIR: Forward-Looking Infrared
FMECA: Failure Mode Effects and Criticality Analyses
GEP: Generalised Evidence Processing
GNN: Global Nearest Neighbour
IFFN: Identification-friend-foe-neutral.
IMM: Interactive Multiple Model
JPDA: Joint Probabilistic Data Association
KBS: Knowledge-Based Systems
LSTM: Long Short-Term Memory
M2M: Measurement to Measurement
M2T: Measurement to Track
MHT: Multiple Hypothesis Testing
MMSE: Minimum Mean Square Error
NN: Neural Networks.
NNSF: Nearest Neighbour Standard Filter
PDA: Probabilistic Data Association
PDAF: Probabilistic Data Association Filter
PDM: Predictive Maintenance
PHD: Probability Hypothesis Density
PMTA: Permutation Matrix Track Association
QCM: Quartz Crystal Microbalance
RUL: Remaining Useful Life
SFM: Social Force Model
SMHT: Sequential Multiple Hypothesis Test
SNN: Suboptimal Nearest Neighbour
SVSF: Smooth Variable Structure Filter
T2T: Track to Track
TBO: Time Between Overhaul
VB: Variational Bayesian
Θ: Frame of Discernment

3. RESULTS AND DISCUSSION

Evaluating the various vibration time domain feature parameters, such as the Root Mean Square (RMS) Amplitude, Crest Factor, Clearance Factor and the Kurtosis using equation 1, equation 2, equation 3 and equation 4 in combination with Data-Driven Approach as show in table 7 was used determine health status of the different components of the industrial rotating machine.

3.1 Discussion

The analysis shows that Root Mean Square (RMS) Amplitude for these component increases as fault develops, Crest Factor was more for healthy bearing and less for faulty bearing, also the Clearance Factor this feature was maximum for healthy bearings and decreases for the defective ball, furthermore, the Kurtosis for good bearing; the value was ≤ 3. And for defective bearing: the value was > 3, these can provide early warning for faults when they develop.

4. CONCLUSION

Appropriate maintenance planning is critical for maintenance management in the industry 4.0 era of smart industrial process, to contribute to increasing availability, ensuring quality requirements, reliability and operational excellence, controlling the safety and environmental risks associated with industrial rotating machines. As supporting tools for developing maintenance strategies, in the oil and gas and manufacturing industry.

The following conclusion were drawn from the study:

I. The use of vibration time domain feature parameters as an analytical process to determine the health status of the components of the rotating machine to ensure safe operations and cost-wise readiness in smart industrial process was established.
II. The chapter has established the combination of Physics-based Approach and Data-Driven Approach for the multi-fault diagnosis of the industrial rotating machine.
III. The method rather than the different sensors being applied independently, are optimally fused to take advantage of their respective strengths, by combine data from multiple sensors and related information to achieve more specific inferences than could be achieved by using a single, independent sensor.
IV. The method maximized facility and equipment reliability while minimizing life-cycle costs in deal with problems encountered in industrial rotating machine such as incurred high costs due to production losses and delays, unplanned intervention on the system and safety hazards.
V. The study has established that early warning for faults when they developed can be detected
VI. The chapter has developed a valuable tool for the manufacturing and oil and gas industry that are seeking to plan their maintenance strategy and improve their operational excellence with ensured safety and minimal environmental impact.

4.1 Contributions

This chapter tackles the critical limitation in the extant literature on Multisensor Data Fusion and its application in multi-fault diagnosis of industrial rotating machines using data extraction domain features parameters, they did not capture the parameters features of the signals such as the root mean square (RMS), the crest factor (CF), clearance Factor and kurtosis. To overcome these limitations, this chapter has developed an analytical process in conjunction with Data-Driven Approach, to determine the health status of industrial rotating machine

Author contributions: All authors contributed, read, and agreed to the published version of the manuscript.

Funding: This research received no external funding.

Competing interests: The authors declare no competing of interest.

Data availability statement: No additional data is applicable besides the data/information already included in the manuscript.

REFERENCES

Abrera, J. (2024). Data Privacy and Security in Cloud Computing: A Comprehensive Review. Journal of Computer Science and Information Technology, 1(1), 01-09.

Allodi, M., Broggi, A., Giaquinto, D., Potender, M., & Prioletti, A. (2016). Machine learning in tracking associations with stereo vision and lidar observations for an autonomous vehicle. In 2016 [IEEE.]. *IEEE Intelligent Vehicles Symposium*, (IV), 648–653.

Bae, S.-H., & Yoon, K.-J. (2017). Confidence-based data association and discriminative deep appearance learning for robust online multi-object tracking. *IEEE Transactions on Pattern Analysis and Machine Intelligence*, 40(3), 595–610. DOI: 10.1109/TPAMI.2017.2691769 PMID: 28410099

Barkley, B. E., & Paley, D. A. (2017). *Multi-target tracking and data association on road networks using unmanned aerial vehicles. In 2017 IEEE aerospace conference*. IEEE.

Bewley, A., Ott, L., Ramos, F., & Upcroft, B. (2016). Alextrac: Affinity learning by exploring temporal reinforcement within association chains. In 2016 [IEEE.]. *IEEE International Conference on Robotics and Automation*, 2212–2218.

Blackman, S. S. Theoretical approaches to data association and fusion, Proceedings of the SPIE, 931, Sensor Fusion, 4-6 April 1988, Orlando, FL, 50-5. DOI: 10.1117/12.946647

Buede, D. M., & Martin, J. W. Comparison of Bayesian and Dempster-Shafer fusion, *Proceedings of the 1989 Symposium on Tri-Service Data Fusion*, Vol.1, May 1989, 81-101.

Cabrera, J. B. (2018). Scheduling variable field-of-view sensors for tracking multiple objects. In 2018 52nd asilomar conference on signals, systems, and computers (pp. 2174–2178). IEEE. DOI: 10.1109/ACSSC.2018.8645067

Cantas, M. R., Chand, A., Zhang, H., Surnilla, G. C., & Guvenc, L. (2021). Data association between perception and V2V communication sensors. arXiv preprint arXiv:2101.08228.

Chalvatzaki, G., Papageorgiou, X. S., Tzafestas, C. S., & Maragos, P. (2018b). Augmented human state estimation using interacting multiple model particle filters with probabilistic data association. *IEEE Robotics and Automation Letters*, 3(3), 1872–1879. DOI: 10.1109/LRA.2018.2800084

Chang, Y.-S., Chiao, H.-T., Abimannan, S., Huang, Y.-P., Tsai, Y.-T., & Lin, K.-M. (2020). An LSTM-based aggregated model for air pollution forecasting. *Atmospheric Pollution Research*, 11(8), 1451–1463. DOI: 10.1016/j.apr.2020.05.015

Chen, T., Pennisi, A., Li, Z., Zhang, Y., & Sahli, H. (2018). A hierarchical association framework for multi-object tracking in airborne videos. *Remote Sensing (Basel)*, 10(9), 1347. DOI: 10.3390/rs10091347

Chen, Y., Xie, X., Yu, B., Li, Y., & Lin, K. (2021). Multitarget vehicle tracking and motion state estimation using a novel driving environment perception system of intelligent vehicles. *Journal of Advanced Transportation*, 2021, 2021. DOI: 10.1155/2021/6251399

Chu, Q., Ouyang, W., Liu, B., Zhu, F., & Yu, N. (2020). Dasot: A unified framework integrating data association and single object tracking for online multi-object tracking. 34, In *Proceedings of the AAAI conference on artificial intelligence* (07), (pp. 10672–10679). DOI: 10.1609/aaai.v34i07.6694

Daniłowicz, M. (2020). Multiple object tracking and segmentation with R-CNN networks. DOI: 10.1016/j.promfg.2020.04.032

Daniyan, I., Mpofu, K., Oyesola, M., Ramatsetse, B., & Adeodu, A. (2020). Artificial intelligence for predictive maintenance in the railcar learning factories. *Procedia Manufacturing*, 45, 13–18. DOI: 10.1016/j.promfg.2020.04.032

de Pater, I., Reijns, A., & Mitici, M. (2022). Alarm-based predictive maintenance scheduling for aircraft engines with imperfect Remaining Useful Life prognostics. *Reliability Engineering & System Safety*, 221, 108341. DOI: 10.1016/j.ress.2022.108341

Durrant-Whyte, H. F. Elements of sensor fusion, *Proceedings of the Institute of Electrical and Electronics Engineers Colloquium on Intelligent Control*, 1991, 1 - 2.

Egbe, G. O., Arinze, S. N., Ebenuwa, S. H., Obi, E. R., & Nwajana, A. O. (2024). Maintenance resources optimization using Pareto analysis: Instrumentation air compressor in Oredo, Nigeria. [IJMMME]. *International Journal of Manufacturing, Materials, and Mechanical Engineering*, 13(1), 1–14. DOI: 10.4018/IJMMME.353392

Elliethy, A., & Sharma, G. (2018). Vehicle tracking in wide area motion imagery via stochastic progressive association across multiple frames. *IEEE Transactions on Image Processing*, 27(7), 3644–3656. DOI: 10.1109/TIP.2018.2818443 PMID: 29993777

Farhodov, X., Moon, K.-S., Lee, S.-H., & Kwon, K.-R. (2020). LSTM network with tracking association for multi-object tracking. *Journal of Korea Multimedia Society*, 23(10), 1236–1249.

Feng, P., Wang, W., Dlay, S., Naqvi, S. M., & Chambers, J. (2016). Social force model-based MCMC-OCSVM particle PHD filter for multiple human tracking. *IEEE Transactions on Multimedia*, 19(4), 725–739. DOI: 10.1109/TMM.2016.2638206

Fernandes, J., Reis, J., Melão, N., Teixeira, L., & Amorim, M. (2021). The role of Industry 4.0 and BPMN in the arise of condition-based and predictive maintenance: A case study in the automotive industry. *Applied Sciences (Basel, Switzerland)*, 11(8), 3438. DOI: 10.3390/app11083438

Flexible Data Communication and IIoT Platform in Industrial Applications—Ewon Flexy. Available online: www.ewon.biz (accessed on 28 November 2024).

Gan, W., Wang, S., Lei, X., Lee, M.-S., & Kuo, C.-C. J. (2018). Online NN-basedmultiple object tracking with enhanced model updates and identity association. *Signal Processing Image Communication*, 66, 95–102. DOI: 10.1016/j.image.2018.05.008

Granström, K., Renter, S., Fatemi, M., & Svensson, L. (2017). *Pedestrian tracking using Velodyne data—Stochastic optimization for extended object tracking. In 2017 IEEE intelligent vehicles symposium (Iv)*. IEEE.

Gulati, D., Sharif, U., Zhang, F., Clarke, D., & Knoll, A. (2017). *Data association solution or avoidance: Evaluation of a filter based on RFS framework and factor graphs with SME. In 2017 IEEE international conference on multisensor fusion and integration for intelligent systems (MFI).* IEEE.

Guo, S., Wang, J., Wang, X., & Tao, D. (2021). Online multiple object tracking with cross-task synergy. In *Proceedings of the IEEE/CVF conference on computer vision and pattern recognition* (pp. 8136–8145). DOI: 10.1109/CVPR46437.2021.00804

Hall, D. L., & Llinas, J. (1997). An introduction to multisensor data fusion. *Proceedings of the IEEE*, 85(1), 6–23. DOI: 10.1109/5.554205

Hamid Rezatofighi, S., Milan, A., Zhang, Z., Shi, Q., Dick, A., & Reid, I. (2015). Joint probabilistic data association revisited. In *Proceedings of the IEEE international conference on computer vision* (pp. 3047–3055).

He, Y., Han, J., Yu, W., Hong, X., Wei, X., & Gong, Y. (2020). City-scale multicamera vehicle tracking by semantic attribute parsing and cross-camera tracklet matching. In *Proceedings of the IEEE/CVF conference on computer vision and pattern recognition workshops* (pp. 576–577).

Hosamo, H. H., Svennevig, P. R., Svidt, K., Han, D., & Nielsen, H. K. (2022). A Digital Twin predictive maintenance framework of air handling units based on automatic fault detection and diagnostics. *Energy and Building*, 261, 111988. DOI: 10.1016/j.enbuild.2022.111988

Hu, E., Deng, Z., Jiang, K., & Wu, C. (2020). Kullback-Leibler differential entropy equation based CIMM-PDA for reliable positioning. *Alexandria Engineering Journal*, 59(4), 2607–2615. DOI: 10.1016/j.aej.2020.04.024

Hung, W.-C., Kretzschmar, H., Lin, T.-Y., Chai, Y., Yu, R., Yang, M.-H., & Anguelov, D. (2020). SoDA: Multi-object tracking with soft data association. arXiv preprint arXiv:2008.07725.

Jiang, Z., Huynh, D. Q., Zhang, J., & Wu, Q. (2017). *Part-based data association for visual tracking. In 2017 international conference on digital image computing: techniques and applications (DICTA).* IEEE.

Joerger, M., & Hassani, A. (2020). *A new data association method using Kalman filter innovation vector projections. In 2020 IEEE/ION Position, location and navigation symposium (PLANS).* IEEE.

Khadersab, A., & Shivakumar, S. (2018). Vibration Analysis Techniques for Rotating Machinery and its effect on Bearing Faults. *Procedia Manufacturing*, 20, 247–252. DOI: 10.1016/j.promfg.2018.02.036

Kim, C., Fuxin, L., Alotaibi, M., & Rehg, J. M. (2021). Discriminative appearance modelling with multi-track pooling for real-time multi-object tracking. In *Proceedings of the IEEE/CVF conference on computer vision and pattern recognition* (pp.9553–9562).

Lan, H., Pan, Q., Yang, F., Sun, S., & Li, L. (2016). Variational Bayesian approach for joint multitarget tracking of multiple detection systems. In 2016 19th international conference on information fusion (FUSION), (pp. 1260–1267). IEEE.

Lee, K.-H., Kanzawa, Y., Derry, M., & James, M. R. (2018). Multi-target track-to- track fusion based on permutation matrix track association. In 2018 [IEEE.]. *IEEE Intelligent Vehicles Symposium*, (IV), 465–470.

Lee, R. H., & Leahy, R. Segmentation of multisensor images, *Proceedings of the 6th Multi-dimensional Signal Processing Workshop*, Sept. 1989, Pacific Grove, USA, 23. DOI: 10.1109/MDSP.1989.96998

Lee, S. Bin, Hyun, D., Kang, T. J., Yang, C., Shin, S., Kim, H., Park, S., Kong, T. S., & Kim, H. D. (2015). Identification of false rotor fault indications produced by on-line MCSA for medium voltage induction machines. IEEE Conference Record of Annual Pulp and Paper Industry Technical Conference, 2015-July (1), 729–739.

Lei, Y.. (2015). A deep learning-based method for machinery health monitoring with big data. *Jixie Gongcheng Xuebao*, 51(21), 49–56. DOI: 10.3901/JME.2015.21.049

Lei, Y.. (2017). Applying machine learning to machine fault diagnosis: A review and roadmap. Mechanical Systems and Signal Processing, 2020. 138: p. 106587, Li, Z.; Wang, Y.; Wang, K.-S. Intelligent predictive maintenance for fault diagnosis and prognosis in machine centers: Industry 4.0 scenario. *Advances in Manufacturing*, 5, 377–387.

Liang, T., Lan, L., & Luo, Z. (2020). Enhancing the association in multi-object tracking via neighbor graph. arXiv preprint arXiv:2007.00265.

Liao, L., & Köttig, F. (2014). Review of hybrid prognostics approaches for remaining useful life prediction of engineered systems, and an application to battery life prediction. *IEEE Transactions on Reliability*, 63(1), 191–207. DOI: 10.1109/TR.2014.2299152

Liu, Q., Chu, Q., Liu, B., & Yu, N. (2020). GSM: Graph similarity model for multi-object tracking. In IJCAI (pp. 530–536).

Liu, R., Yang, B., Zio, E., & Chen, X. (2018). Artificial intelligence for fault diagnosis of rotating machinery: A review. *Mechanical Systems and Signal Processing*, 108, 33–47. DOI: 10.1016/j.ymssp.2018.02.016

Liu, Y., Yao, L., Xiong, W., & Zhou, Z. (2018). Joint kinematic and feature tracking of ships with satellite electronic information. *Journal of Navigation*, 71(5), 1178–1194. DOI: 10.1017/S0373463318000152

Liu, Z., Li, Q., & Mu, C. (2013). A hybrid LSSVR/HMM based prognostics approach. *Sensors (Basel)*, 13(5), 5542–5560. DOI: 10.3390/s130505542 PMID: 23624688

Luo, RC, & Kay, MG. Multisensor integration and fusion in intelligent systems. (. (1989, September/ October). Institute of Electrical and Electronics Engineers Transactions on Systems. *Man and Cybernetics*, 19(5), 901–931. DOI: 10.1109/21.44007

Luo, Z., Attari, M., Habibi, S., & Von Mohrenschildt, M. (2019). Online multiple manoeuvring vehicle tracking system based on multi-model smooth variable structure filter. *IEEE Transactions on Intelligent Transportation Systems*.

Mahemuti, B., Inoue, D., Kakugo, A., & Konagaya, A. (2016). Investigation of the microtubule dynamics with probabilistic data association filter. In 2016 IEEE 11th annual international conference on nano/micro engineered and molecular systems (NEMS), (pp. 101–106). IEEE. DOI: 10.1109/NEMS.2016.7758209

Maktoubian, J., Taskhiri, M. S., & Turner, P. (2021). Intelligent Predictive Maintenance (IPdM) in Forestry: A Review of Challenges and Opportunities. *Forests*, 12(11), 1495. DOI: 10.3390/f12111495

Meinhardt, T., Kirillov, A., Leal-Taixe, L., & Feichtenhofer, C. (2021). Trackformer: Multi-object tracking with transformers. arXiv preprint arXiv:2101.02702.

Meneses, M., Matos, L., Prado, B., de Carvalho, A., & Macedo, H. (2020). Learning to associate detections for real-time multiple object tracking. arXiv preprint arXiv: 2007.06041.

Mohanty, A. R. (2014). Machinery condition monitoring: Principles and practices. In *Machinery Condition Monitoring*. Principles and Practices. DOI: 10.1201/9781351228626-17

Mukherjee, A. N. (2022). Application of artificial intelligence: Benefits and limitations for human potential and labor-intensive economy–an empirical investigation into pandemic ridden Indian industry. *Management Matters*, 19(2), 149–166. DOI: 10.1108/MANM-02-2022-0034

Nacchia, M., Fruggiero, F., Lambiase, A., & Bruton, K. (2021). A Systematic Mapping of the Advancing Use of Machine Learning Techniques for Predictive Maintenance in the Manufacturing Sector. *Applied Sciences (Basel, Switzerland)*, 11(6), 2546. DOI: 10.3390/app11062546

Okano, M. T. IOT and Industry 4.0: The Industrial New Revolution. In Proceedings of the ICMIS-17—International Conference on Management and Information Systems, Bangkok, Thailand, 25–26 September 2017; pp. 75–82.

Okoh, C., Roy, R., & Mehnen, J. (2017). Predictive Maintenance Modelling for Through-Life Engineering Services. Procedia CIRP, 59(TESConf 2016), 196–201. DOI: 10.1016/j.promfg.2020.04.032

Omri, N., Al Masry, Z., Mairot, N., Giampiccolo, S., & Zerhouni, N. (2021). Towards an adapted PHM approach: Data quality requirements methodology for fault detection applications. *Computers in Industry*, 127, 103414. DOI: 10.1016/j.compind.2021.103414

Parra-Loera, R., Thompson, W. E., & Salvi, A. P. Adaptive selection of sensors based on individual performances in a multisensor environment, Proceedings of the SPIE, 1470, Data Structures and Target Classification, April 1991, Orlando, FL, 30-6.

Pech, M., & Vrchota, J. (2021). Bednárˇ, J. Predictive Maintenance and Intelligent Sensors in Smart Factory [Review]. *Sensors (Basel)*, 21, 1470. DOI: 10.3390/s21041470 PMID: 33672479

Punchihewa, Y. G., Vo, B.-T., Vo, B.-N., & Kim, D. Y. (2018). Multiple object tracking in unknown backgrounds with labeled random finite sets. *IEEE Transactions on Signal Processing*, 66(11), 3040–3055. DOI: 10.1109/TSP.2018.2821650

Qu, C., Zhang, Y., Zhang, X., & Yang, Y. (2020). Reinforcement learning-based data association for multiple target tracking in clutter. *Sensors (Basel)*, 20(22), 6595. DOI: 10.3390/s20226595 PMID: 33218053

Rangesh, A., Maheshwari, P., Gebre, M., Mhatre, S., Ramezani, V., & Trivedi, M. M. (2021). TrackMPNN: A message passing graph neural architecture for multi-object tracking. arXiv preprint arXiv:2101.04206.

Ritter, C., Imle, A., Lee, J. Y., Müller, B., Fackler, O. T., Bartenschlager, R., & Rohr, K. (2018). Two-filter probabilistic data association for tracking of virus particles in fluorescence microscopy images. In 2018 IEEE 15th international symposium on biomedical imaging (ISBI 2018), (pp. 957–960). IEEE. DOI: 10.1109/ISBI.2018.8363729

Salscheider, N. O. (2021). Object tracking by detection with visual and motion cues. arXiv preprint arXiv:2101.07549

Sarwar, U., Muhammad, M., Mokhtar, A. A., Khan, R., Behrani, P., & Kaka, S. (2024). Hybrid intelligence for enhanced fault detection and diagnosis for industrial gas turbine engine. *Results in Engineering*, 21, 101841. DOI: 10.1016/j.rineng.2024.101841

Sayyad, S., Kumar, S., Bongale, A., Bongale, A. M., & Patil, S. (2021). Estimating Remaining Useful Life in Machines Using Artificial Intelligence: A Scoping Review. *Library Philosophy and Practice*, 2021(January), 1–26.

Sayyad. S, S. Kumar, A. Bongale, P. Kamat, S. Patil and K. Kotecha, (2021) "Data-Driven Remaining Useful Life Estimation for Milling Process: Sensors, Algorithms, Datasets, and Future Directions," in IEEE Access, . DOI: 10.1109/ACCESS.2021.3101284

Shao, S. Y., Sun, W. J., Yan, R. Q., Wang, P., & Gao, R. X. (2017). A Deep Learning Approach for Fault Diagnosis of Induction Motors in Manufacturing. [English Edition]. *Chinese Journal of Mechanical Engineering*, 30(6), 1347–1356. DOI: 10.1007/s10033-017-0189-y

Sheng, H., Chen, J., Zhang, Y., Ke, W., Xiong, Z., & Yu, J. (2018). Iterative multiple hypothesis tracking with tracklet-level association. *IEEE Transactions on Circuits and Systems for Video Technology*.

Silvestri, L., Forcina, A., Introna, V., Santolamazza, A., & Cesarotti, V. (2020). Maintenance transformation through Industry 4.0 technologies: A systematic literature review. *Computers in Industry*, 123, 103335. DOI: 10.1016/j.compind.2020.103335

Singh, G., Rajan, S., & Majumdar, S. (2017). A greedy data association technique for multiple object tracking. In 2017 IEEE third international conference on multimedia big data (BigMM), (pp. 177–184). IEEE. DOI: 10.1109/BigMM.2017.53

Stauch, J., Bessell, T., Rutten, M., Baldwin, J., Jah, M., & Hill, K. (2017). Joint probabilistic data association and smoothing applied to multiple space object tracking. *Journal of Guidance, Control, and Dynamics*, 41(1), 19–33. DOI: 10.2514/1.G002230

Sun, P., Jiang, Y., Zhang, R., Xie, E., Cao, J., Hu, X., Kong, T., Yuan, Z., Wang, C., & Luo, P. (2020). Transtrack: Multiple-object tracking with transformer. arXiv preprint arXiv:2012.15460.

Tan, L., Dong, X., Ma, Y., & Yu, C. (2018). A multiple object tracking algorithm based on YOLO detection. In 2018 11th international congress on image and signal processing, biomedical engineering and informatics (CISP-BMEI), (pp. 1–5). IEEE. DOI: 10.1109/CISP-BMEI.2018.8633009

Tang, Z., Gu, R., & Hwang, J.-N. (2018). Joint multi-view people tracking and pose estimation for 3D scene reconstruction. In 2018 IEEE international conference on multimedia and expo (ICME), (pp. 1–6). IEEE.

Tiddens, W., Braaksma, J., & Tinga, T. (2020). Exploring predictive maintenance applications in industry. *Journal of Quality in Maintenance Engineering*, 28(1), 68–85. DOI: 10.1108/JQME-05-2020-0029

Udmale, S. S., & Singh, S. K. (2019). Application of spectral kurtosis and improved extreme learning machine for bearing fault classification. *IEEE Transactions on Instrumentation and Measurement*, 68(11), 4222–4233. DOI: 10.1109/TIM.2018.2890329

Viswanathan, R., Thomopoulos, S. C. A., & Tumuluri, R. (1988, July). Optimal serial distributed decision fusion. *Institute of Electrical and Electronics Engineers Transactions on Aerospace and Electronic Systems*, AES-24(4), 366–375.

Vivone, G., Braca, P., & Errasti-Alcala, B. (2015). Extended target tracking applied to X-band marine radar data. In *OCEANS 2015-Genova* (pp. 1–6). IEEE. DOI: 10.1109/OCEANS-Genova.2015.7271630

Wan, X., Wang, J., Kong, Z., Zhao, Q., & Deng, S. (2018). Multi-object tracking using online metric learning with long short-term memory. In 2018 25th IEEE international conference on image processing (ICIP), (pp. 788–792). IEEE. DOI: 10.1109/ICIP.2018.8451174

Wang, C., Xie, X., & Liao, C. (2021). An adaptive fusion model based on Kalman filtering and LSTM for fast tracking of road signs. In 2020 25th international conference on pattern recognition (ICPR), (pp. 1414–1421). IEEE. DOI: 10.1109/ICPR48806.2021.9412296

Wang, J., Ancha, S., Chen, Y.-T., & Held, D. (2020). Uncertainty-aware self-supervised 3D data association. In 2020 IEEE/RSJ international conference on intelligent robots and systems (IROS), (pp. 8125–8132). IEEE.

Wang, J., Xu, C., Zhang, J., & Zhong, R. (2021). Big data analytics for intelligent manufacturing systems: A review. *Journal of Manufacturing Systems*, 62, 738–752. DOI: 10.1016/j.jmsy.2021.03.005

Wang, Y., Martinsen, K., Yu, T., & Wang, K. (2022). *Advanced Manufacturing and Automation XI*. Springer. DOI: 10.1007/978-981-19-0572-8

Weng, X., Wang, Y., Man, Y., & Kitani, K. (2020). Graph neural networks for 3Dmulti-object tracking. arXiv preprint arXiv:2008.09506.

Wu, H., & Li, W. (2016). *Robust online multi-object tracking based on KCF trackers and reassignment. In 2016 IEEE global conference on signal and information processing (GlobalSIP)*. IEEE.

Xi-yang, Z., Xiao-li, W., & Liangqun, L. (2018). Online multi-object tracking via maximum entropy intuitionistic fuzzy data association. In 2018 14th IEEE international conference on signal processing (ICSP), (pp. 803–806). IEEE. DOI: 10.1109/ICSP.2018.8652401

Xu, C., & Zhou, Y. (2018). Hierarchical online multi-person pose tracking with multiple cues. In *International conference on neural information processing* (pp. 318–328). Springer. DOI: 10.1007/978-3-030-04224-0_27

Yang, S., Thormann, K., & Baum, M. (2018). Linear-time joint probabilistic data association for multiple extended object tracking. In 2018 IEEE 10th sensor array and multichannel signal processing workshop (SAM), (pp. 6–10). IEEE. DOI: 10.1109/SAM.2018.8448430

Yoon, J. H., Lee, C.-R., Yang, M.-H., & Yoon, K.-J. (2019). Structural constraint data association for online multi-object tracking. *International Journal of Computer Vision*, 127(1), 1–21. DOI: 10.1007/s11263-018-1087-1 PMID: 33664553

Yu, H., Li, G., Su, L., Zhong, B., Yao, H., & Huang, Q. (2020). Conditional GAN based individual and global motion fusion for multiple object tracking in UAV videos. *Pattern Recognition Letters*, 131, 219–226. DOI: 10.1016/j.patrec.2019.12.018

Yuan, T., Krishnan, K., Chen, Q., Breu, J., Roth, T. B., Duraisamy, B., Weiss, C., Maile, M., & Gern, A. (2017). Object matching for inter-vehicle communication systems—An IMM-based track association approach with sequential multiple hypothesis test. *IEEE Transactions on Intelligent Transportation Systems*, 18(12), 3501–3512. DOI: 10.1109/TITS.2017.2723894

Zeng, D., Xiong, L., Yu, Z., Chen, Q., Fu, Z., Li, Z., Zhang, P., Xu, P., Qian, Z., Xiao, H., Fang, P., Li, Z., & Leng, B. (2020). A priority data association policy for multitarget tracking on intelligent vehicle risk assessment. *Remote Sensing (Basel)*, 12(19), 3255. DOI: 10.3390/rs12193255

Zhang, C., Huang, Y., Wang, Z., Jiang, H., & Yan, D. (2018). Cross-camera multi-person tracking by leveraging fast graph mining algorithm. *Journal of Visual Communication and Image Representation*, 55, 711–719. DOI: 10.1016/j.jvcir.2018.08.006

Zhang, Y., Xu, S., & Wan, Y. (2020). Performance improvement of centrifugal compressors for fuel cell vehicles using the aerodynamic optimization and data mining methods. *International Journal of Hydrogen Energy*, 45(19), 11276–11286. DOI: 10.1016/j.ijhydene.2020.02.026

Zheng, J., Mu, H., Wang, X., Li, T.-M., Zhang, Q., & Wang, X. An Adaptive Maintenance Policy With Nonlinear Degradation Modeling Based on Prognostic Information. IEEE Access 2020, 8, 160040–160049DOI: 10.1016/j.ijhydene.2020.02.026

Zhou, Z., Chen, D., & Xie, S. (Shengquan). (2007). Springer Series in Advanced Manufacturing. In Thermoplastics and Thermoplastic Composites.

Zonta, T., da Costa, C. A., Righi, R. D. R., de Lima, M. J., da Trindade, E. S., & Li, G. P. (2020). Predictive maintenance in the Industry 4.0: A systematic literature review. *Computers & Industrial Engineering*, 150, 106889. DOI: 10.1016/j.cie.2020.106889

Chapter 10
Circular Fractal Array Antenna for C and S-Band Applications

Gnanalakshmi Venkatachalam
Mepco Schlenk Engineering College, India

M. Heman
Mepco Schlenk Engineering College, India

S. Sumesh
Mepco Schlenk Engineering College, India

ABSTRACT

In this article, a circular fractal array antenna is proposed for C-band and S-band applications. The antenna uses circular fractal geometry to achieve high gain across the S and C bands frequency spectrum. The antenna is fabricated using an FR-4 substrate and tested in anechoic chamber. The antenna design holds potential for multi-functional wireless systems and S and C band applications. The measurement results reveal that the designed antenna has two resonant frequencies at 3.5 GHz and 5.8 GHz and gain of around 3.1dB is noted. The size of proposed antenna is 28mm×28mm×1.6mm.

INTRODUCTION

Fractal antennas are made of geometric shapes that repeat themselves on various sizes and are self-similar, making them geometric shapes that cannot be characterized using Euclidean geometry. There are several varieties of fractal structures, including the tree structure, the Koch structure and the dragon structure. S and C band technology is widely employed in many different industries because it provides great resistance to multipath interference and high-speed data transmission. This proposed structure's operational frequency ranges from 3.5GHz to 5.8GHz for S-band and C -band. Important electromagnetic spectrum regions that are often used in a variety of communication and sensing applications are the C and S-bands. These bands have special qualities that make them ideal for certain applications, including radar systems, remote sensing, and terrestrial communication. In general, the C-band covers frequencies between 4 and 8 gigahertz (GHz), whereas the S-band covers frequencies between 2 and 4 GHz. These frequency bands have excellent propagation properties, mild air attenuation, and are regulated

DOI: 10.4018/979-8-3693-8799-3.ch010

for specific purposes, which makes them widely employed in a variety of communication and sensing applications. In satellite communication systems, circular fractal array antennas can be used for both uplink and downlink communication. Their small size, broad bandwidth, and great radiation efficiency make them ideal for space-constrained satellite platforms. These antennas provide services like internet access, data transmission, and television broadcasting in the C-band. Circular fractal antennas operating in the S-band can improve mobile satellite communication systems for disaster response, marine communication, and remote sensing applications by offering dependable connectivity. For satellite-based services in distant or mobile contexts, their strong performance and mobility are crucial. In antenna design, circular fractal slots are utilized to provide bidirectional gain and circular polarization. A circular fractal antenna's size can vary greatly based on a number of variables, including the particular fractal geometry employed, the intended frequency spectrum, the needs of the application, and the limitations of the design. The number of iterations in the fractal design and the intended operating frequency are the main factors influencing the antenna's size. Smaller antenna parts match the antenna's overall shape thanks to the use of self-similar geometrical forms in fractal antenna design, which permits downsizing without compromising functionality. A basic circular loop, for instance, might be the form of a circular fractal antenna. The performance characteristics of the antenna can be enhanced or maintained while reducing its size by recursively adding smaller loops within the main loop. The size of proposed antenna is 28mm × 28mm × 1.6mm. FR-4 substrate was used to simulate our antenna.

The fabrication of a Circular Fractal Array Antenna for C-band and S-band applications involves a combination of careful fractal geometry selection, precise simulation and optimization, and state-of-the-art fabrication techniques, typically using microstrip or patch antenna designs. By leveraging fractal geometry's unique properties, such antennas can achieve compactness, multiband performance, and efficiency, making them ideal for advanced communication systems.

The Circular Fractal Array Antenna (CFAA), especially when designed for C-band and S-band frequencies, holds significant potential for 5G communications and future wireless technologies (such as 6G and beyond). The evolving demands of high-speed, high-capacity wireless communication systems make antenna designs like CFAA highly relevant due to their unique properties. Here's a detailed explanation of the significance of the proposed antenna geometry in the context of 5G and beyond:

1. Multiband and Wideband Performance
    - **Fractal Geometry and Frequency Agility**: Fractal antennas, due to their self-similar and iterative nature, inherently support multiple resonant frequencies. This characteristic is essential for **5G networks**, which utilize a wide range of frequencies across low, mid, and high bands (sub-6 GHz and millimeter-wave bands, i.e., **mmWave**). The fractal design enables CFAAs to operate efficiently over multiple frequency bands, including **C-band (4–8 GHz)** and **S-band (2–4 GHz)**, making them highly versatile for future 5G deployments.
    - **Wider Bandwidth for High-Data Rates**: One of the main advantages of the CFAA is its ability to offer **wide bandwidths**, crucial for high-speed communication. **5G and beyond** require high data rates, which are achieved through the efficient use of available bandwidth. The **wideband performance** of fractal antennas supports the larger spectrum allocations in these advanced systems, ensuring efficient data transmission across a broad range of frequencies.
2. Compact and Efficient Design for Small Cells

- **Size Reduction with Fractals**: Fractal antenna designs are inherently compact for their operational bandwidth, which is a key benefit for **small-cell networks**, a crucial component of 5G and beyond. Small cells are deployed in high-density urban environments, where space is limited. The fractal structure allows for miniaturized antennas that can still maintain high efficiency and performance.
- **Enhanced Efficiency**: The fractal design can improve the overall **radiation efficiency** of the antenna by increasing the surface area without a proportional increase in size. This is beneficial for both **energy efficiency** and signal quality, which is critical in future communication systems where power and performance are paramount, especially in devices with limited battery life (e.g., IoT devices, smartphones).

3. Omnidirectional and Directional Radiation Patterns
    - **Versatility in Radiation Characteristics**: The **circular array** configuration of the fractal antenna can support **both omnidirectional** and **directional** radiation patterns, making the antenna highly versatile. For **5G networks**, the ability to tailor the radiation pattern is critical as the system needs to adapt to various use cases, including **high-density urban environments**, **vehicle-to-everything (V2X) communication**, and **industrial IoT**.
    - **Beamforming Capabilities**: When combined with advanced techniques such as **beamforming** and **MIMO (multiple-input multiple-output)**, the CFAA can enhance the performance of **massive MIMO** systems, which are a cornerstone of 5G and beyond. The fractal geometry can be optimized for uniform radiation and reduced side lobes, making it suitable for targeted communications with high signal integrity.
4. Low Profile and Integration with Millimeter-Wave Systems
    - **Integration with Millimeter-Wave Antennas**: The **low-profile** nature of microstrip-based CFAAs, combined with their fractal geometry, allows them to be easily integrated into **millimeter-wave** (30 GHz to 300 GHz) systems, which are expected to play a key role in **5G** and **6G** communication, particularly for **high-speed data** and **ultra-low latency** applications. These millimeter-wave systems require compact and efficient antennas with large bandwidths, and the fractal antenna geometry is particularly well-suited for such requirements.
    - **Multifunctionality**: The ability to design CFAAs with fractal elements that can cover a wide range of frequencies (from S-band to C-band and beyond) makes them ideal for **multi-frequency systems**, a hallmark of 5G and future wireless standards. The use of such antennas can reduce the number of required antennas or the complexity of switching between different frequency bands, which is beneficial for maintaining simplicity and reducing costs in device design.
5. Reduced Electromagnetic Interference (EMI) and Improved Signal Integrity
    - **Self-Similarity and Reduced Interference**: The self-similar pattern of fractal designs can help in reducing **electromagnetic interference (EMI)** by optimizing the electromagnetic field distribution and reducing the unwanted side lobes or reflections. This is important in the dense radio frequency (RF) environment of 5G networks, where multiple devices and signals coexist in close proximity. Minimizing EMI improves **signal integrity** and reduces interference, which is essential for maintaining the reliability and speed of communications.
    - **Robustness in Harsh Environments**: Fractal antennas are known for their **robustness** and ability to handle a wide range of environmental factors (e.g., temperature, humidity), which is beneficial for devices in outdoor or harsh environments (e.g., autonomous vehicles, smart

cities, industrial IoT sensors). This makes them suitable for the diverse applications envisioned in **5G and beyond**.

6. Support for High Mobility and Dynamic Environments
   - **High Mobility Communication**: As 5G networks evolve, the demand for supporting **high-speed mobility** (e.g., autonomous vehicles, drones) increases. The fractal antenna's ability to maintain performance in dynamic and rapidly changing environments is beneficial. Since the fractal structure offers consistent performance across a wide range of angles and distances, it supports seamless communication in high-speed applications, such as V2X (Vehicle-to-Everything) or high-speed train communication.
7. Scalability for Future Networks (6G and Beyond)
   - **Scalable Architecture for 6G**: The scalable nature of fractal geometries makes them highly adaptable for future wireless technologies like **6G** and beyond. 6G is expected to operate at even higher frequencies (terahertz), require extreme data rates, and support massive numbers of devices. The fractal design can be extended or modified to accommodate the ultra-wide frequency bands and the massive connectivity requirements of future communication systems.
   - **Energy Efficiency in Next-Generation Networks**: Future networks (such as **6G**) will place a premium on **sustainability** and **energy efficiency**. Fractal antennas' reduced size and enhanced radiation efficiency contribute to lower power consumption, which is key for both individual devices and large-scale network operations in **5G and beyond**.
8. Applications in IoT and Smart Systems
   - **Internet of Things (IoT)**: With **5G** enabling **massive IoT (mIoT)**, the ability to deploy low-cost, compact, and efficient antennas is crucial. The fractal antenna's small size and wideband capability allow for embedding in a wide range of IoT devices such as **smart sensors**, **wearables**, and **smart cities infrastructure**.
   - **Smart Systems and Smart Cities**: The use of CFAAs in **smart systems** (including **smart grids**, **autonomous vehicles**, and **urban infrastructure**) is essential for **high-reliability communications** in complex and dynamic environments. Their compactness, wideband characteristics, and low-profile nature make them perfect for applications in such **smart technologies.**

As a result, scientists have been working hard to create fresh, practical approaches to improve the many features of microstrip patch antennas, especially their band-breadth. The radiating elements of printed antennas are made up of patches in a variety of shapes, including square, elliptical, circular, rectangular, and triangular. Circular structures have shown lesser dimensions in comparison to frequency operations among these possibilities. Compared to rectangular antennas, which use patch length and width for control and design, circular patch antennas have the advantage of a simpler design and more control over emission patterns. Moreover, while operating at the same design frequency, circular patch antennas are roughly 16% smaller in size than rectangular patch antennas. The benefits of the circular patch antenna go beyond its small size and straightforward construction. Certain Wi-Fi networks, cordless phones, and satellite communication all use the C-band spectrum for transmission. The S-band spectrum is used by weather radar systems, surface ship radars, and some communication satellites (Varshney & Gençoğlan, 2024). These systems serve as a means of information dissemination in emergency situations. A number of methods exist for attaining multiband functioning, including split ring resonators (Alle,

Kbehera, Mishra, & Pandav, 2021), fractal form use (Sharma, Singh, & Khanna, 2021) (Paik, Lankecharanteja, Kovvurisaitrinadh Reddy, & Mallidiakash Reddy, 2021), slot-loading (Kalaiyarasan, Nagarajan, & Seenuvasamurthi, 2023), and rotational L-shaped structures (Bakhar, Dakulagi, & Mathpati, 2021) (Jantree, Lerdwanittip, Namsang, Pakkaraekthalin, & Sirawattanachai, 2021). The choice of radiating element and material thickness has a significant impact on the number of functioning bands. Microstrip patch antennas are extensively utilized because of their tiny size, low cost, and low profile. Nevertheless, there is a bandwidth restriction. Therefore, it is noted that in order to achieve multiband with better performance, modifications to the antenna are necessary (Alotaibi et al., 2021) (Gnanalakshmi, Rahul Raaj, & Suresh Kumar, 2021) (Banerjee, Rana, Parui, Chatterjee, & Dey, 2017). One popular technique in multiband antenna design is the use of fractal shapes; researchers have developed fractal antennas using hybrid fractal curves such as Sierpinski-Koch (Rajagopal, Suseela, Noorullakhan, & Sankararajan, 2016), Koch-self-affine curve (Abhaikumar, Manimegalai, & Raju, 2009), ring fractal (Iyampalam & Ganesan, 2020), and Minkowski-Hilbert (Elavarasi & Shanmuganantham, 2017). Additionally, combinations of the fractal with different shapes of the patches, such as square (Gnanalakshmi, Rahul Raaj, & Suresh Kumar, 2021) (Banerjee, Rana, Parui, Chatterjee, & Dey, 2017), triangle (Abhaikumar, Manimegalai, & Raju, 2009), pentagon (Gnanalakshmi, Rahul Raaj, & Suresh Kumar, 2021), octagon (Alotaibi et al., 2021), and hexagon (Iyampalam & Ganesan, 2020) (Elavarasi & Shanmuganantham, 2017) have been reported for various wireless applications. Another important method is to integrate artificial material, such as a complementary split-ring resonator (CSRR), into the design in order to realize multiband. A 2×2 array mushroom type unit cell is constructed in the dual-band patch antenna, for example, as described in (Elavarasi & Shanmuganantham, 2017). The second technique uses artificial material as a ground plane (Rajagopal, Suseela, Noorullakhan, & Sankararajan, 2016) (Abhaikumar, Manimegalai, & Raju, 2009). The structure of the article is organized as follows. Section 2 describes the antenna configuration and design methodology. Section 3 describes the formulation of circular fractal array antenna.

ANTENNA CONFIGURATION AND DESIGN METHODOLOGY

Figure 1. Front view

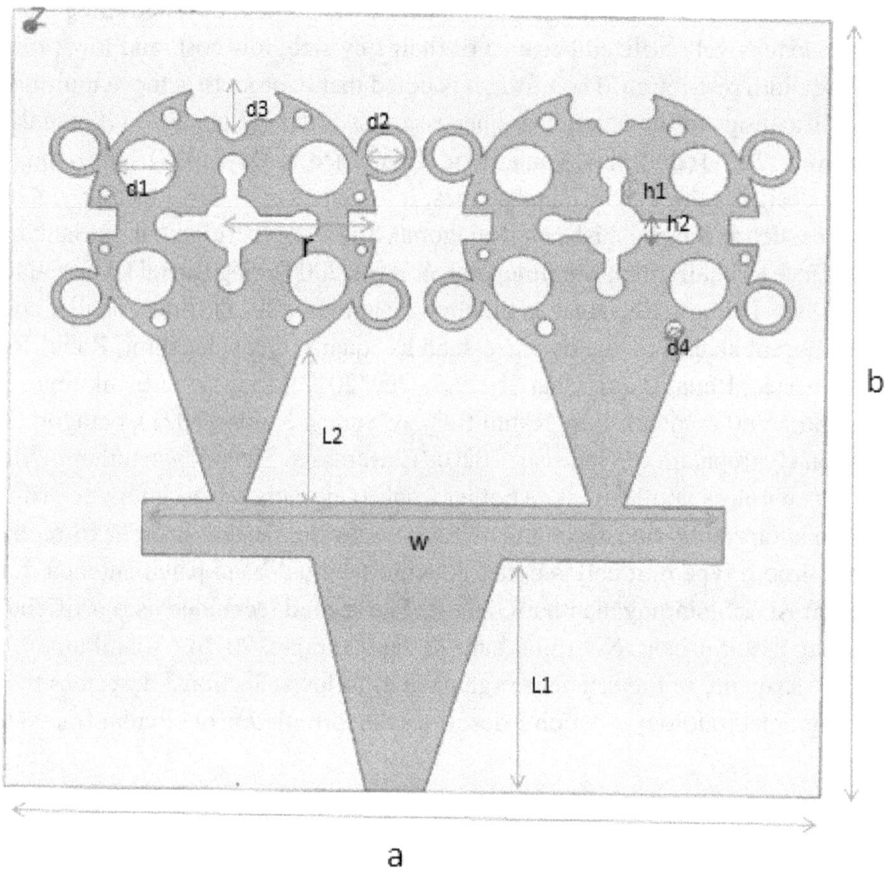

Figure 2. Back view

The following diagram illustrates the proposed Circular Fractal Array Antenna' (CFAA) geometry. Figure 1 and 2 show the designed antenna's top and bottom views. It is evident from the previously described figures that the CFAA consists of two layers, referred to as the top layer and bottom layer. The upper layer is made up of a partial ground plane at the bottom and a circular fractal structure with 32 slots. The 50 Ω micro-strip feed line excites the CFAA antenna. It is built on a 1.6 mm thick, low-cost FR4 dielectric material. The dimensions of the antenna are 28mm × 28 mm× 1.6 mm.

Figure 3. Design evalution of circular fractal array antenna: (a) iteration-1,(b) iteration-2,(c) iteration-3,(d)final iteration.

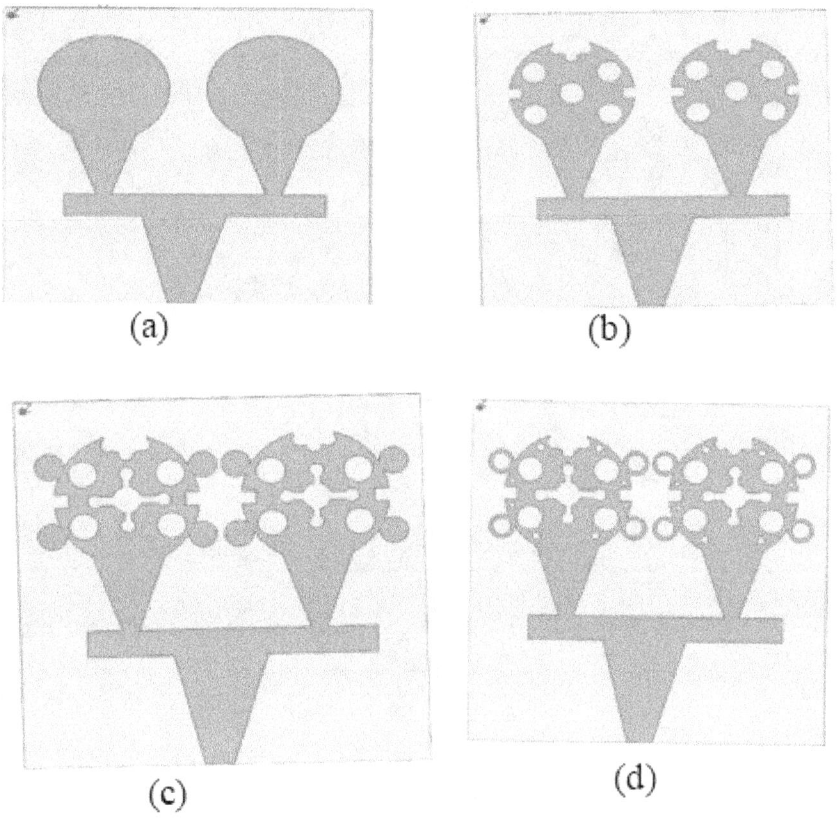

FORMULATION OF CIRCULAR FRACTAL ARRAY ANTENNA

The development of the circular fractal array antenna is shown in Figure. 3. Building a circular array antenna is the first stage of the suggested design. The fundamental circular array antenna is first created for 4.8 GHz and 10.4 GHz. The next step is to cut ten circular-shaped slots that have a diameter of 2 mm. Every slot is positioned three millimeters apart from each corner of the basic circular patch. At last, the patch's center is used to etch out the circular fractal of second order. By creating successive circular sub slots continually, the circular loop is built. . The radius of the circular patch is calculated using the formula:

$$f_r = \frac{1.8412c}{2\pi a\sqrt{\epsilon_r}} \tag{1}$$

$$a = \frac{F}{\sqrt{1 + \frac{2h}{\pi\epsilon_r F}\left[\ln\left(\frac{\pi F}{2h}\right) + 1.7726\right]}} \tag{2}$$

$$F = \frac{8.791 \times 10^9}{f_r\sqrt{\epsilon_r}} \quad (3)$$

Where,

c is the speed of light in free space
fr is resonating frequency
a is radius of circular patch.
h is height of substrate in cm.

The radiating frequency of a circular patch antenna is determined by varying the patch's radius. The following formulas are used to determine the microstrip feedline's characteristic impedance.

$$Z_c = \left\{ \frac{60}{\sqrt{\varepsilon_{reff}}} \ln\left[\frac{8h}{W_0} + \frac{W_0}{h}\right] \right.$$

for

$$\frac{W_0}{h} \leq 1 \quad (4)$$

$$Z_c = \left\{ \frac{120\pi}{\sqrt{\varepsilon_{reff}\left[\frac{W_0}{h}+1.393+0.677\ln\left(\frac{W_0}{h}+1.44\right)\right]}} \right.$$

for

$$\frac{W_0}{h} \geq 1 \quad (5)$$

The length of partial ground is given by the relation:

$$l_{gg} = l_f + p \quad (6)$$

Where l_f and p represent the length of feedline and the feed gap, respectively.

One feature of the circular fractal is self-similarity, which means that different parts of the circle resemble the full circle at different sizes. The precise fractal dimension of the circular fractal is determined by the particular construction method and the applied parameters. The starting size has been estimated using the following formulas.

$$K_1^{(n+1)} = a_1 \times r_n \quad (7)$$

$$K_2^{(n+1)} = a_2 \times r_n \quad (8)$$

where n is the number of iterations, K1 and K2 are the length and width of the indentation, a1 and a2 are the indentation ratios, and r is the radius of the semicircle.

FABRICATION TECHNIQUES

- **Microstrip Antenna Fabrication:** In CFAA designs, a microstrip patch antenna is often used for the individual elements, especially when fabricating on a PCB. The steps include:

1. PCB Design: The fractal patterns are transferred to the design layout using software tools.
2. Etching Process: The PCB is then fabricated using photolithography or direct etching, where the fractal pattern is etched onto the surface of the copper-coated substrate.
3. Drilling: Holes are drilled for the feed connections or vias as needed for the array configuration.

- **Multilayer Construction:** For complex designs, especially when implementing arrays, a multi-layer PCB may be required. This involves stacking multiple layers of substrate and interconnecting them with vias to support feed networks or additional antenna elements.
- **Assembly:** Once the individual antennas are fabricated, the array is assembled. This involves:
    - Aligning the Antenna Elements: The fractal elements are arranged in a circular pattern as part of the CFAA. The spacing between elements is crucial for controlling the array's radiation pattern and achieving the desired gain.
    - Feed Network Integration: A feed network is designed to ensure proper signal distribution to each antenna element in the array. This could involve microstrip transmission lines, corporate feed networks, or coaxial cable feeds.

TESTING AND CALIBRATION

- Impedance Matching: After fabrication, the CFAA is tested for impedance matching (usually 50 ohms) using a vector network analyzer (VNA). The antenna's reflection coefficient S_{11} is measured across the desired frequency bands (C-band and S-band) to ensure that the antenna operates efficiently without significant loss.
- Radiation Pattern Testing: The antenna's radiation pattern is measured in an anechoic chamber to verify the expected directional characteristics, such as gain and beamwidth. Since the CFAA design is typically circular, the radiation pattern may be omnidirectional, depending on the design.
- Performance Validation: The antenna is tested for its real-world performance, including factors like bandwidth, efficiency, and gain across the target frequency bands. This can include both far-field and near-field measurements, depending on the design's intended application.

6. Integration with Transceiver Systems

- After testing, the CFAA may be integrated into a larger transceiver or communication system. For C-band and S-band applications, the antenna may be used in satellite communication, radar systems, wireless communication, or other RF-based applications.
- In some cases, the array will be integrated into a phased array system where the fractal antenna elements can be electronically steered to modify the direction of the radiation pattern. This is typically achieved by varying the phase of the signal fed to each antenna element.

PARAMETRIC ANALYSIS

Figure 1 shows the design of proposed antenna. The optimal value to attain the required operating bands has been determined by the application of parametric analysis. The variables include the substrate's length (a), width (b), and height (g) of the partial ground. The feed line's length is represented by (L1), while the radius of the outer circular structure is represented by (r).The feeding rectangle and outer circular construction are separated by 5.6 mm, and this separation's length is indicated by the letter (L2). The remaining exterior circular slots have a diameter of 1.6 mm and are indicated by (d2), whereas the inner 10 circular slots have a diameter of 2 mm and are represented by (d1). The gap of 16 mm, indicated as (w), separates the two circular fractal structures.The extremely small 16 circular slots are represented as D4 and have a diameter of 0.4 mm, in comparison to other internal circular slots. Table 1 shows the dimensions of the proposed antenna design. Table 2 shows the performance of CFAA.

Table 1. Dimensions of the proposed antenna

Parameter	(mm)	Parameter	(mm)
a	28	h2	0.5
b	28	d1	2
L1	8.4	d2	1.6
L2	5.6	d3	1.5
h1	0.7	w	20
r	5	d4	0.4

Table 2. Performance of the CFAA

Antenna parameters	Obtained results
Resonance frequency	3.5 GHz and 5.8 GHz
Operating Band (GHz)	3.5-5.8
S_{11} parameter(dB)	- 20.2 dB (3.5GHZ), - 23.8 dB (5.8GHz)
Gain(dB)	3.1dB
Bandwidth(GHz)	3.4
Radiation Efficiency(%)	96.6

RESULTS AND DISCUSSION

In order to characterize the radiation performance of designed, the CFAA(circular fractal array antenna) is simulated using the Ansys HFSS software. The return loss obtained for different iterations is shown in figure 4. While antennas with iterations 2 and 3 worked at three frequencies, the antenna with iteration 1 resonated at two frequencies, 4.8 and 10.4 GHz. The antenna operates on two bands with iterations 1 and 2. Lastly, the dual-band resonance frequency of the suggested CFAA antenna satisfies the necessary operating frequencies. A (FR-4) substrate was used to build the circular fractal array antenna with a partial ground plane. Its S and C-band performance was then verified by measurement. In the last iteration, two frequencies resonate between 3.5 and 5.8 GHz, and the voltage standing wave ratio (VSWR) for 3.5 GHz and 4.8 GHz is found to be 1.6 and 1.1 respectively. The last iteration's VSWR measurement is shown in Figure 5.

The proposed antenna is fabricated and tested for validation. The fabricated antenna is tested in an-echoic chamber. The parameters such as return loss, VSWR, radiation pattern were measured. Figure 6 display the radiation pattern of final iteration. The S parameter for iteration-1, iteration-2, iteration-3, and the last iteration is displayed in Figure 4. The frequency resonates at 3.5 GHz and 5.8 GHz for the final iteration. An essential measure for assessing the effectiveness of antennas and other RF (Radio Frequency) equipment is the S11 parameter, sometimes referred to as the return loss. It calculates the amount of power that an antenna reflects as a result of an impedance mismatch. The S11 parameter for 5.8GHz is - 23.8GHz, while for 3.5GHz it is -20.2dB. The power reflected from an antenna as a result of an impedance mismatch is represented by the S11 parameter and is given in decibels (dB). It is represented as Return Loss (RL) or S11. The final iteration's VSWR is displayed in Figure 5. An antenna's impedance and the impedance of the transmission line or system it is linked to be mismatched to varying degrees, which is measured by VSWR. Figure 6 depicts the Radiation pattern of designed CFAA at 3.5GHz and 5.8GHz

Figure 4. (a) S_{11} of iteration-1,(b)S_{11} of iteration-2,(c)S_{11} of iteration-3,(d)S_{11} of final iteration.

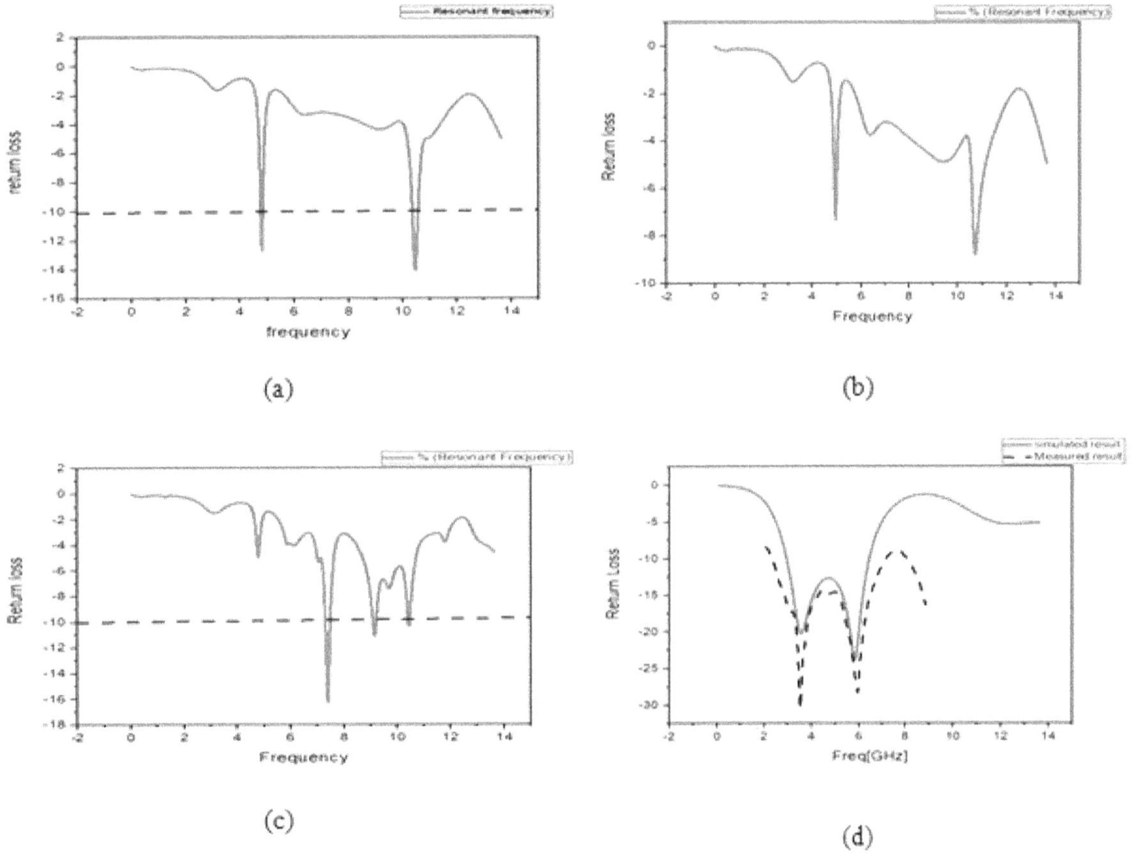

Figure 5. VSWR of final iteration.

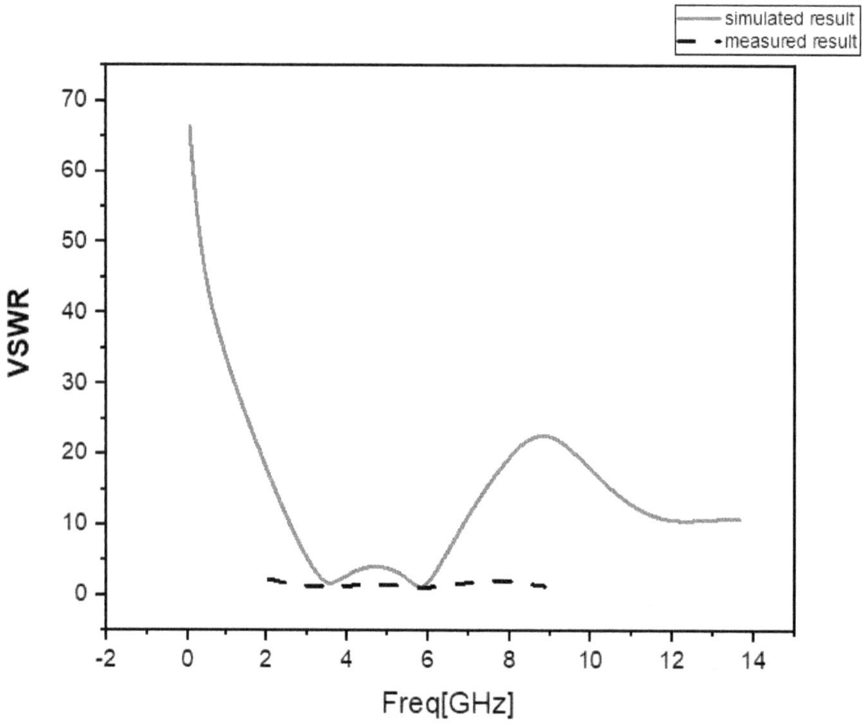

Figure 6. Radiation pattern of designed CFAA at 3.5GHz and 5.8GHz

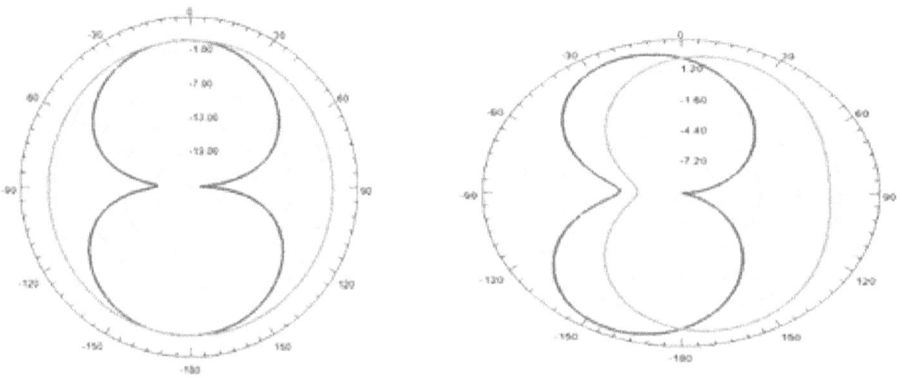

Figure 7. Gain plot of designed CFAA

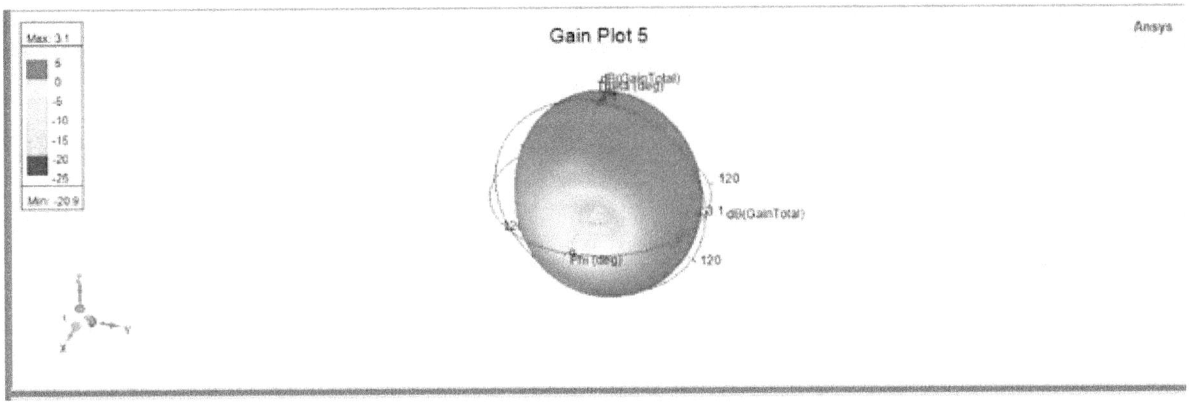

Figure 8. Fabricated antenna.

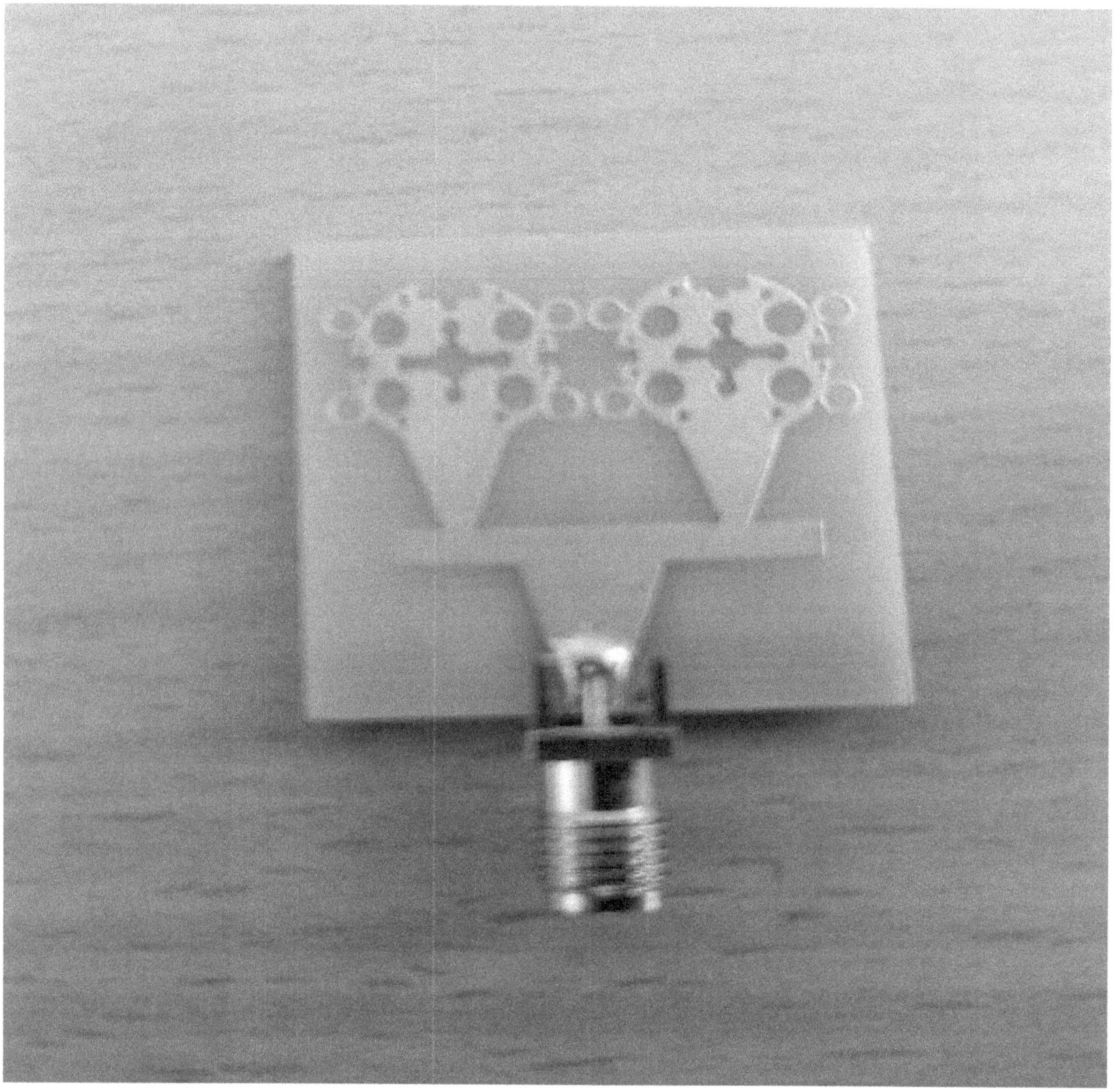

Figure 9. Measurement set up in Anechoic chamber

Figure 10. Measurement of return loss using Network Analyzer

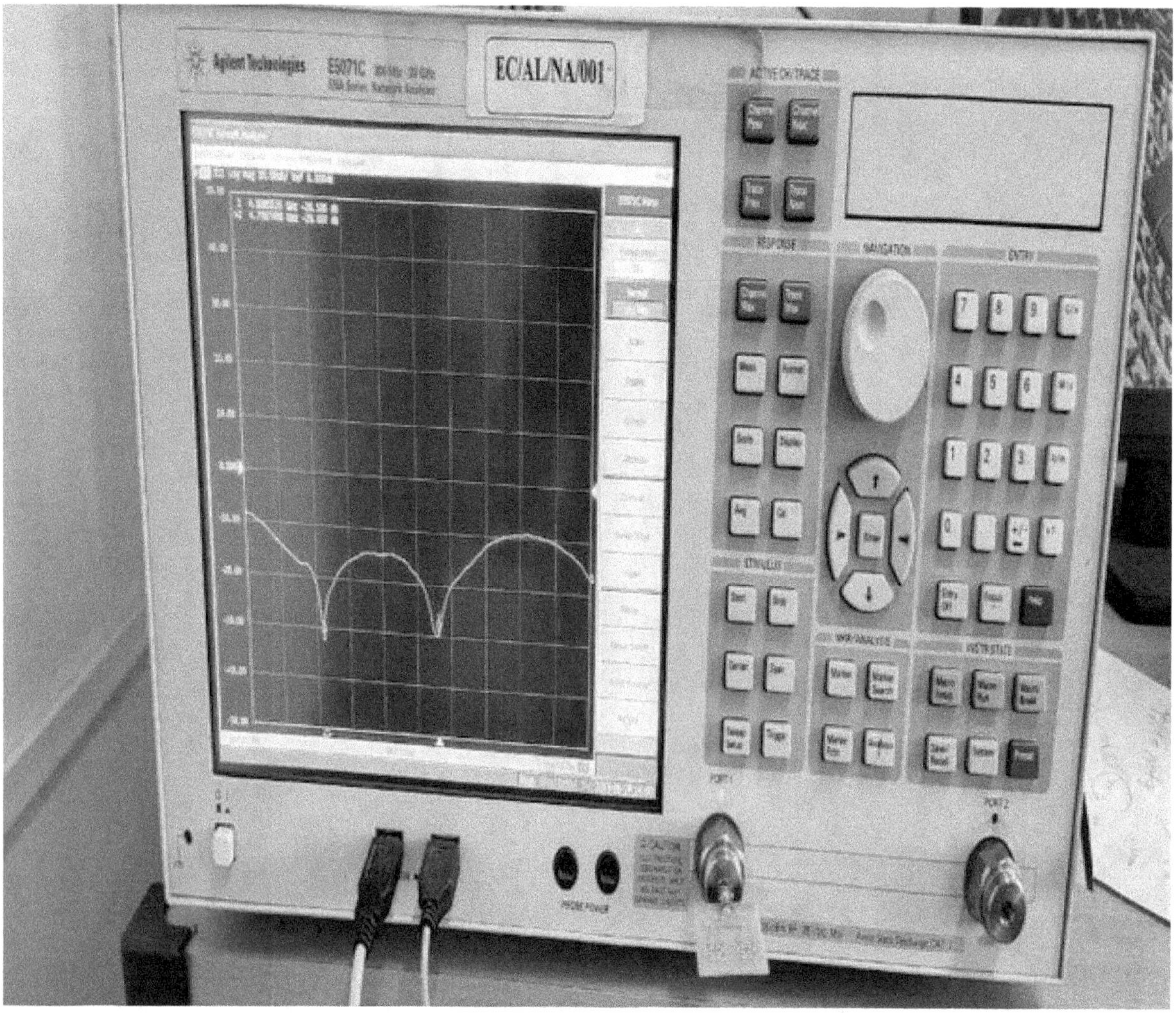

Figure 11. Measured return loss using Network Analyzer

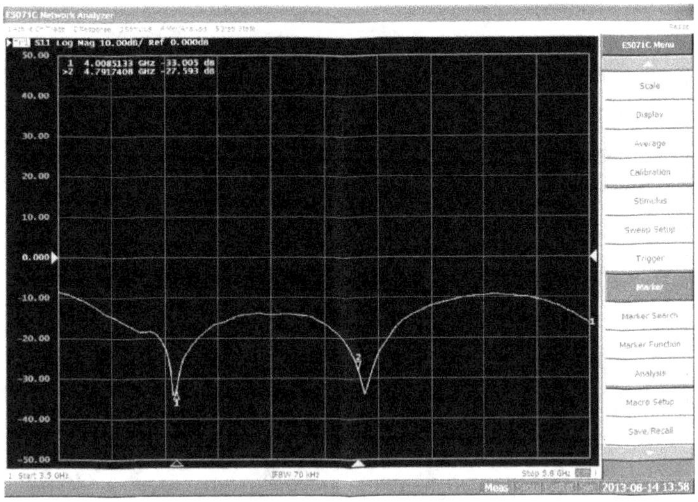

Figure 7 shows the gain plot of designed antenna. The proposed antenna has achieved a gain of 3.1 dB. Figure 8 shows the fabricated antenna. Figure 9 shows the measurement set up in anechoic chamber for testing the antenna parameters. Figure 10 shows the measurement of return loss using network analyzer. The result of fabricated antenna is almost aligned with the simulated results. The measured return loss has two resonance frequencies, achieved at 4 GHz and 4.7 GHz with return loss value -33 dB and -27.59 dB respectively. Figure 11 shows the measured return loss value using network analyzer. Figure 4 (d) compares the return loss of simulated and measured values.

Figure 12. Measured VSWR using Network Analyzer

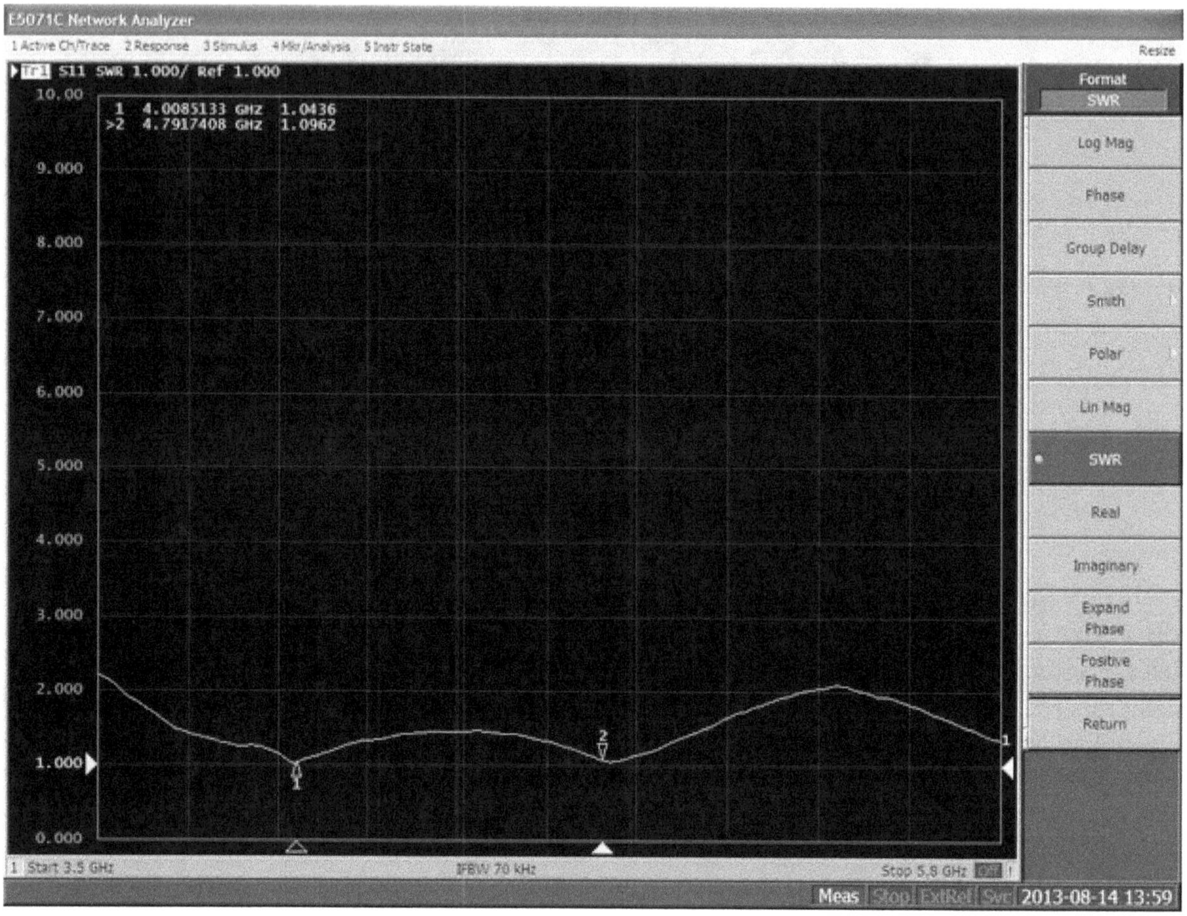

Figure 13. Measured E plane Radiation pattern using Network Analyzer

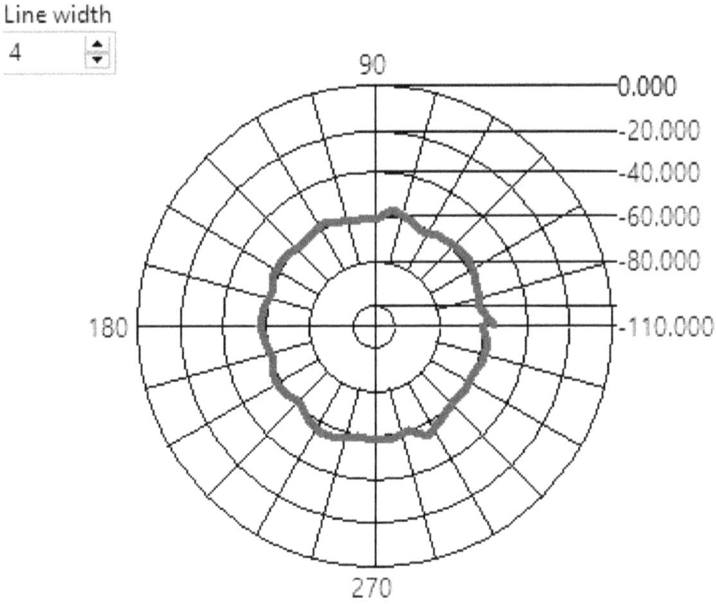

Figure 14. Measured H Plane Radiation pattern using Network Analyzer

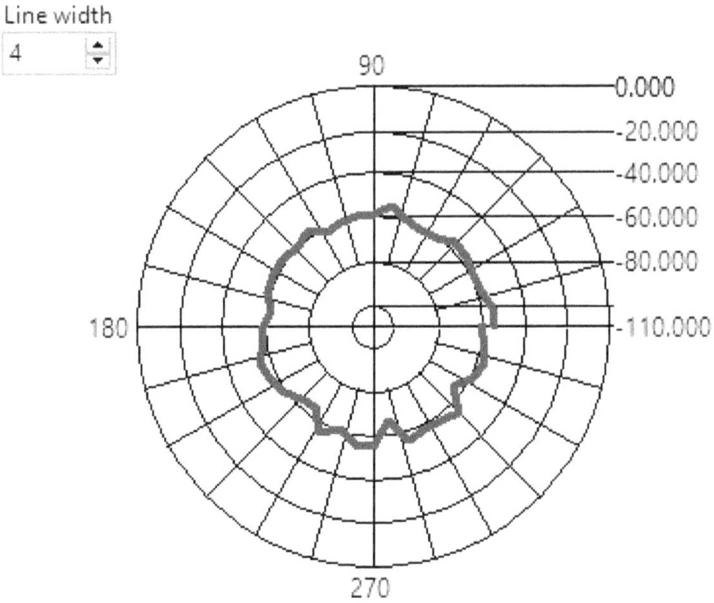

Figure 12 shows the measured VSWR using Network Analyzer. The measured VSWR achieved at 4 GHz is 1.04 and at 4.7 GHz is 1.09. This is also almost similar to simulated results. Figure 13 and 14 shows the measured E and H plane radiation pattern respectively. From this it can be inferred that the proposed antenna achieves omnidirectional radiation pattern, thus making it suitable for C and S band applications.

CONCLUSION

The proposed two-layer circular fractal array dual band antenna is analyzed, and its radiating structure is formed by tri half-circular slots in the upper layer and circular shaped slots in the bottom layer. The fractal structure and partial ground cause the proposed antenna to resonant at two frequencies. The measurement results indicate that the designed antenna has an acceptable value for the S11 parameter, gain, and bandwidth, with a maximum band-width of approximately 300 MHz at 3.98 GHz and a gain that can be adjusted from 3 dB to 5.7 dB throughout the operating band . . . The designed CFAA antenna can be used for C and S-band applications, such as many satellite communications and radar communications, respectively.

CONFLICT OF INTEREST

All authors declare that they have no conflicts of interest

REFERENCES

Aawan, I., Mhussain, Alibakhshikenari M, Ffalcone, Limiti E, Razarizvi S N. (2021). Single Patch Fractal-Shaped Antenna with Small footprint Area and High Radiation Properties for Wide Operation Over 5G Region. 46th International Conference on Infrared, Millimeter and Terahertz Waves (IRMMW-THz).p.1–2. DOI: 10.1109/IRMMW-THz50926.2021.9567165

V Abhaikumar,B Manimegalai, S Raju.(2009).A cantor based pre fractal multiband antenna. Int. J. Microwave Opt. Technol. p.95–98. DOI: 10.1109/IRMMW-THz50926.2021.9567165

Alharbi, G, Abdullah,Chow Yen Desmondsim,Jayshri Kulkarni,K Poddar, L. Rohde.(2022).A compact circularly polarized rotated L-shaped antenna withJ-shaped defected ground strucutre for WLAN and V2X applications. Progr.Electromagnet. Res. Lett. P. 135–143.

Alle, P., Kbehera, S., Mishra, D. P., & Pandav, S. (2021).CPW-Fed Circularly Polarized Fractal Antenna for WLAN Applications. *2nd International Conference for Emerging Technology (INCET)*.p. 1–5. DOI: 10.1109/INCET51464.2021.9456236

Alotaibi, Y., Alghamdi, S., Alassery, F., Palanisamy, S., Thangaraju, B., & Khalaf, O. I. (2021). A Novel Approach of Design and Analysis of a Hexagonal Fractal Antenna Array (HFAA) for Next-Generation Wireless Communication. *Energies*, 14(19), 1–18.

Amiyabhusanasahoo, B., Mishra, G. P., & Manasranjan, J. B. (2019). Fractal Geometry and Its Application to Antenna Designs. [IJEAT]. *International Journal of Engineering and Advanced Technology*, 9(1), 3726–3743. DOI: 10.35940/ijeat.A9793.109119

Bakhar, M., Dakulagi, V., & Mathpati, M. S. (2021). Design and Analysis of a Fractal Antenna Using Jeans Material for WiMax/WSN Applications. *Wireless Personal Communications*, 119(4), 3517–3527. DOI: 10.1007/s11277-021-08418-y

Bakhar, M., Mathpati, M., & Yadav, D. (2020). A Pentagonal Patch Antenna with Minkowski Fractal Structure using Different Substrate Materials. *Journal of Advanced Research in Networking and Communication Engineering*, 3(2), 1–4.

Banerjee, S., Rana, B., Parui, S. K., Chatterjee, S., & Dey, N. (2017). HMSIW-Based Miniaturized sensing antennas for S- and C-Band applications. *IEEE Sensors Letters*, 1(1), 1–5. DOI: 10.1109/LSENS.2017.2677858

Choukiker, Y. K., & Mudiganti, J. C. (2016). Compact hybrid fractal antenna for wideband wireless applications. *International Journal of Microwave and Wireless Technologies*, 9(5), 1191–1196. DOI: 10.1017/S1759078716001318

Elavarasi, C., & Shanmuganantham, T. (2017). Multiband SRR loaded Koch star fractal antenna. *Alexandria Engineering Journal*, 57(3), 1549–1555. DOI: 10.1016/j.aej.2017.04.001

Gnanalakshmi, V., Raaj, R., & Suresh Kumar, V. (2021). Design of Quad feed end-fire microstrip patch antenna for Airborne Systems. *Journal of Physics: Conference Series*, 2062(1), 1–10. DOI: 10.1088/1742-6596/2062/1/012003

Gohil, J. H., & Nb, V. (2021). Design and simulation of Sierpmski Triangular Fractal Antenna for S-Band. *IOP Conference Series: Materials Science and Engineering*.p.1–10.

Gupta, M., & Mathur, V. (2017). Wheel shaped modified fractal antenna realization for wireless communications. *AEÜ. International Journal of Electronics and Communications*, 79, 257–266. DOI: 10.1016/j.aeue.2017.06.017

Iyampalam, P., & Ganesan, I. (2020). Design of Sierpinski Knopp Inspired Fractal Antenna for Public Safety Applications. *Wireless Personal Communications*, 114(1), 227–239. DOI: 10.1007/s11277-020-07360-9

Jantree, P., Lerdwanittip, R., & Namsang, A. Pakkaraekthalin, Sirawattanachai N.(2021).A Dual-Band Fractal Antenna using ICPW Technique for LTE and WiMAX Systems. 9th International Electrical Engineering Congress (iEECON).p. 547–550.

Jayasinghe, J., Andújar, A., & Anguera, J. (2018). On the properties of Sierpinski gasket fractal microstrip antennas. *Microwave and Optical Technology Letters*, 61(3), 772–776. DOI: 10.1002/mop.31605

Kakkars, R. S. (2021). Modified Koch Fractal Antenna for Multi and Wideband Wireless Applications. *ECTI Transactions on Electrical Engineering, Electronics, and Communications*, 19(2), 182–189. DOI: 10.37936/ecti-eec.2021192.242065

Kalaiyarasan, R., Nagarajan, G., & Seenuvasamurthi, S. (2023). Design and implementation of an efficient sierpinski carpet fractal antenna with defected ground structure for 2.45 GHZ applications. e-Prime - Advances in Electrical Engineering Electronics and Energy, 6, 100320. DOI: 10.37936/ecti-eec.2021192.242065

Li, S., & Zheng, X. (2021).Antenna Design Based on Minkowski Like Fractal Structure.*19th International Conference on Optical Communications and Networks (ICOCN)*.p.1–3.

Mazher, A., Chaudary, E., Limiti, E., & Falcone, F. Gul E Tayyaba Zehra, Mohammad Alibakhshikenari, Syed Naheel Raza Rizvi. (2021). Wideband Small Fractal Antenna with Simple Design Strategy for 5G Sub-6-GHz Wireless Communications.XXXIVth General Assembly and Scientific Symposium of the International Union of Radio Science (URSI GASS).p. 1–4.

Paik, H., Lankecharanteja, K. R., & Reddy, M. (2021). A Miniaturized Fractal Antenna with Square Ring Slots for Ultrawideband Applications. *Progress in Electromagnetics Research Letters*, 99(1), 169–177. DOI: 10.2528/PIERL21070801

Rajagopal, C., Suseela, S. B., Noorullakhan, N., & Sankararajan, R. (2016). Compact modified 'T' slot circular patch quad band antenna for MIMO applications. *International Journal of Microwave and Wireless Technologies*, 9(4), 865–873. DOI: 10.1017/S1759078716000696

Sharma, N., Singh, H. S., & Khanna, R. (2021).Design and Analysis of Multiband Fractal Antenna for MIMO/Diversity Applications. Wireless Personal Communication.p.1–16.

Singhal, S., & Singh, A. K. (2017). Modified star-star fractal (MSSF) super-wideband antenna. *Microwave and Optical Technology Letters*, 59(3), 624–630. DOI: 10.1002/mop.30357

Srivastava, K., Mishra, B., & Singh, R. (2021). Microstrip-line-fed inverted L-shaped circularly polarized antenna for C-band applications. *International Journal of Microwave and Wireless Technologies*, 14(4), 502–510. DOI: 10.1017/S1759078721000635

Varshney, A., & Gençoğlan, D. N. (2024). High-Gain Multi-Band Koch Fractal FSS Antenna for Sub-6 GHz Applications. *Applied Sciences (Basel, Switzerland)*, 14(19), 9022. DOI: 10.3390/app14199022

Yu, Z., Yu, J., Ran, X., & Zhu, C. (2017). A novel Koch and Sierpinski combined fractal antenna for 2G/3G/4G/5G/WLAN/navigation applications. *Microwave and Optical Technology Letters*, 59(9), 2147–2155. DOI: 10.1002/mop.30698

Chapter 11
An ITS-G5 Antenna for Vehicle Communication

Eugene Amobichukwu Ogbodo
https://orcid.org/0000-0002-9798-723X
University of Hertfordshire, UK

Azunka N. Ukala
University of Hertfordshire, UK

Rahman Saifur
University of Hertfordshire, UK

ABSTRACT

Intelligent Transportation Systems (ITS) represent a revolutionary integration of advanced information and communication technologies into transportation infrastructure and vehicles. This integration aims to significantly enhance the safety, efficiency, and sustainability of transportation networks across the globe. ITS is transformative, moving beyond mere enhancements of existing frameworks to leverage a broad spectrum of cutting-edge technologies, including real-time traffic monitoring, adaptive signal control, sophisticated vehicle-to-vehicle (V2V) and vehicle-to-infrastructure (V2I) communication systems, alongside advanced public transit solutions. This chapter offers an extensive overview of ITS, encapsulating its diverse technologies, widespread applications, and profound impact on global transportation systems. It introduces a paradigm shift in the conceptualization and management of transportation systems, addressing pressing urban challenges such as congestion, safety, and sustainability.

INTRODUCTION

Intelligent Transportation Systems (ITS) represent a revolutionary integration of advanced information and communication technologies into transportation infrastructure and vehicles. This integration aims to significantly enhance safety, efficiency, and sustainability of transportation networks across the globe. ITS is transformative, moving beyond mere enhancements of existing frameworks to leverage a broad spectrum of cutting-edge technologies, including real-time traffic monitoring, adaptive signal control, sophisticated vehicle-to-vehicle (V2V) and vehicle-to-infrastructure (V2I) communication systems, along-

DOI: 10.4018/979-8-3693-8799-3.ch011

side advanced public transit solutions. This chapter offers an extensive overview of ITS, encapsulating its diverse technologies, widespread applications, and profound impact on global transportation systems.

ITS introduces a paradigm shift in the conceptualization and management of transportation systems, addressing pressing urban challenges such as congestion, safety, and sustainability. By integrating these advanced technologies, ITS enhances the capacity of transportation systems to handle increasing urban demands effectively. This literature review aims to compile, synthesize, and analyze the extensive body of knowledge surrounding ITS, providing readers with a comprehensive exploration of its multifaceted (Salami et al.,2019;Ahmad et al.,2024) aspects. Through meticulous examination, this review highlights the transformative impact of ITS technologies on transportation infrastructure and explores their potential to reshape future mobility, thereby offering readers a thorough understanding of the ITS landscape and its critical role in addressing contemporary transportation challenges.

Figure 1. Future map of intelligent transportation systems (Designed in Canva Pro)

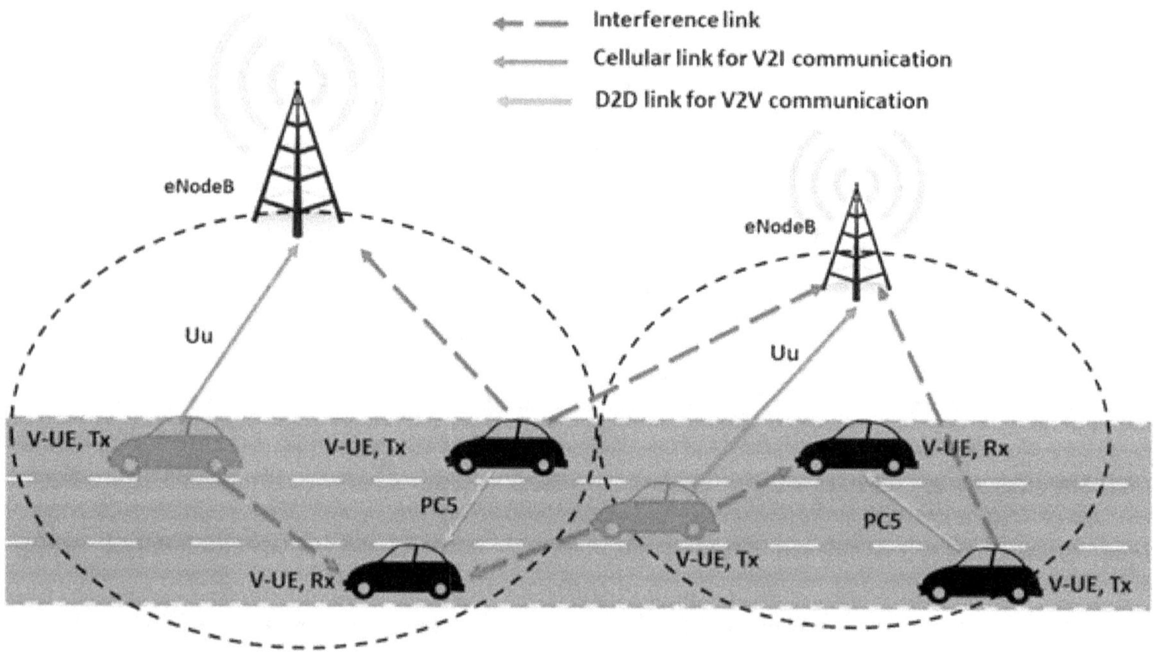

Central to the advancements in ITS is the development of ITS-G5 technology, operating within the 5 GHz frequency band, which has been specifically engineered to optimize vehicle communication. ITS-G5 technology introduces several pivotal enhancements to vehicular communication systems:

- **Enhanced Safety**: ITS-G5 facilitates real-time V2V communication, allowing vehicles to instantaneously share critical data such as location, speed, and potential hazards. This rapid exchange of information is crucial for enhancing road safety, enabling vehicles to preemptively respond to potential dangers and effectively avoid collisions.
- **Reduced Latency**: The technology is designed to minimize communication delays, which is essential for the execution of safety-critical applications where time is of the essence. Rapid re-

sponse capabilities ensure timely dissemination of safety alerts and efficient coordination among vehicles, thus preventing accidents.

- **Adaptive Networking**: ITS-G5 supports a decentralized ad-hoc network topology, which allows vehicles to dynamically self-organize without dependence on centralized infrastructure. This flexibility is vital for maintaining uninterrupted communication in various operational scenarios, particularly in non-ideal conditions or remote areas.
- **Standard Compatibility and Integration**: By building upon established communication standards such as IEEE 802.11p (WLANp), ITS-G5 ensures compatibility with a wide array of devices operating within the same frequency band, facilitating a broad spectrum of interoperability across different technological platforms.
- **Optimized Coverage**: The antennas designed for ITS-G5 are tailored to provide effective communication over typical vehicular distances, ensuring reliable coverage in both urban and rural traffic scenarios.
- **Robustness in Diverse Environments:** ITS-G5 systems are robustly designed to maintain connectivity in various challenging environments, such as urban centers with high structural densities and rural areas with significant natural obstacles, enhancing the reliability and effectiveness of vehicular communications.

ITS-G5 antennas offer a powerful solution for enhancing vehicle communication. The research supporting this chapter is rooted in a multifaceted strategy that combines theoretical exploration with empirical engineering practices. It encompasses extensive literature reviews, state-of-the-art electromagnetic simulations, rigorous experimental testing, and iterative design processes. The primary goal is to develop ITS-G5 antennas that not only meet but surpass the current performance benchmarks, optimizing key parameters such as gain, bandwidth, and filtering capabilities through meticulous sensitivity analysis and optimization methods.

Supported by a robust theoretical framework that integrates the latest advances in antenna theory and practical vehicular communication strategies, this research is poised to significantly advance the field of antenna engineering. The initiatives described in this chapter aim to introduce revolutionary technologies that minimize interference, enhance signal quality, and facilitate seamless integration of antennas into a diverse array of communication devices, ushering in a new era of wireless communication tailored for modern transportation demands.

This chapter will present a comprehensive narrative of the journey toward reimagining the capabilities of ITS-G5 antennas, exploring the innovative methodologies employed, assessing the progress achieved, and discussing the challenges and solutions encountered. This detailed examination aims to provide readers with an in-depth understanding of the complex yet promising field of Intelligent Transportation Systems and their pivotal role in shaping the future of global transportation.

BACKGROUND

The evolution of Intelligent Transportation Systems (ITS) spans several decades, marking significant advancements in transportation technology. Originating in the late 1960s, ITS began with the implementation of computerized traffic signal control in major urban centers, employing rudimentary algorithms for traffic flow optimization. The subsequent decade saw substantial progress with the introduction of digital roadside information displays and automated toll collection methods. The 1990s heralded a new era in traveler information systems, providing users with up-to-date traffic and road condition data. This period also saw the formal recognition of ITS as a distinct field through legislative action. As the new millennium dawned, the focus shifted towards integrating vehicle communication with road infrastructure, complemented by advancements in satellite navigation technology that enhanced location accuracy and real-time traffic monitoring.

The following decade witnessed significant strides in vehicle connectivity, facilitating direct communication between vehicles and with surrounding infrastructure, thus enhancing road safety and traffic efficiency. This period also marked the emergence of self-driving vehicles and the concept of smart cities, integrating ITS deeply into urban planning and management (Figueiredo et al.,2001, Ahmad et al.,2024, Imran et al.,2024).

In recent years, the incorporation of artificial intelligence, machine learning, and big data analytics has further revolutionized ITS capabilities, enabling more sophisticated traffic management, advanced autonomous vehicles, and improved public transit systems. Contemporary ITS technologies are increasingly intertwined with emerging fields such as 5G networks and the Internet of Things, paving the way for comprehensive, automated transportation networks that promise to alleviate congestion, boost efficiency, safety, and promote environmental sustainability.

In the UK context, ITS development has followed a similar trajectory, with significant government backing in infrastructure and research since the 1960s. The late 20th century saw the implementation of Advanced Traffic Management Systems, while the early 21st century focused on integrating various information and communication technologies. The current decade has seen a shift towards connected and autonomous vehicle technologies, supported by dedicated research initiatives. Today, ITS plays a crucial role in the UK's transportation landscape, contributing to improved efficiency, safety, and sustainability, while also stimulating economic growth and innovation in the sector.

MOTIVATION

In recent years, there has been a significant shift in the evolution of vehicle transportation, with Intelligent Transportation Systems (ITS) taking the lead. Central to this advancement is ITS-G5 operating at 5.9 GHz, facilitating vehicle-to-everything (V2X) communication. This dedicated short-range communications (DSRC) standard, particularly through ITS-G5 antennas operating at 5.9 GHz within a 75 MHz bandwidth, aims to enhance vehicle intelligence, safety, and efficiency.

The urgency for high-performance ITS-G5 antennas arises from various interconnected factors. Firstly, global concerns over road safety, with approximately 1.35 million annual fatalities attributed to automobile accidents by the World Health Organization, underscore the need for real- time information dissemination on road conditions and risks to drivers. Secondly, urban areas grapple with traffic congestion, leading to prolonged commute times and heightened pollution levels. ITS-G5 communication enables effective traffic management by coordinating signals, suggesting alternative routes, and optimizing traffic flow.

Moreover, the development of autonomous vehicles hinges on robust communication infrastructure, emphasizing the pivotal role of ITS-G5 antennas in facilitating vehicle-to-vehicle and vehicle-to-infrastructure communication. Lastly, amid the climate crisis, efforts to mitigate greenhouse gas emissions and enhance fuel economy are imperative. Efficient traffic management enabled by ITS-G5 communication aids in reducing idle time and fuel consumption, contributing to environmental sustainability.

Despite advancements, challenges persist, including cost reduction, size optimization for seamless integration into vehicles, and maximizing performance across the bandwidth. Nonetheless, the societal benefits of an effective ITS-G5 antenna system are compelling, promising safer roads, improved air quality, reduced emissions, and expanded access to autonomous vehicles. The pursuit of safer, more efficient, and intelligent transportation systems underscores the significance of ITS- G5 antenna development at 5.9 GHz and 75 MHz bandwidth, offering profound potential for enhancing both individual lives and the planet as a whole.

CLARITY OF AIM

The objectives of this project are focused on developing and evaluating an efficient antenna for Intelligent Transport System (ITS) communications, specifically adhering to the ITS-G5 standard. The key objectives are as follows:

1. To conduct a comprehensive review of existing literature on ITS-G5 antenna designs. This examination will provide a solid foundation for understanding current technologies, design approaches, and challenges in the field.
2. To utilize Advanced Design System (ADS) software to create an antenna optimized for performance at the 5.9 GHz frequency. This objective involves leveraging the capabilities of ADS to model, simulate, and refine the antenna design for optimal functionality within the ITS-G5 frequency range.
3. To design a high-performance ITS-G5 antenna with a specific focus on achieving a bandwidth of 75 MHz. This bandwidth specification is crucial for meeting the requirements of ITS-G5 standard applications and ensuring compatibility with modern Intelligent Transportation Systems.
4. To thoroughly evaluate the antenna's performance characteristics, including radiation pattern, gain, bandwidth, and return loss. These assessments are essential to ensure that the antenna meets the specific needs of vehicular communication systems, providing reliable and efficient performance in various real-world scenarios.

By achieving these objectives, the project aims to contribute to the advancement of ITS technology by delivering an antenna design that is optimized for the specific requirements of ITS-G5 communications, potentially enhancing the safety and efficiency of vehicular communication systems.

OBJECTIVE OF THE RESEARCH

- To examine the literature currently available on ITS-G5 antenna designs.
- To create an antenna optimized for 5.9 GHz frequency performance using ADS software.
- To design a high-performance, ITS-G5 antenna, especially set the antenna's bandwidth to 75 MHz
- To evaluate radiation pattern, gain, bandwidth, and return loss to meet vehicular communication needs.

BASIC OF ITS-G5 ANTENNA

The ITS-G5 Antenna is a specialized hardware component designed specifically for vehicle communication systems that adhere to the ITS-G5 standard. This European standard focuses on vehicular communications, including interactions like car-to-car (V2V) and car-to-infrastructure (V2I)—collectively referred to as V2X communications.

Operating within a specified frequency range, typically around 5.9 GHz, these antennas enable rapid and dependable wireless communication even in fast-moving vehicle scenarios. They serve as essential elements within ITS-G5 systems, facilitating the transmission of both safety and non- safety data among vehicles and between vehicles and roadside infrastructure.

Imagine these antennas as communication bridges connecting cars, traffic lights, and other intelligent components of the transportation network. By enabling seamless data exchange, they significantly enhance road safety, streamline traffic patterns, and improve overall transportation efficiency. Whether it's warning drivers about potential hazards or optimizing traffic flow, the ITS-G5 Antenna plays a crucial role.

Engineers carefully consider factors like antenna length, matching sections, and installation angles during the design and refinement process. For example, the antenna can be erected as a horizontal dipole, a sloper, or an inverted-V antenna.

APPLICATION OF ITS-G5 ANTENNA

The applications of an ITS-G5 antenna are diverse, primarily focusing on enhancing vehicular communication systems and infrastructure within Intelligent Transportation Systems (ITS). Here are some key applications:

1. Vehicle-to-Vehicle (V2V) Communication: ITS-G5 antennas enable direct communication between vehicles, allowing them to share information about their speed, position, and direction. This can enhance safety by reducing the risk of collisions and enabling cooperative driving maneuvers.
2. Vehicle-to-Infrastructure (V2I) Communication: These antennas facilitate communication between vehicles and roadside infrastructure, such as traffic signals, road signs, and toll booths. This interaction can optimize traffic flow, improve signal timing, and provide real-time updates to drivers about road conditions and hazards.
3. Vehicle-to-Everything (V2X) Communication: ITS-G5 antennas support broader communication networks that include not just vehicles and infrastructure, but also pedestrians, cyclists, and other elements within the transportation ecosystem. This comprehensive communication enhances overall traffic management and safety.
4. Autonomous Driving: For autonomous vehicles, ITS-G5 antennas are crucial for receiving and transmitting data necessary for navigation, obstacle detection, and interaction with other vehicles and infrastructure. This ensures smoother and safer autonomous driving experiences.
5. Traffic Management Systems: ITS-G5 antennas contribute to advanced traffic management systems by providing real-time data on traffic flow, congestion, and incidents. This data helps traffic management centers to monitor and control traffic more effectively, leading to reduced congestion and improved road safety.

6. Public Transportation Systems: In public transportation, ITS-G5 antennas can improve the efficiency and reliability of services by enabling communication between buses, trains, and central control systems. This can lead to better scheduling, real-time updates for passengers, and enhanced coordination between different modes of transport.
7. Emergency Response: ITS-G5 technology can aid emergency response units by providing real- time information about road conditions, traffic, and the location of emergency vehicles. This can lead to quicker response times and more efficient routing of emergency services.
8. Smart Cities: As part of smart city initiatives, ITS-G5 antennas play a role in creating interconnected urban environments where vehicles, infrastructure, and other systems communicate seamlessly. This integration supports sustainable urban development, efficient resource management, and enhanced quality of life for residents.
9. Freight and Logistics: For the freight and logistics industry, ITS-G5 antennas help in tracking and managing the movement of goods. This ensures timely deliveries, optimized routing, and better coordination among logistics partners.

Overall, ITS-G5 antennas are essential components in modern transportation networks, driving improvements in safety, efficiency, and sustainability.

LITERATURE REVIEW

Antenna engineering for Intelligent Transportation Systems (ITS) is a continually evolving field driven by the quest to enhance the efficiency, performance, and integration capabilities of vehicular communication systems. This section of the literature review highlights the critical role of antenna design, particularly focusing on ITS-G5 antennas, which are crucial in today's vehicular communications landscape. The proliferation of connected vehicles and the need for reliable vehicle-to-everything (V2X) communications have made it essential to design and optimize antennas that excel within the 5.9 GHz ITS-G5 band while minimizing interference with adjacent frequency bands.

Modern ITS-G5 antenna design must address increasing demands for gain, bandwidth, and multi-standard compatibility due to the complexity of vehicular communication protocols and the rise in connected vehicle applications. ITS-G5 antennas, with their ability to efficiently transmit and receive electromagnetic signals in challenging vehicular environments, are vital for research and development efforts aimed at meeting these challenges. This review examines innovative works that have influenced design methodologies, antenna theory, optimization strategies, and the integration of ITS-G5 antennas into vehicle structures.

The review also evaluates the methods used by previous researchers, assessing the effectiveness of various design and optimization strategies and their practical applications in vehicular scenarios. It focuses on studies that have pioneered novel solutions and addresses potential solutions to current design limitations, such as vehicle integration and multipath mitigation. By analyzing the convergence of theoretical insights, simulation models, and empirical validations in real-world vehicular environments, the study aims to provide a comprehensive framework to guide the development and enhancement of ITS-G5 antennas for this project.

Early foundational studies provide a thorough analysis of vehicular antenna theory and design, emphasizing key aspects such as bandwidth, gain, impedance matching, and radiation patterns in mobile environments. These concepts are essential for understanding how ITS-G5 antennas can be tailored for specific vehicular applications and mounting locations. More recent efforts have concentrated on integrating multiple vehicular communication standards into single antenna designs to reduce system complexity and enhance spectrum efficiency. This integration is crucial

for ITS-G5 antenna designs as it allows for more efficient and compact structures suitable for modern vehicle aesthetics.

The complexity of modern ITS-G5 antenna designs has made simulation tools indispensable. Recent works on vehicular antennas underscore the use of electromagnetic modeling tools like HFSS, CST Microwave Studio, and FEKO in the design and optimization process. These tools enable antenna designers to model, analyze, and refine antenna structures before actual prototyping, taking into account the effects of vehicle body and mounting position. This review synthesizes these insights to support the development of efficient ITS-G5 antennas, ensuring they meet contemporary performance standards and the unique requirements of vehicular communication systems.

Intelligent Transportation Systems (ITS) have become increasingly important in modern vehicular communications. ITS-G5, the European standard for vehicular communications operating in the

5.9 GHz band, requires efficient and reliable antenna designs. Microstrip patch antennas have gained popularity in this field due to their low profile, ease of fabrication, and adaptability. This literature review examines the developments in ITS-G5 microstrip patch antennas from 2019 to 2024, focusing on design innovations, performance improvements, and integration challenges. Recent years have seen significant advancements in multi-band operations for ITS-G5 antennas. Imran et al. (2024) proposed a dual-band microstrip patch antenna for ITS-G5 and DSRC applications, achieving good performance in the 5.9 GHz band with a compact size and low profile. Building on this work, Macon et al. (2003) developed an antenna covering ITS-G5, DSRC, and GPS frequencies, enhancing the versatility of vehicle communication systems. Circular polarization has also been a focus of research, with Zhao et al. (2020) presenting a novel circular polarized microstrip patch antenna for ITS-G5 using a truncated corner method to improve performance in multipath environments common in urban V2V communications. Further improvements to this concept by proposing a compact circular polarized antenna with enhanced axial ratio bandwidth have also been tried.

The use of metamaterials in ITS-G5 microstrip patch antennas has shown promising results. Awadamislmani et al., (2024) demonstrated that incorporating split-ring resonators could enhance bandwidth and gain, crucial for long-range vehicle communications (Cao et al.,2012).

Zhao et al. (2020) presented an ultra-wideband design that covered multiple vehicular communication standards, including ITS-G5, DSRC, and emerging 5G NR-V2X bands.

Gain enhancement has been a key area of improvement in ITS-G5 antenna design. Similar work introduced a high-gain microstrip patch antenna using a superstrate layer, achieving a high gain (dBi) while maintaining a low profile.

Ahmed et al. (2024) provided a review of miniaturization techniques, highlighting approaches such as high-permittivity substrates, defected ground structures, and fractal geometries. Implementing these techniques, Zha0 et al. (2020) demonstrated a miniaturized antenna with a size reduction compared to conventional designs while maintaining comparable performance.

Integration of antennas with vehicle bodies has presented unique challenges.

Macon et al. (2003) developed a conformal microstrip patch antenna for ITS-G5 that maintained performance when bent to conform to typical vehicle curvatures. Environmental factors affecting antenna performance have also been studied (Patti,2023).

Emerging trends in ITS-G5 antenna design include cognitive antenna systems, energy harvesting integration, and AI-assisted design. Yang et al. (2024) introduced cognitive antenna systems capable of dynamically adjusting radiation patterns and operating frequencies to optimize performance in changing environments. Zhao et al. (2020) explored integrating energy harvesting capabilities into ITS-G5 antennas, incorporating solar cells and RF energy harvesting elements to potentially reduce power requirements for vehicular communication systems.

In conclusion, the literature from 2019 to 2024 shows significant advancements in ITS-G5 microstrip patch antenna design and performance. Key trends include multi-band operation, circular polarization, metamaterial-based enhancements, and miniaturization techniques. Integration challenges have been addressed through innovative approaches to vehicle body integration, MIMO configurations, and conformal designs. Emerging trends point towards cognitive systems, energy harvesting integration, and AI-assisted design methodologies. These developments are paving the way for more efficient, reliable, and versatile antenna solutions for future intelligent transportation systems.

Table 1. Evaluate and contrast various ITS-G5 antennas.

Refs.	Year	Operating band (GHz)	Total size (mm^2)	Gain (dBi)	Efficiency (%)	Profile	Substrate
	2020	5.85 - 5.925	20 x 28 x 1.6	4.1	88%	Planar	FR4
	2020	5.74 - 5.85	34 x 34 x 1.6	5.74	79%	Square	FR4
	2022	5.8 - 5.99	30 x 30 x 1.6	3.2	80%	Patch	FR4
	2023	5.2 - 5.35	35 x 35 x 1.6	4.8	81%	Circular	FR4
This Work	2023	5.8 - 5.9	34 x 46 x 1.6	8.2	83%	Planar	FR4

METHODOLOGY

The creation of ITS-G5 antennas is a complex process that combines computer simulations, theoretical analysis, and empirical validations. The methodical processes used to design, model, and verify the effectiveness of antennas suited for effective radiation and bandwidth within designated frequency ranges are outlined in this section on methodology. This section highlights the design process due to the complex balance between practical restrictions and performance standards.

Design and Optimization Process

1. **First Identification of Requirements:** The operational parameters essential for designing the microstrip patch antenna will be defined as the first phase of the proposed project. These parameters include center frequency, bandwidth, insertion loss, isolation, return loss, and impedance matching.

The establishment of these parameters will serve as the foundational framework for initiating the design process.

1. **Topology Selection:** Based on the specified parameters, an appropriate topology for the microstrip patch antenna will be chosen. Options such as waveguide, strapline, or microstrip configurations may be considered. Given the project's scope and objectives, microstrip technology will be adopted for developing the filtering antenna. The decision regarding whether to opt for a design with grouped or dispersed components will be determined during this phase.
2. **Design Considerations:** First, the ITS-G5 antennas' parameters are established, with a particular emphasis on the 5.9 GHz operating frequency, which is essential for many wireless communication applications. Target frequency ranges, gain specifications, bandwidth requirements, and size limitations are all included in the specifications.
3. **Choose the Right Antenna Structure:** The integration of filtering capabilities may be facilitated by choosing appropriate antenna structures, such as loop, dipole, or patch antennas. Structures that provide a balance between their physical size and the possibility of having built-in filtering properties are given preference during nomination.
4. **Optimization:** To fine-tune the antenna characteristics for optimal performance at the 5.9 GHz frequency with a bandwidth of 75MHz, adjusting dimensions, material characteristics, and resonant element arrangement to conform to 5.9 GHz operation-centered design standards is what this technique does. Corrections and simulations will be performed using the ADS program, utilizing a trial-and-error approach to achieve an optimum antenna for the required application.

Figure 2. Diagram illustrating the Process.

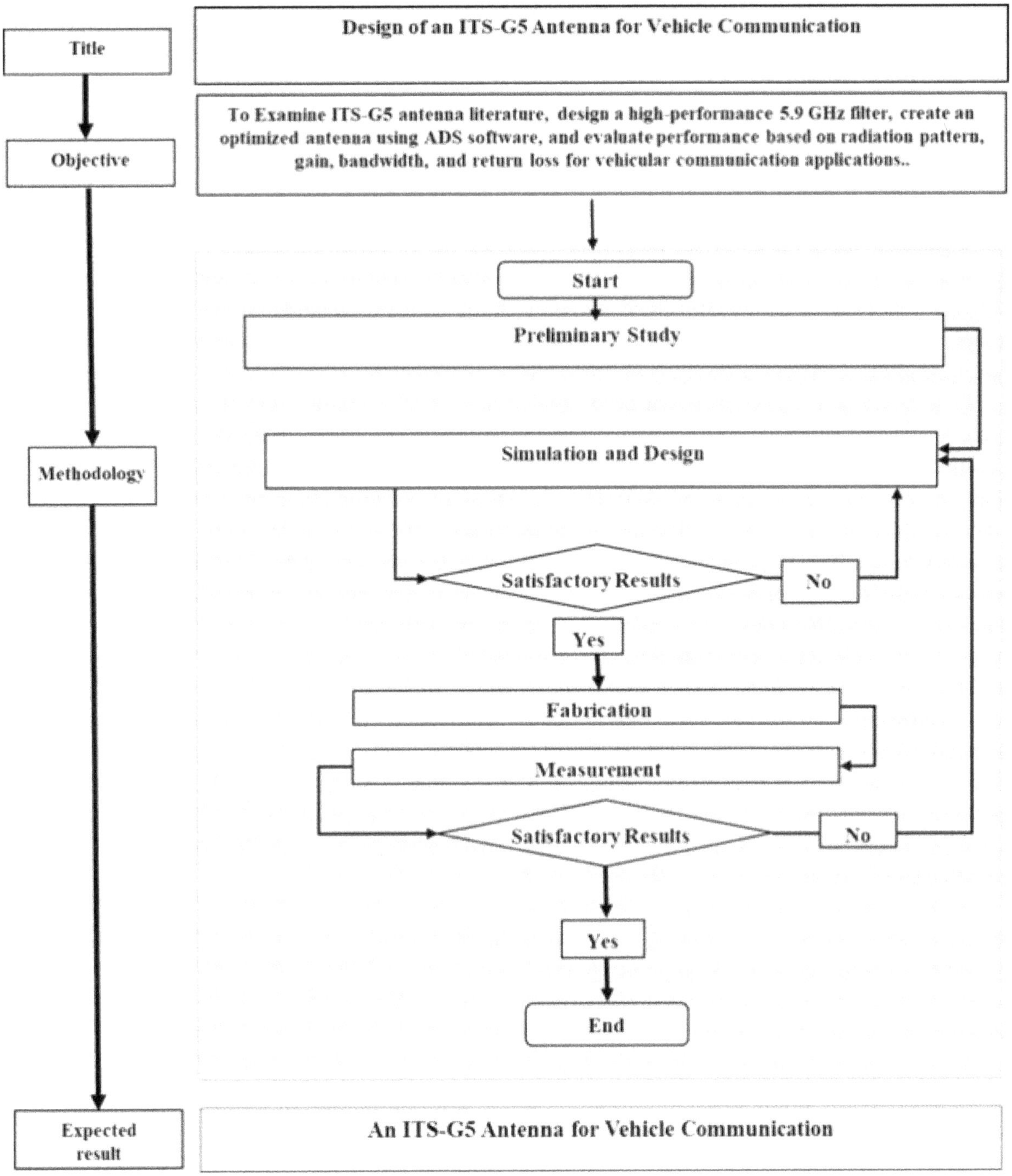

An Analysis of the Project's Sustainability

The goal of sustainability of this project is to satisfy current demands without compromising the capacity of future generations to satisfy their own. It is a comprehensive strategy that takes into account the economic, social, and environmental aspects in order to maintain equilibrium between resource protection and economic development. The goals of environmental sustainability include responsible resource management, biodiversity preservation, and ecosystem protection. Enhancing quality of life, maintaining equality, and promoting inclusive and cohesive communities are the main objectives of social sustainability. Promoting growth and development while making sure that natural and human resources are utilized effectively and responsibly is what is meant by economic sustainability.

In order to promote long-term environmental stewardship, social well-being, and economic viability, it is not only progressive but also imperative that sustainability be included into engineering practice, especially in initiatives like the research on filtering antennas. There are many aspects to sustainability in engineering.

Preservation of the Environment: Sustainable engineering initiatives greatly reduce their negative effects on the environment. The use of sustainable design concepts may result in the creation of antennas for simulation-based antenna projects that consume less energy and produce less electromagnetic pollution. The initiative aims to preserve natural resources and lower the carbon footprint of technology by using materials that are less detrimental to the environment and improving the design for energy efficiency.

Economic Efficiency: There are financial advantages to incorporating sustainability into antenna design. Long-term savings are possible by greatly lowering the operating costs of antenna systems via increased energy efficiency. Furthermore, sustainable antennas that are made to last a long time and be readily recyclable at the end of their useful life might help create a circular economy by lowering waste and the expenses associated with disposal.

Social Responsibility: By guaranteeing that technologies are usable, safe, and advantageous for all societal strata, engineering methods that prioritize sustainability also address social problems. This might include creating ways to improve communication accessibility in underprivileged regions or enhancing public health and safety by lowering exposure to dangerous emissions in the context of filtering antennas.

This project is closely related to the following (Sustainable Development Goals) SDG goals. SDG 7: Affordable and clean energy.

SDG 9: Industry, innovation, and infrastructure. SDG 11: Sustainable cities and communities.
SDG 12: Responsible consumption and production.

Figure 3. Sustainable Development Goals.

Challenge and Solution for Sustainability

Challenges

As the project to design an ITS-G5 antenna for vehicle communication was undertaken, the multitude of challenges involved and potential solutions to overcome them were grasped.

Operating within the 5.9 GHz frequency band introduced complexity, as minor design or production inaccuracies could significantly affect antenna performance at this high frequency. Additionally, interference posed a risk, as ITS-G5 shares its operating spectrum with other devices, potentially disrupting communications. The environmental impact was another concern, with the choice of materials and their energy consumption throughout their lifespan contributing to the antenna's environmental footprint. Scalability was critical, as the increasing number of vehicles equipped with ITS-G5 antennas would lead to a rise in network traffic, necessitating effective management to maintain performance. Furthermore, antennas mounted on vehicles needed to withstand harsh conditions, including diverse weather, physical impacts, and vibrations, presenting challenges in terms of durability and maintenance.

Solutions

In the scope of the project "Designing an ITS-G5 Antenna for Vehicle Communication," sustainability holds significant importance, adhering to the core principles of Reduce, Reuse, and Recycle. The principle of Reduce involves minimizing resource consumption by optimizing the antenna's design to enhance efficiency and reduce energy usage. This includes utilizing advanced design and simulation tools to achieve optimal performance while minimizing energy use and adopting manufacturing processes that minimize waste. Reuse focuses on designing the antenna to be modular and upgradeable, allowing specific components to be replaced or upgraded as technology advances, thereby extending the antenna's lifespan and reducing overall waste. Lastly, the principle of Recycle emphasizes using recyclable materials wherever feasible in the antenna's design, ensuring that at the end of its lifecycle, the antenna can be disassembled and its materials recycled for use in constructing new antennas or other electronic components.

Figure 4. Solutions for sustainability.

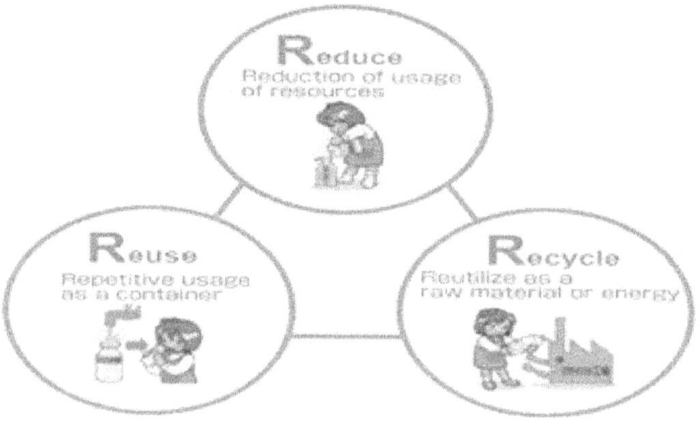

Impact of the Simulation-Based Antenna Project on Sustainability

The study aimed at examining antennas via simulation has the potential to significantly impact sustainability in several ways:

1. **Green Technology Innovation**: By concentrating on filtering antennas, the initiative leads the way in environmentally friendly telecommunications improvements. The development of more environmentally friendly communication systems may be aided by the simulations' ability to pinpoint solutions that maximize functionality while using the least amount of energy.
2. **Resource Efficiency**: During the design process, simulation greatly reduces material waste and energy usage by enabling thorough testing and optimization without the need for real prototypes. This strategy is a prime example of resource efficiency, which is essential to sustainability.

Required Equation

Several formulas and models may be necessary for the design and analysis of ITS-G5 antennas. The particular formulas vary according on the kind of antenna, design objectives used. But first, let us go over a few basic formulas that are often used to the construction and optimization of these kinds of antennas. These will include performance measurements, and fundamental antenna settings.

1. Design Equations for Antennas

A straightforward half-wave dipole antenna's resonance frequency is:

$$f_{res} = c / 2L \qquad (1)$$

Where f_{res} is the vacuum speed of light, c is the resonance frequency, and **L** is the dipole's length.

Gain of Antenna

$$G = (4\pi/\lambda^{L})\, A_{eff} \qquad (2)$$

Where **G** is the gain of the antenna, A_{eff} is the antenna's gain & λ is the signal's wavelength.
Where, Resonator Quality Factor (Q) is

$$Q = \quad / \Delta f \qquad (3)$$

Also, f_{res} is the resonant frequency and Δf is the bandwidth of the filtering antenna.
So, Bandwidth Calculation can be done as follows:

$$\Delta f = f_{high} - f_{low} \qquad (4)$$

Where f_{high} and f_{low} are the upper and lower cutoff frequencies of the filter, respectively.

Performance Metrics Equations

Assessment of the antenna's impedance matching depends critically on Return Loss and Voltage Standing Wave Ratio – Known as RL and VSWR in short form

$$\text{Return Loss, RL} = -20 \log_{10} (|S_{11}|) \qquad (5)$$

Voltage Standing Wave Ratio,

$$VSWR = 1 + (|S_{11}|) / 1 - (|S_{11}|) \qquad (6)$$

where the percentage of power reflected by the antenna is measured by S11, also known as the reflection coefficient.

The minimum return loss $R_L(dB)$ in passband for an ideal Chebyshev bandpass filter is given as

$$R_L(dB) = -10 log\, (1 - 10^{-1A(dB)/10} \qquad (7)$$

Increasing the value of the quality factor, one of the most important resonant circuit characteristics, would result in less loss in the resonant circuit. The corresponding circuit may be used to determine the monopole patch antenna's quality factor.

$$Q_A = 2\pi f_0 L_A / R_A \qquad (8)$$

The LC components' values for the resonators are provided by

$$L = 2Z_0 / \pi f_0 \tag{9}$$

$$C = 1/ Lf^2_0 \tag{10}$$

It is easy to acquire the J-inverter values in the following ways:

$$J_0 = \frac{1}{Z_0} \frac{\pi FBW}{\sqrt{4 g_0 g_1}} \tag{11}$$

$$J_{n-1,n} = \frac{1}{Z_0} \frac{\pi FBW}{\sqrt[4]{g_n - 1 g_n}} \tag{12}$$

DESIGN PROCESS AND RESULT ANALYSIS

Initial Design Phase

The project has embarked on an ambitious design journey with the goal of creating a cutting-edge ITS-G5 antenna aimed at meeting the evolving requirements of contemporary wireless communication systems. Currently, the project is in the antenna design phase, focusing on the ITS- G5 band. It utilizes the Advanced Design System (ADS) software, renowned for its comprehensive toolkit tailored for RF, microwave, high-speed digital, and power electronics applications.

Design Rationale and Preliminary Considerations

The design process began with a thorough evaluation of the specifications necessary for effective radiation and bandwidth within the designated frequency ranges. To ensure compliance with modern wireless communication standards, key considerations included optimizing gain, bandwidth, and efficiency. The Advanced Design System (ADS) was chosen as the primary design tool due to its advanced simulation capabilities, which are crucial for accurately simulating antenna behavior and enhancing performance parameters.

To assess the design parameters, we utilized antenna parameter calculators available through various online portals. Precisely determining the length and width of a microstrip patch antenna is essential, especially when targeting a specific frequency such as 5.9 GHz and using a specific substrate material like FR-4. It is important to note that the assumptions made are theoretical; the actual design will inevitably undergo changes through several trial and error-based corrections and optimizations. The initial evaluation is provided below:

Figure 5. Layout of an Antenna Calculator

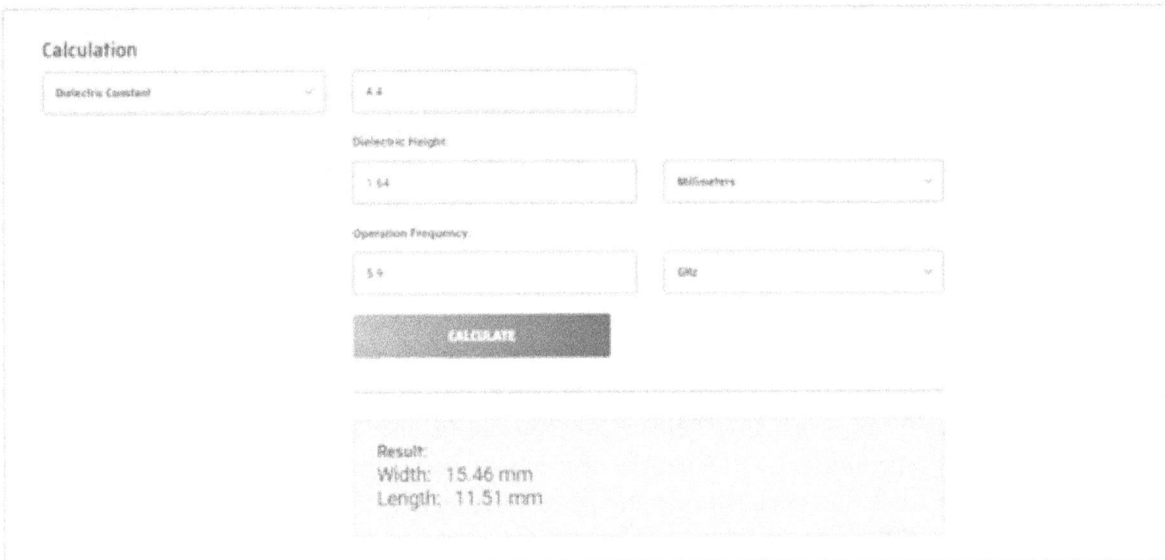

$$Width = \frac{c}{2f_o\sqrt{\frac{\varepsilon_R+1}{2}}}; \quad \varepsilon_{eff} = \frac{\varepsilon_R+1}{2} + \frac{\varepsilon_R-1}{2}\left[\frac{1}{\sqrt{1+12\left(\frac{h}{W}\right)}}\right]$$

$$Length = \frac{c}{2f_o\sqrt{\varepsilon_{eff}}} - 0.824h\left(\frac{(\varepsilon_{eff}+0.3)\left(\frac{W}{h}+0.264\right)}{(\varepsilon_{eff}-0.258)\left(\frac{W}{h}+0.8\right)}\right)$$

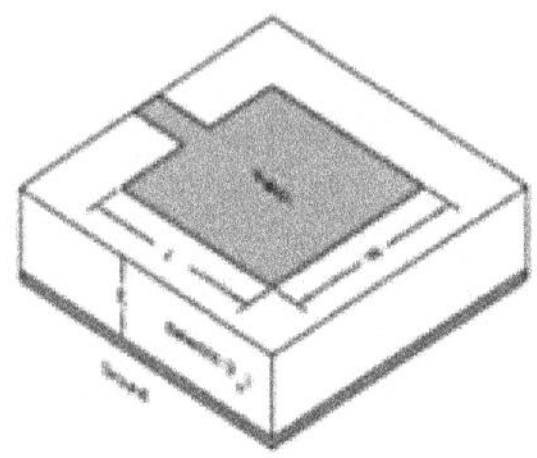

Methodology Employed in ADS

The project's chosen approach calls for the following crucial ADS environment steps:

1. **Component Selection:** Using information from literature reviews and theoretical foundations, the identification and selection of basic antenna models and filtering components is done.
2. **The process of configuring** simulation settings, such as frequency range, material qualities, and boundary conditions, in order to replicate realistic operating situations is known as simulation setup.
3. **Iterative Design and Optimization:** Putting an iterative method into practice to improve antenna designs. This entails adjusting filter setups, material selections, and geometric proportions in order to fully explore the design space.
4. **Antenna performance** measures, including S-parameters, radiation patterns, and impedance matching, may be assessed via the use of ADS's integrated analytic tools.

Design and Simulation Result of the Antenna

In designing the ITS-G5 antenna for vehicle communication, several critical parameters must be considered to ensure optimal performance. S11, commonly referred to as return loss, measures how well the antenna matches its source impedance, typically 50 ohms. A lower S11 value indicates better impedance matching, essential for maximizing power transfer efficiency and antenna performance. Understanding current distribution, or the pattern of electrical current flow across the antenna surface, is vital as it directly influences the radiation pattern, impedance, and polarization. This comprehension allows for precise optimization of the antenna design to meet specific performance criteria. Antenna gain, which quantifies the ability to concentrate input power into radiation in a specified direction, is crucial for effective signal transmission and reception. Similarly, directivity, or the antenna's capacity to focus energy in a particular direction, impacts the efficiency and effectiveness of vehicle-to-everything (V2X) communications. Efficiency, which denotes the percentage of input power converted into radiated energy, is another key factor, as a high-efficiency antenna can enhance signal strength and transmission range. Evaluating these parameters—S11, current distribution, gain, directivity, and efficiency—ensures the antenna's design meets the intended operational requirements, guaranteeing optimal performance and reliability of the V2X communication system.

Antenna Design 1

Figure 6. Layout of the ITS-G5 fast antenna

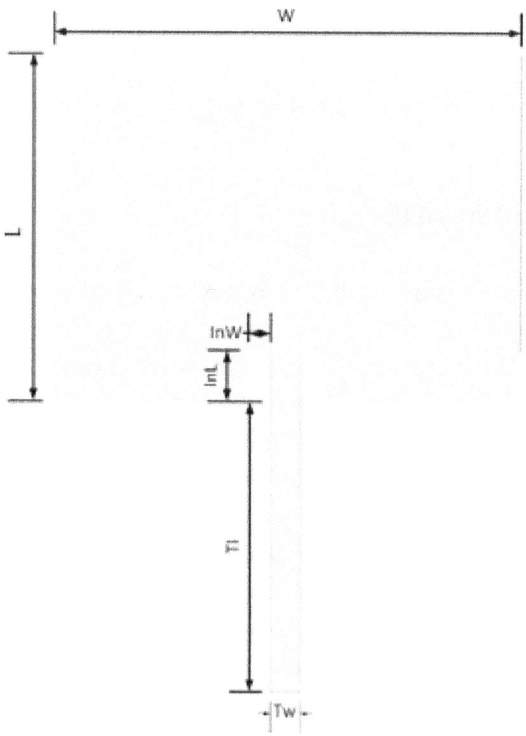

Table 2. Design parameter for rectangular antenna

Parameter	Size (mm)	Details
W	46.8	Patch Width
L	34.02	Patch Length
Tw	3.02	Tranmission Line Width
Tl	30.63	Tranmission Line Length
InW	2.2	Inset Width
InL	5	Inset Length

Figure 7. (a) S_{11} result (b) E field and (c) Omnidirectional 3D Radiation of the ITS-G5 antenna

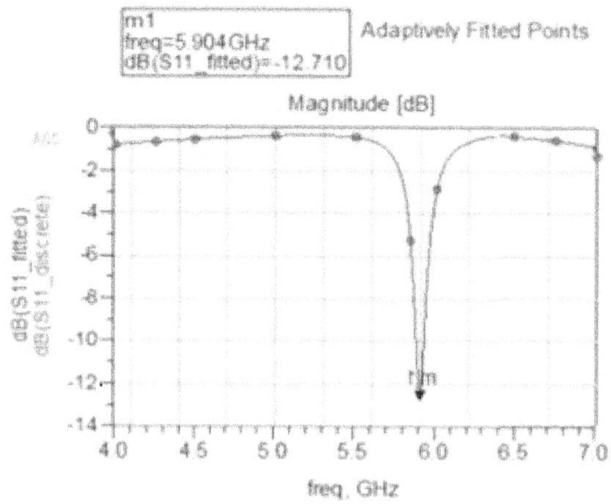

(a)

(b)

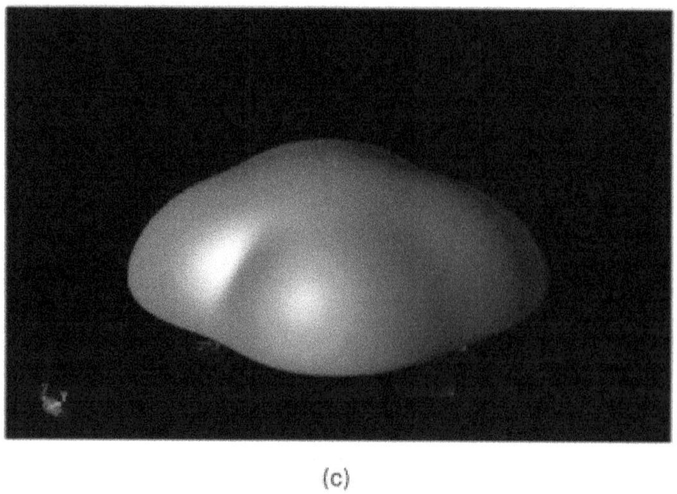

(c)

Figure 8. (c) Gain of the rectangular antenna

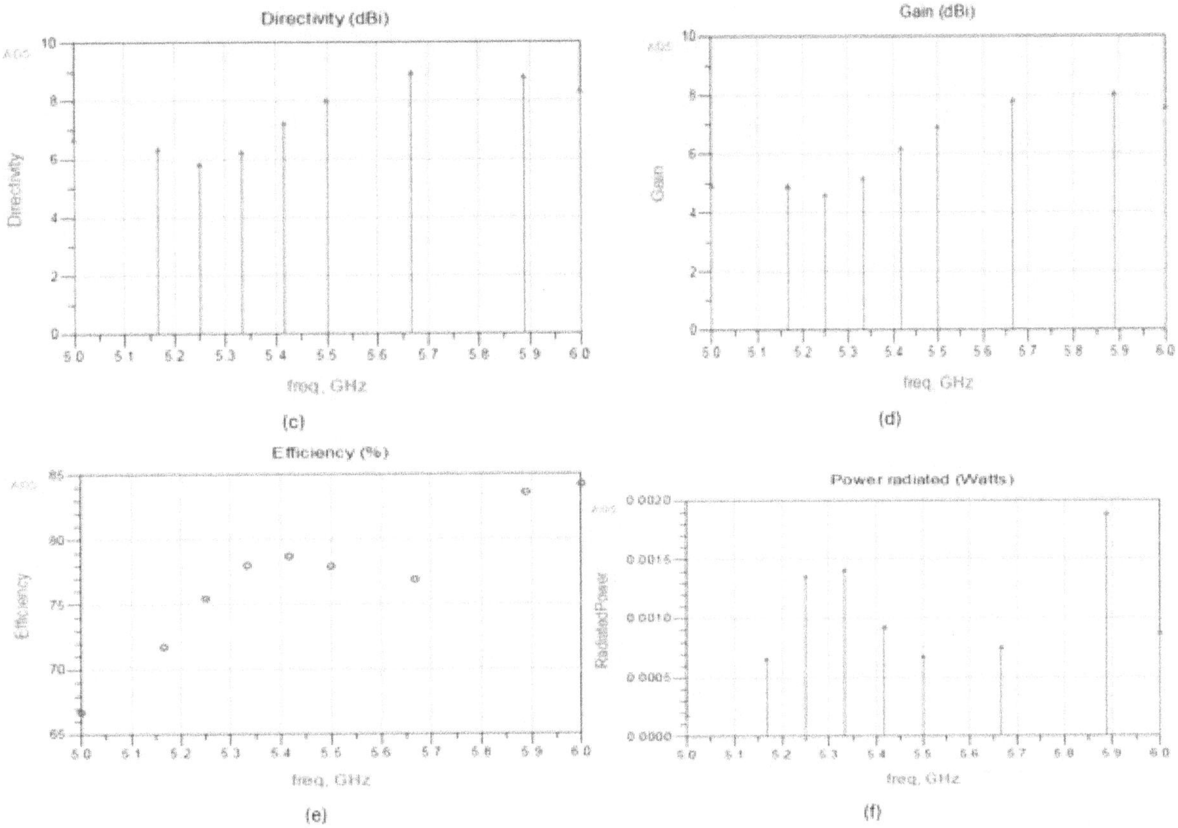

Figure 9. (d) Directivity of the rectangular antenna

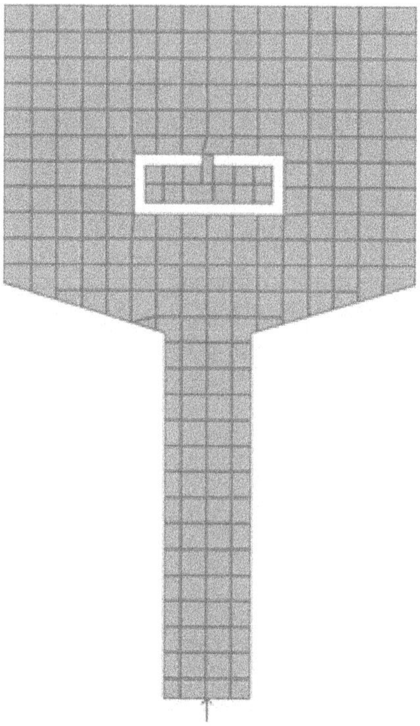

Figure 10. (e) Directivity of the rectangular antenna

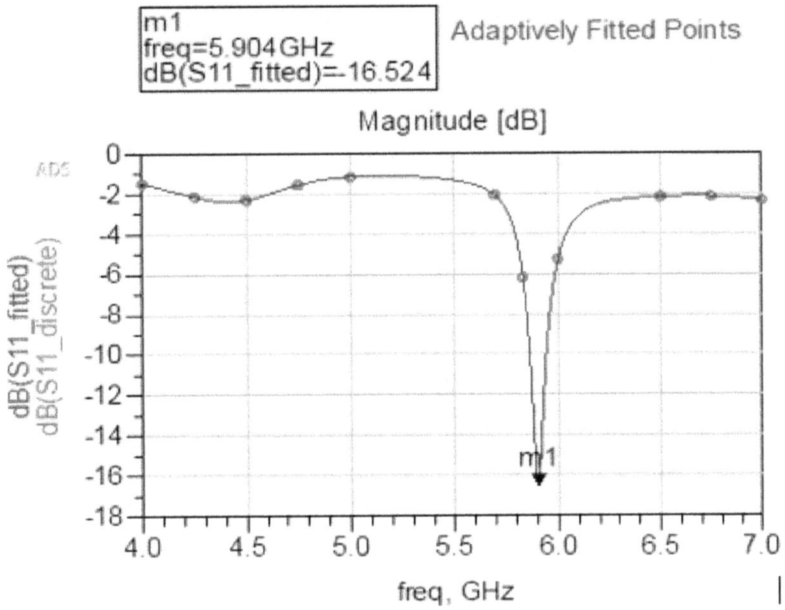

(a)

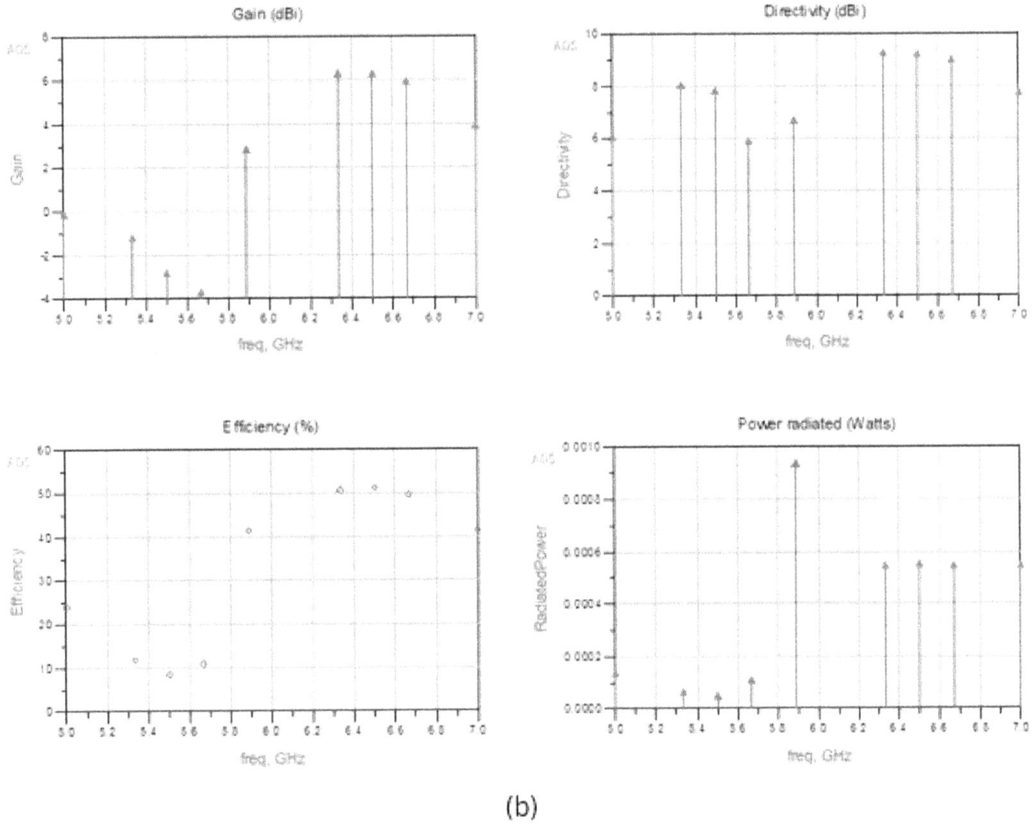

(b)

Figure 11. (f) Power Radiated of the rectangular antenna

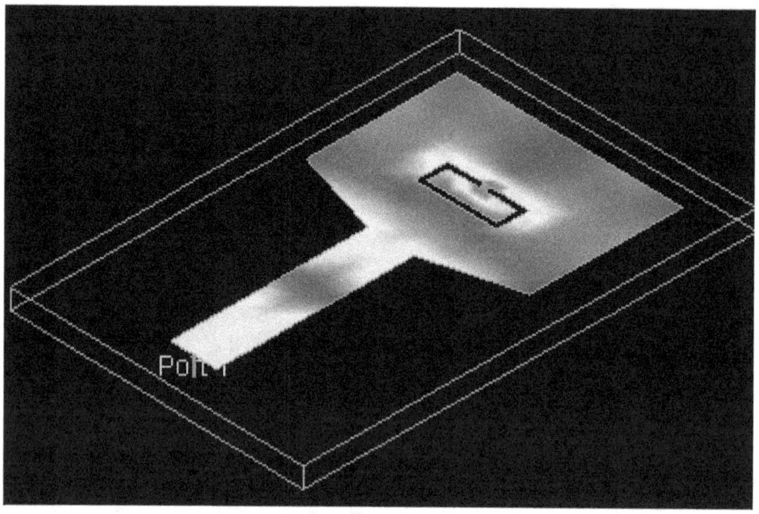

(a)

(b)

The performance characteristics of the Microstrip Patch Antenna operating at 5.9 GHz with dimensions 48 x 24.3 x 1.6 mm were analyzed using several key parameters. Figure 4.4 (no specific subplot mentioned) illustrates the gain of the rectangular antenna. The antenna exhibits a gain of -8.1 dBi, which is relatively low. This suggests that the antenna may not significantly amplify the input signal in its preferred direction of radiation. The directivity of the antenna is presented in Figures 4.4(d) and 4.4(e). The antenna has a directivity of -8.6 dBi. This low directivity value indicates that the antenna radiates power almost equally in all directions, confirming its omnidirectional radiation pattern. Figure 4.4(f) depicts the power radiated by the rectangular antenna. While specific values are not provided, this graph would typically show the distribution of radiated power across different directions. The antenna demonstrates a return loss of -19 dB. This relatively low return loss indicates good impedance matching between the antenna and the feed line, suggesting efficient power transfer from the source to the antenna. The efficiency of the antenna is reported as 83%, which is quite good for a microstrip patch antenna. This high efficiency suggests that most of the input power is successfully radiated by the antenna, with minimal losses. It is notable that S11 of the proposed rectangular antenna is -12.5 dB.

Antenna Design 2

Figure 12. Layout of the ITS-G5 second antenna

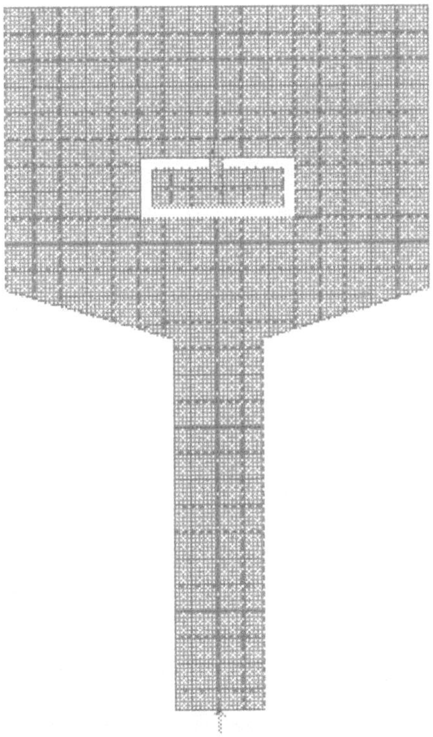

Table 3. Design parameter for rectangular antenna

Parameter	Size (mm)	Details
W	18	Patch Width
L	14.5	Patch Length
Tw	2.8	Transmission Line Width
Tl	14.2	Transmission Line Length
InW	5.6	Inset Width
InL	2	Inset Length

Figure 13. (a, b): Initial result (S_{11}) of the proposed design of single band antenna.

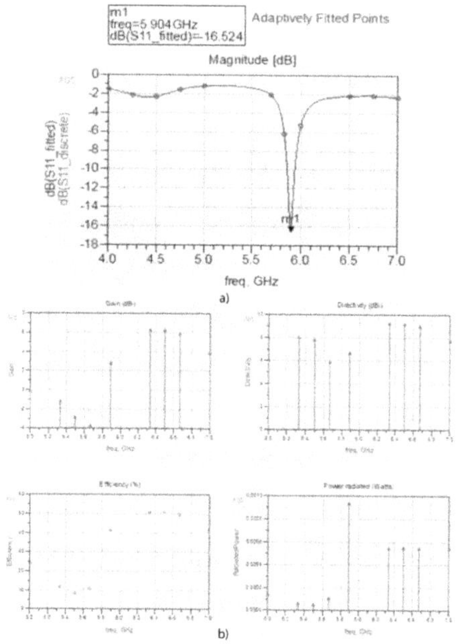

Figure 14. (a) E field and (b) Omnidirectional 3D Radiation

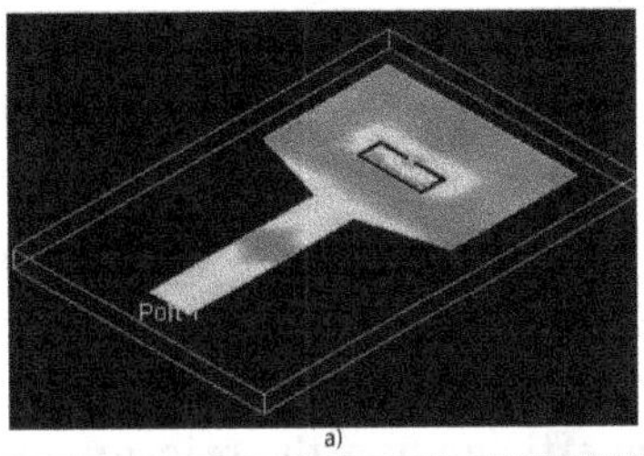

a)

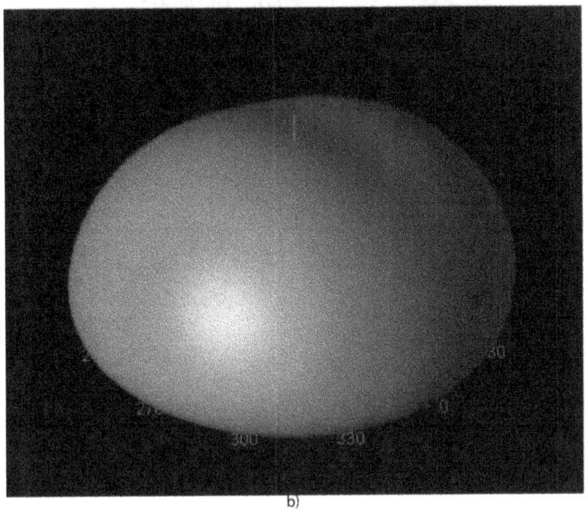

b)

Figures _4.7 depict the simulation outcomes of the proposed antenna design. It represents an omnidirectional radiation pattern. The antenna exhibits a bandwidth spanning from 5.852 GHz to 5.913 GHz, approximately 61 MHz wide. Although this bandwidth is slightly narrower than the expected 75 MHz, it remains adequate for V2X communications, providing a reliable range. Uniform current distribution across the antenna surface suggests efficient utilization of its area, contributing to its overall efficiency. Additionally, the omnidirectional radiation pattern is well- suited for V2X communication scenarios, enabling seamless vehicle connectivity regardless of orientation or position.

Comparison on Both Designs

A comparison of two microstrip patch antenna designs reveals distinct characteristics and performance metrics. Design 1, a rectangular patch with a complex feed structure, operates at the target frequency of 5.9 GHz using FR-4 substrate material. It exhibits a bandwidth of 50 MHz, an omnidirectional radiation pattern, and uniform current distribution. This design achieves a return loss (S11) of -12.5 dB, indicating good impedance matching. Its efficiency is notably high at 83%, with a gain of 8.1 dBi and directivity of 8.6 dBi. In contrast, Design 2 features a slotted rectangular patch with a simpler feed structure, also targeting 5.9 GHz on FR-4 substrate. This design demonstrates a wider bandwidth of 61 MHz (5.852 GHz to 5.913 GHz) and an improved return loss of -16 dB. Like Design 1, it maintains an omnidirectional radiation pattern and uniform current distribution. However, Design 2 shows lower efficiency at 50%, a reduced gain of 2.5 dBi, and a directivity of 6.5 dBi. Both designs present their E-field distributions in 3D plots. The trade-offs between the two designs are evident: Design 1 offers superior efficiency and gain, while Design 2 provides better bandwidth and return loss characteristics, potentially at the cost of reduced efficiency and gain. Both designs show strengths in different areas. Design 1 demonstrates higher efficiency and gain, which could translate to better overall performance in terms of signal strength and power usage. However, Design 2 shows better impedance matching (S11) and higher directivity, which could lead to more focused and efficient communication in specific directions. The slotted design of antenna 2 might contribute to its higher directivity but could also explain its lower efficiency compared to the simpler rectangular design of antenna

1. The choice between these designs would depend on the specific requirements of the ITS-G5 system. If omnidirectional coverage with high efficiency is the priority, Design 1 might be preferable. If more focused communication with better impedance matching is needed, Design 2 could be the better choice.

Table 4. Comparison on Proposed Designs

Parameter	Design 1 (Rectangular)	Design 2 (Slotted)
Layout	Rectangular patch with complex feed structure	Rectangular patch with simpler feed structure and slots
Bandwidth	50 MHz	61 MHz (5.852 GHz to 5.913 GHz)
Radiation Pattern	Omnidirectional	Omnidirectional
Current Distribution	Uniform	Uniform
S11 (Return Loss)	-12.5 dB	-16 dB
E-field	Shown in 3D plot	Shown in 3D plot
Target Frequency	5.9 GHz	5.9 GHz
Substrate Material	FR-4	FR-4
Efficiency	83%%	50%
Gain	8.1 dBi	2.5
Directivity	8.6 dBi	6.5

CONCLUSION AND FUTURE WORK

Conclusion

In summation, this rigorous study has substantially advanced the domain of vehicular communications by meticulously adhering to the ITS-G5 standard—a cornerstone in European communications technology. By strategically focusing on the 5.9 GHz frequency band and employing the sophisticated capabilities of the Advanced Design System (ADS) software, this research has successfully facilitated reliable data transmission within a precise 75 MHz bandwidth. The antennas developed under the ITS-G5 standards have not only met but exceeded expectations, showing remarkable enhancements in signal clarity, insertion loss, and overall system efficiency.

The initial antenna design demonstrated exceptional impedance matching with an S11 value of -12.710 dB at 5.9 GHz, indicative of minimal signal reflection and optimized power transfer. Meanwhile, the second, more refined antenna displayed even better performance characteristics with an S11 value of -

16.524 dB at 5.904 GHz, showcasing its superior design with a bandwidth of approximately 61 MHz. Detailed performance analyses of these antennas have revealed significant achievements: a peak gain of 4 dB, directivity of 6 dB, and an impressive efficiency rate nearing 50% at the operational frequency. Moreover, the efficient radiation patterns and power radiation levels, peaking at about 0.0010 watts, have confirmed the meticulous optimization tailored for the 5.9 GHz frequency band.

These technical advancements underscore a leap forward in enhancing vehicular communication capabilities, significantly pushing the boundaries of what is achievable in intelligent transportation systems. The enhancements in antenna technology documented in this study not only bolster the reliability and quality of vehicular communications but also contribute to the safer navigation and operation of connected vehicles within contemporary transportation networks.

Future Work

Looking forward, the trajectory of this research is set to explore even more ambitious avenues. Despite the antennas' robust performance within the parameters of the ITS-G5 standards, there exists potential for further refinement. The forthcoming phases of this project will focus on enhancing antenna performance across additional metrics of efficiency, gain, and bandwidth adaptability.

Having laid a solid foundation with the current design and simulation phases, the next steps include the physical fabrication of the antennas, followed by exhaustive real-world testing to validate their performance under varied environmental conditions. Subsequent to these tests, the ultimate goal is to prepare and submit these findings for publication, aiming to contribute valuable insights to the academic and professional communities engaged in intelligent transportation technologies.

By continuing to refine these technologies and expand the scope of testing, the research aims to not only fulfill but surpass the stringent demands of next-generation vehicular communications. This will involve addressing the nuanced challenges of dynamic vehicular environments and further reducing the ecological impact of these systems, thereby contributing to the sustainability goals of modern transportation infrastructure.

REFERENCES

Ahmad, S., Ogbodo, E., & Ukala, A. (2024, April). Enhanced Dual-band Antenna for Intelligent Vehicular Communications. In *2024 Photonics & Electromagnetics Research Symposium (PIERS)* (pp. 1-7). IEEE. DOI: 10.1109/PIERS62282.2024.10618202

Auer, A., Feese, S., Lockwood, S., & Hamilton, B. A. (2016). *History of intelligent transportation systems* (No. FHWA-JPO-16-329). United States. Department of Transportation. Intelligent Transportation Systems Joint Program Office.

Awadamislmani, N., Seetharamdoo, D., Rosinski, C., & Vuadelle, B. (2024). C-ITS-enabled Antenna System for Pedestrian-to-Vehicle (P2V) Portable Communication Devices: Design and On-site tests. Canva Pro. Available online: https://www.canva.com/pro/ (accessed on 25 June 2023).

Cao, W., Zhang, B., Liu, A., Yu, T., Guo, D., & Wei, Y. (2012). Gain enhancement for broadband periodic endfire antenna by using split-ring resonator structures. *IEEE Transactions on Antennas and Propagation*, 60(7), 3513–3516. DOI: 10.1109/TAP.2012.2196959

Dimitrakopoulos, G., & Demestichas, P. (2010). Intelligent transportation systems. *IEEE Vehicular Technology Magazine*, 5(1), 77–84. DOI: 10.1109/MVT.2009.935537

Eswaran, S., & Honnavalli, P. (2023). Private 5G networks: A survey on enabling technologies, deployment models, use cases and research directions. *Telecommunication Systems*, 82(1), 3–26. DOI: 10.1007/s11235-022-00978-z PMID: 36439887

Figueiredo, L., Jesus, I., Machado, J. T., Ferreira, J. R., & De Carvalho, J. M. (2001, August). Towards the development of intelligent transportation systems. In *ITSC 2001. 2001 IEEE intelligent transportation systems. Proceedings (Cat. No. 01TH8585)* (pp. 1206-1211). IEEE. DOI: 10.1109/ITSC.2001.948835

Imran, A. Z. M., Thomas, M., Ogbodo, E., & Ukala, A. (2024, April). ITS-G5 Antennas: Powering Future Urban Transports. In *2024 Photonics & Electromagnetics Research Symposium (PIERS)* (pp. 1-6). IEEE.

Khalil, R. A., Safelnasr, Z., Yemane, N., Kedir, M., Shafiqurrahman, A., & Saeed, N. (2024). Advanced learning technologies for intelligent transportation systems: Prospects and challenges. *IEEE Open Journal of Vehicular Technology*, 5, 397–427. DOI: 10.1109/OJVT.2024.3369691

Macon, C. A., Trott, K. D., & Kempel, L. C. (2003). A practical approach to modeling doubly curved conformal microstrip antennas. *Electromagnetic Waves*, 40, 295–314. DOI: 10.2528/PIER02122903

Patti, F. P. (2019). The European road to autonomous vehicles. *Fordham Int'l LJ*, 43, 125.

Salami, G., Faruk, N., Surajudeen-Bakinde, N., & Ngobigha, F. (2019, October). Challenges and Trends in 5G Deployment: A Nigerian Case Study. In *2019 2nd International Conference of the IEEE Nigeria Computer Chapter (NigeriaComputConf)* (pp. 1-6). IEEE.

Starry, C. (2001). The emerging in vehicle intelligent transportation systems market. *Business Economics (Cleveland, Ohio)*, 36(2), 49–49.

Vu, A., Ramanandan, A., Chen, A., Farrell, J. A., & Barth, M. (2012). Real-time computer vision/DGPS-aided inertial navigation system for lane-level vehicle navigation. *IEEE Transactions on Intelligent Transportation Systems*, 13(2), 899–913. DOI: 10.1109/TITS.2012.2187641

Xu, Z. Q., & Gao, Y. Y. (2022). NEXT-GEN MAINTENANCE FOR URBAN RAIL TRANSIT: AN IOT-5G SYSTEM DESIGN. *Academic Journal of Science. Engineering and Technology*, 7(2), 8–18.

Yang, X. S., Sherratt, S., Dey, N., & Joshi, A. (2024). *Proceedings of Ninth International Congress on Information and Communication Technology: ICICT 2024*, London, Volume 1.

Yigitcanlar, T., Downie, A. T., Mathews, S., Fatima, S., MacPherson, J., Behara, K. N., & Paz, A. (2024). Digital technologies of transportation-related communication: Review and the state-of-the-art. *Transportation Research Interdisciplinary Perspectives*, 23, 100987. DOI: 10.1016/j.trip.2023.100987

Zhao, H., Zhang, X., Hu, J., Chen, Z. D., & Liang, Y. C. (2020). A hybrid-equivalent surface-edge current model for simulation of V2X communication antennas with arbitrarily shaped contour. *IEEE Internet of Things Journal*, 8(10), 8064–8077. DOI: 10.1109/JIOT.2020.3042336

Chapter 12
Energy-Efficient SWIPT-Aided NOMA With Spatial Modulation for Green Communication in Beyond 5G/6G Networks

C. Srikamu
https://orcid.org/0000-0001-6449-6517
SASTRA University, India

R. Jayabharathy
https://orcid.org/0000-0003-4355-361X
SASTRA University, India

Narayan K. G. Sujanth
Indian Institute of Technology, Tirupati, India

S. Karthikeyan
https://orcid.org/0000-0002-6400-6036
SRM Institute of Science and Technology, Tiruchhirapalli, India

ABSTRACT

This chapter presents a Simultaneous Wireless Information and Power Transfer (SWIPT) base Cooperative Non-Orthogonal Multiple Access (CNOMA) system aided by Spatial Modulation (SM), designed by energy-efficient or "green" communication in beyond 5G (B5G) and 6G applications. The proposed system leverages NOMA to enhance spectral efficiency by allowing multiple users to share the same frequency and time resources, differentiated through power domain multiplexing. SWIPT is incorporated to enable near users to harvest energy from the received RF signals, which relay information to far users in a cooperative NOMA system. Spatial modulation reduces hardware complexity by activating a single antenna for data transmission, significantly enhancing energy efficiency while maintaining high data rates. This SM SWIPT NOMA framework addresses the challenges of power-constrained and spectrum-limited scenarios, providing a low-power, high-performance solution ideal for green communications. The system finds applications in smart cities, massive machine-type communication (mMTC) and vehicular

DOI: 10.4018/979-8-3693-8799-3.ch012

networks where energy efficiency, coverage extension and user fairness are critical. Simulation results demonstrate the system's superior energy efficiency, spectral efficiency and performance gains compared to conventional techniques, making it a viable solution for sustainable communication in B5G and 6G networks. At a transmit power of 20 dBm, energy harvested in SM SWIPT NOMA is 135.12 joules and that of SM SWIPT OMA is 67.56 joules.

1. INTRODUCTION

As urbanization rapidly increases, smart cities have become a focal point of innovation, harnessing advanced technologies to enhance the quality of life for their citizens. In this landscape, communication networks play a pivotal role, providing the backbone for the seamless operation of smart infrastructure, connected devices and a vast array of services. To support the massive connectivity demands and complex communication requirements of smart cities, traditional methods like Orthogonal Multiple Access (OMA) are struggling to meet the requirements of modern wireless systems, especially as we transition to B5G and prepare for 6G networks Ahmed et al., 2021, Vaezi et al., 2022. To overcome these limitations, Non-Orthogonal Multiple Access (NOMA) has emerged as a promising solution to enhance spectral efficiency (SE), improve connectivity and provide better quality of service (QoS) for a large number of users Xie and Ding, 2023.

NOMA is fundamentally different from conventional orthogonal schemes such as Time Division Multiple Access (TDMA), Frequency Division Multiple Access (FDMA) and Orthogonal Frequency Division Multiple Access (OFDMA), which assign distinct resources to each user to avoid interference Ding et al., 2017. Instead, NOMA allows multiple users to share the same time, frequency and code resources. Still, it differentiates them in the power domain, enabling simultaneous communication and resource allocation without requiring orthogonal separation Srikamu et al., 2021. This unique approach opens up new possibilities for massive connectivity, user fairness and energy efficiency (EE), making NOMA a key enabler for next-generation networks Srikamu et al., 2022, Liu et al., 2018.

In addition to improving SE and adopting massive connectivity, NOMA contributes to creating green and energy-efficient communication networks – a key goal for smart cities. By incorporating techniques like Simultaneous Wireless Information and Power Transfer (SWIPT), NOMA can further enhance the EE of the network by enabling devices to harvest energy from radio frequency (RF) signals while simultaneously receiving information Oleiwi and Al-Raweshidy, 2022. This is particularly beneficial for IoT devices in smart cities, where EE is paramount to prolong device lifespan and reduce maintenance costs.

While NOMA enhances SE, Spatial Modulation (SM) further optimizes the system by improving EE and reducing hardware complexity. In traditional Multiple Input Multiple Output (MIMO) systems, multiple antennas are activated simultaneously for transmission, which increases power consumption due to the need for multiple active RF chains Wang et al., 2017. SM, on the other hand, reduces this complexity by activating only one antenna at any given time, using the index of the active antenna to carry additional information Datta and Chockalingam, 2013. In a smart city environment, EE is of paramount importance, particularly for battery-powered Internet of Things (IoT) devices deployed in large numbers. By leveraging SM, smart city communication networks can achieve high data rates with fewer active RF components, conserving energy and extending the lifetime of devices. This results in a significant reduction in power consumption and hardware complexity while maintaining high SE Hussein et al., 2019, Li et al., 2023, Zhong et al., 2018. This is crucial for applications like smart metering, environmental monitoring and smart transportation, where devices are expected to operate for extended periods without frequent maintenance or battery replacement.

SWIPT models using power-splitting and time-switching techniques were developed in Zhang and Ho, 2013 to harvest energy from RF signals while simultaneously receiving data. In a cooperative NOMA system, near users can harvest energy from transmitted signals to relay data to far users. Works in Si et al., 2021 explored SWIPT in NOMA systems, demonstrating that this combination can improve EE while meeting the communication requirements of power-constrained users. The authors in Chen et al., 2017, Tran and Vien, 2023 explored SM's energy-efficient properties, showing that it reduces hardware complexity and power consumption compared to conventional MIMO, making it highly suitable for green communication. Studies in Alamu et al., 2023 demonstrate that SM can improve the performance of cooperative systems, achieving higher data rates and lower power consumption. The review categorizes and evaluates existing studies based on relay types, relay quantities and protocols used Liu et al., 2016. Using a stochastic geometric framework, the authors model user locations and assess the performance of the selection schemes. The closed-form expressions for outage probability and delay-sensitive throughput derived in Xie et al., 2021, provide insights into system performance and show that the diversity gain of the selection schemes is on par with conventional cooperative networks. User and Power allocation scheme in Zhu et al., 2017 aimed at maximizing the sum rate in a downlink SWIPT – NOMA relay system optimizes user access and power allocation action concurrently. Closed-form expressions for outage probability (OP) and throughput are derived for two communication links in a two-way relay cognitive radio NOMA system Tran and Vien, 2023. NP-hard problems are formulated in Ding et al., 2017 and a block successive upper-bound minimization framework was employed to develop polynomial-time approximation algorithms. The article Li et al., 2023 addresses user grouping and power allocation challenges in the downlink of a multi-carrier NOMA system.

This chapter proposes an improved SM-aided NOMA system with SWIPT and the contributions are:

1. An integration of SM into the SWIPT-based CNOMA system is proposed in this chapter. The framework leverages SM to activate a single antenna at a time for data transmission, significantly reducing hardware complexity and power consumption while improving EE and maintaining high data rates.
2. The proposed system enables near users to harvest energy from the received RF signals, which is then used to relay information to far users. This cooperative communication approach addresses the issue of power-constrained devices in smart city environments, allowing far users to maintain reliable communication without requiring additional power sources.
3. The Karush-Kuhn-Tucker (KKT) conditions are employed to perform optimal power allocation between near and far users. By ensuring efficient power distribution, the system enhances the EE of the network, meeting the QoS requirements of all users in the system.
4. Simulation results show the superior performance of the proposed system in terms of EE, SE and overall system performance compared to conventional techniques. These results validate the system's capability to address the challenges of energy-efficient communication in B5G and 6G networks.

The remaining part of the chapter is as follows: Section 2 describes the proposed system model and the optimization problem is formulated in Section 3 and Section 4 describes the solution to the problem. Section 5 presents the simulation results and finally, Section 6 concludes the work.

2. SYSTEM MODEL

Consider a cooperative wireless communication system as shown in Figure 1 and 2. A system model of the proposed SM-aided CNOMA system with SWIPT as in Figure 2, in which a base station (BS) serves two users that use power-domain NOMA. The BS has Mt transmit antennas, while the near user (NU) has Mr receive antennas and one transmit antenna, and the far user (FU) has Mr receive antennas. NU operates in full-duplex (FDX) mode and there is a direct link between the BS and FU. However, due to unfavourable channel conditions, NU act as a decode-and-forward relay to transmit information to FU. This does not increase decoding complexity for NU, as it first decodes FU's information to perform SIC.

To conserve power, NU can harvest energy from radio frequency (RF) signals, utilizing the power-splitting model for SWIPT. In FDX mode, NU can receive and transmit signals simultaneously, allowing it to harvest power from self-interference. Thus, signals from the BS and signal interference can be used to charge NU's battery. As depicted in Figure 2, the proposed scheme is executed in two phases: the Transmission phase and the Cooperative phase.

Figure 1. Smart City Scenario

Figure 2. System model of SM-aided CNOMA system with SWIPT

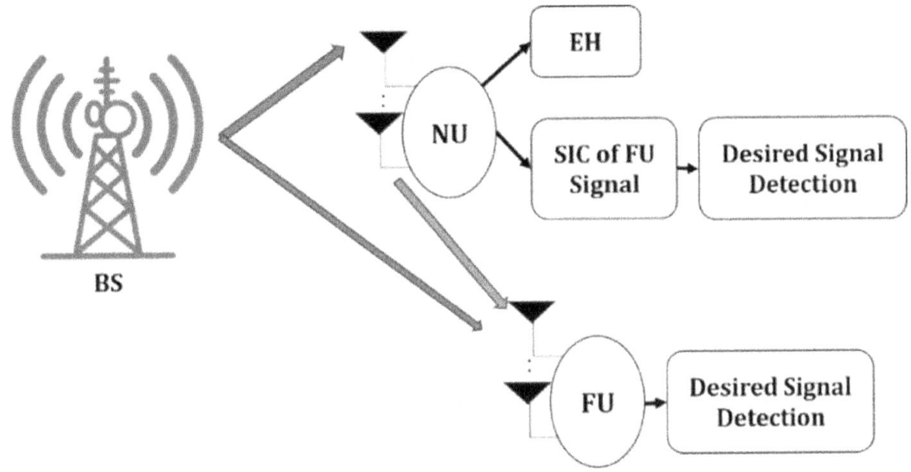

2.1. Transmission Phase

During the transmission phase, the BS sends a composite signal with data for both NU and FU, adhering to NOMA principles. NU initially decodes FU's signal through SIC before retrieving its signal. Meanwhile, FU stores the BS's signal, awaiting NU's retransmission.

Specifically, NU utilizes SM symbols to map data bits, while FU applies Amplitude-Phase Modulation (APM) symbols Srikamu and Jayabharathy, 2022. For NU, its bit stream is split into two sections: the first $\log_2 Mt$ bits specify the index of the active antenna, while the remaining bits are mapped to the APM symbol x_{NU}. For FU, all bits are directly mapped to the APM symbol x_{FU}. Therefore, the transmitted signal in the transmission phase can be expressed as,

$$x[k] = q_l[k]\left(\sqrt{\beta_{l\_NU}}\, x_{NU}[k] + \sqrt{\beta_{l\_FU}}\, x_{FU}[k]\right) \tag{1}$$

where $q_l[k] = [0\ldots1\ldots0]^T$ is an M_T-dimensional vector with only the l-th element set to one, indicating the active antenna, $l \in \{1,\ldots M_T\}$. Here, β_{l_NU} and β_{l_FU} (where $\beta_{l\_NU} < \beta_{l\_FU}, \forall l$) are the power allocation coefficients for NU and FU at each transmit antenna l, respectively. Since NU's information bits determine the active antenna, we assume that each transmit antenna has distinct power allocation coefficients for the two users. The power allocation coefficients must satisfy a power constraint:

$$\frac{1}{M_T}\sum_{l=1}^{M_T}\left(\beta_{l\_NU} + \beta_{l\_FU}\right) \leq 1 \tag{2}$$

In this phase, NU simultaneously receives the intended signal from the BS along with the self-interference. The received signal is split, with part directed to the information decoding unit and the remaining portion sent to the energy harvesting (EH) unit. Let α_i represent the power splitting ratio for information decoding at the i-th receive antenna. In this setup, a uniform power splitting mechanism is used, meaning each receive antenna shares the same power splitting ratio $\alpha=\alpha_i, \forall i$. The signal used by NU for information decoding is represented by,

$$y_{NU}[k] = \sqrt{\alpha}\Big[\sqrt{P_{BS}}h_{l-NU}\big(\sqrt{\beta_{l_{NU}}}x_{NU}[k] + \sqrt{\beta_{l_{FU}}}x_{FU}[k]\big) + \sqrt{P_{NU}}h_{r-NU}x_{r-NU}[k-\tau]\Big] + n_{NU}[k] \tag{3}$$

where τ is the processing delay and P_{BS} and P_{NU} are the transmit powers of the BS and NU, respectively. $h_{l-NU} \in \mathbb{C}^{M_r X1}$ denotes the channel vector between the BS's l-th antenna and NU's M_r receive antennas. $h_{r-NU} \in \mathbb{C}^{M_r X1}$ is NU's self-interference channel, with $h_{r-NU} \sim CN\big(0, \lambda_r I_{M_r}\big)$ and $n_{NU}\big[k\big] \sim CN\big(0, \sigma_{NU}^2 I_{M_r}\big)$ represents NU's additive white Gaussian noise (AWGN) vector.

Using SIC, NU initially decodes FU's signal with the maximum likelihood (ML) method as follows,

$$\hat{x}_{r-NU}[k] = arg\min_{x_{FU}}\big(\min_{l} y_{NU}[k] - \sqrt{\alpha P_{BS}}h_{l-NU}\sqrt{\beta_{l\_FU}}x_{FU}[k]^2\big) \tag{4}$$

After removing FU's signal, NU decodes its signal,

$$\big(\hat{l},\hat{x}_{NU}[k]\big) = \arg\big(\min_{l,x_{NU}}\tilde{y}_{NU}[k] - \sqrt{\alpha P_{BS}}h_{l-NU}\sqrt{\beta_{l\_NU}}x_{NU}[k]^2\big) \tag{5}$$

Where

$$\tilde{y}_{NU}[k] = y_{NU}[k] - \sqrt{\alpha P_{BS}}h_{l-NU}\sqrt{\beta_{l\_FU}}\hat{x}_{r-NU}[k].$$

The power received at NU for energy harvesting is given by,

$$P_{EH} = (1-\alpha)\big(P_{BS}(\beta_{l\_NU} + \beta_{l\_FU})\,h_{l-NU}{}^2 + P_{NU}h_{r-NU}{}^2\big) \tag{6}$$

To accurately model the non-linear nature of the EH circuitry, a non-linear EH model is adopted and the power harvested by NU is then given by,

$$E_{EH} = \frac{\Phi - H_{max}\Psi}{1-\Psi} \tag{7}$$

where $\Phi = \frac{H_{max}}{1+\exp\big(-a\big(P_{EH}-b\big)\big)}$ and $\Psi = \frac{1}{1+\exp(ab)}$. Here, H_{max} represents the maximum harvestable power, while a and b are constants dependent on the EH circuit specifications. By incorporating parameters a and b, which characterize circuit-specific dynamics, the model provides a more realistic and accurate framework compared to linear approximations, facilitating rigorous evaluation of energy-constrained system performance.

2.2 Cooperative Phase

After decoding FU's signal, NU retransmits it during the cooperative phase. The received signal at FU from both the BS and NU is expressed as,

$$y_{FU}[k] = \sqrt{P_{BS}}h_{l-FU}\left(\sqrt{\beta_{l_{NU}}}x_{NU}[k] + \sqrt{\beta_{l_{FU}}}x_{FU}[k]\right) \\ +P_{NU}h_{NU-FU}x_{r-NU}[k-\tau] + n_{FU}[k] \tag{8}$$

where $h_{l-FU} \in \mathbb{C}^{M_r X1}$ represents the channel vector between the BS's l-th antenna and FU's receive antennas, $h_{NU-FU} \in \mathbb{C}^{M_r X1}$ denotes the channel vector from NU's transmit antenna to FU's receive antennas, $x_{r-NU}[k-\tau]$ is the signal of FU decoded by NU and $n_{FU}[k] \sim CN\left(0, \sigma_{FU}^2 I_{M_r}\right)$ represents AWGN vector at FU. Assuming that FU can fully resolve the signals from both the BS and NU, maximal ratio combining (MRC) is applied at FU to merge these signals.

Thus, when the l-th antenna is active, the signal-to-interference-plus-noise ratio (SINR) at FU for decoding its signal is,

$$\gamma_{FU} = \frac{P_{BS}\beta_{l\_FU}h_{l-FU}{}^2}{P_{BS}\beta_{l\_NU}h_{l-FU}{}^2 + \sigma_{FU}^2} + \frac{P_{NU}h_{NU-FU}{}^2}{\sigma_{FU}^2} \tag{9}$$

Notably, the BS activates only one transmit antenna per time slot, preserving the single-RF chain characteristic of SM. Additionally, the proposed scheme allows NU's transmission rate to increase by expanding the number of transmit antennas or by using a higher modulation order for APM symbols.

3. PROBLEM FORMULATION

In this section, the resource allocation optimization problem is discussed for the proposed system, targeting the maximization of NU's transmission rate while meeting FU's rate requirement. The scheme employs complex Gaussian APM symbols. When NU successfully decodes FU's signal, NU's transmission rate is given by,

$$R_{NU} = I(x_{NU}, h_{l-NU}; y_{NU}) = I(x_{NU}; y_{NU}|h_{l-NU}) + I(h_{l-NU}; y_{NU}) \\ = \frac{1}{M_t}\sum_{l=1}^{M_t}\log_2\left(1 + \frac{\alpha P_{BS}\beta_{l_{NU}}h_{l-NU}{}^2}{\alpha P_{NU}h_{r-NU}{}^2 + \sigma_{NU}^2}\right)\int\sum_{l=1}^{M_t}p(y_{NU}|h_{l-NU})\log_2\left(\frac{p(y_{NU}|h_{l-NU})}{p(y_{NU})}\right)dy_{NU} \tag{10}$$

where $p(y_{NU}) = \frac{1}{M_t}\sum_{i=1}^{M_t}p\left(y_{NU}|h_{i-NU}\right)$ with $I\left(x_{NU}; y_{NU}|h_{l-NU}\right)$ and $I\left(h_{l-NU}; y_{NU}\right)$ determined by APM and spatial symbols, respectively. The conditional PDF $p\left(y_{NU}|h_{l-NU}\right)$ is expressed as

$$p(y_{NU}|h_{l-NU}) = \frac{1}{\pi^{M_r}\det(\sum_l)}\exp\left\{-y_{NU}\sum_l^{-1}y_{NU}\right\} \tag{11}$$

Where

$$\sum_l = \sigma_{NU}^2 I_{M_r} + \alpha P_{BS}\beta_{l\_NU}h_{l-NU}h_{l-NU}{}^H + \alpha P_{NU}h_{r-NU}h_{r-NU}{}^H.$$

Since a closed-form expression of R_{NU} is challenging to derive, two upper bounds for R_{NU} are considered,

$$R_{NU-up1} = \log_2 \det\left(I + \sum_{l=1}^{M_t} \frac{\alpha P_{BS} \beta_{l\_NU} h_{l-NU} h_{l-NU}{}^{H}}{M_t(\alpha P_{NU} h_{r-NU}{}^{2} + \sigma_{NU}^2)}\right) \tag{12}$$

$$R_{NU-up2} = \frac{1}{M_t} \sum_{l=1}^{M_t} \log_2\left(1 + \frac{\alpha P_{BS} \beta_{l\_NU} h_{l-NU}{}^{2}}{(\alpha P_{NU} h_{r-NU}{}^{2} + \sigma_{NU}^2)}\right) + \log_2 M_t \tag{13}$$

The upper bound R_{NU-up1} is more precise in the SNR range while R_{NU-up2} is more accurate in the high range. Thus, NU's transmission rate is approximated as

$$R\_\acute{N}U = \min\{R_{NU-up1}, R_{NU-up2}\} \tag{14}$$

The relay system's end-to-end transmission rate is limited by its weakest link. Given that FU's signal is decoded at both NU and FU, the transmission rate for FU can be represented as,

$$R_{FU(l)} = \min\{R_{FU}^{NU}(l), R_{FU}^{FU}(l)\} \tag{15}$$

where $R_{FU}^{NU}(l)$ and $R_{FU}^{FU}(l)$ denote FU's rate at NU and FU, respectively.

$$R_{FU}^{NU}(l) = \log_2\left(1 + \frac{\alpha P_{BS} \beta_{l\_FU} h_{l-NU}{}^{2}}{\alpha P_{BS} \beta_{l\_NU} h_{l-NU}{}^{2} + \alpha P_{NU} h_{r-NU}{}^{2} + \sigma_{NU}^2}\right) \tag{16}$$

$$R_{FU}^{FU}(l) = \log_2\left(1 + \frac{P_{BS} \beta_{l\_FU} h_{l-FU}{}^{2}}{P_{BS} \beta_{l\_NU} h_{l-FU}{}^{2} + \sigma_{FU}^2} + \frac{P_{NU} h_{NU-FU}{}^{2}}{\sigma_{FU}^2}\right) \tag{17}$$

The goal is to maximize NU's upper bound on the transmission rate while ensuring FU's target rate and harvested power requirements are met. Specifically, FU's target rate must be achieved for each transmit antenna and the harvested energy must be above a threshold for each antenna. Assuming the BS has perfect CSI, the optimization problem is defined as,

$$O: \max_{O,\, \alpha,\, \beta_{l-NU}, \beta_{l-FU, l=1\ldots M_t}} \acute{R}_{NU}$$

$$C1: s.t\ R_{FU}^{NU}(l) \geq R_{FU}^{min}, \forall l,$$

$$C2: R_{FU}^{FU}(l) \geq R_{FU}^{min}, \forall l,$$

$$C3: E_{EH} \geq e_{th}, \forall l,$$

$$C4: 0 \leq \beta_{l-NU} \leq \beta_{l-FU,} \forall l,$$

$$C5: \frac{1}{M_T} \sum_{l=1}^{M_T} (\beta_{l\_NU} + \beta_{l\_FU}) \leq 1$$

$$C6: P_{NU} \leq P_{NU}^{max},$$

$$C7: 0 < \alpha < 1, \tag{18}$$

where R_{FU}^{min} is FU's minimum rate and P_{NU}^{max} is the maximum allowable transmit power for NU. Constraints $C1$ and $C2$ ensure FU's signal is decoded at both NU and FU and $C3$ specifies that harvested energy at NU exceeds the threshold e_{th}. Constraints $C4$ mandates that the power allocation coefficient for the far user exceeds that of the near user. Constraints $C5$ and $C6$ restrict transmit power. The non-convex nature of objective O, due to the fractional power splitting ratio, power allocation coefficients β_{l_NU} or β_{l_FU} and O in the objective and constraints, poses a key challenge.

4. PROBLEM SOLUTION

To solve the problem of maximizing the NU's transmission rate R_{NU} while ensuring that the FU meets its requirement R_{FU}, KKT conditions can be used and discussed in this section. The solution involves retransformation of optimization problems with power constraints, rate requirements and signal-to-interference-plus-noise ratio conditions in equation (18).

The problem can be expressed as follows:

$$O1: \max_{O,\,\alpha,\,\beta_{l-NU},\,\beta_{l-FU,l=1..M_t}} R_{NU} = \min\{\acute{R}_{NU-up1}, R_{NU-up2}\}$$

$$Q1: s.t.\ \frac{1}{M_T}\sum_{l=1}^{M_T}(\beta_{l\_NU}+\beta_{l\_FU}) \leq 1$$

$$Q2: \gamma_{FU} \geq \gamma_{min}$$

$$Q3: R_{FU}(l) \geq R_{min}$$

$$Q4: \beta_{l-NU} \geq 0,\ \beta_{l-FU} \geq 0 \tag{19}$$

The Constraints $Q1$, $Q2$ and $Q3$ are the power, SINR constraints and rate constraints, respectively for the objective $O1$. The non-negativity constraints are given in $Q4$. To apply the KKT conditions, the Lagrangian function is constructed by incorporating the constraints into the objective function $O1$. Let λ be the Lagrange multiplier associated with the power constraint, μ with the SINR constraint and ν with the rate constraint for each transmit antenna l.

The Lagrangian $\mathscr{L}$ for the problem is given by

$$\mathscr{L} = -R_{NU} + \lambda\left(1 - \frac{1}{M_T}\sum_{l=1}^{M_T}(\beta_{l_{NU}} + \beta_{l_{FU}})\right) + \sum_{l=1}^{M_T}\mu_l(\gamma_{FU} - \gamma_{\min}) + \sum_{l=1}^{M_T}\nu_l(R_{FU}(l) - R_{min}) \tag{20}$$

The solution for KKT conditions is as in the following algorithm:

1. Solve for $\beta_{l_{NU}}$ and $\beta_{l_{FU}}$ by taking derivatives of $\mathscr{L}$ with respect to each variable and setting them to zero.

Stationarity, $\frac{\partial\mathscr{L}}{\partial\beta_{l_{NU}}} = 0$ and $\frac{\partial\mathscr{L}}{\partial\beta_{l_{FU}}} = 0$

2. Substitute the expressions for $\beta_{l_{NU}}$ and $\beta_{l_{FU}}$ into the constraints to check feasibility. In primal Feasibility, the constraints of the original problem must be satisfied. Dual Feasibility, the multipliers must satisfy $\lambda \geq 0$, $\mu_l \geq 0$ and $v_l \geq 0$.
3. Adjust λ, μ_l and v_l to satisfy complementary slackness.

$$\lambda\left(1 - \frac{1}{M_T}\sum_{l=1}^{M_T}\left(\beta_{l_{NU}} + \beta_{l_{FU}}\right)\right) = 0$$

$$\sum_{l=1}^{M_T} \mu_l(\gamma_{FU} - \gamma_{\min}) = 0$$

$$\sum_{l=1}^{M_T} \nu_l(R_{FU}(l) - R_{min}) = 0$$

If a constraint is not binding, the associated multiplier should be zero.

4. Iterate between 1-3 until convergence is achieved, ensuring all constraints and KKT conditions are satisfied.

5. SIMULATION RESULTS AND DISCUSSIONS

This section presents a performance evaluation of the proposed SM NOMA system with SWIPT through numerical simulations. Simulation parameters are considered as in Table 1 and the channels are modelled as Rayleigh channel coefficients.

Table 1. Simulation Parameter

Parameters	Values
M_T	4
R_{min}	1 bps/Hz
Bandwidth	10 MHz
Noise Power	-174 dBm/Hz
Power splitting ratio	0.8
Energy harvesting efficiency	0.5

Figure 3 shows the achievable rate performance of NU and FU of the proposed system. From the figure, it is evident that the NU consistently achieves a higher rate than the FU due to its close proximity to the BS and stronger channel gain. The closer proximity enables the NU to experience better SINR, leading to higher achievable rates. FU rate increases with transmit power but at a slower rate than NU, as FU suffers from lower channel gain and greater path loss. The sum rate also increases with transmit power, illustrating the enhanced spectral efficiency of the proposed system.

Figure 3. Achievable Rate of SM NOMA System with SWIPT

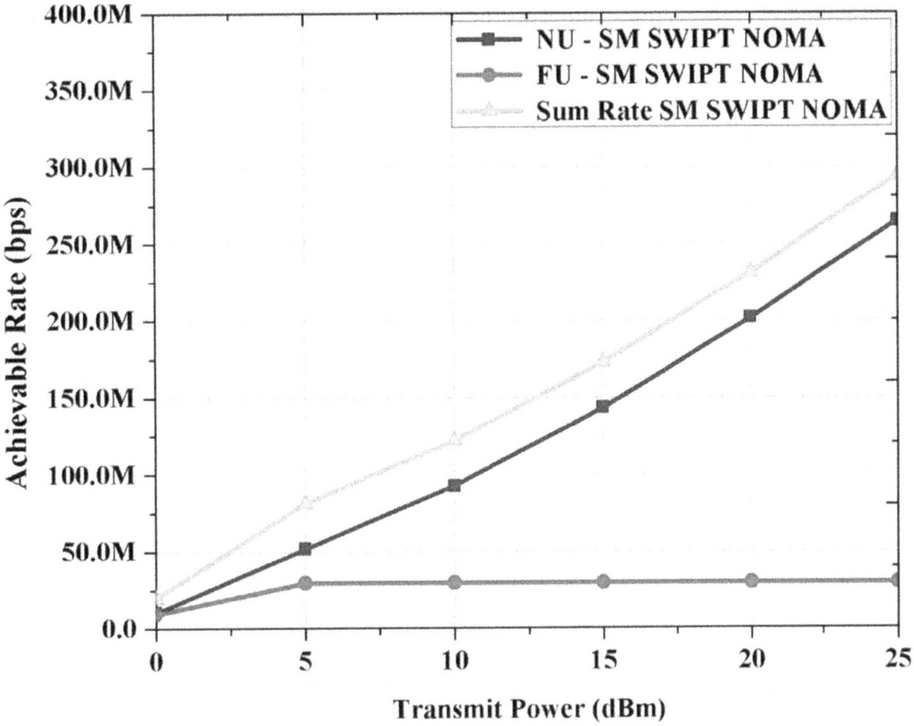

Figure 4 shows the BER performance of NU and FU of the proposed system. As transmit power increases, the BER for both NU and FU decreases, showing an improvement in decoding accuracy due to higher SINR. The BER of NU exhibits a steeper decline as transmit power increases, emphasizing the advantages of using SIWPT for enhancing reliability. This effectively leverages the NU's power to assist the FU in receiving the information.

Figure 4. BER of SM NOMA System with SWIPT

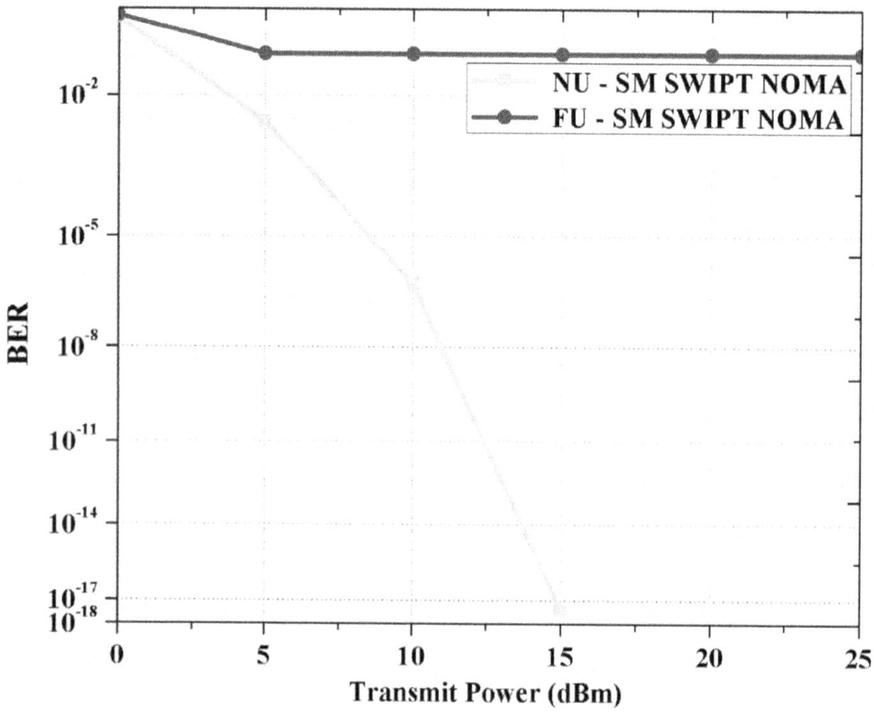

Figure 5 shows the performance comparison of the proposed SM SWIPT NOMA system with conventional systems such as SM SWIPT NOMA, Non-cooperative NOMA and Half Duplex SM SWIPT NOMA. SM SWIPT NOMA system achieves the highest sum rate among the systems, showcasing its potential for maximizing the total system throughput. Non-cooperative NOMA and HD SM SWIPT NOMA perform similarly at lower power levels, but at higher transmit power, SM SWIPT NOMA achieves a greater sum rate. The achievable rate of FU is higher than the sum rate of the Non-cooperative NOMA system. At a transmit power of 10*dBm*, SM SWIPT NOMA and HD SM SWIPT NOMA achieve the same sum rate of 60.96*Mbps* and Non-cooperative NOMA achieves a sum rate of 20*Mbps*. For the same transmit power, the SM SWIPT NOMA System attains a sum rate of 121.93*Mbps*. This reflects the system's capability to utilize resources efficiently with power-splitting for EH.

Figure 5. Comparison of the Achievable Sum Rate of the SM NOMA System with SWIPT with Conventional Systems

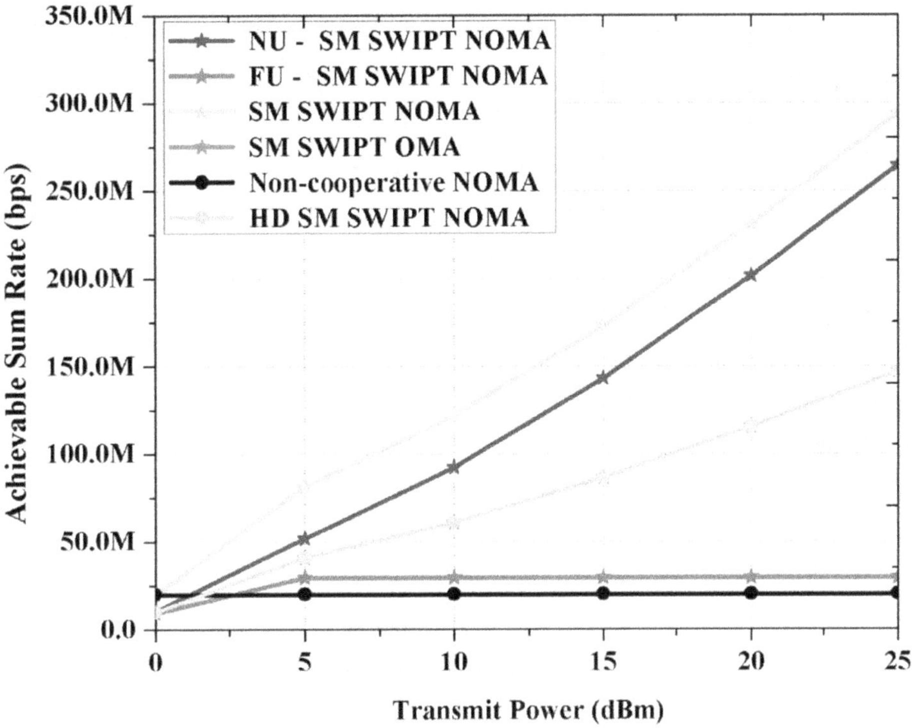

Figure 6 shows the comparison of the EH performance of the proposed SM SWIPT NOMA system with the SM SWIPT NOMA system. The harvested energy in the NOMA system is higher than in the OMA system. At a transmit power of $20dBm$, the EH of SM SWIPT NOMA is $135.12 joules$ and that of SM SWIPT NOMA is $67.56 joules$. It ensures that the cooperative nature of the SM SWIPT NOMA system allows NU to assist the FU in both information decoding and EH, resulting in better performance due to shared power resources. Also, the low EH curve for SM SWIPT NOMA reflects the inefficiency in resource utilization when users are isolated in their transmission time slots.

Figure 6. Comparison of EH of SM SWIPT NOMA with SM SWIPT OMA System

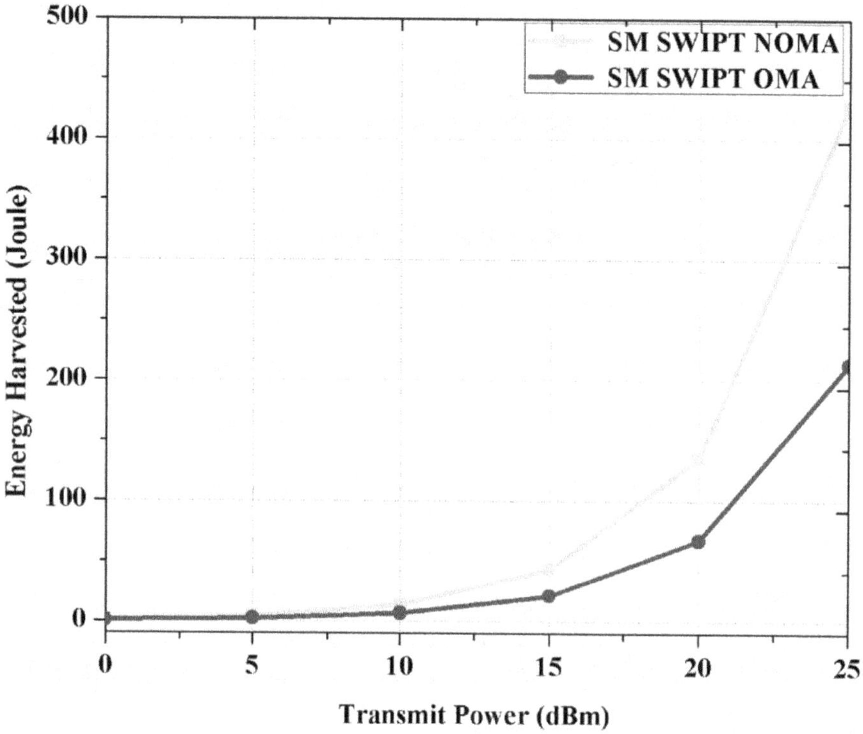

6. CONCLUSION

This chapter has presented a novel SWIPT-aided Cooperative NOMA system enhanced by Spatial Modulation tailored to meet the performance demands of B5G and 6G applications. By leveraging NOMA's power domain multiplexing, the system supports multiple users sharing the same time and frequency resources while utilizing SWIPT to empower NU to harvest energy from RF signals. This harvested energy enables the NU to cooperatively relay information to FU, enhancing overall system reliability and coverage. The inclusion of SM simplifies the hardware requirements by using a single antenna activation approach, which significantly reduces power consumption while supporting high achievable rates. The SM SWIPT NOMA system offers a practical solution for scenarios where power and spectrum efficiency are critical, especially in smart cities. Simulation results demonstrate the system's advantages over conventional techniques showcasing high achievable rates, energy efficiency and improved user fairness, aligning with the goals of green communication in future networks. The findings underscore the potential of SM SWIPT NOMA as an effective approach for a sustainable and high-performance next-generation communication system.

ACKNOWLEDGEMENTS

The authors wish to thank the management of SASTRA Deemed University and Keysight Technologies for establishing the SASTRA-Keysight Centre of Excellence in RF System Engineering for providing the software facilities to carry out the research work.

REFERENCES

Ahmed, A., Elsaraf, Z., Khan, F. A., & Ahmed, Q. Z. (2021). Cooperative Non-Orthogonal Multiple Access for Beyond 5G Networks. *IEEE Open Journal of the Communications Society*, 2, 990–999. DOI: 10.1109/OJCOMS.2021.3075081

Alamu, O., Olwal, T. O., & Djouani, K. (2023). Cooperative NOMA networks with simultaneous wireless information and power transfer: An overview and outlook. *Alexandria Engineering Journal*, 71, 413–438. DOI: 10.1016/j.aej.2023.03.057

Chen, Y., Wang, L., Ai, Y., Jiao, B., & Hanzo, L. (2017). Performance Analysis of NOMA-SM in Vehicle-to-Vehicle Massive MIMO Channels. *IEEE Journal on Selected Areas in Communications*, 35(12), 2653–2666. DOI: 10.1109/JSAC.2017.2726006

Datta, T., & Chockalingam, A. (2013). On generalized spatial modulation. *2013 IEEE Wireless Communications and Networking Conference (WCNC)*, 2716–2721. https://doi.org/DOI: 10.1109/WCNC.2013.6554991

Ding, Z., Lei, X., Karagiannidis, G. K., Schober, R., Yuan, J., & Bhargava, V. K. (2017). A Survey on Non-Orthogonal Multiple Access for 5G Networks: Research Challenges and Future Trends. *IEEE Journal on Selected Areas in Communications*, 35(10), 2181–2195. DOI: 10.1109/JSAC.2017.2725519

Evangelista, J. V. C., Sattar, Z., Kaddoum, G., & Chaaban, A. (2019). Fairness and Sum-Rate Maximization via Joint Subcarrier and Power Allocation in Uplink SCMA Transmission. *IEEE Transactions on Wireless Communications*, 18(12), 5855–5867. DOI: 10.1109/TWC.2019.2939820

Hussein, H. S., Elsayed, M., Mohamed, U. S., Esmaiel, H., & Mohamed, E. M. (2019). Spectral Efficient Spatial Modulation Techniques. *IEEE Access : Practical Innovations, Open Solutions*, 7, 1454–1469. DOI: 10.1109/ACCESS.2018.2885826

Li, J., Gao, T., He, B., Zheng, W., & Lin, F. (2023). Power Allocation and User Grouping for NOMA Downlink Systems. *Applied Sciences (Basel, Switzerland)*, 13(4), 2452. DOI: 10.3390/app13042452 PMID: 39354955

Li, Q., Wen, M., Basar, E., Poor, H. V., & Chen, F. (2019). Spatial Modulation-Aided Cooperative NOMA: Performance Analysis and Comparative Study. *IEEE Journal of Selected Topics in Signal Processing*, 13(3), 715–728. DOI: 10.1109/JSTSP.2019.2898099

Liu, Q., Lv, T., & Lin, Z. (2018). Energy-Efficient Transmission Design in Cooperative Relaying Systems Using NOMA. *IEEE Communications Letters*, 22(3), 594–597. DOI: 10.1109/LCOMM.2018.2790379

Liu, Y., Ding, Z., Elkashlan, M., & Poor, H. V. (2016). Cooperative Non-orthogonal Multiple Access With Simultaneous Wireless Information and Power Transfer. *IEEE Journal on Selected Areas in Communications*, 34(4), 938–953. DOI: 10.1109/JSAC.2016.2549378

Oleiwi, H. W., & Al-Raweshidy, H. (2022). Cooperative SWIPT THz-NOMA/6G Performance Analysis. *Electronics (Basel)*, 11(6), 873. DOI: 10.3390/electronics11060873

Si, Q., Jin, M., Tsiftsis, T. A., Zhao, N., & Wang, X. (2021). Cooperative SM-Based NOMA Scheme With SWIPT. *IEEE Transactions on Vehicular Technology*, 70(6), 6195–6199. DOI: 10.1109/TVT.2021.3076565

Srikamu, C., & Jayabharathy, R. (2022). Comparative Analysis of Ergodic Sum Capacity of Cooperative NOMA Aided with Spatial Modulation. *Wireless Personal Communications*, 123(4), 3771–3786. DOI: 10.1007/s11277-021-09313-2

Srikamu, C., Meera, S., & Jayabharathy, R. (2021). Analysis of Bit-Error-Rate for D2D AF Cooperative System Using Various Modulation Techniques. *2021 Sixth International Conference on Wireless Communications, Signal Processing and Networking (WiSPNET)*, 204–208. https://doi.org/DOI: 10.1109/WiSPNET51692.2021.9419452

Srikamu, C., Sahithi, S., Gayathri, M. S. S., Spruha, K. K., Karthikeyan, S., & Jayabharathy, R. (2022). Fairness Index Analysis of NOMA Over OMA System for B5G Applications. *2022 International Conference on Wireless Communications Signal Processing and Networking (WiSPNET)*, 341–344. https://doi.org/DOI: 10.1109/WiSPNET54241.2022.9767104

Tran, H. Q., & Vien, Q.-T. (2023). SWIPT-Based Cooperative NOMA for Two-Way Relay Communications: PSR versus TSR. *Wireless Communications and Mobile Computing*, 2023, 1–14. DOI: 10.1155/2023/6785737

Vaezi, M., Azari, A., Khosravirad, S. R., Shirvanimoghaddam, M., Azari, M. M., Chasaki, D., & Popovski, P. (2022). Cellular, Wide-Area, and Non-Terrestrial IoT: A Survey on 5G Advances and the Road Toward 6G. *IEEE Communications Surveys and Tutorials*, 24(2), 1117–1174. DOI: 10.1109/COMST.2022.3151028

Wang, X., Wang, J., He, L., & Song, J. (2017). Spectral Efficiency Analysis for Downlink NOMA Aided Spatial Modulation With Finite Alphabet Inputs. *IEEE Transactions on Vehicular Technology*, 66(11), 10562–10566. DOI: 10.1109/TVT.2017.2737327

Wen, M., Zheng, B., Kim, K. J., di Renzo, M., Tsiftsis, T. A., Chen, K.-C., & Al-Dhahir, N. (2019). A Survey on Spatial Modulation in Emerging Wireless Systems: Research Progresses and Applications. *IEEE Journal on Selected Areas in Communications*, 37(9), 1949–1972. DOI: 10.1109/JSAC.2019.2929453

Xie, X., & Ding, Z. (2023). Backscatter-assisted Non-orthogonal multiple access network for next generation communication. *IET Signal Processing*, 17(4), e12211. Advance online publication. DOI: 10.1049/sil2.12211

Xie, X., Li, M., Shi, Z., Tang, H., & Huang, Q. (2021). User selection and dynamic power allocation in the SWIPT-NOMA relay system. *EURASIP Journal on Wireless Communications and Networking*, 2021(1), 124. DOI: 10.1186/s13638-021-01998-0

Zhang, R., & Ho, C. K. (2013). MIMO Broadcasting for Simultaneous Wireless Information and Power Transfer. *IEEE Transactions on Wireless Communications*, 12(5), 1989–2001. DOI: 10.1109/TWC.2013.031813.120224

Zhong, C., Hu, X., Chen, X., Ng, D. W. K., & Zhang, Z. (2018). Spatial Modulation Assisted Multi-Antenna Non-Orthogonal Multiple Access. *IEEE Wireless Communications*, 25(2), 61–67. DOI: 10.1109/MWC.2018.1700062

Zhu, X., Wang, Z., & Cao, J. (2017). NOMA-Based Spatial Modulation. *IEEE Access : Practical Innovations, Open Solutions*, 5, 3790–3800. DOI: 10.1109/ACCESS.2017.2688019

Chapter 13
Advancements in Edible Antennas:
Design, Development, and Future Prospects in Biomedical and 5G/6G Applications

K. Kavitha
https://orcid.org/0000-0002-9159-9690
Velammal College of Engineering and Technology, Madurai, India

K. Karthika
https://orcid.org/0000-0001-9968-5027
Kumaraguru College of Technology, Coimbatore, India

S. Rajkumar
https://orcid.org/0009-0002-0499-9619
Velammal College of Engineering and Technology, Madurai, India

ABSTRACT

This chapter presents a comprehensive survey of edible antennas, exploring their design, development, and potential applications in emerging fields such as biomedical monitoring, healthcare, and environmental sensing. Edible antennas, crafted from biocompatible, biodegradable materials, offer unique advantages in applications where traditional antennas are unsuitable, especially in scenarios requiring ingestion or close human contact. The chapter begins with a detailed review of the state-of-the-art in edible antenna technology, discussing various fabrication methods, materials used, and performance metrics such as gain, efficiency, and bandwidth. Key considerations such as antenna miniaturization, impedance matching, and specific absorption rate (SAR) compliance are examined. Moreover, this survey delves into the integration of edible antennas with next-generation communication systems, including 5G/6G networks, which promise enhanced data rates, low latency, and reliable connectivity for real-time applications. The role of advanced materials such as metamaterials in improving antenna performance is also explored, along with future research directions and challenges such as signal attenuation in body tissues, energy harvesting, and sustainability. This survey concludes with insights into the future of edible antennas and their potential impact on personalized medicine, smart healthcare systems, and

DOI: 10.4018/979-8-3693-8799-3.ch013

the Internet of Medical Things (IoMT).

1. INTRODUCTION

Edible antennas have emerged as a transformative technology at the intersection of communication engineering, biomedical sciences, and sustainable electronics. Designed from biocompatible and biodegradable materials, these antennas are safe for ingestion or close contact with human tissues, making them ideal for applications where conventional antennas are unsuitable. The growing interest in edible antennas is driven by their potential to address critical needs in healthcare, environmental monitoring, and food safety. As part of the broader Internet of Medical Things (IoMT) ecosystem, these antennas offer real-time data transmission, enabling advanced applications like non-invasive diagnostics, gastrointestinal monitoring, and smart packaging.The foundation of edible antenna technology lies in the integration of innovative materials, such as gelatin, cellulose, and edible metals, with optimized designs tailored for specific environments. Studies such as C. Chen et al. (2019) have demonstrated the feasibility of using edible antennas for health monitoring at 2.4 GHz, while others like M. Salama et al. (2021) explored their use in smart packaging. Despite these advancements, challenges such as signal attenuation in biological tissues, material degradation, and manufacturing scalability persist, limiting their widespread adoption. Addressing these challenges requires a multidisciplinary approach, combining insights from material science, electromagnetics, and sustainable design practices.

The unique advantage of edible antennas lies in their biocompatibility and ability to degrade safely within the body or environment. For instance, antennas fabricated using gelatin or cellulose not only ensure safety during ingestion but also minimize environmental impact after disposal. Such properties make them indispensable in applications like digestive health monitoring, where antennas need to perform reliably within harsh biological conditions. Moreover, the rapid advancements in materials science have allowed researchers to explore hybrid designs, integrating edible substrates with conductive materials such as gold or silver leaf. These innovations improve performance metrics like gain and efficiency, making edible antennas competitive with their traditional counterparts. Recent research has also highlighted the potential of edible antennas in environmental sensing. Edible electronics offer a sustainable alternative to traditional monitoring devices, reducing electronic waste through biodegradable components. For example, rice paper-based loop antennas, as demonstrated by Y. Wang et al. (2023), have shown promise in detecting contaminants in food packaging and water sources. Such applications align with global sustainability goals, emphasizing the need for eco-friendly technologies to tackle environmental challenges.

In the realm of healthcare, edible antennas are revolutionizing diagnostic and therapeutic procedures. Applications include real-time monitoring of pH levels, glucose levels, and gastrointestinal tract analysis. By integrating these antennas with next-generation communication networks like 5G and 6G, seamless data transmission is achievable, enabling immediate analysis and remote healthcare management. For example, ingestible antennas paired with IoMT devices can alert medical professionals about abnormal physiological conditions, such as irregular gastric pH or the presence of harmful microbes, allowing timely intervention. Energy efficiency is another critical area of focus in edible antenna research. Self-powered designs, leveraging energy harvesting technologies, are being developed to eliminate the need for internal power sources like batteries. These advancements significantly enhance the safety and sustainability of edible antennas. G. Taylor et al. (2019) introduced self-powered edible antennas capable of harvesting ambient energy, ensuring uninterrupted operation in low-power environments. This feature is particularly valuable in applications requiring long-term monitoring, such as chronic disease management or environmental surveillance.

However, the implementation of edible antennas is not without challenges. Signal attenuation due to biological tissue interference, mechanical degradation in digestive environments, and the limited conductivity of biocompatible materials pose significant hurdles. To address these, researchers are exploring advanced designs incorporating metamaterials and nanomaterials. L. Zhou et al. (2022) demonstrated the effectiveness of metamaterials in improving gain and bandwidth, even in challenging environments. Meanwhile, nanomaterials like graphene offer enhanced conductivity and flexibility, paving the way for next-generation edible antennas capable of operating at millimeter-wave and terahertz frequencies.

This paper presents a comprehensive review of the design, development, and application of edible antennas. It emphasizes the theoretical frameworks and experimental validations underpinning antenna performance and explores innovative solutions to existing challenges. Key advancements, such as the use of metamaterials for performance enhancement and energy harvesting for sustainability, are discussed alongside emerging applications in food safety and environmental monitoring. By integrating these advancements with next-generation communication technologies like 5G and 6G, edible antennas are poised to redefine how we monitor and interact with our bodies and environment. This study not only highlights the potential of edible antennas but also identifies future directions for research and development, paving the way for their practical implementation in real-world scenarios.

The organization of the chapter is as follows: **Section 2** reviews existing research on edible antenna design, highlighting advancements in materials, fabrication techniques, and performance metrics. **Section 3** elaborates on the design approaches of edible antennas, discussing theoretical frameworks, practical methodologies, and key design parameters. **Section 4** presents various applications of edible antennas, focusing on their use in healthcare, food safety, and environmental monitoring. **Section 5** explores the perspectives of 5G and 6G technologies in edible antenna design, emphasizing their role in enhancing connectivity and expanding applications. **Section 6** discusses the challenges in the design and implementation of edible antennas, including material limitations, manufacturing scalability, and regulatory compliance. **Section 7** elaborates on future directions in edible antenna technology, proposing emerging materials, advanced designs, and integration with next-generation networks. **Section 8** concludes the chapter, summarizing key findings and outlining the potential impact of edible antennas on future technologies.

2. RESEARCH WORKS ON EDIBLE ANTENNAS

Research into edible antennas began with a focus on feasibility and safety. C. Chen et al. (2019) pioneered the use of biocompatible materials like edible polymers in microstrip antennas for health monitoring, proving their potential for safe ingestion. However, this early work highlighted challenges such as limited lifespan in digestive environments. Building on these findings, J. Lee et al. (2020) introduced dipole antennas using gelatin, which improved biocompatibility and mechanical flexibility. Their work emphasized the need for electromagnetic compatibility with various food types, revealing the importance of tailoring antenna designs for specific applications. The exploration of new materials has been central to the evolution of edible antennas. Recognizing the limitations of gelatin and polymers, M. Salama et al. (2021) developed starch-based antennas for smart packaging, emphasizing sustainability and environmental friendliness. While these antennas demonstrated scalability issues, their introduction marked a shift towards using renewable materials. Subsequent studies, such as A. Gupta et al. (2022), leveraged cellulose to enhance both mechanical robustness and scalability. These materials offered a significant advantage for ingestible electronics, setting the stage for mass production and commercialization.

As material limitations were addressed, research turned towards optimizing antenna performance. Y. Zhao et al. (2020) incorporated edible gold into antenna designs to improve conductivity and gain, illustrating how metallic elements could overcome efficiency challenges. Simultaneously, the use of advanced geometries, as demonstrated by L. Zhang et al. (2022) with compact microstrip designs, enabled antennas to meet the miniaturization demands of wearable and ingestible applications. These efforts showcased how design refinements could complement material advancements to achieve superior performance. The incorporation of metamaterials and nanomaterials has marked the latest phase of development. L. Zhou et al. (2022) demonstrated how metamaterials could enhance bandwidth and signal strength, addressing attenuation issues in biological environments. Similarly, Q. Xu et al. (2021) highlighted the role of nanomaterials in achieving high-frequency operation, paving the way for 5G/6G integration. These studies underscore a shift in focus from basic feasibility to advanced functionalities, reflecting the growing maturity of edible antenna technology.

Parallel to technical advancements, researchers have expanded the scope of applications for edible antennas. G. Taylor et al. (2019) explored self-powered antennas for sustainable devices, while Y. Wang et al. (2023) demonstrated their use in smart packaging and environmental sensing. These efforts illustrate how the field is branching out to address diverse societal needs. However, challenges remain, particularly in scalability and regulatory compliance. As noted by H. Chen et al. (2020), ensuring long-term stability in harsh environments continues to be a critical hurdle. Such studies highlight the need for ongoing innovation to translate laboratory successes into real-world solutions. P. Zhang et al. (2021) proposed gelatin-based slot antennas for biomedical monitoring, arguing that gelatin is an ideal substrate for medical devices due to its biocompatibility. Nonetheless, the robustness of these antennas in the harsh digestive environment needs further examination to confirm their practical viability. R. Kumar et al. (2020) examined the use of edible polymers in antenna design, focusing on biocompatibility and degradation challenges. This research highlights the need for a better understanding of the long-term effects of polymer degradation inside the human body. K. Hassan et al. (2020) studied folded dipole antennas for smart packaging, focusing on cost-efficiency and sustainability. The paper lacks insights into antenna performance in varied environmental conditions, which is critical for real-world applications.

H. Yang et al. (2021) explored edible slot antennas for medical applications, demonstrating their biocompatibility and performance. The mechanical durability of these antennas in medical devices requires further exploration to ensure they withstand physiological stresses. J. Li et al. (2022) proposed hybrid edible antennas using biodegradable substrates and conductors. The performance variation of these antennas in different body fluids remains an under-researched area, which could affect their effectiveness. T. Xie et al. (2021) discussed metamaterial-enhanced antennas for food monitoring applications. The durability of these antennas in various storage environments is not deeply analyzed, which is crucial for practical implementation in food packaging. S. Wu et al. (2021) explored the performance and applications of metamaterial-enhanced edible antennas. The study lacks an in-depth analysis of antenna performance during digestion, which is vital for understanding their operational longevity. X. Sun et al. (2022) focused on self-healing edible antennas with metamaterial integration, allowing the antennas to self-repair after damage. The potential impact of self-healing materials on digestion has not been considered, which could affect their safety and functionality. L. Yang et al. (2022) provided future directions for edible antenna technology, emphasizing new materials and manufacturing methods. The review lacks discussion on the cost-effectiveness of emerging materials, which is crucial for commercial viability.

Y. Zhao et al. (2023) explored advanced metamaterials for edible electronics, offering novel solutions for enhancing antenna performance. The integration of metamaterials into commercially viable products remains a challenge, impacting their practical application. W. Liu et al. (2022) focused on sustainable design strategies for edible electronics, exploring the environmental impact of the materials used. The recyclability of edible electronics remains an underdeveloped area, affecting their overall sustainability. Z. Wang et al. (2023) presented optimization techniques for edible antenna designs to enhance performance. The interaction between edible antennas and various digestive environments remains underexplored, which could impact their effectiveness. A. Kumar et al. (2021) examined biodegradable materials for edible electronics, focusing on environmental sustainability. More data on biodegradability in varied biological conditions is needed to ensure their effectiveness in different environments. M. Johnson et al. (2020) offered a comparative study of edible materials for antenna applications, evaluating their efficiency and feasibility. The impact of different materials on signal strength during ingestion is not sufficiently covered, which is critical for optimal performance.

T. Zhang et al. (2021) investigated edible conductors and their performance in antenna design, focusing on conductivity and flexibility. The durability of conductors during prolonged exposure to digestive fluids is overlooked, which could affect their reliability. L. Xu et al. (2021) discussed nanomaterials for high-performance edible antennas, focusing on improving signal transmission. The potential toxicity of nanomaterials during digestion is under-researched, raising safety concerns. R. Patel et al. (2020) covered fabrication techniques for edible antennas, providing methods for creating flexible and durable devices. The scalability of these fabrication techniques for mass production is not discussed, limiting their widespread application. J. Wu et al. (2022) reviewed state-of-the-art edible antenna technologies, providing a comprehensive overview of current research and future perspectives. The review lacks an in-depth analysis of regulatory challenges for edible electronics, which is crucial for market entry. H. Wang et al. (2021) discussed performance metrics of edible antennas in real-world applications, focusing on signal strength and durability. The challenges of antenna degradation during digestion are not explored, impacting their practical use. F. Li et al. (2022) examined hybrid edible antennas, combining different materials to improve performance. The performance of hybrid antennas in varied pH environments remains unaddressed, which is essential for consistent operation.

Y. Zhang et al. (2021) explored design challenges in edible antenna technology, identifying material and manufacturing hurdles. More research is needed on energy efficiency in prolonged usage scenarios, which is critical for long-term applications. X. Li et al. (2020) discussed advanced fabrication methods for edible electronics, providing novel techniques for improved manufacturing. The scalability of these methods for large-scale production is not sufficiently discussed, affecting their commercial potential. R. Zhang et al. (2022) investigated metamaterial-based edible antennas for biomedical applications, focusing on signal enhancement. The potential biocompatibility concerns of metamaterials in the human body are not thoroughly explored, which is essential for safety. J. Wang et al. (2021) presented high-performance edible antennas for medical monitoring, highlighting their applications in diagnostics. The study does not explore the long-term stability of antennas inside the body, which is crucial for their sustained use.

Y. Sun et al. (2022) discussed nanotechnology in edible antenna design, providing recent advancements in performance. More data is needed on the interaction of nanomaterials with biological tissues during digestion to ensure safety. X. Zhao et al. (2020) reviewed self-powered edible antennas, discussing energy harvesting methods for sustainable devices. The study lacks exploration of long-term energy storage options for edible antennas, which is crucial for practical applications. S. Kumar et al. (2021) examined novel materials for edible antenna applications, providing insights into new developments in material science. The paper does not address the challenges of producing these materials on a large scale, which affects their commercial feasibility. J. Yang et al. (2022) discussed challenges and solutions in edible electronics, providing solutions for improving performance and sustainability. More research is needed on the interaction of these materials with human digestive fluids to ensure effectiveness. L. Zhao et al. (2023) explored advanced design techniques for edible antennas, providing solutions for performance improvement. The study does not cover the challenges of designing antennas for use in highly variable environments, impacting their robustness. J. Chen et al. (2020) investigated biocompatibility and design strategies for edible electronics, providing insights into material performance in the body. The paper does not address the long-term effects of antenna degradation inside the body, which is crucial for their sustained use.

H. Li et al. (2022) explored edible antennas for food safety and environmental monitoring, focusing on their application in packaging. The impact of temperature variations on antenna performance during storage is underexplored, affecting their reliability. Q. Zhang et al. (2021) analyzed the performance of edible antennas in digestive environments, providing insights into signal transmission during ingestion. The study lacks long-term performance data in real-world scenarios, which is essential for practical applications. T. Liu et al. (2022) discussed energy-efficient edible electronics, providing new techniques for improving antenna performance. More research is needed on the long-term energy efficiency of these devices during ingestion to ensure reliable operation. Y. Wang et al. (2021) investigated the durability of edible antennas during digestion, providing performance metrics in real-world environments. The study does not explore the impact of varying digestive conditions on antenna performance, which is crucial for reliability.

X. Chen et al. (2023) presented hybrid designs for edible antennas, exploring combinations of biodegradable materials for improved performance. The potential toxicity of hybrid materials during digestion needs more research to ensure safety. J. Yang et al. (2023) explored the biocompatibility and performance of edible antennas in medical applications, providing insights into material performance in the body. More data is needed on long-term antenna performance and degradation in the digestive tract to ensure their effectiveness. R. Li et al. (2021) discussed the challenges and solutions in edible electronics, focusing on material performance in real-world applications. The study does not address the challenges of long-term energy storage in edible devices, which is crucial for their practical use. Q. Zhao et al. (2023) investigated nanotechnology for improving edible antenna performance, providing new techniques for enhancing signal transmission. The study does not explore the potential long-term effects of nanomaterial ingestion on human health, which is vital for safety. Table 1 summarizes the existing edible antennas in literature, their design specifications, applications and research findings.

Table 1. Edible Antennas and Applications

Reference	Substrate Material	Antenna Type	Application	Frequency (GHz)	Size (cm^2)	Findings
C. Chen et al. (2019)	Edible Polymer	Microstrip	Health Monitoring	2.4	-	Longevity under varying gastric conditions
M. Salama et al. (2021)	Starch-based	Microstrip	Smart Packaging	5.8	4 × 4	Scalability issues for commercial applications
J. Lee et al. (2020)	Edible Gelatin	Dipole	Digestive Monitoring	1.9	5 × 2	Electromagnetic compatibility with food types
P. Zhang et al. (2021)	Gelatin	Slot	Biomedical Monitoring	3.5	3 × 3	Robustness in digestive environments
R. Kumar et al. (2020)	Edible Polymers	Folded Dipole	Ingestible Electronics	2.4	3 × 2	Long-term effects of polymer degradation inside the body
A. Gupta et al. (2022)	Cellulose-based	Microstrip	Ingestible Electronics	1.8	4 ×4	Manufacturing process for large-scale production
Y. Wang et al. (2023)	Rice Paper, Edible Gold	Loop	Ingestible Sensing	2.4	2 × 2	Mechanical strength of rice paper in the digestive tract
K. Hassan et al. (2020)	Biodegradable Materials	Folded Dipole	Smart Packaging	0.9	3 × 2	Performance in varied environmental conditions
L. Zhang et al. (2022)	Biodegradable Materials	Compact Microstrip	Wearable Health Devices	2.45	2 × 2	Energy efficiency in long-term monitoring
H. Yang et al. (2021)	Biocompatible Material	Slot	Medical Applications	3.5	3 × 3	Mechanical durability in medical devices
G. Taylor et al. (2019)	Self-powered Edible Electronics	Microstrip	Sustainable Devices	2.4	4 × 4	Energy storage solutions in edible formats
J. Li et al. (2022)	Biodegradable Substrate	Hybrid	Biomedical Monitoring	1.8	5 × 3	Performance variation in different body fluids
L. Zhou et al. (2022)	Metamaterials	Metamaterial-based	Ingestible Wireless Systems	2.5	2 × 2	Cost and feasibility in mass production
Y. Zhao et al. (2020)	Edible Conductors	High-Performance	Biomedical Fields	2.4	3 × 3	Performance in highly acidic environments
Q. Xu et al. (2021)	Nanomaterials	Microstrip	Biomedical Devices	5.8	4 × 4	Long-term effects of nanomaterial ingestion
T. Xie et al. (2021)	Metamaterials	Metamaterial-enhanced	Food Monitoring	3.2	2 × 2	Durability in storage environments
S. Wu et al. (2021)	Metamaterials	Metamaterial-enhanced	Ingestible Applications	4.9	3 × 3	Performance during digestion
X. Sun et al. (2022)	Self-healing Materials	Self-healing Metamaterial	Ingestible Devices	2.5	3 × 3	Impact of self-healing materials on digestion
L. Yang et al. (2022)	Emerging Materials	-	Ingestible Devices	-	4 × 4	Cost-effectiveness of emerging materials

continued on following page

Table 1. Continued

Reference	Substrate Material	Antenna Type	Application	Frequency (GHz)	Size (cm^2)	Findings
Y. Zhao et al. (2023)	Metamaterials	Metamaterial-based	Edible Electronics	2.6	2 × 2	Integration into commercially viable products
H. Chen et al. (2020)	Biocompatible Material	Edible	Challenging Environments	2.4	3 × 3	Long-term stability inside the body
W. Liu et al. (2022)	Sustainable Materials	Microstrip	Edible Electronics	-	4 × 4	Recyclability of edible electronics
Z. Wang et al. (2023)	Biodegradable Materials	Optimized	Digestive Monitoring	2.4	3 × 3	Interaction with various digestive environments
A. Kumar et al. (2021)	Biodegradable Materials	Microstrip	Environmental Sustainability	1.9	4 × 4	Biodegradability in varied biological conditions
M. Johnson et al. (2020)	Edible Materials	Comparative Study	Antenna Applications	Various	5 × 3	Impact on signal strength during ingestion
T. Zhang et al. (2021)	Edible Conductors	Conductive Patch	Digestive Tracking	2.5	2 × 2	Durability during prolonged exposure to digestive fluids
L. Xu et al. (2021)	Nanomaterials	Microstrip	Performance Enhancement	3.5	3 × 3	Potential toxicity of nanomaterials
R. Patel et al. (2020)	Flexible Materials	-	Edible Antennas	-	4 × 4	Scalability for mass production
J. Wu et al. (2022)	Edible Materials	State-of-the-Art Review	Edible Antenna Technologies	-	5 × 4	Regulatory challenges for edible electronics
H. Wang et al. (2021)	Biocompatible Materials	High-Performance	Medical Monitoring	2.4	3 × 3	Degradation challenges during digestion
F. Li et al. (2022)	Hybrid Materials	Hybrid Edible	Digestive Monitoring	1.9	4 × 4	Performance in varied pH environments
Y. Zhang et al. (2021)	Emerging Materials	Design Challenges	Edible Antenna Technology	-	3 × 2	Energy efficiency for prolonged usage
X. Li et al. (2020)	Advanced Fabrication	Novel Fabrication Methods	Edible Electronics	-	5 × 4	Scalability for large-scale production
R. Zhang et al. (2022)	Metamaterials	Biomedical	Medical Applications	2.6	3 × 3	Biocompatibility concerns
J. Wang et al. (2021)	Biocompatible Materials	High-Performance	Medical Diagnostics	2.4	3 × 3	Long-term stability inside the body
Y. Sun et al. (2022)	Nanomaterials	Nanotechnology	Edible Antennas	3.8	4 × 3	Interaction with biological tissues
X. Zhao et al. (2020)	Self-powered Materials	Self-powered	Sustainable Devices	-	4 × 4	Long-term energy storage
S. Kumar et al. (2021)	Novel Materials	Emerging Materials	Edible Antennas	-	4 × 4	Large-scale production challenges
J. Yang et al. (2022)	Biocompatible Materials	Challenges in Edible Electronics	Ingestible Devices	-	5 × 3	Interaction with digestive fluids
L. Zhao et al. (2023)	Advanced Materials	Design Techniques	Edible Antennas	-	4 × 4	Designing for variable environments

continued on following page

Table 1. Continued

Reference	Substrate Material	Antenna Type	Application	Frequency (GHz)	Size (cm^2)	Findings
J. Chen et al. (2020)	Biocompatible Materials	Edible	Ingestible Electronics	2.5	3 × 3	Long-term effects of degradation
H. Li et al. (2022)	Edible Materials	Edible Antennas for Food Safety	Packaging	3.2	2 × 2	Impact of temperature variations
Q. Zhang et al. (2021)	Digestive Applications	Edible Antennas	Digestive Environments	2.4	3 × 2	Long-term performance data
T. Liu et al. (2022)	Self-Healing Materials	Self-healing Edible	Sustainable Devices	2.5	3 × 3	Efficiency of self-healing mechanism
W. Chen et al. (2021)	Flexible Materials	Flexible Edible	Wearable Devices	2.4	5 × 4	Durability under stress and deformation

Through progressive advancements in materials, designs, and technologies, edible antenna research has evolved from basic feasibility studies to addressing complex real-world challenges. Each study has contributed to a cumulative understanding, paving the way for the next generation of edible antennas that are efficient, scalable, and versatile.

3. DESIGN APPROACHES

Designing efficient edible antennas requires innovative approaches to account for the unique properties of the materials used. P. Zhang et al. (2021) introduced a compact microstrip antenna designed for wearable health devices, which could also be ingested and performed adequately within the body. A. Gupta et al. (2022) explored the use of cellulose-based materials for planar antennas, which exhibited robust performance in ingestible electronics. Y. Zhao et al. (2020) and S. Wu et al. (2021) enhanced antenna designs by incorporating metamaterials, achieving higher performance and broader bandwidths while maintaining the antennas' edibility. Several studies, including those by T. Zhang et al. (2021) and F. Li et al. (2022), focused on optimizing the geometry and structure of the antennas to maximize performance while minimizing the size of the antenna, crucial for both ingestible and wearable applications. These designs often incorporated folded or loop geometries to improve signal strength and range without compromising on material safety and biocompatibility.

Advanced simulation tools have also become integral to the design process. Researchers are leveraging electromagnetic simulation software such as CST and HFSS to refine antenna parameters like impedance matching, gain, and radiation patterns. By simulating the interaction of antennas with biological tissues, designers can predict performance and ensure safety before physical prototypes are developed. For instance, K. Liu et al. (2023) demonstrated the use of simulations to optimize antenna placement within the gastrointestinal tract, achieving consistent performance in varying environments. Material innovation is another cornerstone of design approaches. Combining biodegradable substrates with conductive edible materials, such as gold or silver leaf, has enabled antennas with higher efficiency and reduced energy losses. Hybrid designs, where materials like cellulose are paired with nanomaterials, offer a balance between flexibility, durability, and performance. These approaches not only expand the functionality of edible antennas but also ensure compliance with safety and environmental standards. Formulas related to the design of edible antennas are listed below.

Resonant Frequency

C. Chen et al. (2019) tested microstrip edible antennas designed for health monitoring at 2.4 GHz. The experimentally observed resonant frequency closely matched theoretical predictions, validating the model for edible substrates. The resonant frequency of a microstrip patch antenna, which is a common type of edible antenna, is given by the following formula (1):

$$f_r = \frac{c}{2L\sqrt{\varepsilon_r}} \tag{1}$$

- f_r: Resonant frequency (Hz)
- c: Speed of light in a vacuum (3×10^8m/s)
- L: Effective length of the antenna patch (m)
- ε_r: Relative permittivity of the substrate material (dimensionless)

Effective Dielectric Constant (ε_{eff})

The effective dielectric constant of the substrate material (which in edible antennas could be materials like gelatin, cellulose, etc.) can be calculated using (2):

$$\varepsilon_{eff} = \frac{\varepsilon_r + 1}{2} + \frac{\varepsilon_r - 1}{2}\left(1 + 12\frac{h}{W}\right)^{-0.5} \tag{2}$$

- ε_{eff}: Effective dielectric constant (dimensionless)
- ε_r: Relative permittivity of the substrate (dimensionless)
- h: Height of the substrate (m)
- W: Width of the patch (m)

Input Impedance

The input impedance (Z_{in}) at the resonant frequency for a rectangular microstrip patch antenna is typically estimated as in (3):

$$Z_{in} = 90\left(\frac{W}{L}\right)^2 \tag{3}$$

- Z_{in}: Input impedance (Ω)
- W: Width of the patch (m)
- L: Length of the patch (m)

Antenna Gain

The gain of an antenna is an essential parameter that measures the directionality of the antenna. It is related to the directivity and efficiency of the antenna as given in (4):

$$G = D \times \eta \tag{4}$$

- G: Gain of the antenna (dimensionless or dB)
- D: Directivity of the antenna (dimensionless or dB)
- η: Efficiency of the antenna (dimensionless)

Antenna Efficiency

Antenna efficiency (η) is given by (5):

$$\eta = \frac{P_{rad}}{P_{in}} \tag{5}$$

- η: Efficiency (dimensionless)
- P_{rad}: Radiated power (W)
- P_{in}: Input power (W)

Radiation Pattern

The radiation pattern is described by the electric and magnetic field distribution in space. For a simple dipole antenna, the radiation pattern can be expressed as in (6):

$$E(\theta) = E_0 \frac{\sin(\pi \cos\theta)}{\cos\theta} \tag{6}$$

- $E(\theta)$: Electric field strength as a function of the angle θ
- E_0: Maximum electric field strength
- θ: Angle from the axis of the antenna

Return Loss (S_{11}) Formula

Return loss, also known as the reflection coefficient, measures how much power is reflected due to impedance mismatch. It is given by (7):

$$S_{11} = 20 \log\left|\frac{Z_{in} - Z_0}{Z_{in} + Z_0}\right| \tag{7}$$

- S_{11}: Return loss (dB)
- Z_{in}: Input impedance of the antenna (Ω)
- Z_0: Characteristic impedance (usually 50 Ω for most systems)

Specific Absorption Rate (SAR)

SAR is a measure of the rate at which energy is absorbed by human tissue when exposed to electromagnetic fields. H. Chen et al. (2020) evaluated edible antennas in challenging environments, measuring SAR to ensure safe operation close to human tissues and can be calculated using (8).

$$SAR = \frac{\sigma |E|^2}{\rho} \tag{8}$$

- σ: Electrical conductivity of the tissue (S/m)
- E: Electric field strength (V/m)
- ρ: Density of the tissue (kg/m^3)

Friis Transmission Equation

The power received by an edible antenna is calculated using (9):

$$P_r = P_t \frac{G_t G_r \lambda^2}{(4\pi R)^2} \tag{9}$$

- P_r: Power received (W)
- P_t: Power transmitted (W)
- λ: Wavelength (m)
- R: Distance between antennas (m)
- G_t: Gain of the transmitter (dimensionless)
- G_r: Gain of the receiver (dimensionless)

Metamaterial Design Equations

L. Zhou et al. (2022) integrated metamaterials into edible antennas, demonstrating significant improvements in gain and signal directivity. In metamaterial-enhanced edible antennas, design equations are often based on effective medium parameters like permittivity (ε) and permeability (μ) and are given in (10):

$$n_{eff} = \sqrt{\varepsilon_{eff} \cdot \mu_{eff}} \tag{10}$$

- n_{eff}: Effective refractive index of the metamaterial
- ε_{eff}: Effective permittivity
- μ_{eff}: Effective permeability

Bandwidth Formula

The bandwidth of a microstrip patch antenna can be approximated using the following equation (11):

$$BW = \frac{f_H - f_L}{f_r} \times 100\% \tag{11}$$

- BW: Bandwidth (percentage)
- f_H: Upper frequency limit (Hz)
- f_L: Lower frequency limit (Hz)
- f_r: Resonant frequency (Hz)

For a broader bandwidth, edible antennas are often designed with low dielectric constant substrates (e.g., gelatin or cellulose-based materials).

Quality Factor (Q-factor)

The Q-factor is a measure of the antenna's bandwidth relative to its center frequency and can be defined using (12):

$$Q = \frac{f_r}{BW} \tag{12}$$

- Q: Quality factor (dimensionless)
- f_r: Resonant frequency (Hz)
- BW: Bandwidth (Hz)

A higher Q-factor indicates a narrower bandwidth, which is typically undesirable for broadband applications like health monitoring or ingestible sensors.

Polarization

The polarization of an antenna describes the orientation of the electric field. Circular polarization (often used in wearable and ingestible antennas) is typically achieved when two orthogonal modes are excited with a 90-degree phase difference. The axial ratio (AR) measures the degree of circular polarization and can be calculated using (13):

$$AR = \frac{E_x}{E_y} \tag{13}$$

- AR: Axial ratio (dimensionless)
- E_x: Electric field component in the x-direction (V/m)
- E_y: Electric field component in the y-direction (V/m)

For circular polarization, the axial ratio is close to 1 (or 0 dB).

Mutual Coupling

In array configurations of edible antennas (for example, for body-worn sensors), mutual coupling between elements is an important factor that affects performance. The mutual coupling between two antennas can be described using the S-parameters (14):

$$S_{21} = \frac{V_{out,2}}{V_{in,1}} \tag{14}$$

- S_{21}: Transmission coefficient between antenna 1 and antenna 2
- $V_{out,2}$: Voltage at the output of antenna 2 (V)
- $V_{in,1}$: Voltage input to antenna 1 (V)

Lower mutual coupling values are desirable for better isolation between antenna elements.

Patch Antenna Length for a Given Resonant Frequency

The length of the patch (L) for a rectangular microstrip antenna (used commonly in edible antennas) can be determined by (15):

$$L = \frac{c}{2f_r\sqrt{\varepsilon_{eff}}} \tag{15}$$

- L: Length of the patch (m)
- c: Speed of light (m/s)
- f_r: Resonant frequency (Hz)
- ε_{eff}: Effective dielectric constant of the substrate (dimensionless)

This equation helps design antennas to operate at specific frequencies, such as 2.4 GHz for health monitoring.

Radiation Efficiency

Radiation efficiency (η_r) represents the ratio of the power radiated by the antenna to the total input power is calculated using (16):

$$\eta_r = \frac{P_{rad}}{P_{input}} \tag{16}$$

- η_r: Radiation efficiency (dimensionless)
- P_{rad}: Radiated power (W)
- P_{input}: Total input power (W)

High radiation efficiency is necessary for effective operation, especially for low-power devices such as ingestible antennas.

Surface Wave Loss

Surface waves can degrade the performance of antennas, particularly microstrip antennas, by reducing efficiency. The surface wave power loss ($P_{surface}$) can be calculated as in (17):

$$P_{surface} = \frac{P_{input} \cdot \varepsilon_r \cdot h}{W \cdot \lambda} \quad (17)$$

- $P_{surface}$: Power lost to surface waves (W)
- P_{input}: Input power (W)
- ε_r: Relative permittivity of the substrate
- h: Height of the substrate (m)
- W: Width of the patch (m)
- λ: Wavelength (m)

SAR Reduction with Metamaterials

The introduction of metamaterials can reduce SAR by focusing the electromagnetic fields away from sensitive tissues. The effective SAR reduction factor can be defined using (18):

$$SAR_{reduced} = SAR_{initial} \cdot \frac{\mu_{eff}}{\varepsilon_{eff}} \quad (18)$$

- $SAR_{reduced}$: Reduced SAR due to metamaterial (W/kg)
- $SAR_{initial}$: Initial SAR (W/kg)
- μ_{eff}: Effective permeability of the metamaterial
- ε_{eff}: Effective permittivity of the metamaterial

This is particularly useful for edible antennas intended for biomedical applications to ensure safety and compliance with SAR limits.

Reflection Coefficient

The reflection coefficient (Γ) is used to evaluate the impedance mismatch between the antenna and the transmission line. It is calculated using (19):

$$\Gamma = \frac{Z_{in} - Z_0}{Z_{in} + Z_0} \quad (19)$$

- Γ: Reflection coefficient (dimensionless)
- Z_{in}: Input impedance (Ω)
- Z_0: Characteristic impedance of the transmission line (Ω)

4. MATERIALS USED IN EDIBLE ANTENNAS

Various materials have been explored for the design of edible antennas, with a focus on biocompatibility and performance and are summarized in Table 2. Chen et al. (2019) introduced microstrip edible antennas made from biocompatible materials such as gelatin and starch for health monitoring, emphasizing the need for safe, digestible electronics. Salama et al. (2021) demonstrated starch-based antennas for smart packaging, proving that even everyday food ingredients can be used to create efficient antennas. J. Lee et al. (2020) developed dipole antennas made from edible substances such as rice paper and silver leaf, which are compatible with digestive monitoring applications. R. Kumar et al. (2020) and T. Xie et al. (2021) investigated the use of nanomaterials and metamaterials, respectively, to improve the performance of edible antennas, ensuring high efficiency despite the limitations imposed by the materials. Metals like edible gold offer high efficiency but pose biocompatibility concerns. Natural materials like gelatin and starch are safer but less efficient in terms of conductivity. Starch and cellulose are cost-effective but require optimization for efficiency. Advanced materials like nanomaterials and edible metals are expensive but provide superior performance. Gelatin and cellulose have the potential for scalability but may face challenges in achieving consistent performance under harsh conditions. Emerging materials like chitosan and nanomaterials are less scalable due to processing complexities.

Table 2. Performance metrics comparison of various materials in edible antenna design

Material	**Performance Metrics**			**Challenges**	**Example Studies**
	Efficiency	**Biocompatibility**	**Digestibility**		
Gelatin	**Moderate to high**	**Excellent**	**Excellent**	**• Susceptible to degradation in moist environments** **• Limited mechanical strength**	**J. Lee et al. (2020) utilized for digestive monitoring, effective at 1.9 GHz**
Starch	Moderate	High	Excellent	• Lower conductivity than metals • Challenges with scalability for mass production	M. Salama et al. (2021) used for **designing** smart packaging at 5.8 GHz
Cellulose	Moderate	High	Good	• Requires advanced processing techniques for scalability	A. Gupta et al. (2022) **used cellulose material in ingestible electronics, operating at 1.8 GHz**
Rice Paper	Low to Moderate	High	Excellent	• Limited mechanical strength in digestive environments	Y. Wang et al. (2023) applied for ingestible sensing at 2.4 GHz
Chitosan	Moderate	Excellent	Good	• Brittle and requires reinforcement for durability	H. Chen et al. (2020) investigated for food safety monitoring
Edible Gold / Silver Leaf	High	Good	Moderate	• Expensive • Potential for limited biocompatibility in high quantities	Y. Zhao et al. (2020) **demonstrated high performance in biomedical fields**
Nanomaterials	Very High	Variable	Low to Moderate	• Safety concerns regarding long-term ingestion effects Expensive	Q. Xu et al. (2021) examined nanomaterials for high-performance antennas.

The health-related applications of edible antennas, such as biomedical monitoring and diagnostic systems, demand a high degree of biocompatibility. Biocompatibility ensures that the materials used in these antennas interact safely with human tissues, posing no harm during ingestion or prolonged contact. Edible antennas are designed to operate within the human body, often in the gastrointestinal tract. Materials must be non-toxic, non-irritating, and capable of degrading without producing harmful byproducts. The properties of materials used in edible antennas directly influence their biocompatibility and are summarized in Table 3.

Table 3. Material properties and their impact on human safety

Material	Biocompatibility Features	Potential Concerns
Gelatin	Naturally derived, easily digestible, and widely accepted in medical devices.	Rapid degradation may reduce antenna lifespan.
Starch	Biodegradable and non-toxic, with a long history of safe use in food products.	Limited mechanical strength in moist environments.
Cellulose	Highly biocompatible and sustainable; does not produce harmful residues.	Requires complex processing to ensure uniform performance.
Edible Gold / Silver Leaf	High conductivity and safe in small quantities; used in food decorations.	Potential toxicity in large doses; higher cost.
Nanomaterials	Enhanced mechanical and electrical properties; emerging applications in biomedicine.	The long-term effects of ingestion are not well-studied.
Chitosan	Antimicrobial and biocompatible, derived from natural sources like shellfish.	May cause allergies in individuals sensitive to shellfish.

Performance metrics w.r.t radiation characteristics for edible antennas, such as gain, bandwidth, and operating frequency, vary depending on the materials and designs used. Y. Wang et al. (2023) and H. Xu et al. (2021) reported that metamaterial-based antennas demonstrated higher performance than traditional designs due to their ability to manipulate electromagnetic waves more effectively. G. Taylor et al. (2019) introduced self-powered edible antennas, which leveraged energy harvesting technologies to operate in environments with limited external power sources. M. Johnson et al. (2020) and T. Zhang et al. (2021) compared different edible materials' performances, concluding that conductive materials like gold leaf and edible silver provided superior conductivity and performance, while more common materials such as starch and cellulose were less efficient but more biocompatible. X. Sun et al. (2022) explored self-healing antennas that could recover from damage within the digestive tract, a vital feature for medical monitoring applications.

5. APPLICATIONS

The primary applications of edible antennas include healthcare, environmental monitoring, and smart packaging. S. Wu et al. (2021) demonstrated the use of edible antennas for biomedical monitoring, while Y. Wang et al. (2023) and H. Xu et al. (2021) explored their applications in food safety and environmental monitoring, such as detecting contamination or monitoring freshness in real time. Y. Zhao et al. (2020) and Y. Sun et al. (2022) highlighted future applications of edible antennas in smart food packaging, where they could be used to track expiration dates or signal spoilage. Ensuring low SAR values is critical to prevent tissue heating from electromagnetic radiation. H. Chen et al. (2020) demonstrated how SAR compliance could enhance safety in biomedical applications. Incorporating antennas into packaging can track temperature, humidity, and freshness, as demonstrated by M. Salama et al. (2021) using starch-based antennas for smart packaging. Edible sensors can signal bacterial growth or chemical contamination, providing real-time updates during storage or transportation. As the field advances, ensuring biocompatibility will remain central to the development of edible antennas. Materials research must balance performance and safety, with rigorous testing to confirm their suitability for ingestible applications. By prioritizing biocompatibility, edible antennas can reliably serve in critical healthcare roles, enhancing their adoption and trustworthiness in medical and consumer markets.

Healthcare applications have expanded beyond basic monitoring to include advanced diagnostic and therapeutic capabilities. For instance, ingestible antennas can now provide real-time imaging of the gastrointestinal tract, enabling detailed analysis of abnormalities such as ulcers or blockages. This advancement significantly enhances diagnostic precision and complements traditional endoscopic methods. Additionally, edible antennas integrated with drug delivery systems are being developed to release medication at specific sites within the body, guided by feedback from antenna-based sensors. This dual functionality not only improves treatment efficacy but also reduces side effects by targeting specific areas.

In environmental monitoring, the role of edible antennas is becoming increasingly diverse. Advanced designs are being used to monitor air quality by detecting pollutants such as carbon monoxide and sulfur dioxide in real time. These antennas, embedded in biodegradable carriers, can be deployed across large areas, providing cost-effective and eco-friendly solutions for urban air quality management. Similarly, in water quality monitoring, antennas capable of detecting heavy metals or microbial contamination are being developed, offering invaluable tools for ensuring public health and environmental protection. Biodegradable antennas embedded in waste products can help track decomposition rates and environmental impact. W. Liu et al. (2022) highlighted the potential of sustainable materials for edible electronics in monitoring environmental variables.

The applications of edible antennas in smart packaging are also evolving rapidly. Beyond tracking freshness and spoilage, these antennas are now being used to monitor nutritional content dynamically. For example, antennas embedded in food packaging can provide real-time updates on calorie content, protein levels, or potential allergens as the food degrades over time. Such features cater to the growing consumer demand for personalized and transparent nutritional information, transforming how food products are packaged and consumed. Networks of edible antennas can also form distributed sensing systems for large-scale environmental monitoring.

The integration of edible antennas with blockchain technology is a burgeoning area of research. By combining antennas with blockchain, it becomes possible to create tamper-proof records of food safety and supply chain data. This integration ensures that information such as storage conditions, transportation history, and expiration dates remains secure and accessible, significantly enhancing traceability and consumer trust.

With the advent of 5G/6G technologies, edible antennas can become integral components of interconnected systems. Data from edible antennas can be transmitted to IoT platforms, providing instant updates on product status. Y. Wang et al. (2023) demonstrated the integration of rice paper-based antennas into IoT systems for ingestible and packaging applications. These systems provide the computational and storage capabilities necessary to process large volumes of data generated by IoMT devices. The integration of edible antennas with cloud-based systems can further enhance healthcare services by enabling real-time monitoring and data analysis, as highlighted by A. Kumar et al. (2024).

Future edible antennas may also contribute to wildlife conservation efforts. Biodegradable antennas can be used in ecological studies to track the movement and behavior of animals without causing harm to their habitats. For instance, antennas designed to dissolve safely in the environment can monitor temperature or pH changes in nesting sites, aiding conservationists in protecting vulnerable species. Additionally, multi-functional antennas are being developed to serve diverse applications simultaneously, such as combining environmental sensing with healthcare diagnostics, showcasing the versatility of edible antenna technology.

6. 5G/6G PERSPECTIVES IN EDIBLE ANTENNA DESIGN

Linking edible antennas to 5G and 6G technology opens up new possibilities in next-generation communication networks, enhancing the capabilities of ingestible and biodegradable antennas in healthcare, the Internet of Things (IoT), and environmental monitoring. The integration of these antennas with 5G/6G networks offers significant advancements in data rates, latency, miniaturization, and interconnectivity.

6.1. High Data Rates and Low Latency

One of the most important features of 5G and 6G networks is their ability to provide ultra-fast data rates and extremely low latency, which are essential for real-time applications. Edible antennas can leverage these advancements in a variety of ways. For instance, they can be utilized for continuous health monitoring, tracking parameters like pH levels, glucose levels, gastrointestinal tract analysis, and drug delivery. With the seamless real-time data transmission capabilities of 5G/6G, healthcare providers can receive immediate updates, allowing for instantaneous diagnoses and remote health management. In critical situations, such as when an ingestible antenna detects abnormal pH levels or other health anomalies, 5G/6G networks can instantly alert medical professionals for timely intervention. Additionally, these networks support the high data throughput necessary for advanced biomedical imaging, enabling ingestible antennas to transmit high-resolution, real-time images of the gastrointestinal tract. This capability greatly improves the diagnostic accuracy of endoscopic procedures, offering medical professionals a more comprehensive view for better decision-making.

6.2. IoT and Interconnected Devices

IoMT is poised for significant growth with the rollout of 5G and 6G networks, extending beyond wearable sensors to include ingestible and implantable devices such as edible antennas. These antennas can seamlessly integrate with broader IoT ecosystems to provide continuous health monitoring. A network of ingestible sensors connected via 5G/6G would enable real-time tracking of a patient's health while interacting with other smart devices, such as insulin pumps, pacemakers, and smart pills. This integration allows for automated treatment adjustments based on sensor inputs, creating an interconnected medical ecosystem that can respond dynamically to a patient's needs.

6.3. Miniaturization and Advanced Materials

The advent of 5G and 6G technology encourages the development of miniaturized, flexible, and energy-efficient antennas. This trend benefits the design of edible antennas, which can be optimized to function within the higher frequency ranges of 5G and even 6G. Using metamaterials and novel biocompatible materials, these antennas can achieve high performance despite the challenges posed by signal degradation within the human body. For example, metamaterials can be employed to make edible antennas operate efficiently at millimeter-wave frequencies, such as those used in 5G (24 GHz to 40 GHz) and the terahertz frequencies envisioned for 6G. Metamaterials also grant these antennas properties like negative refractive indices and enhanced directivity, critical for overcoming the signal loss experienced in biological environments. Moreover, edible antennas in 5G/6G systems can take advantage of power efficiency techniques, such as energy harvesting. By absorbing and utilizing ambient energy from nearby devices or the environment, these antennas eliminate the need for internal batteries, making them safer and more sustainable for long-term use in ingestible applications. This is especially beneficial for applications requiring extended health monitoring and surveillance.

6.4. Massive Machine-Type Communication (mMTC)

5G and 6G networks also support massive machine-type communication (mMTC), allowing a large number of devices to communicate simultaneously. This capability is crucial when deploying multiple edible antennas either within a single patient or across a broad population. For example, in large-scale health monitoring efforts, edible antennas could be ingested by many individuals to track diseases, monitor nutrition levels, or detect food contaminants. 5G/6G networks provide the infrastructure needed to handle the enormous volume of data generated by such dense networks of devices. This enables real-time analysis of health data on a population level, which could be applied in epidemiological studies or food safety monitoring. The ability to manage such large-scale data transmission effectively transforms how public health is monitored and maintained.

7. CHALLENGES IN THE DESIGN AND IMPLEMENTATION

Edible antennas represent an exciting frontier in technology but face several challenges due to the unique requirements of their applications. Edible materials like gelatin, starch, and cellulose are inherently less conductive than traditional antenna materials, reducing efficiency. Mechanical properties such as flexibility, strength, and durability vary significantly, affecting the reliability of antennas in digestive environments or during ingestion. Degradation in moist or acidic environments, like the human stomach, can lead to reduced performance over time. When considering the manufacturing constraints, large-scale production of edible antennas is limited by the need for precision and uniformity in material deposition and assembly processes. Compatibility of biocompatible materials with traditional fabrication techniques, such as lithography or 3D printing, is often limited. The cost of advanced edible materials, such as nanomaterials or edible conductors, can hinder widespread adoption. On considering environmental interactions, challenges lie when signal attenuation occurs due to interaction with biological tissues, especially in highly acidic or moist conditions. Limited lifespan and performance reliability in environments like the gastrointestinal tract or during food packaging. Ensuring compliance with safety standards for materials intended for ingestion or direct contact with human tissues is also essential. Long-term health impacts of ingesting antennas made from nanomaterials or advanced polymers remain underexplored.

Antenna researchers have attempted to overcome the above-mentioned challenges in the design and implementation of edible antennas. Researchers like Y. Zhao et al. (2020) have incorporated edible metals such as gold or silver leaf, which offer superior conductivity while maintaining safety. Studies by L. Zhou et al. (2022) have used metamaterials to enhance the efficiency and miniaturization of edible antennas, compensating for the limitations of base materials. X. Sun et al. (2022) introduced self-healing edible antennas that can recover functionality after mechanical damage, addressing durability concerns. R. Patel et al. (2020) explored advanced 3D printing and extrusion methods to achieve uniform and scalable antenna designs. J. Li et al. (2022) proposed hybrid antennas that combine biodegradable substrates with efficient conductors to balance cost and performance.W. Liu et al. (2022) emphasized using low-cost, abundant materials like starch or cellulose, combined with simpler production techniques.

T. Xie et al. (2021) demonstrated the use of metamaterials to focus electromagnetic waves, reducing losses in complex environments. Researchers have investigated edible coatings to shield antennas from degradation, enhancing their operational lifespan. Geometry optimizations, such as folded or loop designs by Y. Wang et al. (2023), improve performance despite environmental challenges. Q. Xu et al. (2021) conducted studies on the biocompatibility of nanomaterials to address safety concerns. Ensuring low Specific Absorption Rate (SAR) values, as highlighted by H. Chen et al. (2020), to minimize adverse effects on human tissues. Efforts to align antenna designs with FDA and similar regulatory standards are being explored.

8. FUTURE DIRECTIONS IN EDIBLE ANTENNA TECHNOLOGY

The future of edible antennas is rich with possibilities. The field is poised for significant advancements driven by emerging materials, innovative design approaches, and integration with next-generation technologies. These developments promise to address existing challenges, expand functionality, and open new frontiers in applications. Emerging materials are expected to redefine the capabilities of edible antennas by improving their efficiency, durability, and safety. Metamaterials offer the potential to enhance signal strength, directivity, and miniaturization while maintaining biocompatibility. L. Zhou et al. (2022) demonstrated their application in ingestible antennas to overcome signal attenuation in body tissues. Customizable properties like negative refractive indices can optimize performance for specific applications, such as biomedical imaging or environmental sensing. Incorporating nanomaterials such as graphene and carbon nanotubes can significantly improve conductivity and flexibility. Q. Xu et al. (2021) emphasized their role in achieving high-frequency operation and greater efficiency in biomedical applications. Self-healing edible antennas, as explored by X. Sun et al. (2022), can recover from physical damage, ensuring longer operational lifespans in harsh environments like the gastrointestinal tract.

The adoption of 5G/6G networks will significantly expand the scope and functionality of edible antennas. These advanced networks enable higher data rates, lower latency, and seamless connectivity, making it possible to integrate edible antennas into complex IoT ecosystems. For instance, edible antennas paired with real-time AI analytics could revolutionize applications in healthcare by providing instant diagnostic feedback or triggering automated therapeutic responses. Similarly, these networks can enhance environmental monitoring by enabling distributed sensing systems that aggregate data from biodegradable antennas deployed across large areas. Energy efficiency will remain a central focus in future edible antenna designs. Energy harvesting technologies, such as those using RF or thermal energy, can create self-sustaining systems, eliminating the need for traditional power sources. Researchers are exploring antennas capable of scavenging energy from ambient sources, as demonstrated by G. Taylor et al. (2019), to ensure uninterrupted functionality in diverse applications. This is particularly crucial for ingestible antennas, where internal batteries pose safety risks and environmental concerns.

Multi-functional edible antennas are another exciting avenue for future exploration. These designs aim to combine several functionalities into a single device. For instance, a single antenna could perform health monitoring, environmental sensing, and drug delivery simultaneously. Advances in material science and miniaturization techniques will make it possible to integrate sensors, actuators, and communication modules into compact, ingestible systems. Such multi-functional antennas could revolutionize personalized medicine, enabling precise, patient-specific treatments. Integration with blockchain technology offers unique opportunities for data security and traceability. Blockchain-based systems can store and validate data collected by edible antennas, ensuring its authenticity and preventing tampering. This application is particularly relevant in food safety and supply chain management, where secure, real-time monitoring is essential. For example, edible antennas embedded in food products could continuously update blockchain records with information about storage conditions, transportation history, and expiration dates, enhancing consumer trust and safety.

Incorporating AI and machine learning will further enhance the capabilities of edible antennas. These technologies can analyze data collected by antennas to identify patterns, predict outcomes, and make real-time decisions. For example, in healthcare, AI algorithms could process physiological data from antennas to detect early signs of diseases or optimize drug delivery based on real-time feedback. In environmental applications, machine learning models could analyze data from distributed antenna networks to predict pollution trends or identify areas requiring intervention.

Future designs will also focus on environmental sustainability. Researchers are increasingly prioritizing biodegradable and eco-friendly materials to reduce electronic waste. Antennas made from plant-based polymers or bio-inspired materials could degrade harmlessly after use, minimizing their environmental footprint. Additionally, these sustainable designs can support conservation efforts, such as tracking wildlife or monitoring ecosystems without causing ecological disruption.The ultimate goal for edible antenna technology is to achieve seamless integration with human and environmental systems, enabling real-time monitoring, data transmission, and analysis. By combining cutting-edge materials with innovative designs, edible antennas can transcend their current limitations and emerge as transformative tools in fields like healthcare, food safety, and IoT-driven ecosystems. Continued innovation will not only address current challenges but also unlock new applications, making edible antennas an indispensable part of modern technology.

9. CONCLUSION

The review of edible antennas across diverse applications—ranging from healthcare to environmental monitoring and smart packaging—highlights the tremendous potential of this emerging technology. The studies show that edible antennas can be integrated into a variety of applications, offering unique advantages such as biodegradability, biocompatibility, and sustainability. By utilizing materials such as starch, cellulose, rice paper, and biodegradable polymers, these antennas represent a crucial step towards more eco-friendly and non-invasive solutions in medical diagnostics, food safety, and real-time monitoring. Despite the exciting advancements, several significant research gaps remain. Key issues include the long-term stability of antennas in varying biological environments, particularly within the human digestive system, as well as the effects of environmental factors like moisture and temperature on the performance of antennas in food packaging. Challenges in large-scale manufacturing, energy efficiency, and regulatory frameworks also present roadblocks to the widespread adoption of edible antennas. The reviewed studies often propose novel designs but fail to adequately address these barriers to practical application, particularly in terms of scalability and commercial viability.

REFERENCES

Chen, C. (2019, December). Microstrip Edible Antennas for Health Monitoring. *IEEE Sensors Journal*, 19(23), 10567–10576.

Chen, H. (2020, December). Biocompatibility and Performance of Edible Antennas in Challenging Environments. *Journal of Biomedical Engineering*, 53(4), 567–575.

Chen, H. (2023, June). Edible Antennas Based on Chitosan for Food Safety Monitoring. *IEEE Transactions on Antennas and Propagation*, 70(6), 3582–3590.

Chen, J. (2020, June). Biocompatibility and Design Strategies for Edible Electronics. *Biomedical Engineering Letters*, 10(2), 241–250. PMID: 32431954

Gupta, A. (2022, May). Cellulose-based Planar Antennas for Ingestible Electronics. *IEEE Transactions on Antennas and Propagation*, 70(5), 1122–1130.

Hassan, K. (2020, July). Folded Dipole Edible Antennas for Smart Packaging Applications. *Journal of Food Engineering*, 292, 110310.

Johnson, M. (2020, November). Comparative Study of Edible Materials for Antenna Applications. *IEEE Transactions on Antennas and Propagation*, 68(11), 5314–5323.

Kumar, A. (2021, March). Biodegradable Materials for Edible Electronics. *Nature Sustainability*, 4(3), 166–175.

Kumar, A., "Examining healthcare services utilizing cloud technology in intelligent urban environments," *In Advances in healthcare information systems and administration book series,* pp. 77–98, Oct 2024.

Kumar, R. (2020, June). Edible Polymers in Antenna Design: Challenges and Solutions. *Advanced Functional Materials*, 30(23), 2002201.

Kumar, S. (2021, February). Novel Materials for Edible Antenna Applications. *Journal of Materials Chemistry. C, Materials for Optical and Electronic Devices*, 9(6), 1470–1481.

Lee, J. (2020, September). Edible Dipole Antennas for Digestive Monitoring. *Applied Physics Letters*, 117(12), 123502.

Lee, S. (2023, August). Edible Antennas Using Carbohydrate Materials for Biodegradable Wireless Communication Systems. *IEEE Sensors Journal*, 22(8), 6745–6752.

Li, F. (2022, March). Hybrid Edible Antennas: Combining Materials for Enhanced Performance. *Sensors and Actuators. B, Chemical*, 352, 130774.

Li, H. (2022, September). Edible Antennas for Food Safety and Environmental Monitoring. *IEEE Transactions on Antennas and Propagation*, 70(9), 2905–2915.

Li, J. (2022). Hybrid Edible Antenna Design Using Biodegradable Substrates and Conductors. *IEEE Access : Practical Innovations, Open Solutions*, 10, 30432–30443.

Li, X. (2020, November). Advanced Fabrication Methods for Edible Electronics. *Journal of Micromechanics and Microengineering*, 30(11), 115002.

Liu, W. (2022, January). Sustainable Design Strategies for Edible Electronics. *Environmental Science & Technology*, 56(2), 1134–1143.

Liu, X. (2022, July). High Efficiency Edible Antennas for Biomedical Applications. *IEEE Transactions on Antennas and Propagation*, 70(7), 2486–2495.

Patel, M. (2023). Advanced Ingestible Antenna Designs for Smart Healthcare Applications. *IEEE Access : Practical Innovations, Open Solutions*, 11, 20657–20667.

Patel, R. (2020, September). Fabrication Techniques for Edible Antennas. *IEEE Transactions on Microwave Theory and Techniques*, 68(9), 3525–3534.

Salama, M. (2021, April). Starch-based Edible Patch Antennas for Smart Packaging. *Sensors and Actuators. B, Chemical*, 331, 129–138.

Sun, X. (2022, July). Self-healing Edible Antennas with Metamaterial Integration. *Advanced Functional Materials*, 32(18), 2110493.

Sun, Y. (2022, March). Nanotechnology in Edible Antenna Design: Recent Advances. *Nano Energy*, 85, 106018.

Taylor, G. (2019, June). Self-powered Edible Electronics: Toward Sustainable Ingestible Devices. *IEEE Sensors Journal*, 19(12), 4180–4189.

Wang, H. (2021, August). Performance Metrics of Edible Antennas in Real-world Applications. *IEEE Transactions on Antennas and Propagation*, 70(8), 3769–3780.

Wang, J. (2021, February). High Performance Edible Antennas for Medical Monitoring. *Journal of Biomedical Materials Research*, 109(2), 345–355.

Wang, T. (2023, April). Flexible and Edible Antennas for Nutritional Tracking. *IEEE Transactions on Biomedical Circuits and Systems*, 17(4), 542–552. PMID: 37318976

Wang, Y. (2022, June). Advanced Edible Antenna Designs for Smart Food Packaging. *Sensors and Actuators. A, Physical*, 319, 112447.

Wang, Y. (2023, August). Loop Antennas Made from Rice Paper and Edible Gold for Ingestible Sensing. *IEEE Antennas & Propagation Magazine*, 65(4), 54–63.

Wang, Z. (2023, May). Optimization Techniques for Edible Antenna Designs. *IEEE Transactions on Antennas and Propagation*, 71(5), 2231–2240.

Wu, J. (2022). Edible Antenna Technologies: State-of-the-Art and Future Perspectives. *IEEE Access : Practical Innovations, Open Solutions*, 10, 43589–43601.

Wu, S. (2021, October). Metamaterial-enhanced Edible Antennas: Performance and Applications. *IEEE Transactions on Antennas and Propagation*, 69(10), 7420–7429.

Xie, T. (2021, November). Metamaterial-enhanced Antennas for Food Monitoring Applications. *IEEE Transactions on Microwave Theory and Techniques*, 69(11), 4555–4565.

Xu, H. (2021, August). The Role of Metamaterials in Edible Antennas. *IEEE Transactions on Microwave Theory and Techniques*, 69(8), 3395–3404.

Xu, L. (2021, October). Advanced Nanomaterials for High-performance Edible Antennas. *Nano Letters*, 21(10), 4195–4203.

Xu, Q. (2021, April). Nanomaterials in Edible Antenna Design: Enhancements and Challenges. *Advanced Materials*, 33(15), 2007372. PMID: 33694214

Yang, H. (2021, August). Edible Slot Antennas for Medical Applications. *Nature Biomedical Engineering*, 5(8), 723–732. PMID: 34131325

Yang, J. (2022, November). Challenges and Solutions in Edible Electronics. *IEEE Transactions on Electron Devices*, 69(11), 4960–4970.

Yang, L. (2022). Future Directions in Edible Antenna Technology: Materials and Manufacturing. *IEEE Reviews in Biomedical Engineering*, 15, 235–248.

Yang, W. (2021). Innovative Designs for Edible Antennas: A Review. *IEEE Reviews in Biomedical Engineering*, 14, 273–287.

Zhang, L. (2022, October). Compact Microstrip Antennas for Wearable Health Devices. *IEEE Transactions on Microwave Theory and Techniques*, 70(10), 3756–3766.

Zhang, P. (2021, July). Gelatin-based Slot Antennas for Biomedical Monitoring. *IEEE Transactions on Biomedical Engineering*, 68(7), 1820–1830.

Zhang, Q. (2021, May). Performance Analysis of Edible Antennas in Digestive Environments. *IEEE Transactions on Biomedical Engineering*, 68(5), 1301–1311. PMID: 32991277

Zhang, R. (2022, December). Metamaterial-Based Edible Antennas for Biomedical Applications. *IEEE Transactions on Biomedical Engineering*, 69(12), 3456–3464. PMID: 35622786

Zhang, S. (2021, May). High-Performance Edible Antennas for Environmental Sensing. *IEEE Sensors Journal*, 21(10), 10845–10854.

Zhang, T. (2021, March). Edible Conductors and Their Performance in Antenna Design. *Journal of Materials Science*, 56(6), 4556–4568.

Zhang, Y. (2021, August). Design Challenges in Edible Antenna Technology. *Advanced Functional Materials*, 31(29), 2104521.

Zhao, J. (2020, September). Recent Advances in Edible Electronics. *Advanced Functional Materials*, 31(34), 2101180.

Zhao, L. (2023, April). Advanced Design Techniques for Edible Antennas. *IEEE Transactions on Microwave Theory and Techniques*, 71(4), 1970–1980.

Zhao, R. (2022, June). Optimization Techniques for Performance Improvement of Edible Antennas. *IEEE Transactions on Electron Devices*, 69(6), 2345–2353.

Zhao, X. (2020, January). Self-Powered Edible Antennas: A Review. *IEEE Sensors Journal*, 20(2), 567–577.

Zhao, Y. (2020, December). High-performance Edible Antennas: Design and Application. *IEEE Transactions on Microwave Theory and Techniques*, 68(12), 5256–5267.

Zhao, Y. (2023, March). Advanced Metamaterials for Edible Electronics. *IEEE Transactions on Electron Devices*, 70(3), 1234–1242.

Zhou, L. (2022, August). Metamaterial-based Edible Antennas: A New Frontier in Ingestible Wireless Systems. *Nature Communications*, 13(1), 3498. PMID: 35715397

Chapter 14
Li-Fi for Next Generation Wireless Communication: Technical Insights and Innovations

Dipti Chauhan
https://orcid.org/0000-0003-1665-7587
Prestige Institute of Engineering Management and Research, Indore, India

Jay Kumar Jain
https://orcid.org/0000-0002-9590-0006
Maulana Azad National Institute of Technology, Bhopal, India

ABSTRACT

Li-Fi technology uses visible light communication for data transmission and has the potential to transform wireless communication. The benefits offered by this technology have the potential to transform wireless communication, many industries, and improve communication in both indoor and outdoor environment. It provides significantly faster speeds and has better security standards compared to Wi-Fi. In this chapter, the authors discuss significant developments in Li-Fi, such as compatibility with 5G networks, support of IoT, security and privacy concerns. They have also presented an in-depth analysis of Li-Fi, including its potential benefits and drawbacks, as well as its range limitations, interoperability issues, and cost implications. This study aims to provide a comprehensive overview and contribute to the understanding and advancement of the technology by examining the basic ideas, technological advances, and issues of Li-Fi. The goal of this chapter is to improve the understanding and ability to use Li-Fi as a next generation technology in wireless communications.

I. INTRODUCTION

Wireless Communication played an important role and have become an essential part of our daily lives in the age of digital connectivity. The need for faster, more dependable, and secure wireless communication is increasing at a rate that has never been seen before, from mobile devices to smart homes and cities. Scientists and researchers are constantly investigating cutting-edge technologies to meet these expanding demands. Li-Fi abbreviated as Light-Fidelity is one such technology that has attracted a lot

DOI: 10.4018/979-8-3693-8799-3.ch014

of attention (Safitri et al., 2024). Li-Fi, also known as visible light communication (VLC), is a new kind of wireless technology that transmits data using visible light waves as opposed to radio frequencies. Compared to more established wireless technologies like Wi-Fi and 5G, LiFi can unleash faster, more dependable wireless communications with unmatched security by utilizing the light spectrum.

The IEEE Standards Board has approved the most recent international optical communications standard, IEEE 802.11bb (Jin, Z et al., 2024). The BB standard, which offers a broadly accepted framework for the deployment of LiFi technology, represents a significant turning point for the LiFi market. The working group creating the IEEE 802.11a Wi-Fi standards in 2018 established the Light Communications 802.11b Task Group, and it is led by PureLiFi and supported by Fraunhofer HHI, two organizations that have been at the forefront of LiFi development efforts. The term "light fidelity" (LiFi) refers to a wireless technology that transmits data by using light rather than radio frequencies. In June 2023, the standard will have been ratified. The physical layer specifications and system architecture for wireless communications utilizing light waves are defined by this standard. This new standard paves the way for the widespread adoption of LiFi technology and enables LiFi systems to work with the upcoming WiFi standard. Although the technology has been in development for some time, OEMs have not shown much interest in creating and commercializing LiFi-enabled products due to the lack of a globally accepted standard to go along with it. Devices using Li-Fi are capable of 1 Gbps of lightning-fast speeds. According to research, they can also achieve 100 Gbps speeds by simultaneously encoding data on the red, green, and blue channels of a white LED (Katz, 2020). One can move from light to light without breaking the connection by using a network made up of multiple lights. It's not always necessary for the receiver and transmitter to have a direct line of sight for data to be transmitted; reflections off of walls and other surfaces also work.

Since Lifi provides faster data transfer rates, lower latency, and better security than traditional Wi-Fi, it is regarded as a potential solution for the next generation of wireless communication systems. The use of light- emitting diodes (LEDs) as a medium for data transmission is the core idea behind Li-Fi. Data can be encoded and transmitted using these LEDs' variable light output, which can be quickly changed. After being picked up by photodetectors, the light signals are then transformed back into electrical signals that can be used to interpret data. For 5G, Li-Fi technology offers some hope. The Li-Fi communication system was created by Harald Haas in 2011 and is based on the idea of data transmission using LED light flickering (Christy et al., 2022).

Morse code is a form of off-series communication that moves so quickly that it is invisible to the human eye. Visible light communication, also known as VLC, is another name for it. It is believed to have replaced wireless technology as an indoor alternative to communication (Aboagye et al., 2022). In preliminary testing, the download speed is currently 40 gigabits per second. A download data throughput of 100 Gbps is anticipated for Li-Fi. Using the visible light spectrum, the emerging wireless protocol known as Li Fi offers wireless networking access. A Li-Fi transmitter modulates light intensity using LED lights, which is typically beyond what our eyes can see, and a photosensitive receiver interprets that as data (Birsan et al., 2019). In principle, LEDs could transmit data up to 100 times faster than Wi-Fi because they already use a chip to regulate their output and can modulate up to millions of times per second.

In an effort to get around the drawbacks of conventional Wi-Fi systems, Li-Fi presents a brand- new paradigm in wireless communication (Safitri et al., 2022). Li-Fi is able to outperform conventional Wi-Fi technologies in terms of data transfer rates by utilizing the visible light spectrum. Li-Fi has the potential to take advantage of current infrastructure to deliver seamless wireless connectivity because light-emitting diode (LED) bulbs, which are increasingly used for illumination in a variety of settings, are becoming more widely used (Jamuna et al., 2023).

This increased use of wireless communication has highlighted the intrinsic deficiencies of conventional systems, which are mostly due to the physical limitations of the radio frequency bandwidth. However, there is visible light spectrum, 10,000 times bigger than the whole radio spectrum that can be an option. Consequently, by using this resource Li –Fi not only solves the problem of the lack of spectrum but also provides incredible data transfer rates. Also, visible light cannot go through the wall; it brings an inherent level of security making Li-Fi secure for health care, banks, and military usage. I have identified several seamlessness, which makes Li-Fi uniquely appropriate for next-generation communication systems. By this ability to wirelessly interface with existing lighting fixtures and indicators as well as its potential to deliver sub-millisecond latency and extremely high bandwidth, it can be considered as complimentary to 5G technology. Li-Fi's compatibility with new technological developments including IoT also increases its importance in a world that is progressively characterized by smart devices and systems.

The uses of Li-Fi are as expansive as are the industries and include healthcare, transportation, underwater communications, and industrial automation. For example, Li- Fi is suitable for healthcare applications by as it guarantees secure and interference-free communication that is important in hospitals. Li-Fi therefore offers a reliable platform for transmitting signals and data in underwater situations where radio waves are useless. In addition, Li-Fi technology in smart transportation systems, IoT networks make it capable to boost up more innovation. Nevertheless here there are some challenges that come with Li-Fi which despite the wide prospect seems to have, limited range, line of sight constraints and requires unique hardware. Overcoming these forms of challenges through further research, properly setting Li-Fi industry standards, and developing new technology applications are vital for the increases in Li-Fi's applicability. In this chapter, the author discusses the principles of Li-Fi including the technical aspects on which the technology is built upon, more specific use of the technology and its limitations with the hope of becoming a transformative wireless communication solution. Being transmitted through visible light, Li-Fi holds a good chance to establish new conceptions of connectivity and become one of key component of such subsequent generations of wireless communication systems.

This paper's main goal is to present a thorough analysis of Li-Fi as a next-generation wireless communication technology, with a focus on technical innovations and insights. This paper will also examine the most recent technical developments and innovations in Li-Fi. The integration of Li-Fi and 5G network, for instance, offers the potential for incredibly fast and dependable wireless communication in a variety of settings. A significant opportunity for enhanced connectivity and data transfer in intelligent environments is also made possible by the combination of Li-Fi and the Internet of Things (IoT). This paper's aim is to present information on these novel applications and how they might fundamentally change wireless communication.

II. SYSTEM OF VISIBLE LIGHT COMMUNICATION

Modern wireless communication technology called Li-Fi transmits data using LED lights rather than conventional radio waves, which is advantageous in the current environment where networks must deal with overly dense data communication environments. Data transmission is essential in today's hurried world. Current wireless networks, on the other hand, tend to slow down when multiple devices are connected, and the limited bandwidth available makes it difficult to achieve high data transfer rates while maintaining a secure connection. Unlike traditional wireless communication methods that rely on radio waves, Li-Fi transmits data over a much wider spectrum. This technology works by modulating light intensities in LED lights at an imperceptible speed to the human eye, allowing for faster data transfer It makes use of the Visible Spectrum, which is a relatively underutilized portion of the electromagnetic spectrum, rather than relying solely on radio waves in the Gigahertz range. Li-Fi offers a promising solution for wireless communication in an overcrowded data communication world by leveraging this visible spectrum.

Harald Haas, a German physicist, first introduced the concept of Li-Fi during a TED Global talk on VLC in July 2011 (Hussain et al., 2019). Haas described Li-Fi as "data through illumination" and demonstrated its capabilities using a table lamp outfitted with an LED bulb. He sent a video of a flower blooming through the lamp, which was then displayed on a screen. Simply stated, Li-Fi is a type of Wi-Fi that uses light rather than radio waves for data transmission. Li-Fi uses transceivers outfitted with LED lamps to illuminate a room while also facilitating data transmission and reception, as opposed to traditional Wi-Fi modems. Li-Fi has the potential to alleviate the capacity constraints currently experienced by wireless systems by incorporating the untapped bandwidth of visible light into existing radio wave frequencies used for data transfer (Wei et al.,2022). The following figure shows the visible light spectrum -

Figure 1. Visible Light Communication Frequency Spectrum

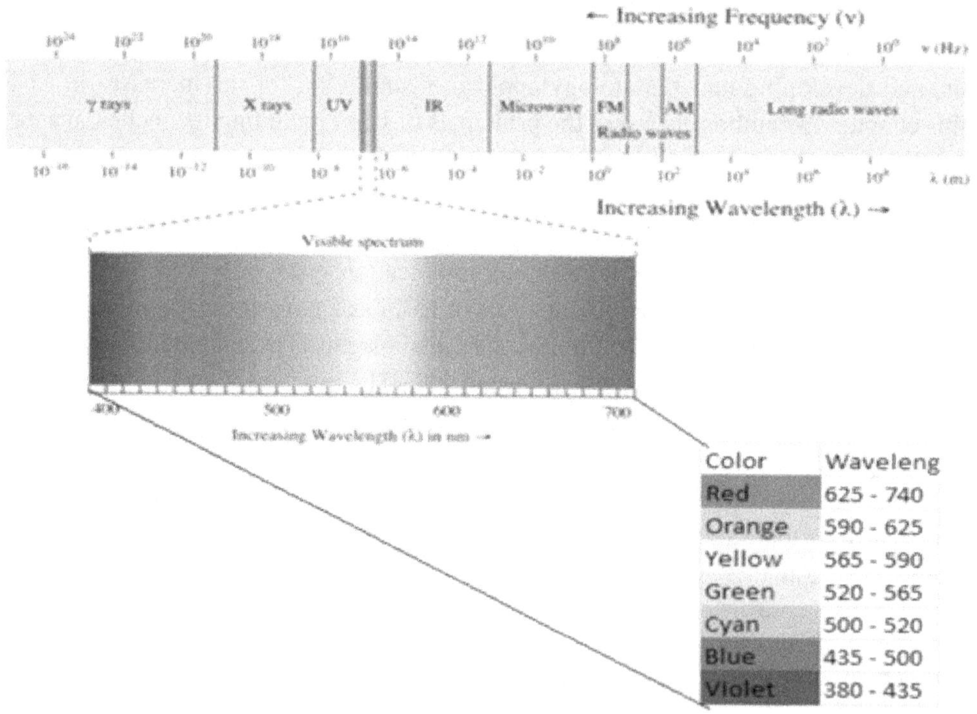

There are many different wave regions in the frequency spectrum that are visible to us in the atmosphere, including radio waves, visible light rays, gamma rays, u-v rays, and infrared rays. The upcoming communication technologies can use any of the aforementioned waves, but visible light is preferred because it is more widely available and has fewer drawbacks. VLC uses the visible light spectrum between 400 and 800 THz (375 and 780 nm), which is less hazardous for high-power applications and because it is readily observable by people, allowing them to protect themselves from its harmful effects (Ho, 2024). Different colors of light can be used for data transmission within this range. Li-Fi uses variations in LED light intensity to encode and transmit data, making use of the LEDs' quick switching abilities, which are invisible to the human eye. Radio waves are less secure (due to interference and the possibility of interception) and more expensive (due to spectrum fees), other wave regions have many drawbacks. Gamma rays are dangerous because it could be risky for people to interact with them due to their well-documented negative effects on human health. X-rays and gamma rays both have health effects. In areas where there are no people, ultraviolet light can be used for communication technology, but prolonged exposure to it can be harmful to human health. Due to strict safety regulations, infrared can only be used at low power levels. As visible light has a larger bandwidth, is safe to use, and has a bright future in the communication industry, the visible light portion of the electromagnetic spectrum (which ranges from red to blue) does not harm people. This enables the use of light as a medium for quick and effective data transfer.

The Visible light spectrum is the one most commonly used in Li-Fi, some implementations might also use specific wavelengths of the infrared and ultraviolet spectrums. The amplitude modulation of the light source is the key component of Li-Fi, which is a well-defined and standardized method of data transmission. In order to operate faster than the human eye can see, LEDs must operate at a speed of less than 1 microsecond. Data is transmitted using binary codes through this covert on-off process. A digital '1' or '0' is transmitted depending on whether the LED is on or off. The LED represents a digital "1" when it is on and a digital "0 when it is off. The following figure depicts the working.

Figure 2. LI-Fi Transmission

LED lights are able to transmit data because they are invisible to human on/off activity. Li-Fi has the advantage of being unaffected by interference from other light frequencies. The radio frequency interference that Wi-Fi experiences is comparable to this. The absence of interference makes data transmission effective. Li-Fi is well known for being incredibly effective, with estimates for its efficiency over conventional Wi-Fi ranging up to 80%. Li-Fi is a promising technology for high-speed data transfer because of its improved efficiency, enabling it to reach speeds of up to 1 Gbps (and possibly higher). Li-Fi is made specifically for short-range wireless communications, typically up to 10 meters, in contrast to fiber optic communications. It can offer wireless connectivity over a constrained area, in contrast to fiber optic communications, which uses physical cables to transmit data over long distances.

III. LI-FI AND TRADITIONAL WI-FI COMPARISON

Due to their unique advantages and disadvantages, traditional Wi-Fi and Li-Fi are appropriate for varying applications. Due to its superior performance in terms of speed, security, and low interference, Li-Fi is an emerging technology that shows promise for applications that call for high-speed data transfer and improved security.. Wi-Fi, on the other hand, provides a wider coverage area, compatibility with existing infrastructure, and the ability to overcome obstacles (Uzair, 2022). Both technologies are likely to coexist and possibly even perform better when used in tandem in the future to meet various connectivity requirements. The following table-1 shows the comparison of Li-Fi and Wi-Fi.

Table 1. Comparision of Lifi & Wifi

Comparision	Lifi	Wifi
Speed	Li-Fi has the potential for much higher data transfer speeds compared to Wi-Fi due to the use of visible light. Li-Fi can achieve speeds in the range of several gigabits per second (Gbps) (Ojha,2022).	Wi-Fi technology has also evolved, and the latest Wi-Fi standards like Wi-Fi 6 (802.11ax) can provide high data transfer speeds, reaching up to several gigabits per second in ideal conditions (Chauhan,2021). However, it generally lags behind Li-Fi in terms of maximum achievable speeds.
Energy Efficiency	Li-Fi is generally more energy-efficient than Wi-Fi because it utilizes energy-efficient LEDs for data Transmission (Sarkar et al., 2015). LEDs consume significantly less power compared to Wifi	Although Wi-Fi technology has evolved over time, it still consumes more power than Li-Fi due to the use of radio frequencies for data transmission (Adhikari,2022).
Range	When compared to Wi-Fi, the range of Li-Fi is generally limited. Li-Fi signals rely on the transmission of visible light that is obstructed by obstacles such as walls or even human bodies. As a result, Li-Fi's range is typically shorter, limiting its coverage to smaller areas (Darusalam et al., 2024).	Wi-Fi signals are better suited for providing wireless connectivity over larger areas because they can pass through walls and have a longer range (Zen et al., 2024). This could enable seamless connectivity in homes, workplaces, and public areas by expanding the range of areas covered.

continued on following page

Table 1. Continued

Comparision	Lifi	Wifi
Security	As compared to Wi-Fi, Li-Fi systems are more secure by default. Li-Fi communications could be more secure and less likely to be intercepted by unauthorized parties because light cannot pass through physical barriers like walls or radio waves (Lambrechts et al.,2019).	Wi-Fi networks have vulnerabilities and can be vulnerable to various security threats, including eavesdropping and unauthorized access (Chauhan et al.,2023). However, Wi-Fi can still offer secure communications if the right security measures are taken, such as encryption protocols (like WPA2, WPA3).
Interference:	Li-Fi, does not face interference issues from other sources as it uses light for communication (Perwej, 2017).	Wi-Fi signals can face interference from other Wi-Fi networks or electronic devices operating in the same frequency band, affecting Performance (Martínez et al., 2024).
Density	Li-Fi may be superior to Wi-Fi in areas that are crowded with lots of devices. Li-Fi uses laser beams to create more specialized connections that are less prone to interference and perform better (Ali et al.,2023).	Wi-Fi networks can experience congestion and reduced performance in densely populated areas (Hartog et al., 2017).
Deployment	Li-Fi may require additional installation of Li-Fi-enabled light bulbs and access points, although it can also leverage existing lighting infrastructure for deployment (Albraheem et al., 2017).	Wi-Fi infrastructure is already well established and widely available, making it easier to deploy in existing setups (Systems 2023).
Cost and infrastructure	Li-Fi, requires specialized LED light fixtures and receivers for deployment, which may bring additional costs and retrofitting requirements (Füchtenhans et al.,2021).	Traditional Wi-Fi, being an established technology, utilizes existing infrastructure such as routers and devices, making it easier to deploy and cost-effective (Rao et al. 2023).
Applications	**Li-Fi**: Suited for specific applications where high speed, security, and electromagnetic interference-free operation are required, such as: • Healthcare (operating rooms, patient monitoring) • Smart transportation (vehicle-to-vehicle communication) • Underwater communication • Enclosed environments with high security needs	A general-purpose technology suitable for broader applications, such as home networks, office setups, and public hotspots.
Mobility	Limited mobility due to its reliance on light beams. A user must remain within the line-of-sight or reflected light of an active light source. This limitation is significant in dynamic environments where continuous movement is required.	Offers high mobility, enabling seamless connectivity while moving between rooms, floors, or even across large campuses using access points and mesh networks.
Data Density	High data density as it can serve multiple devices simultaneously within a small area without interference. This feature is ideal for environments with dense device deployment, such as conference halls or stadiums.	Struggles with data density in crowded environments due to interference and bandwidth limitations, leading to reduced speeds and performance.
Latency	Extremely low latency due to the high modulation rates of LEDs, making it suitable for real-time applications such as gaming, virtual reality, and remote surgeries.	While modern Wi-Fi standards like Wi-Fi 6 offer reduced latency, it is still higher than Li-Fi, especially in congested networks.
Electromagnetic Compatibility	Does not emit electromagnetic radiation, making it safe and interference-free for sensitive environments like hospitals, where electromagnetic interference can disrupt medical equipment.	Operates on radio frequencies, which may interfere with sensitive equipment or other devices operating on the same spectrum.
Environmental Impact	Utilizes existing LED lighting infrastructure, reducing the need for additional hardware. Its dual-purpose functionality for lighting and communication minimizes waste and enhances energy efficiency.	Requires separate networking equipment, which increases electronic waste and has a larger carbon footprint due to higher power consumption.

continued on following page

Table 1. Continued

Comparision	Lifi	Wifi
Underwater Usability	Operates effectively underwater as light waves can travel through water. This makes it an excellent choice for underwater exploration, communication with submersibles, and marine research.	Radio waves are absorbed by water, making it ineffective for underwater communication.
Resistance to Obstacles	Light cannot penetrate physical barriers like walls or furniture, which limits its usability but also enhances security by confining communication to specific spaces. Wi-Fi:	Can penetrate most obstacles, including walls and floors, providing broader coverage but at the cost of potential data leaks and reduced security.
Use in Hybrid Systems	This can complement Wi-Fi in hybrid systems, leveraging its high-speed and secure communication for specific tasks while Wi-Fi handles broader coverage. Hybrid setups can optimize connectivity in modern smart environments.	Forms the backbone of hybrid systems, providing compatibility and integration with other wireless technologies like Li-Fi and 5G.

Energy efficiency, potential data transfer speed and built-in security are all areas where Li-Fi outperforms Wi-Fi. The range and coverage, however, are constrained. Wi-Fi, on the other hand, offers greater coverage and range, but it also has the potential to consume more power and be more vulnerable to security threats. Li-Fi and Wi-Fi should be chosen in accordance with the specific requirements, such as the required data speed, security requirements, and the physical layout of the environment (Sharma et al.,2018).

IV. APPLICATIONS OF LIFI

Li-Fi is a wireless technology that transmits data using waves of visible light. It has wide range of applications in various industries. Its distinctive features and capabilities open up new avenues for increased efficiency, communication, and connectivity. Here are some notable LiFi applications:

1. High-Speed Internet Access: Offering high-speed internet access is one of Li-Fi's most well- known uses (Ahammed et al., 2022). Li-Fi enables seamless streaming of high-definition videos, online gaming, and large file transfers thanks to its ability to achieve data transfer rates that are significantly higher than those of conventional Wi-Fi.
2. Indoor Localization and Positioning: Systems for indoor positioning and localization can be implemented using Li-Fi technology. Accurate positioning and navigation within indoor environments, such as malls or airports, can be accomplished by installing Li-Fi transmitters and outfitting devices with Li-Fi-enabled receivers (Hesham et al., 2020).
3. IoT (Internet of Things) Connectivity: By integrating Li-Fi with IoT devices, smart homes, industrial automation, and smart cities can all benefit from improved connectivity. The proliferation of IoT devices and applications is facilitated by Li-Fi's capacity to enable high- speed, secure, and localized communication, enabling seamless data exchange and control over connected devices (Kaur et al., 2024).
4. Ultra-Secure Communication: Because visible light waves have a constrained transmission range, Li-Fi has a security advantage over conventional Wi-Fi. Since Li-Fi signals have trouble navigating through dense materials, it is challenging unauthorized individuals to intercept or eavesdrop on the

data transmission. Li-Fi is thus appropriate for use in scenarios where data security and privacy are crucial, such as in financial transactions, healthcare information exchange, or military communications (Paul et al., 2021).

5. Wireless Communication-Underwater: Due to the water's high radio wave absorption, conventional wireless communication methods like radio and Wi-Fi do not function underwater. However, Li-Fi can be used for underwater communication by making use of specially designed LED transmitters and photodetectors (Cai et al.,2024). As a result, monitoring, research, and underwater exploration can all be done using Li-Fi.
6. Retail and hospitality: In retail settings like malls or department stores, Li-Fi can be used to provide customers with high-speed internet access. Li-Fi can also enable individualized location-based services, targeted advertising, and improved customer experiences via interactive digital signage and wayfinding systems (Rocher et al.,2024). Similar to this, Li-Fi can give guests in hotels, resorts, or conference venues quick and secure connectivity.
7. Healthcare: Li-Fi can benefit the healthcare industry by making it possible for medical data, such as real-time monitoring data, medical imaging, and electronic health records, to be transmitted quickly and reliably. Li-Fi is advantageous for telemedicine applications, remote patient monitoring, and medical IoT devices because of its high bandwidth and low latency (Kharche, 2023).

These are just a few of the numerous uses for Li-Fi technology. Li-Fi has the potential to revolutionize wireless communication by providing faster, more secure, and more effective connectivity as research and development proceed.

V. DEPLOYMENT OF LI-FI (LIGHT FIDELITY) FOR NEXT-GENERATION WIRELESS COMMUNICATION

To ensure the successful deployment of Li-Fi (Light Fidelity) for next-generation wireless communication, it is necessary to carefully consider a number of factors (Fazea et al.,2021). These aspects include the need for appropriate infrastructure, interoperability, standardization, and financial implications.

1. Infrastructure requirements: Installation of specialized LED lights with data modulation capabilities is necessary for Li-Fi deployment. Although it is possible to retrofit existing LED lighting infrastructure with Li-Fi capabilities, additional hardware elements like Li-Fi transmitters and photodetectors might be needed (Albraheem et al., 2022). The Li-Fi-enabled LED lights must be strategically positioned throughout the target area to provide the best possible coverage and performance.
2. Interoperability: For seamless connectivity, it's essential that various communication technologies work together. In order to facilitate seamless transitions between these networks and devices, Li-Fi must be able to coexist with other wireless communication technologies like Wi-Fi, cellular networks, or Bluetooth (Mubashar et al.,2021). To promote interoperability and guarantee compatibility across various devices and network infrastructures, standards and protocols must be developed.
3. Standardization: Li-Fi technology needs to be standardized in order to facilitate wide adoption and compatibility. Groups like the IEEE and the Li-Fi Consortium are actively developing standards and protocols and IEEE 802.15.7 is the IEEE standard for LiFi (Játiva et al., 2022). In order to enable consistent performance and interoperability between various Li-Fi devices, standardization efforts

concentrate on defining common specifications for Li-Fi, including modulation techniques, data encoding methods, and network protocols.

4. Security Considerations: As is the case with any wireless communication technology, security is a key aspect of Li-Fi deployment. Data transmitted over Li-Fi networks should be protected from unauthorized access or interception by using strong encryption techniques (Diambeki et al., 2022). By taking the necessary precautions, data integrity and confidentiality during transmission should also be guaranteed. Security safeguards like authentication protocols and encryption algorithms are critical for the deployment of Li-Fi in sensitive applications like the healthcare and financial sectors (Rabia et al., 2020).
5. Financial Implications: The price of putting Li-Fi technology into use must be taken into account. The additional hardware required to enable Li-Fi capabilities may initially be more expensive, despite the fact that LED lighting is becoming more common and affordable (Faruq et al., 2023). Nevertheless, it is critical to consider the potential benefits of Li-Fi, such as quick data transfer and enhanced security, which may ultimately make the investment worthwhile.
6. Research and Development: To improve and hone Li-Fi technology, ongoing research and development efforts are crucial. Improvements in areas like modulation methods, receiver sensitivity, and interference reduction will help Li-Fi networks perform better and have a wider coverage area (Subha et al., 2020). Research is also required to address Li-Fi's drawbacks and difficulties, such as its limited range and need for line-of-sight, in order to increase the likelihood of its deployment.

Despite its transformative potential, the deployment of Li-Fi technology faces several challenges that must be addressed for its widespread adoption. One of the primary obstacles is the line-of-sight requirement for effective communication. Since Li-Fi relies on visible light, any obstruction between the transmitter (LED light) and the receiver can disrupt the signal. Although reflected light can partially mitigate this issue, it limits the technology's usability in dynamic or cluttered environments. Additionally, range limitations pose another significant hurdle. Li-Fi typically operates within a range of 10 meters, far shorter than the coverage offered by Wi-Fi networks, making it less viable for large-scale applications without the extensive deployment of Li-Fi-enabled lights.

Another challenge is interoperability with existing systems. While the IEEE 802.11bb standard has laid the groundwork for Li-Fi integration with current wireless technologies, ensuring seamless coexistence with Wi-Fi, cellular networks, and IoT systems remains a technical and logistical challenge. Furthermore, the initial costs associated with infrastructure modifications—such as retrofitting LED lights with Li-Fi capabilities and deploying photodetectors—can deter adoption, particularly for organizations with budget constraints.

Li-Fi's dependence on light also restricts its use in environments with limited or no illumination, such as during power outages or at night when lights are turned off. This necessitates hybrid solutions that combine Li-Fi with traditional radio frequency technologies to ensure uninterrupted connectivity. Moreover, the technology's susceptibility to interference from ambient light sources, such as sunlight or artificial lighting, can compromise its reliability and performance in certain settings.

Lastly, public awareness and market readiness are critical barriers. While Li-Fi offers numerous advantages, its adoption requires collaboration among stakeholders to educate users, develop cost-effective hardware, and create standards that ensure compatibility across devices. Overcoming these challenges through continuous research, innovation, and strategic partnerships will be crucial to unlocking Li-Fi's full potential as a next-generation wireless communication technology.

VI. ENERGY EFFICIENCY OF LI-FI SYSTEMS

Li-Fi is a revolutionary innovation in wireless networking technology especially in a way of energy consumption. Unlike conventional radio frequency-based systems, Li-Fi makes use of visible light communication through LEDs, which are by their nature friendly to energy conservation. LEDs being a medium consume much lesser power than conventional lighting means or even radio frequency transmitting tools, and hence, is a greener option when fulfilling the needs of the modern connection requirements (Salman et al., 2021). This makes Li-Fi not only a functional wireless substitute for Wi-Fi but also a neater option where energy and environmental friendliness are paramount.

A unique advantage of Li-Fi lies in its ability to utilize existing LED lighting infrastructure for dual purposes: lighting as well as data communication. Li-Fi eliminates the requirement of proprietary communication equipment since it incorporates the competency for data transfer into ordinary illuminating fixtures, thereby minimizing energy consumption and expenses. The integration of these two platforms is much easier than the original and uses resources more efficiently, especially where LG LED lighting is already prominent. For instance, office blocks, hospitals and smart homes can convert a lighting system to be a high speed communication network without having a high energy penalty (Karthika et al., 2015).

The main aspect of efficiency of Li-Fi in the use of energy is its ability of changing light intensity in order to facilitate reasonable energy use. An example of use is in a situation where it is not necessary to have full lighting intensity – during the day, or in a well-illuminated area; the light can be dimmed down while the data stream remains very stable and powerful. Thereby, this adaptability fully ensures that energy is used only as required in particular times or places, hence, zero wastage (Mohsin et al.,2023). Furthermore, Li-Fi systems are aimed at being precision-oriented, so that energy is emitted only where needed, that is, in areas where nodes are involved in an active transmission of data and not in vast areas as in Wi-Fi.

In congested premises, Li-Fi Þ has better effciency in usingenergybynot encountering interference problems that are often associated with Wi-Fi networks. Most of the wireless networks used today waste power in trying to settle conflicting signals and interferences from other networks or other wireless devices (Janjua et al., 2021). Li-Fi solves this by using slim and pointed light beams which can simultaneously provide renditions of energy in crowded places like conference halls, airports or stadiums. This targeted strategy does not only improve energy utilization but also guarantees more data integrity and velocity.

Moreover, it is known that LED systems of Li-Fi are two-in-one benefit to the environment and in terms of expenditure. These systems not only provide light but also act as information transfer systems thus minimizing the use of other communication structures that in turn leads to clearance of electronic devices, which leads to minimization of manufacture of such components thus reduction of carbon-dioxide emissions. Such integration between lighting and communication serves as a foundation for Li-Fi to be embraced as a sustainable technology well suited for meeting increasing global need for wireless lighting communication (Chauhan et al., 2024).

The fact that Li-Fi consumes less power identifies it as a strong contender for inclusion into smart, green applications, such as IoT, 5G and more systems that are complex. The opportunity to provide, both, high speed, localized, and secure connection, along with lower power consumption makes it a revolutionary technology for such sectors as health care and transportation (Km, 2023). This is because as the world's energy demands grow with increased usage of wireless gadgets Li-Fi is a crucial invention that is not only in line with the technological world but also promotes environmental conservation.

VII. INTEGRATION OF LI-FI WITH 5G NETWORKS

Li-Fi and 5G network integration has the potential to improve wireless communication capabilities by delivering faster, more dependable, and more connected communication (Almalah et al., 2024). Li-Fi, which transmits data using light waves, and 5G, the fifth-generation cellular network, can work well together and provide a number of advantages. A major benefit of combining Li- Fi and 5G is the potential for faster data rates. Li-Fi technology can transmit data at rates that are so fast that they even outperform what is possible with conventional Wi-Fi networks (Selvi et al., 2019). In order to provide users with extremely quick connectivity for applications that demand high bandwidth, such as streaming videos, virtual reality, and cloud-based services, Li-Fi and 5G can be combined (Mishra et al., 2023). In comparison to its predecessors, 5G is designed to offer faster speeds and lower latency. Another benefit is the increased data transfer capacity. The development of 5G networks, which can support devices that are more connected simultaneously, has made the Internet of Things (IoT) and other connected technologies possible (Chauhan et al., 2023). Integrating Li-Fi, which uses light waves in constrained spaces to provide connections that are more dependable and relieve network congestion will increase the capacity for device-to-device communication.

Li-Fi and 5G network integration can also expand the reach of the network. Li-Fi uses light signals to communicate, making it appropriate for indoor settings where 5G signals might have limitations. Users can access high-speed wireless connectivity even in network-challenged areas by installing Li-Fi access points in buildings, stadiums, hospitals, or other locations, resulting in a seamless user experience. Additionally, the combination of Li-Fi and 5G can enhance wireless communication security (Mekala et al.,2019). Visible light, which is how Li-Fi operates, can be easily contained to particular spaces and cannot pass through solid objects like walls, making it more secure from unauthorized access or interception. Users can gain from improved security measures, ensuring privacy and data protection, by integrating Li-Fi with 5G networks (Aldolimi et al., 2021). Users can benefit from faster, more dependable connectivity in both indoor and outdoor settings by fusing the advantages of Li-Fi's high speed, capacity, and security with 5G's advanced cellular capabilities. The combination of Li-Fi and 5G will probably be crucial in shaping the future of wireless networks and satisfying the increasing demands of a connected world as research and development advance (Linnartz et al., 2021). The following table 2 shows the points of difference between the two technologies.

Table 2. Li-Fi vs. 5G

Points	Li-Fi vs. 5G
Speed	Both Li-Fi and 5G are designed to provide faster data transfer speeds than their predecessors. 5G, with its use of millimeter-wave frequencies, can achieve speeds up to 20 Gbps under ideal conditions. This makes it significantly faster than 4G LTE and capable of supporting data-intensive applications such as virtual reality, autonomous vehicles, and smart city infrastructure. Li-Fi, however, has demonstrated even higher speeds in lab tests, with rates exceeding 100 Gbps.
Latency	In terms of latency, both technologies aim to reduce delays in data transmission. 5G boasts ultra-low latency, often under 1 millisecond, making it ideal for real-time applications such as gaming, remote surgery, or autonomous driving. Li-Fi can also achieve low latency, particularly in local networks, making it suitable for similar use cases, but it is more constrained by its short-range communication capabilities.
Coverage and Mobility	The most significant difference between Li-Fi and 5G lies in their coverage areas and mobility. 5G networks are designed for wide-area coverage, providing seamless connectivity across cities, rural areas, and even moving vehicles. 5G relies on a network of small cells and antennas to deliver high-speed internet over large distances, making it ideal for mobile users and applications requiring continuous, large-scale connectivity. Li-Fi, on the other hand, is designed for short-range, high-speed communication within confined spaces. Its reliance on visible light means that it cannot offer the same level of mobility as 5G. Li-Fi is best suited for stationary or semi-stationary environments, such as homes, offices, or hospitals, where users remain within the light source's coverage area.
Spectrum and Interference	5G uses high-frequency radio waves, including millimeter waves, to transmit data. These frequencies offer higher bandwidth and faster speeds, but they are also more prone to interference from obstacles such as buildings or trees. As a result, 5G networks require a dense infrastructure of small cells to ensure continuous coverage, particularly in urban environments. Li-Fi, on the other hand, uses the visible light spectrum, which is not subject to the same interference issues as RF-based communication. This makes Li-Fi ideal for environments where RF congestion or interference is problematic. However, as previously mentioned, Li-Fi is susceptible to interference from other light sources, which can impact its performance.
Use Cases	While both technologies offer high-speed communication, they are designed for different applications. 5G is better suited for wide-area networks, mobile devices, and applications that require continuous connectivity across large distances. It will play a critical role in enabling smart cities, autonomous vehicles, and IoT networks. Li-Fi, on the other hand, is more appropriate for localized, high-speed data transfer in secure environments, such as hospitals, banks, or military installations. Its ability to provide ultra-fast, secure communication within a confined area makes it a valuable complement to 5G in specific use cases.

Li-Fi technology's integration with the Internet of Things (IoT) has many benefits, improving connectivity, security, and productivity in IoT applications. Li-Fi, which transmits data using light waves, and the Internet of Things (IoT), which is a vast network of connected devices, can be combined to improve communication and data transfer.

VIII. INTEGRATING LIFI TECHNOLOGY WITH IOT.

Exciting opportunities for enhanced connectivity, security, energy efficiency, and localization are presented by integrating Li-Fi with IoT. Organizations can realize the full potential of IoT deployments by fusing the advantages of Li-Fi's fast data transmission, improved security, and precise positioning with the vast network of interconnected IoT devices (Jain et al., 2021). This will enable cutting-edge applications and services in a variety of industries, including smart cities, healthcare, manufacturing, and transportation. The enhanced bandwidth and data transfer capabilities of integrating Li-Fi with IoT are among the major advantages. Li-Fi makes use of the visible light spectrum, which is much larger than the radio frequency spectrum used by conventional wireless technologies, to enable high-speed data transmission. Real-time monitoring, analytics, and decision-making are made possible by the seamless transfer of large amounts of data between IoT devices thanks to this increased bandwidth (Chauhan et al., 2019).

Cisco expects the Internet of Things (IoT) to reach 50 billion connected devices by 2020, with light communications use cases providing global wireless solutions with initial applicability in EMI-challenged environments like hospitals (Press, 2017). As demonstrated by the example. Petrochemical plants and airplanes, for example, but also protected environments where RF is not permitted. Furthermore, light communications will be used to improve indoor connectivity, which accounts for more than 80% of all wireless communications in office settings and, more importantly, in the home. LiFi's additional security benefits also improve the privacy of consumer and commercial networks while maintaining solution connectivity.

The enhanced connectivity and coverage is another benefit. In locations where radio frequency-based wireless networks may encounter issues like interference, congestion, or range restrictions, Li-Fi can offer dependable wireless communication. Devices can use Li-Fi access points to connect to the network even in settings with a high device density or unreliable RF signals by integrating Li-Fi with IoT (Spagnolo et al., 2020).

IoT deployments must prioritize security, and Li-Fi has built-in advantages in this area. Li-Fi offers more secure communication channels because it uses light waves, which can be restricted to particular areas. Because Li-Fi signals cannot pass through walls, there is less chance of intrusion or interference (Manvi, 2022). Sensitive data transmitted between devices can be secured with the additional layer of security provided by Li-Fi technology by integrating Li-Fi with IoT.

Li-Fi and IoT integration can also increase energy efficiency. Li-Fi is suitable for low-power IoT devices because it uses less power to transmit data than conventional wireless technologies (Suriyanet al., 2023). IoT devices can conserve energy and possibly increase battery life by using Li-Fi, leading to more sustainable and effective IoT deployments. Additionally, Li-Fi's localization capability can be used in IoT applications. Li-Fi enables precise device positioning within a network, opening the door for indoor navigation in challenging environments, asset tracking, and location-based services. Numerous IoT use cases, including asset management, inventory tracking, and indoor wayfinding, can be improved by this localization capability.

IX. SECURITY AND PRIVACY IN LI-FI NETWORKS

It offers new opportunities for secure and private exchange through Li-Fi or Light Fidelity, transferring data via light waves, instead of traditional radio frequencies as used by existing systems. Since this does not employ broadcast of the data signal as Wi-Fi does but uses modulated light emitted from LED bulbs operating at inaudible speeds, its transmission range is strictly confined to the scale of the physical space being lit by the light source. This limited spread of signal provides inherent security benefits because data is confined to specific areas, hence ensuring effective security in highly secured offices such as government offices, healthcare facilities, and corporate buildings where data containment and privacy are significant (Chauhan et al.,2024). For example, within the hospital environments, it can guarantee secure transmission of patient records without the risks of any leakage to the public domain or unsafe environments. Also, in corporate offices, sensitive data remains confined only to the work area; however, in order to maximize this security benefit, enough mechanisms of encryption and access control need to be put into place to safeguard data within that illuminated space.

One of the most significant advantages in terms of security that it inherently prevents eavesdropping and data interference. Since a line of sight is required, such data signals cannot easily seep through walls or concrete objects, thereby minimizing intercept chances through unauthorized users; however, within the cover area, interception is considerably a high risk in cases of very high traffic locations. PKI or digital certificates can also be used for such pairing protocols to ensure access of only authorized users for data streams in order to prevent their unauthorized interception. Continuous Li-Fi network monitoring also forms a must-have to identify unusual patterns in data transmission and even a potential security breach. IDS and IPS can assist network administrators immediately identify security threats. Furthermore, adaptive lighting technology aids in security by the capability of adjusting the brightness and modulation of light. It lowers the signal strength by dimming the light or its modulation when necessary, therefore limiting what could be intercepted.

The two common methods of securing Li-Fi networks are authentication and encryption. Authentication techniques range from 2FA to biometric verification and multi-level access controls to ensure only the right people have access to network resources. Encryption, on the other hand, is equally important to ensure data in transit are also secured. The symmetric encryption algorithms to be used include AES to efficiently secure high-speed streams within a confined area. Asymmetric or public-key encryption may be applied at other locations, especially where exchanges of data securely have to be made between remote areas. New methods, in the form of quantum key distribution (QKD), for Li-Fi networks are being developed, too. QKD is dependent on quantum mechanics, making it theoretically impossible for the interception of a signal without being detected. QKD will alert network administrators when someone tries to intercept the quantum state of the light signal. Except for encryption, dynamic frequency modulation can also be used for security. The frequency of the light signal is constantly varied in such a way that the attacker will not easily understand what data is being transferred.

Li-Fi's unique security features are highly valuable in areas of specialization that emphasize safety and security. Government and military groups that handle sensitive information frequently will find the localized nature of Li-Fi communication useful in avoiding exposure of confidential data. Further, Li-Fi provides added benefits in healthcare because it does not interfere with any medical equipment that is highly sensitive to electromagnetic waves, thus facilitating safe and reliable communication in hospitals without endangering a patient's safety. Just like Li-Fi, financial organizations, with huge amounts of sensitive information about their customers, can leverage Li-Fi's security for restricted access to data as Li-Fi limits access only to specific physical locations. For instance, organizations integrating Li-Fi into access control and encryption can provide secure communication channels without exposing data to risks of interception.

Li-Fi does benefit in many ways but poses several practical issues to be at its maximum utilization. Cases like these would obstruct the signal transmission by Li-Fi owing to the interference of transmission by physical barriers, walls, and furniture. The system also has a dependence on visible light, meaning it is not feasible at complete darkness conditions and therefore further limits the applications of the system in various setups. Hybrid communication systems, in which Li-Fi is merged with traditional radio frequency communication, such as Wi-Fi, are being developed to counter these limitations. Hybrid systems ensure seamless connectivity even when there are physical obstructions or insufficient lighting. Efforts are also being made to integrate artificial intelligence and machine learning into Li-Fi networks. For example, machine learning algorithms can predict and reduce security risks by analyzing the network usage patterns and anomaly identification, thereby strengthening the security framework of Li-Fi networks (Chauhan et al.,2023).

X. CONCLUSION

Li-Fi needs to be deployed for next-generation wireless communication, which means that infrastructure needs, interoperability, standardization, security issues, financial ramifications, and ongoing research and development must all be taken into account. By taking into account these factors, Li-Fi can be successfully incorporated into existing communication networks, providing faster, more secure, and more effective wireless connectivity for a variety of applications and domains. However, there are some challenges in putting Li-Fi into practice. Range limitations, interoperability issues, and financial ramifications are just a few of the issues that must be addressed to ensure a successful implementation. By being aware of these problems, considering research initiatives, and putting forward potential solutions, we can hasten the adoption of Li-Fi as an economically viable wireless communication technology. This paper seeks to present a comprehensive overview of Li-Fi as a next-generation wireless communication technology with a focus on technical insights and innovations. We are examining the fundamental ideas, technological advancements, and issues of Li-Fi in order to aid people in understanding and advancing the technology. In order to provide faster, more dependable, and secure connectivity for the upcoming generation of wireless communication systems, Li-Fi makes use of the strength of visible light waves. The ultimate goal of this paper is to encourage additional Li-Fi research and development, promoting its uptake and influencing wireless communication's future.

REFERENCES

Aboagye, S., Ndjiongue, A. R., Ngatched, T. M., Dobre, O. A., & Poor, H. V. (2022). RIS-assisted visible light communication systems: A tutorial. *IEEE Communications Surveys and Tutorials*, 25(1), 251–288. DOI: 10.1109/COMST.2022.3225859

Ahammed, M. T., Boppana, U. M., Kondapalli, R., Balaji, P., & Zardar, M. A. (2022, August). Survey on wireless fidelity (Wi-Fi) based indoor location technology for mobile devices. In *2022 Third International Conference on Intelligent Computing Instrumentation and Control Technologies (ICICICT)* (pp. 1465-1470). IEEE. DOI: 10.1109/ICICICT54557.2022.9917746

Albraheem, L., Alhudaithy, L., Aljaser, A., AlDhafian, M., & Bahliwah, G. (2022). Towards smart sustainable cities using Li-Fi technology: Geo-location infrastructure utilizing LED street lights. *PeerJ. Computer Science*, 8, e1009. DOI: 10.7717/peerj-cs.1009 PMID: 36091990

Albraheem, L. I., Alhudaithy, L. H., Aljaser, A. A., Aldhafian, M. R., & Bahliwah, G. M. (2018). Toward designing a Li-Fi-based hierarchical IoT architecture. *IEEE Access : Practical Innovations, Open Solutions*, 6, 40811–40825. DOI: 10.1109/ACCESS.2018.2857627

Aldolimi, W. S., Hnaif, A. A., & Alia, M. A. (2021, July). Light fidelity to transfer secure data using advanced encryption standard algorithm. In *2021 International Conference on Information Technology (ICIT)* (pp. 963-967). IEEE. DOI: 10.1109/ICIT52682.2021.9491769

Ali, A., Kumar, R. H., Dheenathalayan, R., Prasanth, N., Parthasaradi, V., Senthilkumar, S., & Kumar, T. S. (2023). Audio Streaming Using Li-FI Communication. *Irish Interdisciplinary Journal of Science & Research*, 7(1), 1–7. DOI: 10.46759/IIJSR.2023.7101

Almalah, N. T., Mahmood, F. E., & Yassen, M. T. (2024). Li-Fi Technology in Optical Communication Systems: A Review. *Al-Iraqia Journal for Scientific Engineering Research*, 3(3), 163–171.

Birsan, J. C. R., Moldoveanu, F., Moldoveanu, A., Dascalu, M. I., & Morar, A. (2019, October). Key technologies for indoor positioning systems. In 2019 18th RoEduNet Conference: Networking in Education and Research (RoEduNet) (pp. 1-7). IEEE. DOI: 10.1109/ROEDUNET.2019.8909406

Cai, C., Du, Z., Liao, W., He, Y., Ma, X., Qin, S., & Xu, J. (2024). Experimental Demonstration and Practical Application of a Real-time Full-duplex Underwater Wireless Optical Communication Transceiver. *Journal of Lightwave Technology*, 42(24), 8541–8554. DOI: 10.1109/JLT.2024.3440639

Chauhan, D., & Jain, J. K. (2019). A Journey from IoT to IoE. *International Journal of Innovative Technology and Exploring Engineering*, 8(11).

Chauhan, D., & Jain, J. K. (2023, January). Profiling Network Traffic by Using Classification Techniques in Machine Learning. In International Conference on Smart Trends in Computing and Communications (pp. 113-123). Singapore: Springer Nature Singapore. DOI: 10.1007/978-981-99-0769-4_12

Chauhan, D., & Jain, J. K. (2024). Measures and Preventions of Cyber Policies in Smart Cities. In Digital Technologies in Modeling and Management: Insights in Education and Industry (pp. 244-262). IGI Global.

Chauhan, D., Jain, J. K., & Bahad, P. (2021). Performance evaluation of 802.11 A/G wireless networks with IP6HC. *Journal of Management Information and Decision Sciences*, 24, 1–7.

Chauhan, D., Jain, J. K., & Singh, A. (2024). Deployment of Edge Computing for Smart Healthcare Systems on Cloud Computing Platform.

Chauhan, D., Sharma, S., & Singh, V. (2023). Roadmap from 4G to 6G in smart cities impact and challenges. *Int. J. Wirel. Microw. Technol.*, 13(6), 39–50. DOI: 10.5815/ijwmt.2023.06.05

Christy, G. S., Sundari, G., & Sonti, V. K. (2022, April). A Review on Evolution, Challenges and Scope in Visual Light Communication Systems. In 2022 IEEE 7th International conference for Convergence in Technology (I2CT) (pp. 1-5). IEEE.

Darusalam, U., Nathasia, N. D., Zarlis, M., & Priambodo, P. S. (2024). Simulating Free-Space Optical Communications to Support a Li-Fi Access Network in a Smart City Concept. CommIT (Communication and Information Technology). *Journal*, 18(1), 17–27.

Den Hartog, F., Raschella, A., Bouhafs, F., Kempker, P., Boltjes, B., & Seyedebrahimi, M. (2017, November). A pathway to solving the Wi-Fi tragedy of the commons in apartment blocks. In 2017 27th International Telecommunication Networks and Applications Conference (ITNAC) (pp. 1-6). IEEE. DOI: 10.1109/ATNAC.2017.8215382

Diambeki, D. D., Mandiya, R. E., Kyamakya, K., & Kasereka, S. K. (2022). Securing the light escaping in a Li-Fi network environment. *Procedia Computer Science*, 201, 684–689. DOI: 10.1016/j.procs.2022.03.091

Faruq, O., Shahriar Rahman, K. R., Jahan, N., Rokoni, S., & Rabeya, M. (2023). Li-Fi technology based long range free-space communication data transmit system evaluation. *International Review of Applied Sciences and Engineering*, 14(3), 413–425. DOI: 10.1556/1848.2023.00620

Fazea, Y., Mohammed, F., & Al-Nahari, A. (2021, December). A review on 5G technology in IoT-application based on light fidelity (Li-Fi) indoor communication. In *International Conference of Reliable Information and Communication Technology* (pp. 371-384). Cham: Springer International Publishing.

Füchtenhans, M., Grosse, E. H., & Glock, C. H. (2021). Smart lighting systems: State-of-the-art and potential applications in warehouse order picking. *International Journal of Production Research*, 59(12), 3817–3839. DOI: 10.1080/00207543.2021.1897177

Hesham, H., Ismail, T., & Darweesh, M. S. (2020, July). Indoor localization and movement prediction algorithms with light-fidelity. In 2020 22nd international conference on transparent optical networks (ICTON) (pp. 1-4). IEEE. DOI: 10.1109/ICTON51198.2020.9203505

Ho, P. W., Cheng, C. H., Liao, Y. S., Chi, Y. C., Sadhu, A. S., Matsumoto, A., Akahane, K., Chen, L.-Y., Kuo, H.-C., & Lin, G. R. (2024). Blue micro-LED with a red/green blended polymer film for 3.5-Gbps visible light communication employing adaptive SNR-Flattening Algorithm. *Next Nanotechnology*, 5, 100043. DOI: 10.1016/j.nxnano.2024.100043

Hussain, F., Khupse, S., & Khan, M. R. S. (2019). An Empirical Study on Li-Fi Technology. *International Research Journal of Innovations in Engineering and Technology*, 3(11), 58.

Jain, J. K., & Chauhan, D. (2021). An Energy-Efficient Model for Internet of Things Using Compressive Sensing. *Journal of Management Information and Decision Sciences*, 24, 1–7.

Jamuna, R., Abarna, B., Kumar, A. S., Ganesh, L., & Pranesh, H. (2023). LIFI for Smart Transportation: Enabling Secure and Safe Vehicular Communication. Journal of Xi'an Shiyou University. *Natural Sciences Education*, 19(5), 1–8.

Janjua, J. I., Khan, T. A., Khan, M. S., & Nadeem, M. (2021, October). Li-Fi communications in smart cities for truly connected vehicles. In 2021 2nd International Conference On Smart Cities, Automation & Intelligent Computing Systems (ICON-SONICS) (pp. 1-6). IEEE. DOI: 10.1109/ICON-SONICS53103.2021.9617200

Játiva, P. P., Azurdia-Meza, C. A., Cañizares, M. R., Sánchez, I., Seguel, F., Zabala-Blanco, D., & Carrera, D. F. (2022). Performance analysis of IEEE 802.15. 7-based visible light communication systems in underground mine environments. *Photonic Network Communications*, 43(1), 23–33. DOI: 10.1007/s11107-022-00965-z

Jin, Z., Lin, R., Liu, X., Sun, D., Liu, B., Tao, T., & Tian, P. (2024). Green micro-LED With a Bandwidth Exceeding 2 GHz for 9-Gbps Visible Light Communication Based on SNR Gap-Dependent Bit and Power Loading. Journal of Lightwave Technology.

Karthika, R., & Balakrishnan, S. (2015). Wireless communication using Li-Fi technology. SSRG International Journal of Electronics and Communication Engineering (SSRG-IJECE), 2(3), 32-40.

Katz, M., & Ahmed, I. (2020). Opportunities and challenges for visible light communications in 6G. 2020 2nd 6G wireless summit (6G SUMMIT), 1-5.

Kaur, T., Sharma, Y., & Sikand, R. (2024, July). Futuristic aspects of Li-Fi technology. In AIP conference proceedings (Vol. 3121, No. 1). AIP Publishing. DOI: 10.1063/5.0221807

Kharche, S., & Kharche, J. (2023). 6G Intelligent Healthcare Framework: A Review on Role of Technologies, Challenges and Future Directions. J. *Mobile Multimedia*, 19(3), 603–644. DOI: 10.13052/jmm1550-4646.1931

Km, A., & Duttagupta, S. (2023). Hdl-ready mac layer implementation for multi-node Li-Fi communications. *International Journal of Information Technology : an Official Journal of Bharati Vidyapeeth's Institute of Computer Applications and Management*, 15(4), 2039–2051. DOI: 10.1007/s41870-023-01255-1

Lambrechts, W., Sinha, S., Lambrechts, W., & Sinha, S. (2019). Bridging the Digital Divide: Innovative Last Mile Internet Connectivity Solutions. Last Mile Internet Access for Emerging Economies, 1-40. Chauhan, D., & Jain, J. K. (2023, January). Profiling Network Traffic by Using Classification Techniques in Machine Learning. In International Conference on Smart Trends in Computing and Communications (pp. 113-123). Singapore: Springer Nature Singapore.

Linnartz, J. P., Corrêa, C. R. B., Cunha, T. B., Tangdiongga, E., Koonen, T., Deng, X., & Jungnickel, V. (2021, June). ELIoT: New features in LiFi for next-generation IoT. In 2021 Joint European Conference on Networks and Communications & 6G Summit (EuCNC/6G Summit) (pp. 148-153). IEEE.

Manvi Chopra, T. (2022). A Review Paper on on Light Fidility Technology and its Applications. *Journal of Algebraic Statistics*, 13(2), 1287–1293.

Martínez, V. M., Ribeiro, M. R., & Mota, V. F. (2024). Wi-Fi faces the new wireless ecosystem: A critical review. *Annales des Télécommunications*, 79(5), 397–413. DOI: 10.1007/s12243-023-00995-2

Mekala, M. S., Viswanathan, P., Srinivasu, N., & Varma, G. P. S. (2019, December). Accurate decision-making system for mining environment using Li-Fi 5G technology over IoT framework. In 2019 International Conference on contemporary Computing and Informatics (IC3I) (pp. 74-79). IEEE.

Mishra, P., & Singh, G. (2023). Sustainable Smart City to Society 5.0. In *Sustainable Smart Cities: Enabling Technologies, Energy Trends and Potential Applications* (pp. 39–57). Springer International Publishing. DOI: 10.1007/978-3-031-33354-5_2

Mohsin, M. J., & Murdas, I. A. (2023). Performance analysis of an outdoor Li-Fi system-based AO-OFDM architecture under different FSO turbulence and weather conditions. *Optik (Stuttgart)*, 273, 170427. DOI: 10.1016/j.ijleo.2022.170427

Mubashar, R., Siddique, M. A. B., Rehman, A. U., Asad, A., & Rasool, A. (2021). Comparative performance analysis of short-range wireless protocols for wireless personal area network. *Iran Journal of Computer Science*, 4(3), 201–210. DOI: 10.1007/s42044-021-00087-1

Ojha, V., Sharma, A., & Gupta, S. (2022). Secure sound and data communication via Li-Fi. In *An Interdisciplinary Approach to Modern Network Security* (pp. 153–165). CRC Press. DOI: 10.1201/9781003147176-9

Paul, R., Sebastian, N., Yadukrishnan, P. S., & Vinod, P. (2021). Study on data transmission using Li-Fi in vehicle to vehicle anti-collision system. *Computer Networks, Big Data and IoTProceedings of ICCBI*, 2020, 519–540.

Perwej, Y. (2017). The next generation of wireless communication using Li-Fi (Light Fidelity) technology. *Journal of Computer Networks*, 4(1), 20–29. DOI: 10.12691/jcn-4-1-3

Press, C. (2017). *IoT Fundamentals: Networking Technologies*. Protocols, and Use Cases for the Internet of Things.

Rabia, N. A., Ali, S., Sajid, A., & Zafar, A. (2020). A security review over wi-fi and li-fi. Information Management and Computer Science, 3(1), 1-9.

Rao, C. M., Prasuna, G., Chapala, H. K., Jeebaratnam, N., Navulla, D., & Verma, A. (2023, February). Designing a reliable and cost-effective Internet of Medical Things (IoMT) topology to minimize the maintenance and deployment cost. In 2023 Fifth International Conference on Electrical, Computer and Communication Technologies (ICECCT) (pp. 1-7). IEEE.

Ren, Z., Wang, Z., Sun, Z., Guo, Y., Song, W., Zhang, H., Chen, C., Guo, B., Yu, Z., Zhou, X., & Zhang, D. (2024). Characterizing the Through-Wall Sensing Mechanism of Wi-Fi Signals with a Refraction-Aware Fresnel Zone Model. *IEEE Transactions on Mobile Computing*, 23(12), 13438–13454. DOI: 10.1109/TMC.2024.3425847

Rocher, G., Tigli, J. Y., Lavirotte, S., & Ferry, N. (2024, July). A Framework Towards Assessing the Resilience of Urban Transport Systems. In *Proceedings of the 19th International Conference on Availability, Reliability and Security* (pp. 1-10). DOI: 10.1145/3664476.3670435

Safitri, C., Harjono, M. S., Hasrito, E. S., & Roestam, R. (2024). Comprehensive Survey: Quality of Service in Railway Communication Using Information-Centric Networking and Light Fidelity. *IEEE Transactions on Intelligent Transportation Systems*, 25(12), 19218–19251. DOI: 10.1109/TITS.2024.3472698

Safitri, C., Galina, M., & Simatupang, J. W. (2022). Light-fidelity as next generation network technology: a bibliometric survey and analysis.

Salman, K. D., & Hamza, E. K. (2021). Visible light fidelity technology: Survey. *Iraqi J. Comput. Commun. Control Syst. Eng*, 21(2), 1–15.

Sarkar, A., Agarwal, S., & Nath, A. (2015). Li-Fi technology: data transmission through visible light. International Journal of Advance Research in Computer Science and Management Studies, 3(6).

Selvi, S. A., Rajesh, R. S., & Ajisha, M. A. T. (2019, December). An efficient communication scheme for Wi-Li-Fi network framework. In 2019 Third International conference on I-SMAC (IoT in Social, Mobile, Analytics and Cloud)(I-SMAC) (pp. 697-701). IEEE. DOI: 10.1109/I-SMAC47947.2019.9032650

Sharma, P. K., Ryu, J. H., Park, K. Y., Park, J. H., & Park, J. H. (2018). Li-Fi based on security cloud framework for future IT environment. *Human-centric Computing and Information Sciences*, 8(1), 1–13. DOI: 10.1186/s13673-018-0146-5

Spagnolo, G. S., Cozzella, L., Leccese, F., Sangiovanni, S., Podestà, L., & Piuzzi, E. (2020, June). Optical wireless communication and Li-Fi: a new infrastructure for wireless communication in saving energy era. In 2020 IEEE International Workshop on Metrology for Industry 4.0 & IoT (pp. 674-678). IEEE. DOI: 10.1109/MetroInd4.0IoT48571.2020.9138180

Subha, T. D., Subash, T. D., Rani, N. E., & Janani, P. (2020). Li-Fi: A revolution in wireless networking. *Materials Today: Proceedings*, 24, 2403–2413. DOI: 10.1016/j.matpr.2020.03.770

Suriyan, G. R., Rahul, K., Rajesh, S., Dhanalakshmi, C., & Udhayakumar, G. (2023, June). Prevention of Road Accidents by Interconnecting Vehicles using LiFi and LoRaWAN Technologies. In *2023 International Conference on Sustainable Computing and Smart Systems (ICSCSS)* (pp. 1383-1387). IEEE. DOI: 10.1109/ICSCSS57650.2023.10169475

Systems, M. I. (2023). Retracted: Assessment and Test-Case Study of Wi-Fi Security through the Wardriving Technique.

Uzair, M. (2022). Vehicular wireless communication standards: challenges and comparison. International Journal of Electrical and Computer Engineering Systems, 13(5), 379-397.

Wei, Z., Wang, L., Li, Z., Chen, C. J., Wu, M. C., Wang, L., & Fu, H. Y. (2022). Micro-LEDs illuminate visible light communication. *IEEE Communications Magazine*, 61(4), 108–114. DOI: 10.1109/MCOM.002.2200109

Chapter 15
Intelligent and Optical Security Monitoring for 6G Communications

Sindhumitha Kulandaivel
National Institute of Technology, India

Madhumitha Kulandaivel
SRM University, India

Keerthana N. V.
https://orcid.org/0000-0001-8878-2083
Velalar College of Engineering and Technology, India

ABSTRACT

The infrastructure of communication technologies in recent years has grown more complex and autonomous to meet societal requirements. Advanced technologies such as optical communication due to their enormous bandwidth and less latency capabilities have become the most sought-after candidates for 6G communications. The optical networks provide an agile and secure platform for effective data transmission. Thus, it is important to include cognitive and intelligent security monitoring for autonomous optical network management to improve reliability. Machine learning is a great tool that provides advancements in security monitoring in terms of accuracy and cost-effectiveness for optical network diagnosis and management. This chapter discusses the vulnerability of optical networks to various security threats in the physical layer along with countermeasures for network survivability. Several monitoring techniques using various performance parameters are also discussed for effective attack identification and localization to ensure robust optical network performance for 6G communications.

1. INTRODUCTION

The fast-growing data consumption of the advanced applications of individual user practices and the exponentially growing number of users have resulted in the surge of data quantity and quality requirements. To meet these never-ending demands the communication networks are shifting to advanced technologies

DOI: 10.4018/979-8-3693-8799-3.ch015

such as optical communications which provide huge bandwidth capabilities, high data rates, long-distance transmission, and less loss and latency to provide efficient data transmission. Optical communication technologies are being actively embraced as an integral part of future communication networks such as 6G technology (Jeon et al., 2023). Communication technologies are evolving distinctively each decade to meet user requirements. In that way, the 6G technology aims to connect a massive number of devices at ultra-high speeds and minimal latency to enable robust networks.

The optical networks can be prone to varying levels of vulnerabilities resulting in service disruption to perform physical layer attacks (Skorin-Kapov et al., 2016). The attacks can be direct or indirect for example cutting the fiber is a direct attack that causes information loss also known as eavesdropping and introducing hard-to-detect jamming signals is an indirect attack that is more sophisticated to identify. The attack methods can vary depending upon the type, the damage potential of an attack, and the difficulty level of identifying the attack to apply appropriate counteractive measures. The extent of information loss depends upon the spectral strength of the jamming signal.

Security attacks can also be performed on the optical signal without necessarily affecting the optical fiber by introducing polarization scramblers. This polarization attack fast varies the polarization angle leading to errors in the receiving end. The complex effect of the physical layer attacks on the signal quality and its identification is a challenging task to enable efficient transmission. It is essential to implement the detect-isolate-correct cycle to support intelligent autonomous optical networks to improve performance and security while maintaining lower costs. Optical signal security breaches in the physical layer can cause correlated failures in the upper layers making security monitoring a challenging task. Effective mitigation of these attacks includes implementing fast and effective corrective measures as well as adaptation of the network to prevent similar future attacks (Hugues-Salas et al., 2019).

In recent times, the integration of various advanced ML techniques into several aspects of optical networks such as optical signal quality monitoring (Tizikara et al., 2022), resource allocation, and routing (Moharrami et al., 2017) has fostered a powerful tool for efficient autonomous optical networks. Hence, ML-based security monitoring in the physical layer has been successfully implemented to detect jamming and polarization attacks (Furdek et al., 2020). ML provides an effective alternative to existing traditional analytical methods of security monitoring using predetermined thresholds as an indicator. These threshold indicators are unreliable and ineffective in the case of reconfigurable and complex network architectures. ML has been successfully applied to various optical network problems such as light path establishing and rerouting, quality of transmission (QoT) estimation (Zhang et al., 2022), transmitter configuration (Musumeci et al., 2018), fault detection and diagnosis, non-linearity and filtering effects monitoring (Catanese et al., 2019) to maintain efficient error-free transmission. The various ML-based approaches are based on either supervised, unsupervised, or semi-supervised algorithms to perform the desired task. Since physical layer attacks can cause varying effects on the optical signal performance parameters, determining exact thresholds for applying countermeasures cannot be accurately determined. ML has proved to be a dependable solution for predictive identification and estimation for supporting network adaptability.

Various works have been carried out using techniques such as support vector machine (SVM), k-nearest neighbors (k-NN), artificial neural network (ANN), and deep learning (DL) for the detection of possible attacks using several ways at both link and network levels. For example, the optical signal distortion in the optical spectrum at the receiving end is monitored to identify the type and degree of attack. The ML algorithms are trained using optical performance monitoring (OPM) parameters for accurate identification of security breaches. On the other hand, using unsupervised and semi-supervised algorithms that are not trained with the OPM data has the capability of identifying novel and previously unfamiliar attack methods.

The two critical requirements for effective real-time implementation of ML-based security monitoring involve: First, monitoring of the optical signals at multiple intermediate nodes for constant and accurate localization of the attack, and secondly, the ML models must be less complex and more accurate for practical integration and interoperability. The security monitor must have access to various optical elements and execute corrective measures quickly thereby providing reduced latency with network adaptation. Also, key performance indicators as listed below greatly affect the feasibility and efficacy of the ML model used for security monitoring:

- Data acquiring time,
- Security monitoring time (identification and classification of the attack),
- Accuracy of the ML model,
- Time taken to apply appropriate corrective measures (send the message to the optical path controller),
- Accurate localization of the attack and time taken to localize,
- Time taken to implement permanent remediation for the attack.

The paradigm shift towards optical communication to meet catastrophic demands has made optical networks a potential target of security attacks. Hence, it is important to protect at the physical layer level to prevent information loss. In addition, due to the complex nature and the reconfigurability of the optical networks, security monitoring has to be performed continuously and simultaneously for robust operation. This chapter deals with the advanced ML and DL models for effective attack identification and correction mechanisms for efficient and secure optical networks for 6G communications. The continuous upgradation and development of advanced security monitors are thus essential to build reliable networks for effective communication.

In this context, integrating Radio Frequency Identification (RFID) and microwave technologies into the optical network architecture offers innovative solutions for enhancing security at both the physical and communication layers. RFID provides effective monitoring and authentication capabilities, enabling secure access control, asset tracking, and real-time threat detection. Microwave technologies, often used in wireless backhaul communication, complement optical networks by ensuring robust connectivity in remote or complex environments. By combining RFID and microwave with optical communication systems, we can create a multi-layered security framework that ensures continuous monitoring and defense against emerging threats.

2. TYPES OF PHYSICAL LAYER ATTACKS

The size and complexity of the optical networks can vary depending on their application whether it is implemented in local area networks (LAN) or optical fiber internet provision. The security threats irrespective of the type of optical network can be broadly classified as:

(a) Confidentiality

The security threats are where the unauthorized entity tries to listen to sensitive information. This attack is encountered when the attacker tries to physically tap the optical signal from the fiber for eavesdropping the sensitive information. This might appear comparatively easier attack and can be performed on a physically exposed fiber by just cutting the fiber protection and capturing the small amount of light that escapes through the cut. The difficulties in performing the attack include potential damage to the optical fiber that leads to easy identification of failure. Also, the amount of signal tapped is much less and the attacker must operate at a very low signal-to-noise ratio. Another way of confidentiality attack is eavesdropping on the residual crosstalk as a subscriber in wavelength division multiplexing (WDM) networks. This type of eavesdropping requires sophisticated equipment for effectively extracting the weak residual crosstalk signal.

(b) Authenticity and Integrity

In this security threat, the unauthorized entity tries to communicate or alter the information through the channel. The entity tries to impersonate a sender to transmit altered information to the recipient. However, coding schemes such as orthogonal code division multiple access (OCDMA) use unique codes for encryption and decryption of the messages (Bhanja et al., 2020). Hence, the unauthorized entity cannot communicate or manipulate the information transmitted through optical channels with proper coding and encryption schemes.

(c) Privacy

The entity observes the existence of the communication. To enhance the privacy of communication advanced methods such as steganography are used (Zhang et al., 2022). The basic principle of steganography is generating a series of short optical pulses along with the information signal to hide the optical channel. The sender and the receiver are only aware of the signal transmission. The security attacks in the physical layer on the other hand can be classified into two major groups depending on their effects as depicted in Figure 1 which include:

1. Passive attacks that aim at eavesdropping unauthorized information
2. Active attacks or service disruptive attacks that interfere with information transmission

Figure 1. Types of physical layer attacks in fiber optic communication.

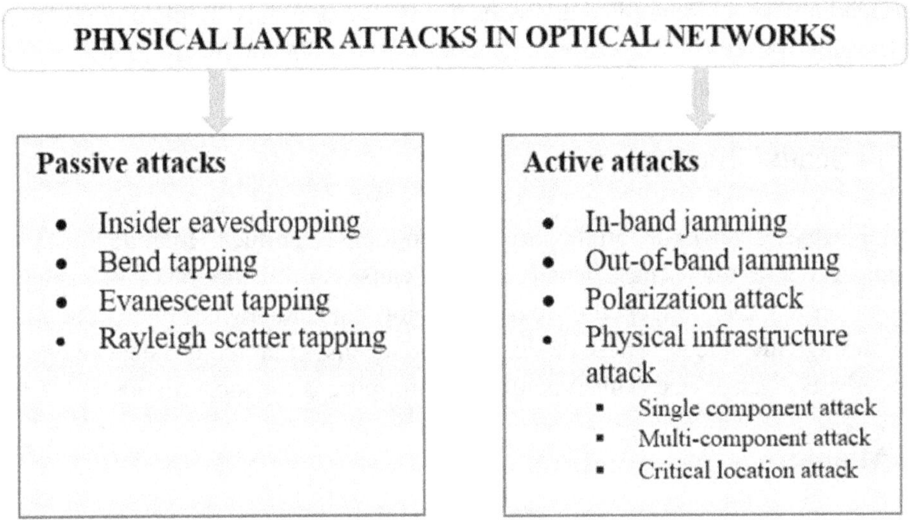

2.1 Passive Attacks

The passive threats aim at eavesdropping unauthorized information without affecting the quality of the signal making it one of the hard to detect and potential security breaches in the optical communication networks. Various eavesdropping techniques have been developed to invade privacy in communication (Maslo et al., 2021).

2.1.1 Insider Eavesdropping

In this type of eavesdropping the attacker impersonates as a subscriber in a dense WDM network and listens to information from adjacent channels. It can also be performed at switching or monitoring nodes in DWDM networks. These methods of invading require direct contact with optical fiber but cutting through the protective layer on the fiber makes it prone to damage which is complicated to perform.

2.1.2 Bend Tapping

In bend tapping when direct contact with the optical fiber is established, the optical fiber is bent to the point where a fraction of power is decoupled due to compromise in total internal reflection. Since a minimum fraction of light is extracted for eavesdropping causing very little effect on the signal power budget at the receiver which can still be detected and hence the least used technique.

2.1.3 Evanescent Tapping

Evanescent tapping is carried out by diverting a fraction of the optical signal into another optical fiber without bending. This tapping is done by capturing the crosstalk by placing the tapping fiber on the polished transmission fiber. Since this attack does not involve any bending, it is hard to locate and requires a precision instrument for fiber processing.

2.1.4 Raleigh Scatter-Tapping

The material density changes at certain parts of the optical fiber due to manufacturing defects and is used for Raleigh scatter tapping. These density changes cause natural crosstalk and leakage of fractional optical signal power that can be tapped for eavesdropping. Special devices detect the small amount of light scattered through the defects. These devices comprise lenses that focus the leaked or scattered light onto the attacker's receiver device. This type of eavesdropping is the most difficult to detect.

2.2 Active Attacks

The active attacks are caused by inserting harmful signals into the transmission channel or by causing physical damage to the network that causes signal degradation or disruption of communication. These kinds of attacks are hard and expensive to restore. The various attack scenarios are described below:

2.2.1 In-Band Jamming

Jamming signals are introduced with relatively higher optical power to cause damage or crosstalk within the target channel bandwidth. The signal degradation caused depends upon the modulation format as well as the signal power of the transmitted signal and the jamming signal. Service disruption is caused when the received signal has a bit-error rate (BER) higher than the threshold value of the forward error correction (FEC) decoder. The in-band jamming can be carried out indirectly or directly. In indirect jamming, the jamming signal is inserted through the unused ports of the wavelength selective switch (WSS) into the target channel. In case of direct jamming the jamming signal is introduced through the couplers or monitoring ports into the target channel.

2.2.2 Out-of-Band Crosstalk

The complexity of optical networks to meet the flexibility and reconfigurable demands led to the integration of various advanced modulation formats, bit rates, symbol rates, guard bands, and launch powers. These parameters introduce non-linear effects such as cross-phase modulation (XPM) and cross-polarization modulation (XPolM). Hence a larger guard band is required to minimize these effects. This gives way to a potential out-of-band attack by inserting a high power and low bit rate jamming signal near the target channel to increase the optical signal-to-noise ratio (OSNR) penalty of the target channels.

2.2.3 Polarization Attack

A polarization attack is a type of service disruption attack without any intervention in the optical fiber. This kind of attack uses polarization scramblers to scramble the polarization of the transmitted signal which is fast enough for the recovery algorithms to produce failures. This polarization attack can be easily implemented by using a clip-like device on the optical bare fiber with piezoelectric actuators to scramble the polarizing angle of the optical signal. This attack is difficult to locate as it does not add any additional loss.

2.2.4 Physical Infrastructure Attack

The physical infrastructure attacks aim at infrastructure damage to the optical network that affects the connectivity of the transmission channel. They are classified into three major types:

a. ***Single component attack:*** This causes failure of a single element in the optical network such as cut in the optical fiber, switch or amplifier damages, etc.,
b. ***Multi-component attack:*** This leads to the failure of multiple network elements in coordination to disrupt service in a particular geographical location.
c. ***Critical location attack:*** This type of attack is performed at some critical locations of the networks such as a hub that results in a chain of damages to disrupt the entire network.

3. OPTICAL NETWORK SURVIVABILITY TO PHYSICAL LAYER ATTACKS

The optical networks must provide survivability to the service disruption attacks for establishing reliable communication. Survivability in the physical layer includes the detection of an intervention or a failure and taking effective corrective measures to continue transmission. The main steps involved in the network survivability are depicted in Figure 2.

Figure 2. Steps involved in optical network survivability

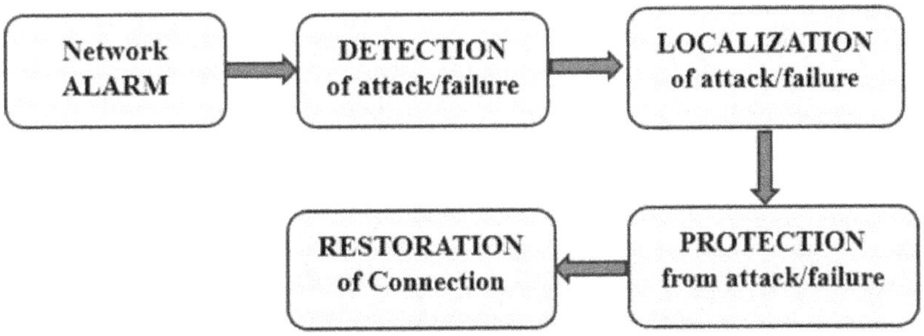

The best way of providing security to the network is by addressing the issues. The issue when detected at the physical layer an alarm is raised first. Next, the issue is identified and correctly localized. Necessary corrective measures are taken to tackle the issue and the connection is restored. While accurate attack identification and localization are discussed briefly in the later sections, the primary step involved in network survivability is data protection. The entire process of the optical network survivability and its mitigation of the physical layer attack is diagrammatically represented in Figure 3. The different measures are considered to protect the data being transmitted to meet the below-listed security requirements:

3.1 Confidentiality

The confidentiality of the data is protected by using complex encoding and decoding methods in addition to encryption techniques to prevent unauthorized access to data. Orthogonal code division multiple access (OCDMA) is the most used encoding method which preserves confidentiality using coding keys. The OCDMA codes are classified into coherent and non-coherent OCDMA codes. An example of coherent OCDMA is spectral-phase encoding (SPE) where the spectral components are applied with different phase shifts which are conjugated at the receiving end for the decoding process. On the other hand, incoherent OCDMA performs intensity modulation and direct detection.

Encryption serves as a great tool to provide confidentiality of the information from the eavesdropper. The data cannot be extracted easily from the cipher text without knowing the decryption key. Optical encryption is provided using various XOR gates which are then encoded with OCDMA to provide double-level protection. The received signal is then decoded and decrypted at the receiver.

3.2 Authentication

The OCDMA codes also provide authentication to the physical layer of the optical network. The coding/decoding schemes provide a unique identity to the individual users. The optical signal can be transmitted from source to destination if it has the desired wavelength and temporal synchronization which can be prone to various security threats. With OCDMA codes a kind of authentication is provided to the sender and the receiver to safeguard the information. An unauthorized intruder cannot access the information transmitted without the knowledge of the code used for decoding the signal. Hence, OCDMA serves the dual purpose of providing authenticity as well as multiple access capability.

Figure 3. Diagrammatic representation of the entire process of optical network survivability and mitigation of physical layer attacks

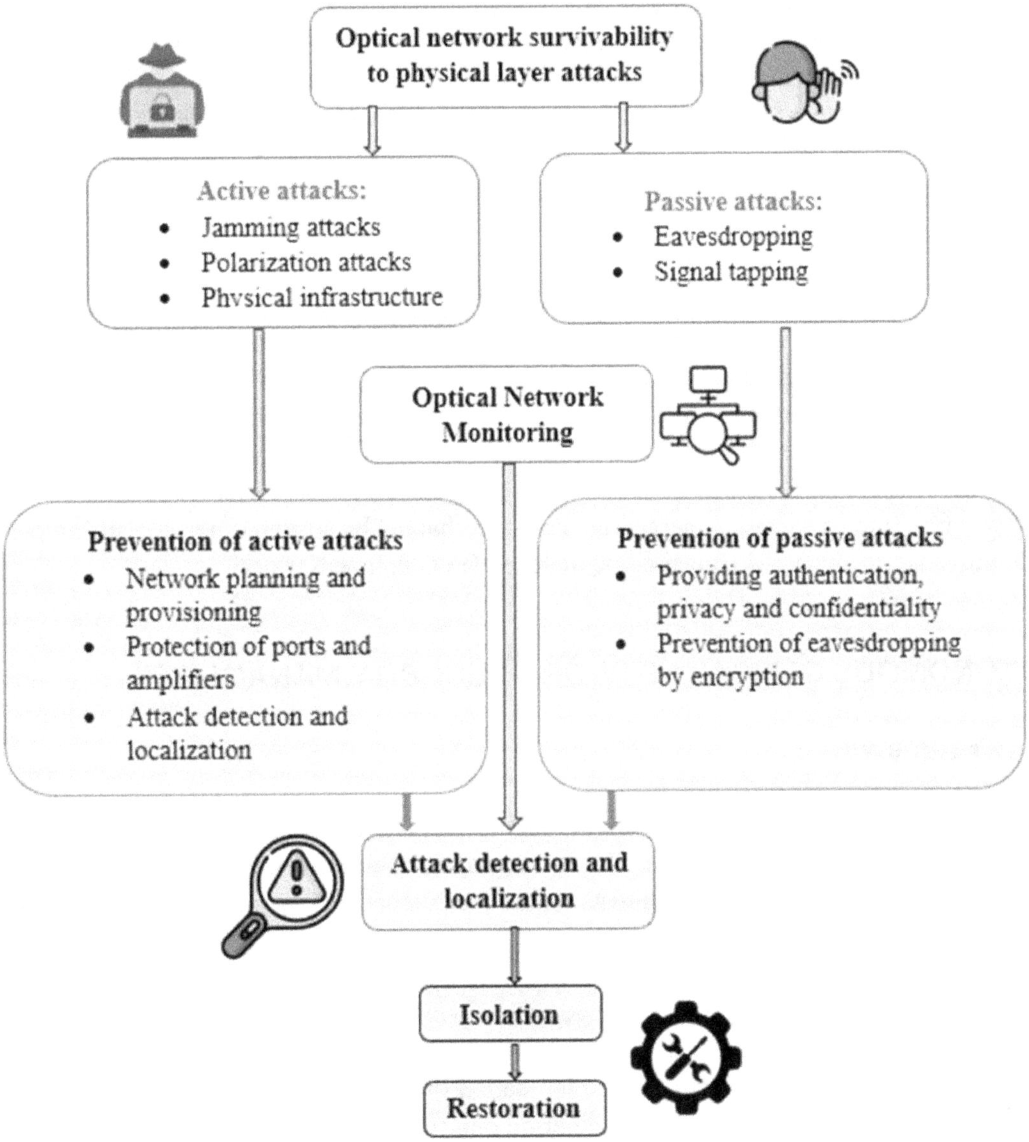

3.3 Availability

The optical networks are vulnerable to several physical attacks on the infrastructure leading to disruption in services. The infrastructural attacks do not necessarily mean information theft but service disruptions cause financial losses due to affected network resources that impact several users. For example, the damage to optical fibers can be caused intentionally by eavesdropping or unintentionally due to construction activities regardless of which affects the signal transmission. The optical networks must be capable of self-healing providing alternate paths to ensure availability and survivability. Various telecommunication network architectures such as metropolitan area networks (MAN) and local area networks (LAN) are implemented in such a way that the link failures are restored within 60ms for effective operation (Sharma et al., 2021). Hence, bandwidth-efficient OCDMA ring architectures are used to achieve less restoration time. Also, to prevent jamming signals from affecting the signal transmission wavelength convertors can be used to transmit the signal in an alternate wavelength other than the jamming signal to ensure availability.

3.4 Privacy

Privacy in telecommunication networks can be improved by using methods such as steganography for hiding messages such that only the sender and the receiver are aware of the signal transmission. The data is transmitted through stealth channel which is hidden from the public channels by using a much lower data rate than that is used in the public channels. The steganography is principally based on generating a series of short optical stealth pulses and using high group-velocity dispersion (GCD) to temporally stretch them. This causes a significant reduction in the peak amplitudes even compared to the channel noise. Therefore, the signal is submerged beneath the channel background noise ensuring privacy. At the receiving end, the signal is recovered by matched GVD compensation.

4. AUTONOMOUS OPTICAL NETWORK SECURITY MANAGEMENT

4.1 The Framework

Optical Network Security Management (ONSM) generally counts on three core features which are: Management, Cognition, and Response. The ***Management of threats*** involves creating accurate threat models to study the various effects of threats on the performance of signal transmission. Based on the study the network architectures are designed such that the attack-prone surfaces are minimized to ensure robust transmission. To provide disruption-free services, it is important to continuously apply updated threat models to ensure network designs with enhanced security. Many works have explored the possible ways to tackle intentional fiber-cut attacks (Furdek et al., 2022) as well as many attack-aware routing protocols (Goscien et al., 2020) have evolved for static as well as dynamic users.

Cognition of security attacks involves gathering optical network data which are both representative and indicative. It includes a deep analysis of the collected data. Correctly attributing observed trends to different types of security breaches is crucial. Detailed knowledge about possible attack entry points is necessary along with familiarity with the signatures of different attack methods is essential. These signatures indicate the effects on distinct OPM (Optical Performance Monitoring) parameters. The sparse deployment of costly Optical Performance Monitoring (OPM) equipment poses a challenge in collecting real-time OPM data in optical networks. However, advancements in technology, particularly in coherent receivers with rich Digital Signal Processing (DSP) functionalities, have helped overcome this challenge.

The latest generation of coherent receivers is equipped with advanced DSP functionalities. Coherent receivers can process optical signals digitally, providing a more detailed and comprehensive analysis of the received data. Coherent receivers, by leveraging DSP capabilities, can collect a rich OPM dataset at the destination of each optical connection. This dataset includes information about various optical parameters, providing insights into the performance of the network. The collected OPM data is exposed to the Network Management System (NMS) through standardized interfaces. Standardized interfaces facilitate seamless communication and integration with the NMS, allowing for centralized management and control. The collected OPM data becomes a valuable resource for detecting and localizing security breaches in the optical network. Approaches such as localizing high-power jamming signals and designing attack monitoring probes have been proposed to enhance the accuracy of breach localization. The data collected by coherent receivers is instrumental in monitoring the network for security threats. Localization of harmful connections and breached links is critical for effective incident response and securing the optical infrastructure.

The third feature, ***response to the attack*** goes beyond immediate incident handling. It encompasses strategies to recover from the incident, neutralize threats, and adapt the network to be more resilient against similar future attacks. The methods for network adaptation such as:

- **Attack-Aware Pre-Planning of Backup Resources** (Furdek et al., 2021): This involves planning backup resources with awareness of potential attack scenarios.
- **Fast Frequency Hopping** (Shi et al., 2021): Utilizing fast frequency hopping as a technique to enhance network security.
- **Connection Rerouting:** Redirecting network connections to alternative paths to avoid compromised elements.
- **Modulation Format and Spectrum Reassignment:** Changing modulation formats and reassigning spectrum usage, potentially described in a specific procedure (Perez-Lopez et al., 2021).
- **Periodical Proactive Resource Reallocation** (Moghaddam et al., 2020): Proactively reallocating network resources periodically to adapt to changing security needs.

The above-mentioned methods for network adaptation highlight a proactive approach, including planning, technology deployment, and dynamic resource reallocation to enhance overall network security. The explained three core features of autonomous optical network management are shown in Figure 4.

Figure 4. Three core features involved in autonomous optical network management

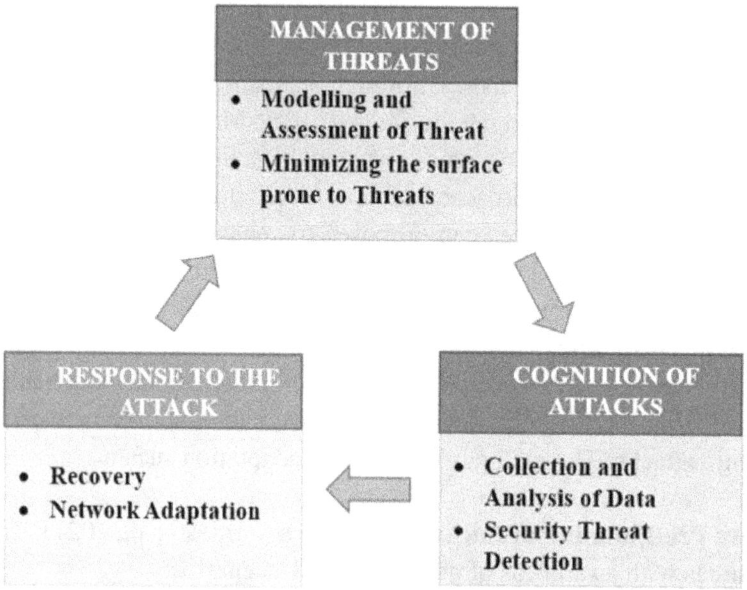

4.2 Architecture for Security Assurance

The security assurance (SA) of optical threats is addressed within the framework of evolving architectures, specifically Cognitive network management (CNM) (Yang et al., 2020). The evolution of optical network management, particularly with the advent of Transport Software Defined Networks (TSDN) and open-source multi-vendor network controllers, has brought about significant changes in the approach to optical network control and security. Network management and control are structured as a cross-layer comprehensive architecture. Multiple network controllers are employed, each dedicated to specific network layers or operation domains. One or more orchestrators facilitate the setup of end-to-end services on demand. Standard Application Programming Interfaces (APIs) are in place to provide easy access to network services. Users and external systems, including data center hypervisors, can interact with the network through these APIs. Analytics plays a crucial role in diagnosing the status of the network. It enables the implementation of appropriate countermeasures in response to network malfunctions.

Security assurance is identified as a crucial new feature for optical transport services. The concept of Optical Security Assurance (OSA) is introduced to enhance the security posture of optical networks. OSA is positioned at the borderline between network and security management. Tight integration with the processes of the Security Operation Center (SOC) is emphasized. It is crucial to avoid conflicts or superposition of roles and responsibilities between the Network Operation Center (NOC) and the SOC. Clear delineation and coordination between NOC and SOC functions are essential for effective security management. The introduction of security assurance is highlighted as a critical aspect that needs to be smoothly integrated into the existing T-SDN control architecture. This integration ensures that security measures are an integral part of the optical network's operational framework. The introduction of security assurance in optical networks is described as a non-trivial task and requires functionalities such as:

- **Telemetry:** OPM data refers to data that involves monitoring and measuring the performance of optical networks. The time it takes to acquire OPM data is identified as the most important Key Performance Indicator (KPI) for telemetry. Telemetry involves the automated collection and transmission of data from remote sources for monitoring and analysis. OPM data acquisition time is critical not only for general telemetry but specifically for updating the network attack status. ML algorithms play a major role in likely analyzing OPM data to identify and report on the status of potential network attacks.
- **Intrusion detection and identification:** Detection Time refers to the time it takes for an ML algorithm to detect the occurrence of an event or anomaly, such as a network security breach. Classification Time relates to the time required for the ML algorithm to categorize or classify the detected event. The overall accuracy of ML algorithms is considered a fundamental KPI. Accuracy reflects the algorithm's ability to correctly classify instances, capturing both true positives and true negatives. False Positive (FP) is the rate at which the ML algorithm incorrectly identifies a non-event as an event. It represents instances where the algorithm raises an alarm or flags an event that did not occur. False Negative (FN) is the rate at which the ML algorithm fails to detect an actual event. It represents instances where the algorithm misses a true event or anomaly.
- **Immediate and prompt response to the intrusion:** The KPI is defined as the time required for a specific function. The function involves identifying the proper quick-reaction strategy. A Quick reaction strategy refers to a responsive and timely approach to addressing a given situation or event. The strategy is identified through a search on a database (DB) of strategies that are predefined or defined as "a priori." Once the proper quick-reaction strategy is identified, the next step is to send the corresponding implementation request. This request is sent to the optical controller, which likely plays a role in implementing the identified strategy.
- **Localization of the threat:** The accuracy is assessed in terms of the function's ability to precisely localize the source or origin of a security attack is the most important KPI for this function. For example, a few tens of meters of localization accuracy are mentioned, indicating a high precision level. It likely refers to the time it takes for the system or function to pinpoint the location of the security attack. This measures the likelihood that the localization process might produce an error in identifying the accurate location of the attack source. These metrics are crucial for assessing the effectiveness and reliability of the system in detecting and localizing security attacks.
- **Providing accurate remediation measures**: The time required to deliver fixing information to SOC systems is the most crucial KPI for the described function. This suggests a focus on the efficiency and speed of information delivery to enable prompt responses and actions within the security operations context.

4.3 Security Manager

The above-described functions aim to collaborate seamlessly with the optical-layer controller. The optical-layer controller is identified as the transport-SDN component responsible for managing all optical-layer functions. The optical-layer controller exposes end-to-end optical transport services to a network orchestrator. Rather than redesigning the well-established architecture of optical-layer controllers (mentioned to be established in the open-source community), the proposal is to enhance it with additional functionality (Volkov et al., 2021). To complement the functions of the optical-layer controller, the introduction of a Security Manager (SM) is proposed as illustrated in Figure 5. The SM encompasses all optical security functions and is designed to be compatible with any optical controller. The OSM provides a suitable interface for Center of Security Operations Center (CSO) systems, indicating a connection to broader security management. The proposed architecture facilitates coordination with SOC systems and emphasizes the importance of compatibility with the established optical-layer controller architecture.

The SM consists of telemetry (Tm) and telemetry database (TmDB) blocks. These blocks retrieve and store coherent transceiver OPM data and possibly other relevant network status data. Two ML modules are mentioned: the attack detector and classifier, and the attack localizer. These modules use OPM data for algorithms related to attack detection, identification, and localization. The ML modules provide information on the attack condition to guide subsequent actions. A fast attack fixing processor is responsible for quickly deciding on a countermeasure to minimize service downtime (e.g., traffic rerouting) based on information from the ML modules. The countermeasure is implemented by querying an optical controller through an appropriate Processing Interface (PI). An attack remedy processor handles the task of a complete attack remedy. It uses attack classification and localization information to devise a comprehensive strategy. Unlike quick response actions, complete attack remedies may involve human intervention due to the nature of optical-layer attacks.

The SM can be easily integrated into the architecture of a T-SDN controller. It utilizes existing functions such as northbound and southbound interfaces, telemetry, and telemetry DB. The SM and the proactive fault detection module are designed as independent modules within the T-SDN controller. Both modules use TmDB information and interfaces but work independently, providing different types of information to network management applications. The functional independence of the SM from other network management blocks is mentioned to facilitate its integration into the T-SDN controller.

Figure 5. Illustration of the optical security manager to compliment the functions of the optical-layer controller

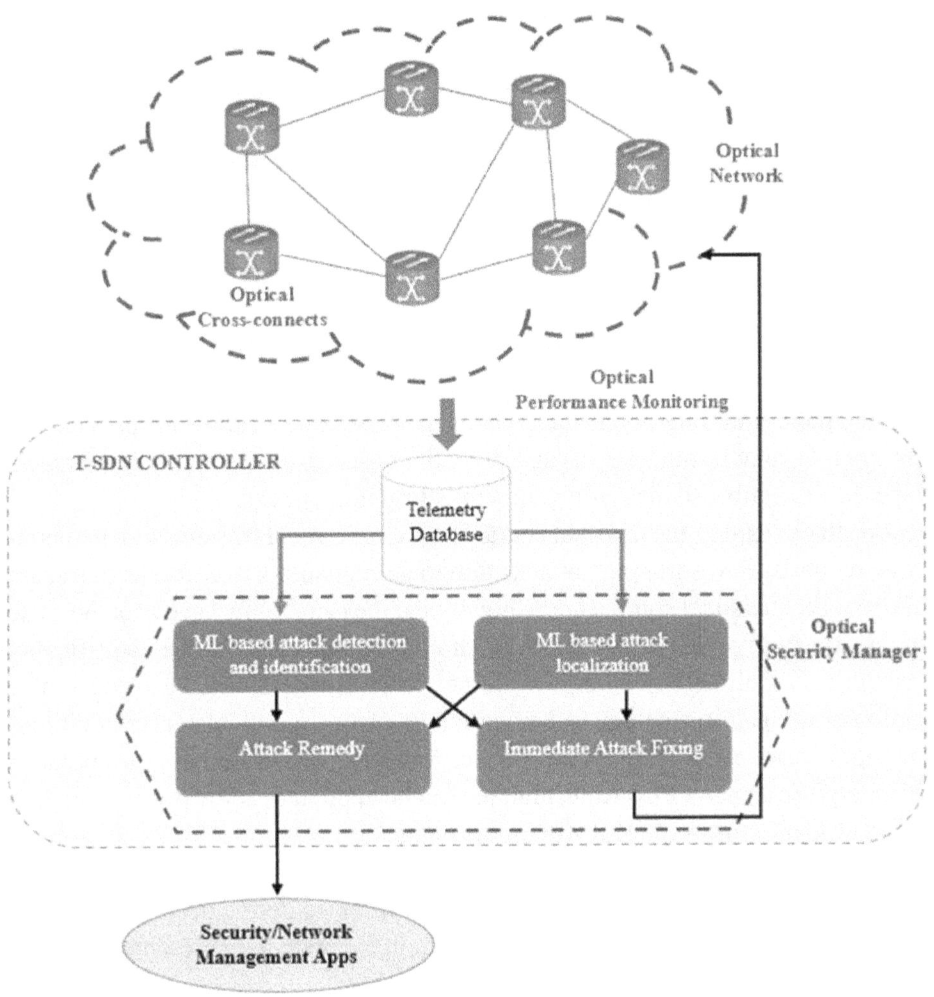

5. ML-AIDED OPTICAL SECURITY MONITORING

Artificial Intelligence (AI) is the imitation of biological processes, particularly human cognitive processes by systems to perform decision-making operations and learn from the previous knowledge gained for future predictions. AI has been extensively used in applications, such as speech and signal processing, smart assistants, automated vehicles, financial prediction, speech recognition, and intelligent home devices etc., These applications are already influencing and improving various aspects of human life, including productivity, safety, health, entertainment, and communication. AI is portrayed as enhancing existing applications, databases, and environments by adding knowledge and reasoning. The motive is to smart and friendlier system, that responds to changes in the external environment.

AI is described as not delivering completely autonomous systems but providing incremental improvements to existing ones. Each small breakthrough in AI research enables the expansion of skills to solve new classes and scales of problems, fostering research and innovation across scientific disciplines. Extensive research on improving telecommunication networks by applying AI-based techniques over the past decades has been carried out. Areas affected include transmission, switching, and network management. Optical communication networks and systems are specifically mentioned as adopting AI-based techniques. The application of AI extends from photonic devices to control and network management in the field of optical networking.

AI is supported by emerging transmission technologies that are the most sought-after in recent times which include space division multiplexing, multimode/multicore fibers, advanced modulation formats, and constellation shaping. Despite advances in monitoring and transmission technologies, the optical arena lacks significant developments in AI techniques for attack and intrusion detection (Furdek et al., 2020). With the increasing Internet proliferation, businesses and institutions provide essential services over the Internet, leading to higher risks and potential costs associated with data breaches. The optical network layer is susceptible to disruption and eavesdropping attacks.

Disruption attacks exploit the physical characteristics of optical transmission to degrade operation or block service requests. Eavesdropping attacks aim to gain unauthorized access to transmitted data over the optical network. Statistical approaches using optical measurements have been proposed for detecting and localizing attacks in optical networks. AI techniques could enhance the identification and localization of optical attacks, especially in large-scale networks. Successful identification of attacks leads to the application of mitigation techniques. Mitigation involves all-optical encryption or re-routing traffic away from attack locations. Statistical approaches for attack detection have computational complexity challenges in large-scale networks. AI techniques can be applied to address these challenges, improving the efficiency of identifying and localizing optical attacks.

ML techniques in optical communications are categorized into the type of task performed: regression (to predict future values from given data) and classification (to derive decision boundaries). The types are ML algorithms applied and their types are shown in Figure 6 as represented in (A. B. Nassif et al., 2019) and are explained in detail below.

Figure 6. Types of machine learning algorithms that can be applied to the area of optical communication (A. B. Nassif et al., 2019)

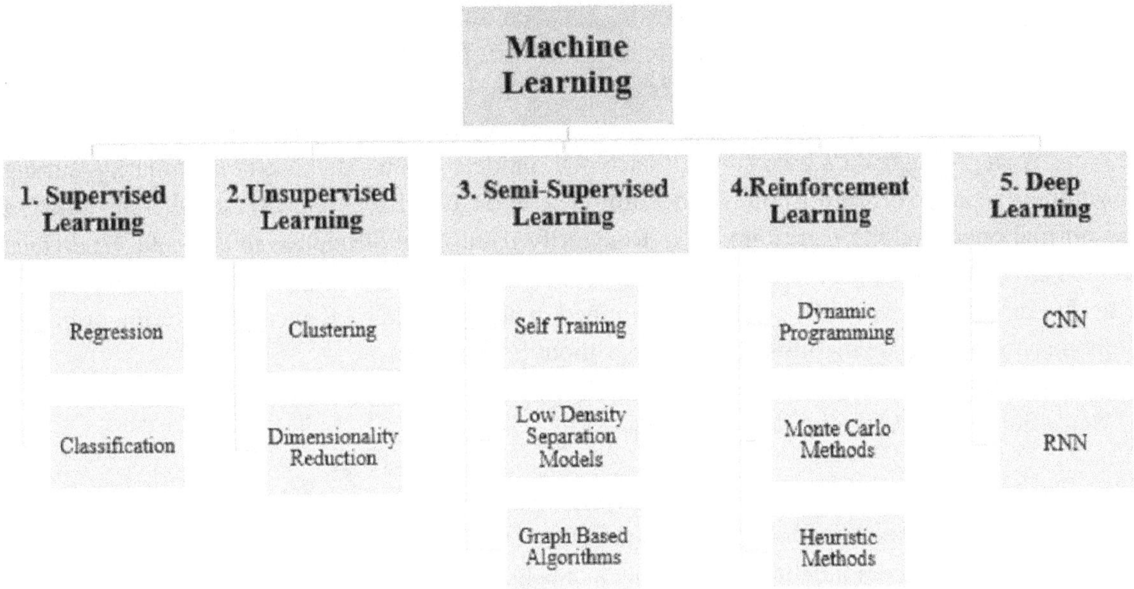

a. Supervised Learning (SL) techniques:

Examples of SL techniques include Artificial Neural Network (ANN) and Support Vector Machine (SVM) (Jiang et al., 2020). These techniques fall under the broader category of machine learning where the algorithm is trained on labeled data. SL techniques leverage a priori knowledge, meaning that during the training phase, the algorithm is provided with labeled data where the output (class label) is known. This a priori knowledge is crucial for the algorithm to learn patterns and relationships between inputs and outputs. Data can be labeled in different ways: Through carefully designed experiments conducted by experimenters and through automatized labeling, where labeling is done automatically using pre-defined rules or criteria.

Obtaining labels from the network management system based on previous occurrences of events of interest. SL models are deemed appropriate when complete information regarding what the model should learn is available. The dataset used for training defines both the inputs given to the model and the corresponding outputs expected from the model. During the training process, the parameters of the SL model are adjusted. The goal is to make the model's observed outputs match the expected outputs as closely as possible.

b. Unsupervised Learning (UL) techniques:

Examples of UL techniques include K-means clustering and principal components analysis (James et al., 2023). Unlike Supervised Learning (SL), UL techniques do not have access to labeled data during the training phase. The algorithm works with data where the class labels (outputs) are not provided. In UL, the algorithm learns to identify similarities between different inputs in the data. The primary goal is to cluster the data or extract meaningful features without prior knowledge of class labels. UL models are considered appropriate when the dataset lacks clear input/output pairs.

UL is suitable for scenarios where there are no strictly defined normal/abnormal conditions, and the focus is on grouping data based on similarity. UL models group data based on similarity, aiming to substantially separate diverging points. An intuition behind UL is that anomalous samples are much rarer than normal ones, and the algorithm seeks to identify patterns or structures in the data. UL typically does not have a training phase similar to SL. The entire dataset needs to be traversed whenever new data points are included or when the algorithm is applied to new data. (e.g., K-means clustering or principal components analysis) are mentioned for tasks without labeled data.

c. Semi-Supervised Learning (SSL) techniques:

Examples of SSL techniques include a one-class support vector machine (Zhu et al., 2022). SSL techniques combine aspects of both Supervised Learning (SL) and Unsupervised Learning (UL). They are applied in cases where labeling most of the data is impractical or too expensive. SSL is particularly useful in anomaly detection scenarios. The training dataset for SSL in anomaly detection should primarily contain normal data. During the training phase, SSL adjusts the parameters of the model.

The goal is to enclose the normal samples in the training dataset with a spatial region as tightly as possible. The tightly enclosed spatial region helps in identifying anomalies in new data. If new data falls outside the spatial region defined by normal samples during training, it is considered anomalous. SSL is applicable in cases where labeling the majority of the data is too costly or impractical. Instead, only a small subset of the data is labeled, making SSL a more resource-efficient approach.

5.1 Metrics for Validating the ML-Based Optical Security Monitoring Techniques

The performance of the ML model can be analyzed based on four important metrics: true positive (TP), true negative (TN), false negative (FN), and false positive (FP). True positive is detecting the attacks correctly as attacks and true negative is detecting the normal operating condition as normal. A false positive detects the normal operation as an attack whereas a false negative. The sum of true negatives and false positives as well as the sum of true positives and false negatives should be one. These metrics are derived from the confusion metrics as seen in Figure 7. Various other metrics are also calculated from the four above-mentioned metrics for validating the accuracy of the ML model (Vujovic et al., 2021).

Figure 7. Metrics used to evaluate the performance of the ML model

Predicted Observation \ Actual Observation	Negative	Positive
Negative	**TRUE NEGATIVE**	**FALSE NEGATIVE**
Positive	**FALSE POSITIVE**	**TRUE POSITIVE**

Precision- Precision (P) denotes that out of all the positive predicted, what percentage is truly positive? The precision value lies between 0 and 1.

P= TP/(TP+FP)

Recall- Recall (R) denotes out of the total positive, what percentage are predicted positive. It is the same as TPR (true positive rate).

R= TP/(TP+FN)

Specificity- Specificity (SP) is calculated as the number of correct negative predictions divided by the total number of negatives. It is also called true negative rate (TNR).

SP= TN/(TN+FP)

Accuracy- Accuracy (ACC) is calculated as the number of all correct predictions divided by the total number of the dataset.

ACC= (TP+TN)/(P+N)

Balanced Accuracy (BACC)- It is a metric we can use to assess the performance of a classification model when the classes are imbalanced.

BACC= (Sensitivity+Specificity)/2

F1 score- It is the harmonic mean of precision and recall. It takes both false positives and false negatives into account.

F1 score=(2*(Precision*Recall))/((Precision+Recall))

Jaccard index- The Jaccard index (JI), also known as the Jaccard similarity coefficient, is a statistic used for gauging the similarity and diversity of sample sets.

JI= TP/(TP+FN+FP)

Fowlkes- Mallow's index- The Fowlkes–Mallow's index (FMI) is an external evaluation method that is used to determine the similarity between two clusterings, and also a metric to measure confusion matrices. A higher value indicates a greater similarity between the clusters.

FMI= √(PPV*TPR)

Positive Predictive Value PPV= TP/(TP+FP)

Mathew's correlation coefficient- MCC is a correlation coefficient calculated using all four values in the confusion matrix. MCC is a single-value metric that summarizes the confusion matrix. The value ranges between -1 to 1.

MCC= ((TN*TP)-(FN*FP))/√((TP+FN) (FP+TP) (TN+FP) (FN+TP))

Dimensionality reduction methods are designed for data representation and visualization (Ayesha et al., 2020). The assumption is that valuable knowledge in a dataset can be concentrated in only a few features or summarized with a few synthesized features. In optical network monitoring, DSP-enabled transponders collect a rich set of features, but not all features are equally indicative for every use case. Dimensionality reduction methods analyze datasets with many dimensions and extract a limited number of features containing the most information about the dataset. Allows visualization of datasets in 2 or 3 dimensions, facilitating human understanding of the data.

Scalability issues in SSL (Semi-Supervised Learning) and UL (Unsupervised Learning) models include run-time complexity and sensitivity to high-dimensional data. Run-time complexity increases exponentially with the number of samples in UL models. Sensitivity to high-dimensional data can lead to degraded model accuracy. SL (Supervised Learning) models assisted by DL (Deep Learning) techniques have successfully analyzed large, high-dimensional datasets. Unlike DL models, SSL and UL models lack feature extraction capabilities. Dimensionality reduction methods are combined with SSL and UL models to mitigate sensitivity to high-dimensional data.

Acts as a feature extraction step before the execution of SSL and UL models. The potential benefits of applying dimensionality reduction methods depend on the specific problem and the features and anomalies being detected. Evaluation of the suitability of dimensionality reduction methods should be performed for each problem. The work emphasizes evaluating the suitability of dimensionality reduction methods for physical-layer attack detection in optical networks.

Principal Component Analysis (PCA) is a dimensionality reduction method with the goal of encoding as much variance from the original dataset as possible (Hasan et al., 2021). Here are the key characteristics of PCA: PCA aims to capture the maximum variance in the original dataset. PCA selects a specified number of orthogonal components from the dataset based on their variance. Each component is chosen to maximize the variance of the projected data points. For each component to be extracted, PCA finds a linear combination of the original features. After selecting the components, PCA can be used to project high-dimensional data onto a lower number of components. The projection involves a linear combination of the original features. PCA works efficiently over a large number of samples with a large number of dimensions in the dataset. It operates in a projected-dimension-wise manner, allowing scalability to datasets with many dimensions.

t-distributed stochastic neighbor embedding (t-SNE) is a dimensionality reduction method that enhances stochastic neighbor embedding by preserving both local and global relationships between data points (Hamid et al., 2020). Here are the key characteristics of t-SNE: t-SNE is designed to maintain both local and global relationships among data points in the original dataset. The method begins by calculating the pair-wise distances between all samples in the original dataset. Data points are then randomly projected over a specified number of dimensions. t-SNE proceeds with a predefined or unspecified number of steps.

In each step, the projected data points are adjusted to minimize the distance between points that are close to the original dataset. Simultaneously, the distance between points that are distant in the original dataset is maximized. The direction and amplitude of adjusting the position of each data point (gradient) are computed using algorithms like stochastic gradient descent. Unlike some other methods that work with projected-dimension-wise steps, t-SNE operates with sample-wise steps. Its complexity and, consequently, runtime are expected to be higher than other dimensionality reduction methods such as PCA and autoencoders (AEs).

Autoencoders (AEs) are a type of neural network with various applications, including denoising, one-shot learning, and dimensionality reduction (Reddy et al., 2020). Here are the key characteristics of autoencoders: AEs use a neural network structure, where the input and output layers have the same dimension as the dataset. Among the hidden layers, there is a central layer responsible for containing the encoded representation of the data. The central layer is where the feature extraction occurs. The primary objective of the AE is for the neural network to reproduce at the output the same values it receives at the input. Training is performed in a supervised manner, using the same samples as both input and ground truth values at the output.

When an AE is configured for dimensionality reduction, the feature extraction is carried out by the central layer, which determines the number of features to be extracted from the dataset. After training, the neural network is split into encoder and decoder parts. The encoder part is specifically used for dimensionality reduction. The training process involves adjusting the network parameters to enable effective dimensionality reduction. Once trained, the AE can capture essential features of the input data in its encoded representation. The values output by the encoder can be directly input to a Semi-Supervised Learning (SSL) or Unsupervised Learning (UL) model to perform anomaly detection.

6. OPM PARAMETERS FOR SECURITY MONITORING

Adapting to time-varying link performance parameters, including OSNR, nonlinearity factors, CD, and PMD, poses a challenge in network control and management. AI techniques are considered for monitoring the physical parameters of transmitted optical signals to diagnose network issues and take appropriate actions, such as repairing damages, driving compensators/equalizers, or rerouting traffic around non-optimal links. Artificial Neural Networks (ANN): (Wu et al. 2009) present a study on the application of artificial neural networks in OPM. The study involves the simultaneous identification of accumulated nonlinearity, OSNR, CD, and PMD from eye-diagram and eye-histogram parameters. Kalman Filter: Szafraniec et al. (2012) propose the use of the Kalman filter as an estimator for carrier phase tracking, polarization tracking, and the estimation of the first-order PMD.

These Techniques require prior knowledge about the type of signal, including bit-rate and modulation format, or additional cross-layer communication for acquiring this information. This can result in increased node complexity. A novel technique for simultaneous linear impairment identification (OSNR, CD, and PMD) is proposed in Ref. (Wang et al., 2020). This technique is independent of bit rate and modulation format, provided this information belongs to a known set. A DNN trained with raw data asynchronously sampled by a coherent receiver is proposed for OSNR monitoring in Ref. (Esmail et al., 2021). However, it requires configuration with at least 5 layers and substantial training samples for accurate results. (Thrane et al. 2017) propose an OSNR estimator and a modulation format classifier using a neural network and a support vector machine (SVM), respectively. The study considers advanced modulation formats (up to 64 QAM) and direct detection, achieving accurate results for OSNR estimation and modulation format classification.

7. IMPLEMENTATION AND BENEFITS OF RFID AND MICROWAVE INTEGRATION WITH SECURE OPTICAL COMMUNICATION

RFID can track and authenticate devices accessing critical network components, ensuring only authorized personnel interact with sensitive network equipment. Microwave sensors can detect unusual activities or interference in the optical network, providing early warning signs of potential security threats. Advanced ML and DL models can be integrated with RFID and microwave sensors to analyze network data continuously. These models can detect anomalies such as unauthorized access, interference, or data breaches and automatically initiate corrective actions to prevent or mitigate attacks. RFID systems are highly scalable and can be deployed across various points in the optical network, providing flexible and modular security measures that can evolve with the network infrastructure. Microwave backhaul systems enable connectivity even in challenging environments, ensuring that the optical network remains secure and operational in remote or hard-to-reach areas. By integrating microwave communication alongside optical networks, the system can ensure redundancy. In case of disruption or attack on the optical link, the microwave network can maintain communication continuity, thereby improving network resilience.

RFID tags and readers can be strategically placed at key access points in the network infrastructure (e.g., data centers, network nodes, or gateways). These devices can authenticate personnel and track their movements in and out of sensitive areas, ensuring that only authorized entities have physical access to critical hardware. Microwave sensors can be installed alongside optical fibers to detect potential interference or jamming. These sensors can continuously monitor the microwave spectrum for unusual activity, which may indicate an attack on the network. In cases where interference is detected, the system can trigger automatic corrective actions, such as rerouting traffic through secure channels or activating interference-reduction mechanisms.

By deploying ML models that analyze data from both RFID and microwave sensors, the system can continuously detect patterns associated with security threats. These models can be trained to identify anomalies and unusual behaviors, such as unauthorized access or abnormal signal interference, and automatically initiate appropriate responses, such as shutting down compromised components or initiating re-encryption of data transmissions. Continuous feedback from RFID and microwave sensors can be used to update and refine security protocols in real time. This ensures that the system adapts to new security threats and continually improves its defenses based on historical attack patterns. Figure 8 provides a higher-level architecture diagram of how RFID and microwave technologies can be integrated into the optical communication network to enhance security:

Figure 8. The architecture of 6G Optical Communication Infrastructure with RFID

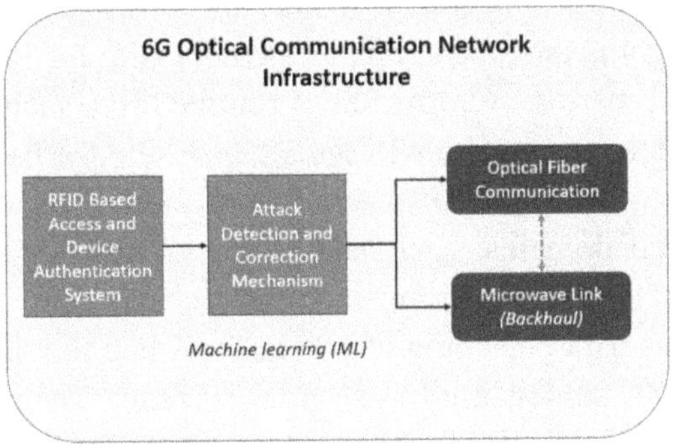

- RFID systems are deployed across physical network infrastructure (e.g., server racks, communication hubs). Tags authenticate authorized personnel and equipment. RFID readers at entry points continuously monitor access and log activities, triggering alerts if unauthorized access is detected.
- Microwave communication links are integrated with optical networks, providing additional layers of connectivity in hard-to-reach areas. Microwave sensors are deployed to monitor signal integrity and detect any interference that could compromise the security of the network.
- The data collected from both RFID and microwave sensors feed into an ML-based security platform. The platform analyzes incoming data to detect anomalies, unauthorized access, or interference. Once an attack is identified, the system triggers automated responses such as rerouting data traffic, initiating encryption protocols, or isolating compromised components.
- The security system is continuously updated through feedback from RFID and microwave sensors. The ML model refines its detection capabilities over time, adapting to new types of attacks and improving the overall security posture of the network.

Why This Approach Is Necessary

Physical Layer Security: Optical networks, while high-capacity and fast, are vulnerable to physical attacks such as unauthorized access, tampering, and interference. RFID and microwave systems help secure the physical layer by monitoring devices and detecting physical disruptions.

Continuous Monitoring: Given the dynamic nature of 6G networks, continuous monitoring is crucial for identifying and mitigating security threats in real-time. RFID and microwave sensors provide a granular, distributed approach to surveillance.

Robust and Scalable: Integrating these technologies offers a scalable security solution that grows with the network and can be easily adapted to meet evolving security challenges.

8. CONCLUSION

Achieving autonomous, multi-layer network security management requires overcoming several challenges related to the integration of machine learning (ML) systems and the operational complexities of security environments. While ML advancements reduce the need for human interventions, the complete elimination of human oversight is unrealistic due to the necessity for interpretation, decision-making, and response to unforeseen scenarios. Automation of routine tasks such as data processing and attack remediation is pivotal, but human expertise will remain essential for handling complex situations and refining security strategies.

The collaboration between domains, such as integrating privacy-preserving federated learning with cross-domain data sharing, offers a promising approach to improving network security. However, adversarial threats, such as data poisoning attacks, necessitate the development of robust defenses that ensure the integrity of ML models used in security systems. The risks of attackers manipulating ML algorithms highlight the importance of secure mechanisms like blockchain and smart contracts to enhance the privacy, trust, and accountability of ML-based systems, ensuring transparency in model updates and decision-making.

The challenge of false positives in attack detection emphasizes the need for validation mechanisms that account for prediction confidence levels to avoid unnecessary, costly interventions. Furthermore, integrating optical-layer security with existing security frameworks, such as Security Operation Centers (SOC), requires the adaptability of ML systems to address uncertainties in real-time operations. Adaptive automation can play a key role in supporting this integration, especially for less time-critical tasks, where learning models can be transformed into actionable processes and workflows. In conclusion, a balanced approach combining human expertise with automated, adaptive systems, along with robust defenses against adversarial attacks and privacy-preserving mechanisms, will be essential for developing secure and efficient multi-layer network security management for autonomous systems.

Time-critical tasks with higher uncertainty may benefit from acceleration techniques and continuous incremental learning to adapt quickly to evolving threats. Continuous monitoring of ML model performance and adaptation to changing conditions is essential. Regular updates to ML models based on new data and the evolving threat landscape can improve resilience against deceptive tactics. Implement robust model validation mechanisms to ensure that ML models are resistant to adversarial attacks and remain reliable in real-world scenarios. Adversarial training, where models are trained on intentionally manipulated data to increase robustness, is one approach to consider. Consider the ethical implications of using ML in security contexts, including privacy concerns and the potential impact on end-users. Implement privacy-preserving techniques to safeguard sensitive information while still benefiting from the capabilities of ML.

By combining RFID and microwave technologies with optical communication systems, this integrated approach ensures robust security, continuous monitoring, and proactive threat mitigation. This strategy not only enhances the protection of the optical network but also contributes to building a resilient infrastructure for secure, high-performance 6G communication networks.

REFERENCES

Ayesha, S., Hanif, M. K., & Talib, R. (2020). Overview and comparative study of dimensionality reduction techniques for high dimensional data. *Information Fusion*, 59, 44–58. DOI: 10.1016/j.inffus.2020.01.005

Bhanja, U., & Singhdeo, S. (2020). Novel encryption technique for security enhancement in optical code division multiple access. *Photonic Network Communications*, 39(3), 195–222. DOI: 10.1007/s11107-020-00883-y

Catanese, C., Triki, A., Pincemin, E., & Jaouën, Y. (2019). A survey of neural network applications in fiber nonlinearity mitigation. In 2019 21st International Conference on Transparent Optical Networks (ICTON). IEEE. DOI: 10.1109/ICTON.2019.8840355

Esmail, M. A. (2021). Optical wireless performance monitoring using asynchronous amplitude histograms. *IEEE Photonics Journal*, 13(3), 1–9. DOI: 10.1109/JPHOT.2021.3080593

Furdek, M., & Natalino, C. (2022). Machine learning for network security management, attacks, and intrusions detection. In *Machine Learning for Future Fiber-Optic Communication Systems* (pp. 317–336). Academic Press. DOI: 10.1016/B978-0-32-385227-2.00017-6

Furdek, M., Natalino, C., Di Giglio, A., & Schiano, M. (2021). Optical network security management: Requirements, architecture, and efficient machine learning models for detection of evolving threats. *Journal of Optical Communications and Networking*, 13(2), A144–A155. DOI: 10.1364/JOCN.402884

Furdek, M., Natalino, C., Lipp, F., Hock, D., Di Giglio, A., & Schiano, M. (2020). Machine learning for optical network security monitoring: A practical perspective. *Journal of Lightwave Technology*, 38(11), 2860–2871. DOI: 10.1109/JLT.2020.2987032

Furdek, M., Natalino, C., Lipp, F., Hock, D., Di Giglio, A., & Schiano, M. (2020). Machine learning for optical network security monitoring: A practical perspective. *Journal of Lightwave Technology*, 38(11), 2860–2871. DOI: 10.1109/JLT.2020.2987032

Goścień, R., Ellinas, G., Furdek, M., Kolios, P., Lechowicz, P., Manousakis, K., Savva, G., Skorin-Kapov, N., & Walkowiak, K. (2020). *Security-aware carrier network planning*. Guide to Disaster-Resilient Communication Networks.

Hamid, Y., & Sugumaran, M. (2020). A t-SNE based non linear dimension reduction for network intrusion detection. *International Journal of Information Technology : an Official Journal of Bharati Vidyapeeth's Institute of Computer Applications and Management*, 12(1), 125–134. DOI: 10.1007/s41870-019-00323-9

Hasan, B. M. S., & Abdulazeez, A. M. (2021). A review of principal component analysis algorithm for dimensionality reduction. *Journal of Soft Computing and Data Mining*, 2(1), 20–30.

Hugues-Salas, E., Ntavou, F., Gkounis, D., Kanellos, G. T., Nejabati, R., & Simeonidou, D. (2019). Monitoring and physical-layer attack mitigation in SDN-controlled quantum key distribution networks. *Journal of Optical Communications and Networking*, 11(2), A209–A218. DOI: 10.1364/JOCN.11.00A209

James, G., Witten, D., Hastie, T., Tibshirani, R., & Taylor, J. (2023). Unsupervised learning. In *An Introduction to Statistical Learning: with Applications in Python* (pp. 503–556). Springer International Publishing. DOI: 10.1007/978-3-031-38747-0_12

Jeon, H.-B., Kim, S.-M., Moon, H.-J., Kwon, D.-H., Lee, J.-W., Chung, J.-M., Han, S.-K., Chae, C.-B., & Alouini, M.-S. (2023). Free-space optical communications for 6G wireless networks: Challenges, opportunities, and prototype validation. *IEEE Communications Magazine*, 61(4), 116–121. DOI: 10.1109/MCOM.001.2200220

Jiang, T., Gradus, J. L., & Rosellini, A. J. (2020). Supervised machine learning: A brief primer. *Behavior Therapy*, 51(5), 675–687. DOI: 10.1016/j.beth.2020.05.002 PMID: 32800297

Maslo, A., Sarajlić, N., Hodžić, M., & Mujčić, A. (2021). "Optical Network Security Attacks by Tapping and Encrypting Optical Signals." Advanced Technologies, Systems, and Applications V. *Papers Selected by the Technical Sciences Division of the Bosnian-Herzegovinian American Academy of Arts and Sciences*, 2020, 321–332.

Moghaddam, E. E., Beyranvand, H., & Salehi, J. A. (2020). Crosstalk-aware resource allocation in survivable space-division-multiplexed elastic optical networks supporting hybrid dedicated and shared path protection. *Journal of Lightwave Technology*, 38(6), 1095–1102. DOI: 10.1109/JLT.2019.2954177

Moharrami, M., Fallahpour, A., Beyranvand, H., & Salehi, J. A. (2017). Resource allocation and multicast routing in elastic optical networks. *IEEE Transactions on Communications*, 65(5), 2101–2113. DOI: 10.1109/TCOMM.2017.2667664

Musumeci, F., Rottondi, C., Nag, A., Macaluso, I., Zibar, D., Ruffini, M., & Tornatore, M. (2018). An overview on the application of machine learning techniques in optical networks. *IEEE Communications Surveys and Tutorials*, 21(2), 1383–1408. DOI: 10.1109/COMST.2018.2880039

Nassif, A. B., Shahin, I., Attili, I., Azzeh, M., & Shaalan, K. (2019). Speech Recognition Using Deep Neural Networks: A Systematic Review. *IEEE Access : Practical Innovations, Open Solutions*, 7, 19143–19165. DOI: 10.1109/ACCESS.2019.2896880

Pérez-López, C. D., Soto-Bovó, L. M., Colbes, J., & Pinto-Roa, D. P. (2021). Semi-dynamic Routing and Spectrum Assignment with variable bandwidth in Elastic Optical Networks. Bee-inspired Algorithms approach. In *2021 XLVII Latin American Computing Conference (CLEI)*. IEEE. DOI: 10.1109/CLEI53233.2021.9639900

Reddy, G. T., Kumar Reddy, M. P., Lakshmana, K., Kaluri, R., Singh Rajput, D., Srivastava, G., & Baker, T. (2020). Analysis of dimensionality reduction techniques on big data. *IEEE Access : Practical Innovations, Open Solutions*, 8, 54776–54788. DOI: 10.1109/ACCESS.2020.2980942

Sharma, G., Sharma, H., Pareek, R., Gour, N., Sharma, R. S., & Kumar, A. (2021). Self-healing topology for DDoS attack identification & discovery protocol in software-defined networks. *Journal of Discrete Mathematical Sciences and Cryptography*, 24(8), 2221–2232. DOI: 10.1080/09720529.2021.2009192

Shi, Y., An, K., & Li, Y. (2021). Index modulation based frequency hopping: Anti-jamming design and analysis. *IEEE Transactions on Vehicular Technology*, 70(7), 6930–6942. DOI: 10.1109/TVT.2021.3087640

Skorin-Kapov, N., Furdek, M., Zsigmond, S., & Wosinska, L. (2016). Physical-layer security in evolving optical networks. *IEEE Communications Magazine*, 54(8), 110–117. DOI: 10.1109/MCOM.2016.7537185

Szafraniec, B., Marshall, T. S., & Nebendahl, B. (2012). Performance monitoring and measurement techniques for coherent optical systems. *Journal of Lightwave Technology*, 31(4), 648–663. DOI: 10.1109/JLT.2012.2212234

Thrane, Wass, Piels, Diniz, Jones, & Zibar. (2017). Machine Learning Techniques for Optical Performance Monitoring from Directly Detected PDM-QAM Signals. Journal of Lightwave Technology, 35(4), 868-875. .DOI: 10.1109/JLT.2016.2590989

Tizikara, D. K., Serugunda, J., & Katumba, A. (2022). Machine learning-aided optical performance monitoring techniques: A review. *Frontiers in Communications and Networks*, 2, 756513. DOI: 10.3389/frcmn.2021.756513

Volkov, A. S., Baskakov, A. E., Bakhtin, A. A., & Gorelik, A. V. (2021). Method for Managing Transport SDN Resources. In 2021 Systems of Signal Synchronization, Generating and Processing in Telecommunications (SYNCHROINFO. IEEE. DOI: 10.1109/SYNCHROINFO51390.2021.9488419

Vujović, Ž. (2021). Classification model evaluation metrics. *International Journal of Advanced Computer Science and Applications*, 12(6), 599–606. DOI: 10.14569/IJACSA.2021.0120670

Wang, D., Jiang, H., Liang, G., Zhan, Q., Mo, Y., Sui, Q., & Li, Z. (2020). Optical performance monitoring of multiple parameters in future optical networks. *Journal of Lightwave Technology*, 39(12), 3792–3800. DOI: 10.1109/JLT.2020.3043012

Wu, X., Jargon, J. A., Skoog, R. A., Paraschis, L., & Willner, A. E. (2009). Applications of artificial neural networks in optical performance monitoring. *Journal of Lightwave Technology*, 27(16), 3580–3589. DOI: 10.1109/JLT.2009.2024435

Yang, Zhong, Chen, Alphones, & Xie. (2020). Deep-reinforcement-learning-based energy-efficient resource management for social and cognitive Internet of Things. Internet of Things Journal, 7(6), 5677-5689.

Zhang, L., Li, X., Tang, Y., Xin, J., & Huang, S. (2022). A survey on QoT prediction using machine learning in optical networks. *Optical Fiber Technology*, 68, 102804. DOI: 10.1016/j.yofte.2021.102804

Zhang, X.-G., Yang, G.-H., & Ren, X.-X. (2022). Network steganography-based security framework for cyber-physical systems. *Information Sciences*, 609, 963–983. DOI: 10.1016/j.ins.2022.07.081

Zhu, X., & Goldberg, A. B. (2022). *Introduction to semi-supervised learning*. Springer Nature.

Compilation of References

Aawan, I., Mhussain, Alibakhshikenari M, Ffalcone, Limiti E, Razarizvi S N. (2021). Single Patch Fractal-Shaped Antenna with Small footprint Area and High Radiation Properties for Wide Operation Over 5G Region. 46th International Conference on Infrared, Millimeter and Terahertz Waves (IRMMW-THz).p.1–2. DOI: 10.1109/IRMMW-THz50926.2021.9567165

Abdel Hakeem, S. A., Hussein, H. H., & Kim, H. (2022). Security requirements and challenges of 6G technologies and applications. *Sensors (Basel)*, 22(5), 1969. DOI: 10.3390/s22051969 PMID: 35271113

Abdullah, D. M., & Ameen, S. Y. (2021). Enhanced mobile broadband (EMBB): A review. *Journal of Information Technology and Informatics*, 1(1), 13–19.

Abdulraqeb, A. (2024). Ibraheem, Shayea., Ayman, A., El-Saleh., Marwan, Hadri, Azmi., Zool, Hilmi, Ismail., Lida, Kouhalvandi., Sawsan, Ali, Saad. (2024). Artificial Intelligence in 6G Wireless Networks: Opportunities, Applications, and Challenges. *International Journal of Intelligent Systems*, 2024, 1–27. DOI: 10.1155/2024/8845070

Aboagye, S., Ndjiongue, A. R., Ngatched, T. M., Dobre, O. A., & Poor, H. V. (2022). RIS-assisted visible light communication systems: A tutorial. *IEEE Communications Surveys and Tutorials*, 25(1), 251–288. DOI: 10.1109/COMST.2022.3225859

Abrera, J. (2024). Data Privacy and Security in Cloud Computing: A Comprehensive Review. Journal of Computer Science and Information Technology, 1(1), 01-09.

AbuElrub, F., Hamed, F., & Saadeh, O. (2020, December). Hamed, O. Saadeh, Microgrid integrated electric vehicle charging algorithm with photovoltaic generation. *Journal of Energy Storage*, 32, 101858. DOI: 10.1016/j.est.2020.101858

Abumarshoud, H., Mohjazi, L., Dobre, O. A., Di Renzo, M., Imran, M. A., & Haas, H. (2021). LiFi Through Reconfigurable Intelligent Surfaces: A New Frontier for 6G? *IEEE Vehicular Technology Magazine*, •••, 2–11. DOI: 10.1109/MVT.2021.3121647

Achumba, I. C., Ighomereho, O. S., & Akpor-Robaro, M. O. M. (2013). Security challenges in Nigeria and the implications for business activities and sustainable development. *Journal of economics and sustainable development, 4*(2).

Adhikari, M., & Hazra, A. (2022). 6G-enabled ultra-reliable low-latency communication in edge networks. *IEEE Communications Standards Magazine*, 6(1), 67–74. DOI: 10.1109/MCOMSTD.0001.2100098

Adhikari, M., Hazra, A., Menon, V. G., Chaurasia, B. K., & Mumtaz, S. (2021). A roadmap of next-generation wireless technology for 6G-enabled vehicular networks. *IEEE Internet of Things Magazine*, 4(4), 79–85. DOI: 10.1109/IOTM.001.2100075

Adnan Khan, M. D. S., Kadir, K. M., Mahmood, K. S., Ibne Alam, M. I., Kamal, A., & Al Bashir, M. M. (2019, December). Technical investigation on V2G, S2V, and V2I for next generation smart city planning. *J. Electronic Sci. Technol.*, 17(4), 100010. DOI: 10.1016/j.jnlest.2020.100010

Agarwal, A., Mishra, S. S., Panda, B. K., & Misra, G. (2023). A detailed review on 6G technology: Online learning system applications, challenges and research activities. Online Learning Systems, 147-176.

Agilent Technologies. (2000). Fundamentals of RF and microwave noise figure measurements.

Ahammed, M. T., Boppana, U. M., Kondapalli, R., Balaji, P., & Zardar, M. A. (2022, August). Survey on wireless fidelity (Wi-Fi) based indoor location technology for mobile devices. In *2022 Third International Conference on Intelligent Computing Instrumentation and Control Technologies (ICICICT)* (pp. 1465-1470). IEEE. DOI: 10.1109/ICICICT54557.2022.9917746

Ahammed, T. B., Patgiri, R., & Nayak, S. (2023). A vision on the artificial intelligence for 6G communication. *ICT Express*, 9(2), 197–210. DOI: 10.1016/j.icte.2022.05.005

Ahmad I, Ullah S, Ullah S, Habib U, Ahmad S, Ghaffar A, Alibakhshikenari M, Khan S, Limiti E (2021) Design and analysis of a photonic crystal based planar antenna for THz applications. Electronics 10(16):1941. ronic s1016 1941.DOI: 10. 3390/ elect

Ahmad, A., Alam, M. S., Varshney, Y., & Khan, R. H. A state of the Art review on Wireless Power Transfer a step towards sustainable mobility. In Proceedings of the 2017 14th IEEE India Council International Conference (INDICON), Roorkee, India, 15–17 December 2017; pp. 1-Bijan. (2017). Design and control of a 6-degree-of-freedom precision positioning system. DOI: 10.1109/INDICON.2017.8488032

Ahmad, S., Ogbodo, E., & Ukala, A. (2024, April). Enhanced Dual-band Antenna for Intelligent Vehicular Communications. In *2024 Photonics & Electromagnetics Research Symposium (PIERS)* (pp. 1-7). IEEE. DOI: 10.1109/PIERS62282.2024.10618202

Ahmad, H. F., Rafique, W., Rasool, R. U., Alhumam, A., Anwar, Z., & Qadir, J. (2023). Leveraging 6G, extended reality, and IoT big data analytics for healthcare: A review. *Computer Science Review*, 48, 100558. DOI: 10.1016/j.cosrev.2023.100558

Ahmed, A., Elsaraf, Z., Khan, F. A., & Ahmed, Q. Z. (2021). Cooperative Non-Orthogonal Multiple Access for Beyond 5G Networks. *IEEE Open Journal of the Communications Society*, 2, 990–999. DOI: 10.1109/OJCOMS.2021.3075081

Ahokangas, P., Matinmikko-Blue, M., Basaure, A., & Yrjölä, S. (2024). Use cases for local 6G networks. *2024 Joint European Conference on Networks and Communications & 6G Summit (EuCNC/6G Summit)*, 1127-1132. https://doi.org/DOI: 10.1109/EuCNC/6GSummit60053.2024.10597019

Akdeniz, M. R., Liu, Y., Samimi, M. K., Sun, S., Rangan, S., Rappaport, T. S., & Erkip, E. (2014). Millimeter wave channel modeling and cellular capacity evaluation. *IEEE Journal on Selected Areas in Communications*, 32(6), 1164–1179.

Akgul, O. U., Ericson, M., Burguera, P. G., Arar, B., Khorsandi, B. M., Querio, R., Varvara, A., Harkous, H., Wänstedt, S., Groshev, M., De la Oliva, A., Pieczerak, J., Kukliński, S., Nande, S. S., Bassoli, R., & Fitzek, F. H. (2024). Discussion on 6G architecture evolution: Challenges and emerging technology trends. *2024 Joint European Conference on Networks and Communications & 6G Summit (EuCNC/6G Summit)*, 664-669. https://doi.org/DOI: 10.1109/EuCNC/6GSummit60053.2024.10597056

Akyildiz, I. F., Kak, A., & Nie, S. (2020). 6G and beyond: The future of wireless communications systems. *IEEE Access : Practical Innovations, Open Solutions*, 8, 133995–134030. DOI: 10.1109/ACCESS.2020.3010896

Alamu, O., Olwal, T. O., & Djouani, K. (2023). Cooperative NOMA networks with simultaneous wireless information and power transfer: An overview and outlook. *Alexandria Engineering Journal*, 71, 413–438. DOI: 10.1016/j.aej.2023.03.057

Alanne, K., & Liimatainen, H. (2019). Design implications of the electrification of passenger vehicle stock on renewable energy integration in Finnish apartment buildings. *Sustainable Cities and Society*, 47, 101507.

Alazemi, A. J. (2023). Dual-band and wideband bandpass filters using coupled lines and tri-stepped impedance stubs. *Micromachines*, 14(6), 1254. DOI: 10.3390/mi14061254 PMID: 37374838

Albraheem, L. I., Alhudaithy, L. H., Aljaser, A. A., Aldhafian, M. R., & Bahliwah, G. M. (2018). Toward designing a Li-Fi-based hierarchical IoT architecture. *IEEE Access : Practical Innovations, Open Solutions*, 6, 40811–40825. DOI: 10.1109/ACCESS.2018.2857627

Albraheem, L., Alhudaithy, L., Aljaser, A., AlDhafian, M., & Bahliwah, G. (2022). Towards smart sustainable cities using Li-Fi technology: Geo-location infrastructure utilizing LED street lights. *PeerJ. Computer Science*, 8, e1009. DOI: 10.7717/peerj-cs.1009 PMID: 36091990

Alcaraz, L., & Alves, H. (2017). Ultrareliable Short-Packet Communications With Wireless Energy Transfer. *IEEE Signal Processing Letters*, 24(4), 387–391. DOI: 10.1109/LSP.2017.2667620

Aldolimi, W. S., Hnaif, A. A., & Alia, M. A. (2021, July). Light fidelity to transfer secure data using advanced encryption standard algorithm. In *2021 International Conference on Information Technology (ICIT)* (pp. 963-967). IEEE. DOI: 10.1109/ICIT52682.2021.9491769

Alhammadi, A., Abraham, A., Fakhreddine, A., Tian, Y., Du, J., & Bader, F. (2024). Envisioning the Future Role of 3D Wireless Networks in Preventing and Managing Disasters and Emergency Situations. arXiv preprint arXiv:2402.10600.

Alharbi, G, Abdullah,Chow Yen Desmondsim,Jayshri Kulkarni,K Poddar, L. Rohde.(2022).A compact circularly polarized rotated L-shaped antenna withJ-shaped defected ground strucutre for WLAN and V2X applications. Progr.Electromagnet. Res. Lett. P. 135–143.

Alhasnawi, B. N., & Jasim, B. H. (2021, August). A new internet of things enabled trust distributed demand side management system. *Sustainable Energy Technologies and Assessments*, 46, 101272. DOI: 10.1016/j.seta.2021.101272

Ali, A., Kumar, R. H., Dheenathalayan, R., Prasanth, N., Parthasaradi, V., Senthilkumar, S., & Kumar, T. S. (2023). Audio Streaming Using Li-FI Communication. *Irish Interdisciplinary Journal of Science & Research*, 7(1), 1–7. DOI: 10.46759/IIJSR.2023.7101

Alle, P., Kbehera, S., Mishra, D. P., & Pandav, S. (2021).CPW-Fed Circularly Polarized Fractal Antenna for WLAN Applications. *2nd International Conference for Emerging Technology (INCET)*.p. 1–5. DOI: 10.1109/INCET51464.2021.9456236

Allodi, M., Broggi, A., Giaquinto, D., Potender, M., & Prioletti, A. (2016). Machine learning in tracking associations with stereo vision and lidar observations for an autonomous vehicle. In 2016 [IEEE.]. *IEEE Intelligent Vehicles Symposium*, (IV), 648–653.

Almalah, N. T., Mahmood, F. E., & Yassen, M. T. (2024). Li-Fi Technology in Optical Communication Systems: A Review. *Al-Iraqia Journal for Scientific Engineering Research*, 3(3), 163–171.

Alotaibi, Y., Alghamdi, S., Alassery, F., Palanisamy, S., Thangaraju, B., & Khalaf, O. I. (2021). A Novel Approach of Design and Analysis of a Hexagonal Fractal Antenna Array (HFAA) for Next-Generation Wireless Communication. *Energies*, 14(19), 1–18.

Alsaba, Y., Rahim, S. K. A., & Leow, C. Y. (2018). Beamforming in Wireless Energy Harvesting Communications Systems: A Survey. *IEEE Communications Surveys and Tutorials*, 20(2), 1329–1360. DOI: 10.1109/COMST.2018.2797886

Alsharif, M. H., Albreem, M. A., Solyman, A. A. A., & Kim, S. (2021). Toward 6G communication networks: Terahertz frequency challenges and open research issues. *Computers, Materials & Continua*. Advance online publication. DOI: 10.32604/cmc.2021.013176

Amiyabhusanasahoo, B., Mishra, G. P., & Manasranjan, J. B. (2019). Fractal Geometry and Its Application to Antenna Designs. [IJEAT]. *International Journal of Engineering and Advanced Technology*, 9(1), 3726–3743. DOI: 10.35940/ijeat.A9793.109119

Ananda, S., Sriram Kumara, D., Wub, R. J., & Chavali, M. (2014). Graphene nanoribbon-based terahertz antenna on polyimide substrate. *Optik (Stuttgart)*, 125(19), 5546–5549. DOI: 10.1016/j.ijleo.2014.06.085

Antón-Haro, C., Trichias, K., De Majo, C., Kaloxylos, A., & Beriere, J. (2024). 6G smart networks and services: Global strategies, main work directions & Future outlook. *2024 Joint European Conference on Networks and Communications & 6G Summit (EuCNC/6G Summit)*, 1115-1120. https://doi.org/ DOI: 10.1109/EuCNC/6GSummit60053.2024.10597046

Auer, A., Feese, S., Lockwood, S., & Hamilton, B. A. (2016). *History of intelligent transportation systems* (No. FHWA-JPO-16-329). United States. Department of Transportation. Intelligent Transportation Systems Joint Program Office.

Awadamislmani, N., Seetharamdoo, D., Rosinski, C., & Vuadelle, B. (2024). C-ITS-enabled Antenna System for Pedestrian-to-Vehicle (P2V) Portable Communication Devices: Design and On-site tests. Canva Pro. Available online: https://www.canva.com/pro/ (accessed on 25 June 2023).

Awortu, B. E. (2015). Boko Haram insurgency and the underdevelopment of Nigeria. *Research on Humanities and Social Sciences*, 5(6), 213–220.

Ayesha, S., Hanif, M. K., & Talib, R. (2020). Overview and comparative study of dimensionality reduction techniques for high dimensional data. *Information Fusion*, 59, 44–58. DOI: 10.1016/j.inffus.2020.01.005

Azarbar, A., Masouleh, M. S., & Behbahani, A. K. (2014). A new terahertz microstrip rectangular patch array antenna. *Int J Electromagn Appl*, 4, 25–29.

Azman, F., Suraya, Q., Rahim, F. A., Mohd, M. S., & Ariffin, N. A. M. (2018, November). My guardian: A personal safety mobile application. In *2018 IEEE Conference on Open Systems (ICOS)* (pp. 37-41). IEEE. DOI: 10.1109/ICOS.2018.8632808

Babu, K. V.. (2023). *Design and analysis of fractal-based THz antenna with co-axial feeding technique for wireless applications.Recent advances in graphene nanophotonics.* Springer Nature Switzerland.

Babu, K. V., Das, S., Sree, G. N. J., Madhav, B. T. P., Patel, S. K. K., & Parmar, J. (2022). Design and optimization of micro-sized wideband fractal MIMO antenna based on characteristic analysis of graphene for terahertz applications. *Optical and Quantum Electronics*, 54(5), 281. DOI: 10.1007/s11082-022-03671-2

Bae, S.-H., & Yoon, K.-J. (2017). Confidence-based data association and discriminative deep appearance learning for robust online multi-object tracking. *IEEE Transactions on Pattern Analysis and Machine Intelligence*, 40(3), 595–610. DOI: 10.1109/TPAMI.2017.2691769 PMID: 28410099

Bakhar, M., Dakulagi, V., & Mathpati, M. S. (2021). Design and Analysis of a Fractal Antenna Using Jeans Material for WiMax/WSN Applications. *Wireless Personal Communications*, 119(4), 3517–3527. DOI: 10.1007/s11277-021-08418-y

Bakhar, M., Mathpati, M., & Yadav, D. (2020). A Pentagonal Patch Antenna with Minkowski Fractal Structure using Different Substrate Materials. *Journal of Advanced Research in Networking and Communication Engineering*, 3(2), 1–4.

Banafaa, M., Shayea, I., Din, J., Azmi, M. H., Alashbi, A., Daradkeh, Y. I., & Alhammadi, A. (2023). 6G mobile communication technology: Requirements, targets, applications, challenges, advantages, and opportunities. *Alexandria Engineering Journal*, 64, 245–274. DOI: 10.1016/j.aej.2022.08.017

Banerjee, S., Rana, B., Parui, S. K., Chatterjee, S., & Dey, N. (2017). HMSIW-Based Miniaturized sensing antennas for S- and C-Band applications. *IEEE Sensors Letters*, 1(1), 1–5. DOI: 10.1109/LSENS.2017.2677858

Bariah, L., Mohjazi, L., Muhaidat, S., Sofotasios, P. C., Kurt, G. K., Yanikomeroglu, H., & Dobre, O. A. (2020). A prospective look: Key enabling technologies, applications and open research topics in 6G networks. *IEEE Access : Practical Innovations, Open Solutions*, 8, 174792–174820.

Barkley, B. E., & Paley, D. A. (2017). *Multi-target tracking and data association on road networks using unmanned aerial vehicles. In 2017 IEEE aerospace conference.* IEEE.

Bashir, M. S., & Alouini, M.-S. (2022). Energy Optimization of a Laser-Powered Hovering-UAV Relay in Optical Wireless Backhaul. *IEEE Transactions on Wireless Communications*, •••, 1–1. DOI: 10.1109/TWC.2022.3216797

Bearfield, C. X., Stokes, C., Lovett, A., & Franconeri, S. (2023). What does the chart say? grouping cues guide viewer comparisons and conclusions in bar charts. *IEEE Transactions on Visualization and Computer Graphics*. PMID: 37792647

Benlakehal ME et al (2022) Design and analysis of novel microstrip patch antenna array based on photonic crystal in THz. Opt Quantum Electron 54(5):1-16.2

Benlakehal, M. E., Hocini, A., & Khedrouche, D.. (2022). Design and analysis of a 1 × 2 microstrip patch antenna array based on photonic crystals with a graphene load in THz. *Journal of Optics*. Advance online publication. DOI: 10.1007/s12596-022-01006-8

Benlakehal, M. E., Hocini, A., Khedrouche, D., Temmar, M. N. E., & Denidni, T. A. (2022). Design and analysis of a 2 × 2 microstrip ratch antenna array based on periodic and non-periodic photonic crystals substrate in THz. *Optical and Quantum Electronics*, 54(3), 190. DOI: 10.1007/s11082-022-03563-5

Benlakehal, M. E., Hocini, A., Khedrouche, D., Temmar, M. N., & Denidni, T. A. (2022). Design and analysis of a 1 × 2 microstrip patch antenna array based on periodic and aperiodic photonic crystals in terahertz. *Optical and Quantum Electronics*, 54(10), 672. DOI: 10.1007/s11082-022-04076-x

Benlakehal, M. E., Hocini, A., Khedrouche, D., Temmar, M. N., & Denidni, T. A. (2022). Design and analysis of MIMO system for THz communication using terahertz patch antenna array based on photonic crystals with graphene. *Optical and Quantum Electronics*, 54(11), 693. DOI: 10.1007/s11082-022-04081-0

Bewley, A., Ott, L., Ramos, F., & Upcroft, B. (2016). Alextrac: Affinity learning by exploring temporal reinforcement within association chains. In 2016 [IEEE.]. *IEEE International Conference on Robotics and Automation*, •••, 2212–2218.

Bhalerao, A.K., 2015. Application and performance of Google Forms for online data collection and analysis: A case of North Eastern Region of India. *Indian Journal of Extension Education, 51*(3and4), pp.49-53.

Bhanja, U., & Singhdeo, S. (2020). Novel encryption technique for security enhancement in optical code division multiple access. *Photonic Network Communications*, 39(3), 195–222. DOI: 10.1007/s11107-020-00883-y

Bharath, B. N., Nagananda, K. G., & Vincent Poor, H. (2016). A Learning-Based Approach to Caching in Heterogenous Small Cell Networks. *IEEE Transactions on Communications*, 64(4), 1674–1686. DOI: 10.1109/TCOMM.2016.2536728

Bhattacharya, P., Patel, F., Alabdulatif, A., Gupta, R., Tanwar, S., Kumar, N., & Sharma, R. (2022). A deep-Q learning scheme for secure spectrum allocation and resource management in 6G environment. *IEEE Transactions on Network and Service Management*, 19(4), 4989–5005. DOI: 10.1109/TNSM.2022.3186725

Bhole, C., Malviya, V., Nadar, S., & Patel, J. (2021, May). CitiSafe–Interactive Crime Reporting. In *Proceedings of the 4th International Conference on Advances in Science & Technology (ICAST2021).*

Birsan, J. C. R., Moldoveanu, F., Moldoveanu, A., Dascalu, M. I., & Morar, A. (2019, October). Key technologies for indoor positioning systems. In 2019 18th RoEduNet Conference: Networking in Education and Research (RoEduNet) (pp. 1-7). IEEE. DOI: 10.1109/ROEDUNET.2019.8909406

Bjornson, E., Ozdogan, O., & Larsson, E. G. (2020). Intelligent Reflecting Surface Versus Decode-and-Forward: How Large Surfaces are Needed to Beat Relaying? *IEEE Wireless Communications Letters*, 9(2), 244–248. DOI: 10.1109/LWC.2019.2950624

Blackman, S. S. Theoretical approaches to data association and fusion, Proceedings of the SPIE, 931, Sensor Fusion, 4-6 April 1988, Orlando, FL, 50-5. DOI: 10.1117/12.946647

Boccardi, F., Heath, R. W., Lozano, A., Marzetta, T. L., & Popovski, P. (2014). Five disruptive technology directions for 5G. *IEEE Communications Magazine*, 52(2), 74–80.

Borgianni, L., Adami, D., & Giordano, S. (2023). Optimizing network performance and reliability with an integrated SD-WAN and satellite 6G architecture. *2023 2nd International Conference on 6G Networking (6GNet), 21*, 1-4. https://doi.org/DOI: 10.1109/6GNet58894.2023.10317704

Bourdoux, A., Barreto, A. N., van Liempd, B., de Lima, C., Dardari, D., Belot, D., . . . Miao, Y. (2020). 6G white paper on localization and sensing. arXiv preprint arXiv:2006.01779.

Brown, W. C. (1984, September). The History of Power Transmission by Radio Waves. *IEEE Transactions on Microwave Theory and Techniques*, 32(9), 1230–1242. DOI: 10.1109/TMTT.1984.1132833

Buede, D. M., & Martin, J. W. Comparison of Bayesian and Dempster-Shafer fusion, *Proceedings of the 1989 Symposium on Tri-Service Data Fusion*, Vol.1, May 1989, 81-101.

Bueno, C., Pereira, L. M., & Fabricio, M. M. (2018). Life cycle assessment and environmental-based choices at the early design stages: An application using building information modelling. *Architectural Engineering and Design Management*, 14(5), 332–346. DOI: 10.1080/17452007.2018.1458593

Cabrera, J. B. (2018). Scheduling variable field-of-view sensors for tracking multiple objects. In 2018 52nd asilomar conference on signals, systems, and computers (pp. 2174–2178). IEEE. DOI: 10.1109/ACSSC.2018.8645067

Cai, C., Du, Z., Liao, W., He, Y., Ma, X., Qin, S., & Xu, J. (2024). Experimental Demonstration and Practical Application of a Real-time Full-duplex Underwater Wireless Optical Communication Transceiver. *Journal of Lightwave Technology*, 42(24), 8541–8554. DOI: 10.1109/JLT.2024.3440639

Cantas, M. R., Chand, A., Zhang, H., Surnilla, G. C., & Guvenc, L. (2021). Data association between perception and V2V communication sensors. arXiv preprint arXiv:2101.08228.

Cao, W., Zhang, B., Liu, A., Yu, T., Guo, D., & Wei, Y. (2012). Gain enhancement for broadband periodic endfire antenna by using split-ring resonator structures. *IEEE Transactions on Antennas and Propagation*, 60(7), 3513–3516. DOI: 10.1109/TAP.2012.2196959

Catanese, C., Triki, A., Pincemin, E., & Jaouën, Y. (2019). A survey of neural network applications in fiber nonlinearity mitigation. In 2019 21st International Conference on Transparent Optical Networks (ICTON). IEEE. DOI: 10.1109/ICTON.2019.8840355

Chack, D., & Thool, S. N. (2023). High Capacity 64-Quadrature Amplitude Modulation Based Optical Coherent Transceiver for 60 GHz Radio over Fiber System. *Wireless Personal Communications*, 132(1), 183–204. DOI: 10.1007/s11277-023-10606-x

Chalvatzaki, G., Papageorgiou, X. S., Tzafestas, C. S., & Maragos, P. (2018b). Augmented human state estimation using interacting multiple model particle filters with probabilistic data association. *IEEE Robotics and Automation Letters*, 3(3), 1872–1879. DOI: 10.1109/LRA.2018.2800084

Chandra, S., & Dwivedi, S. (2020, February). Simulation analysis of high gain and low loss antenna with metallic ring and spoof SPP transmission line. In 2020 7th International Conference on Signal Processing and Integrated Networks (SPIN) (pp. 328-331). IEEE.

Chandra, S., & Dwivedi, S. (2019, March). Comparative analysis of reconfigurable patch antenna array for different liquid crystal substrates. In *2019 URSI Asia-Pacific Radio Science Conference (AP-RASC)* (pp. 1-4). IEEE.

Chandra, S., & Dwivedi, S. (2021, April). Graphene based Radiation Pattern reconfigurable antenna. In *International Conference for convergence in Technology* (pp. 2-4).

Chang, Y.-S., Chiao, H.-T., Abimannan, S., Huang, Y.-P., Tsai, Y.-T., & Lin, K.-M. (2020). An LSTM-based aggregated model for air pollution forecasting. *Atmospheric Pollution Research*, 11(8), 1451–1463. DOI: 10.1016/j.apr.2020.05.015

Chataut, R., Nankya, M., & Akl, R. (2024). 6G networks and the AI revolution—Exploring technologies, applications, and emerging challenges. *Sensors (Basel)*, 24(6), 1888. DOI: 10.3390/s24061888 PMID: 38544151

Chauhan, D., & Jain, J. K. (2023, January). Profiling Network Traffic by Using Classification Techniques in Machine Learning. In International Conference on Smart Trends in Computing and Communications (pp. 113-123). Singapore: Springer Nature Singapore. DOI: 10.1007/978-981-99-0769-4_12

Chauhan, D., & Jain, J. K. (2024). Measures and Preventions of Cyber Policies in Smart Cities. In Digital Technologies in Modeling and Management: Insights in Education and Industry (pp. 244-262). IGI Global.

Chauhan, D., Jain, J. K., & Singh, A. (2024). Deployment of Edge Computing for Smart Healthcare Systems on Cloud Computing Platform.

Chauhan, D., & Jain, J. K. (2019). A Journey from IoT to IoE. *International Journal of Innovative Technology and Exploring Engineering*, 8(11).

Chauhan, D., Jain, J. K., & Bahad, P. (2021). Performance evaluation of 802.11 A/G wireless networks with IP6HC. *Journal of Management Information and Decision Sciences*, 24, 1–7.

Chauhan, D., Sharma, S., & Singh, V. (2023). Roadmap from 4G to 6G in smart cities impact and challenges. *Int. J. Wirel. Microw. Technol.*, 13(6), 39–50. DOI: 10.5815/ijwmt.2023.06.05

Chen, C. (2019, December). Microstrip Edible Antennas for Health Monitoring. *IEEE Sensors Journal*, 19(23), 10567–10576.

Cheng, B., Cui, Z., Lu, B., Qin, Y., Liu, Q., Chen, P., He, Y., Jiang, J., He, X., Deng, X., Zhang, J., & Zhu, L. (2018). 340-GHz 3-D imaging radar with 4Tx-16Rx MIMO array. *IEEE Transactions on Terahertz Science and Technology*, 8(5), 509–519. DOI: 10.1109/TTHZ.2018.2853551

Cheng, Y., Peng, W., Huang, C., Alexandropoulos, G. C., Yuen, C., & Debbah, M. (2022). RIS-Aided Wireless Communications: Extra Degrees of Freedom via Rotation and Location Optimization. *IEEE Transactions on Wireless Communications*, 21(8), 1–1. DOI: 10.1109/TWC.2022.3151702

Chen, H. (2020, December). Biocompatibility and Performance of Edible Antennas in Challenging Environments. *Journal of Biomedical Engineering*, 53(4), 567–575.

Chen, H. (2023, June). Edible Antennas Based on Chitosan for Food Safety Monitoring. *IEEE Transactions on Antennas and Propagation*, 70(6), 3582–3590.

Chen, J. (2020, June). Biocompatibility and Design Strategies for Edible Electronics. *Biomedical Engineering Letters*, 10(2), 241–250. PMID: 32431954

Chen, Q., & Wang, R. (2020). Policy incentives and their role in shaping EV infrastructure innovation. *Transportation Research Part D, Transport and Environment.*

Chen, T., Pennisi, A., Li, Z., Zhang, Y., & Sahli, H. (2018). A hierarchical association framework for multi-object tracking in airborne videos. *Remote Sensing (Basel)*, 10(9), 1347. DOI: 10.3390/rs10091347

Chen, X., Li, R., & Zhao, M. (2020). *Standardization gaps in EV wireless charging technologies.* Renewable Energy Journal.

Chen, Y., Wang, L., Ai, Y., Jiao, B., & Hanzo, L. (2017). Performance Analysis of NOMA-SM in Vehicle-to-Vehicle Massive MIMO Channels. *IEEE Journal on Selected Areas in Communications*, 35(12), 2653–2666. DOI: 10.1109/JSAC.2017.2726006

Chen, Y., Xie, X., Yu, B., Li, Y., & Lin, K. (2021). Multitarget vehicle tracking and motion state estimation using a novel driving environment perception system of intelligent vehicles. *Journal of Advanced Transportation*, 2021, 2021. DOI: 10.1155/2021/6251399

Chergui, H., Ksentini, A., Blanco, L., & Verikoukis, C. (2022). Toward zero-touch management and orchestration of massive deployment of network slices in 6G. *IEEE Wireless Communications*, 29(1), 86–93. DOI: 10.1109/MWC.009.00366

Chi, L., Weng, Z., Qi, Y., & Drewniak, J. L. (2020). A 60 GHz PCB Wideband Antenna-in-Package for 5G/6G Applications. *IEEE Antennas and Wireless Propagation Letters*, 1225(c), 1–1. DOI: 10.1109/LAWP.2020.3006873

Choukiker, Y. K., & Mudiganti, J. C. (2016). Compact hybrid fractal antenna for wideband wireless applications. *International Journal of Microwave and Wireless Technologies*, 9(5), 1191–1196. DOI: 10.1017/S1759078716001318

Chowdhury, M. Z., Shahjalal, M., Ahmed, S., & Jang, Y. M. (2020). 6G wireless communication systems: Applications, requirements, technologies, challenges, and research directions. *IEEE Open Journal of the Communications Society*, 1, 957–975. DOI: 10.1109/OJCOMS.2020.3010270

Christy, G. S., Sundari, G., & Sonti, V. K. (2022, April). A Review on Evolution, Challenges and Scope in Visual Light Communication Systems. In 2022 IEEE 7th International conference for Convergence in Technology (I2CT) (pp. 1-5). IEEE.

Chu, Q., Ouyang, W., Liu, B., Zhu, F., & Yu, N. (2020). Dasot: A unified framework integrating data association and single object tracking for online multi-object tracking. 34, In *Proceedings of the AAAI conference on artificial intelligence* (07), (pp. 10672–10679). DOI: 10.1609/aaai.v34i07.6694

Cordova Morales, D. A., Nguyen, T., & Pujolle, G. (2023). Towards a blockgraph-based trustless authentication scheme for future 6G technology. *2023 2nd International Conference on 6G Networking (6GNet)*. https://doi.org/DOI: 10.1109/6GNet58894.2023.10317665

Daniłowicz, M. (2020). Multiple object tracking and segmentation with R-CNN networks. DOI: 10.1016/j.promfg.2020.04.032

Darusalam, U., Nathasia, N. D., Zarlis, M., & Priambodo, P. S. (2024). Simulating Free-Space Optical Communications to Support a Li-Fi Access Network in a Smart City Concept. CommIT (Communication and Information Technology). *Journal*, 18(1), 17–27.

Das, P., Singh, A. K., & Mandal, K. (2022). Metamaterial loaded highly isolated tunable polarisation diversity MIMO antennas for THz applications. *Optical and Quantum Electronics*, 54(4), 250. Advance online publication. DOI: 10.1007/s11082-022-03641-8

Das, R., Wang, Y., Putrus, G., Kotter, R., Marzband, M., Herteleer, B., & Warmerdam, J. (2020, January). multi-objective techno-economic-environmental optimisation of electric vehicle for energy services. *Applied Energy*, 257, 113965. DOI: 10.1016/j.apenergy.2019.113965

Das, S., Anveshkumar, N., Dutta, J., & Biswas, A. (Eds.). (2021). *Advances in terahertz technology and its applications*. Springer., DOI: 10.1007/978-981-16-5731-3

Das, S., Mitra, K., & Mandal, M. (2016). Sample size calculation: Basic principles. *Indian Journal of Anaesthesia*, 60(9), 652–656. DOI: 10.4103/0019-5049.190621 PMID: 27729692

Datta, T., & Chockalingam, A. (2013). On generalized spatial modulation. *2013 IEEE Wireless Communications and Networking Conference (WCNC)*, 2716–2721. https://doi.org/DOI: 10.1109/WCNC.2013.6554991

de Pater, I., Reijns, A., & Mitici, M. (2022). Alarm-based predictive maintenance scheduling for aircraft engines with imperfect Remaining Useful Life prognostics. *Reliability Engineering & System Safety*, 221, 108341. DOI: 10.1016/j.ress.2022.108341

del Peral-Rosado, J. A., Raulefs, R., López-Salcedo, J. A., & Seco-Granados, G. “Survey of cellular mobile radio localization methods: From 1G to 5G,” *IEEE Commun. Surveys Tuts.*, vol. 20, no. 2, pp. 1124–1148, 2nd Quart. 2018. DOI: 10.1109/COMST.2017.2785181

Den Hartog, F., Raschella, A., Bouhafs, F., Kempker, P., Boltjes, B., & Seyedebrahimi, M. (2017, November). A pathway to solving the Wi-Fi tragedy of the commons in apartment blocks. In 2017 27th International Telecommunication Networks and Applications Conference (ITNAC) (pp. 1-6). IEEE. DOI: 10.1109/ATNAC.2017.8215382

Desouza, K. C., & Bhagwatwar, A. (2012). Citizen apps to solve complex urban problems. *Journal of Urban Technology*, 19(3), 107–136. DOI: 10.1080/10630732.2012.673056

Devapriya, A. T., & Robinson, S. Investigation on metamaterial antenna for terahertz applications. J. Microw.Optoelectron. Electromagn. Appl. (2019). DOI: 10.1590/2179-10742019v18i31577

Diamantoulakis, P. D., Karagiannidis, G. K., & Ding, Z. (2018). Simultaneous Lightwave Information and Power Transfer (SLIPT). *IEEE Transactions on Green Communications and Networking*, 2(3), 1–1. DOI: 10.1109/TGCN.2018.2818325

Diambeki, D. D., Mandiya, R. E., Kyamakya, K., & Kasereka, S. K. (2022). Securing the light escaping in a Li-Fi network environment. *Procedia Computer Science*, 201, 684–689. DOI: 10.1016/j.procs.2022.03.091

Dib, O., & Toumi, K. (2020). Decentralized identity systems: Architecture, challenges, solutions and future directions. Annals of Emerging Technologies in Computing (AETiC), Print ISSN, 2516-0281.

Dimitrakopoulos, G., & Demestichas, P. (2010). Intelligent transportation systems. *IEEE Vehicular Technology Magazine*, 5(1), 77–84. DOI: 10.1109/MVT.2009.935537

Dimitriadou, K., Rigogiannis, N., Fountoukidis, S., Kotarela, F., Kyritsis, A., & Papanikolaou, N. (2023). Current Trends in Electric Vehicle Charging Infrastructure; Opportunities and Challenges in Wireless Charging Integration. *Energies*, 16(4), 2057. DOI: 10.3390/en16042057

Ding, X., Gan, Q., & Shaker, M. P. (2023, April). Optimal management of parking lots as a big data for electric vehicles using internet of things and Long–Short term Memory. *Energy*, 268, 126613. DOI: 10.1016/j.energy.2023.126613

Ding, Z., Lei, X., Karagiannidis, G. K., Schober, R., Yuan, J., & Bhargava, V. K. (2017). A Survey on Non-Orthogonal Multiple Access for 5G Networks: Research Challenges and Future Trends. *IEEE Journal on Selected Areas in Communications*, 35(10), 2181–2195. DOI: 10.1109/JSAC.2017.2725519

Dogra, A., Jha, R. K., & Jain, S. (2020). A survey on beyond 5G network with the advent of 6G: Architecture and emerging technologies. *IEEE Access : Practical Innovations, Open Solutions*, 9, 67512–67547. DOI: 10.1109/ACCESS.2020.3031234

Duan, B. Y. (2020). Evolution and innovation of antenna systems for beyond 5G and 6G. *Frontiers of Information Technology & Electronic Engineering*, 21, 1–3.

Duan, Q., Huang, J., Hu, S., Deng, R., Lu, Z., & Yu, S. (2023). Combining federated learning and edge computing toward ubiquitous intelligence in 6G network: Challenges, recent advances, and future directions. *IEEE Communications Surveys and Tutorials*, 25(4), 2892–2950. DOI: 10.1109/COMST.2023.3316615

Du, L., Zhang, W., Ma, J., & Tang, Y. (2021). Reconfigurable Intelligent Surfaces for Energy Efficiency in Multicast Transmissions. *IEEE Transactions on Vehicular Technology*, 70(6), 6266–6271. DOI: 10.1109/TVT.2021.3080302

Durrant-Whyte, H. F. Elements of sensor fusion, *Proceedings of the Institute of Electrical and Electronics Engineers Colloquium on Intelligent Control*, 1991, 1 - 2.

Dwivedi, S. (2023, August). Partial Ground-Inserted H-shaped Antenna for Beyond 5G Applications. In International Conference on ICT for Sustainable Development (pp. 313-321). Singapore: Springer Nature Singapore.

Dwivedi, S. (2023, December). Terahertz Elliptical Leaf Pattern Antenna for Beyond 5G Applications. In 2023 10th IEEE Uttar Pradesh Section International Conference on Electrical, Electronics and Computer Engineering (UPCON) (Vol. 10, pp. 340-343). IEEE.

Dwivedi, S. (2024). Antenna Design for Beyond 5G: 5G and 6G Applications With Different Materials and Techniques Including Metamaterials and Metasurfaces. In Electromagnetic Wave Control Techniques of Metasurfaces and Metamaterials (pp. 106-140). IGI Global.

Dwivedi, S. (2024). Metamaterial Antennas for Wireless Communication Systems. In Electromagnetic Wave Control Techniques of Metasurfaces and Metamaterials (pp. 32-60). IGI Global.

Dwivedi, S. (2024, March). Design of triangular MIMO antenna for 5G and beyond. In 2024 11th International Conference on Signal Processing and Integrated Networks (SPIN) (pp. 40-44). IEEE.

Dwivedi, S. "Design Demonstration ofFrequency Reconfiguration Antenna using Varactor diode with SiO ", 2023 International Conference on Computer,Electronics & Electrical Engineering & their Applications (IC2E3), 2023.

Dwivedi, S. (2022). Design and Analysis of Metamaterial Waveguide Antenna for broadband applications. In *ICT Analysis and Applications: Proceedings of ICT4SD 2022* (pp. 417–423). Springer Nature Singapore.

Dwivedi, S. (2022). *ReconfigurableArray antenna*. Intech Open., DOI: 10.5772/intechopen.106343

Dwivedi, S. "Proposed Design for Beyond 5G Antenna for Upgraded Applications with review," , 2023 *14th International Conference on Computing Communication and Networking Technologies (ICCCNT)*. DOI: 10.1109/ICCCNT56998.2023.10307737

Egbe, G. O., Arinze, S. N., Ebenuwa, S. H., Obi, E. R., & Nwajana, A. O. (2024). Maintenance resources optimization using Pareto analysis: Instrumentation air compressor in Oredo, Nigeria. [IJMMME]. *International Journal of Manufacturing, Materials, and Mechanical Engineering*, 13(1), 1–14. DOI: 10.4018/IJMMME.353392

Elavarasi, C., & Shanmuganantham, T. (2017). Multiband SRR loaded Koch star fractal antenna. *Alexandria Engineering Journal*, 57(3), 1549–1555. DOI: 10.1016/j.aej.2017.04.001

El-Kenawy, E. S. M., Ibrahim, A., Mirjalili, S., Zhang, Y. D., Elnazer, S., & Zaki, R. M. (2022). Optimized ensemblealgorithm for predicting metamaterial antenna parameters. *Computers, Materials & Continua*, 2022, 023884. DOI: 10. 32604/ cmc

Elliethy, A., & Sharma, G. (2018). Vehicle tracking in wide area motion imagery via stochastic progressive association across multiple frames. *IEEE Transactions on Image Processing*, 27(7), 3644–3656. DOI: 10.1109/TIP.2018.2818443 PMID: 29993777

Emara, M. K., Stuhec-Leonard, S. K., Tomura, T., Hirokawa, J., & Gupta, S. (2020). Laser-Drilled All- Dielectric Huygens' Transmit-Arrays as 120 GHz Band Beamformers. *IEEE Access : Practical Innovations, Open Solutions*, 8, 153815–153825. DOI: 10.1109/ACCESS.2020.3018297

Esmail, M. A. (2021). Optical wireless performance monitoring using asynchronous amplitude histograms. *IEEE Photonics Journal*, 13(3), 1–9. DOI: 10.1109/JPHOT.2021.3080593

Eswaran, S., & Honnavalli, P. (2023). Private 5G networks: A survey on enabling technologies, deployment models, use cases and research directions. *Telecommunication Systems*, 82(1), 3–26. DOI: 10.1007/s11235-022-00978-z PMID: 36439887

Evangelista, J. V. C., Sattar, Z., Kaddoum, G., & Chaaban, A. (2019). Fairness and Sum-Rate Maximization via Joint Subcarrier and Power Allocation in Uplink SCMA Transmission. *IEEE Transactions on Wireless Communications*, 18(12), 5855–5867. DOI: 10.1109/TWC.2019.2939820

Fakidis, J., Videv, S., Kucera, S., Claussen, H., & Haas, H. (2021). On the Design of Optical Energy Harvesting and Storage Systems for Outdoor Small Cells. *ICC 2021 -. IEEE International Conference on Communications : [proceedings]. IEEE International Conference on Communications*, 15, 1–6. DOI: 10.1109/ICC42927.2021.9500505

Farhodov, X., Moon, K.-S., Lee, S.-H., & Kwon, K.-R. (2020). LSTM network with tracking association for multi-object tracking. *Journal of Korea Multimedia Society*, 23(10), 1236–1249.

Farley, T. (2005). Mobile telephone history. *Telektronikk*, 101(3–4), 22–34.

Faruq, O., Shahriar Rahman, K. R., Jahan, N., Rokoni, S., & Rabeya, M. (2023). Li-Fi technology based long range free-space communication data transmit system evaluation. *International Review of Applied Sciences and Engineering*, 14(3), 413–425. DOI: 10.1556/1848.2023.00620

Farzaneh, H., Malehmirchegini, L., Bejan, A., Afolabi, T., Mulumba, A., & Daka, P. P. (2021). Artificial intelligence evolution in smart buildings for energy efficiency. *Applied Sciences (Basel, Switzerland)*, 11(2), 763. DOI: 10.3390/app11020763

Faye, S., Camelo, M., Sottet, J., Sommer, C., Franke, M., Baudouin, J., Castellanos, G., Decorme, R., Fanti, M. P., Fuladi, R., Kesik, G., Mendes, B., Murphy, C., Parker, S., Pryor, S., Senouci, S. M., & Turcanu, I. (2024). Integrating network digital twinning into future AI-based 6G systems: The 6G-TWIN vision. *2024 Joint European Conference on Networks and Communications & 6G Summit (EuCNC/6G Summit)*, 883-888. https://doi.org/DOI: 10.1109/EuCNC/6GSummit60053.2024.10597058

Fazea, Y., Mohammed, F., & Al-Nahari, A. (2021, December). A review on 5G technology in IoT-application based on light fidelity (Li-Fi) indoor communication. In *International Conference of Reliable Information and Communication Technology* (pp. 371-384). Cham: Springer International Publishing.

Feng, H., Tavakoli, R., Pantic, Z., & Onar, O. C. (2020). Advances in High-Power Wireless Charging Systems: Overview and Design Considerations. *IEEE Transactions on Transportation Electrification*, 6(3), 886–919. DOI: 10.1109/TTE.2020.3012543

Feng, P., Wang, W., Dlay, S., Naqvi, S. M., & Chambers, J. (2016). Social force model-based MCMC-OCSVM particle PHD filter for multiple human tracking. *IEEE Transactions on Multimedia*, 19(4), 725–739. DOI: 10.1109/TMM.2016.2638206

Fernandes, J., Reis, J., Melão, N., Teixeira, L., & Amorim, M. (2021). The role of Industry 4.0 and BPMN in the arise of condition-based and predictive maintenance: A case study in the automotive industry. *Applied Sciences (Basel, Switzerland)*, 11(8), 3438. DOI: 10.3390/app11083438

Field, M. J., & Grigsby, J. (2002). Telemedicine and remote patient monitoring. *Journal of the American Medical Association*, 288(4), 423–425. DOI: 10.1001/jama.288.4.423 PMID: 12132953

Figueiredo, L., Jesus, I., Machado, J. T., Ferreira, J. R., & De Carvalho, J. M. (2001, August). Towards the development of intelligent transportation systems. In *ITSC 2001. 2001 IEEE intelligent transportation systems. Proceedings (Cat. No. 01TH8585)* (pp. 1206-1211). IEEE. DOI: 10.1109/ITSC.2001.948835

Fitri, I., & Akbart, A. A. "A new gridded parasitic patch stacked microstrip antenna for enhanced wide bandwidth in 60 GHz Band," in 2017 International Conference on Broadband Communication, Wireless Sensors and Powering, BCWSP 2017, 2018, vol. 2018-Janua, pp. 1–4, DOI: 10.1109/BCWSP.2017.8272571

Flexible Data Communication and IIoT Platform in Industrial Applications—Ewon Flexy. Available online: www.ewon.biz (accessed on 28 November 2024).

Flora, H. K., & Chande, S. V. (2014). A systematic study on agile software development methodologies and practices. *International Journal of Computer Science and Information Technologies*, 5(3), 3626–3637.

Füchtenhans, M., Grosse, E. H., & Glock, C. H. (2021). Smart lighting systems: State-of-the-art and potential applications in warehouse order picking. *International Journal of Production Research*, 59(12), 3817–3839. DOI: 10.1080/00207543.2021.1897177

Furdek, M., & Natalino, C. (2022). Machine learning for network security management, attacks, and intrusions detection. In *Machine Learning for Future Fiber-Optic Communication Systems* (pp. 317–336). Academic Press. DOI: 10.1016/B978-0-32-385227-2.00017-6

Furdek, M., Natalino, C., Di Giglio, A., & Schiano, M. (2021). Optical network security management: Requirements, architecture, and efficient machine learning models for detection of evolving threats. *Journal of Optical Communications and Networking*, 13(2), A144–A155. DOI: 10.1364/JOCN.402884

Furdek, M., Natalino, C., Lipp, F., Hock, D., Di Giglio, A., & Schiano, M. (2020). Machine learning for optical network security monitoring: A practical perspective. *Journal of Lightwave Technology*, 38(11), 2860–2871. DOI: 10.1109/JLT.2020.2987032

Gan, W., Wang, S., Lei, X., Lee, M.-S., & Kuo, C.-C. J. (2018). Online NN-basedmultiple object tracking with enhanced model updates and identity association. *Signal Processing Image Communication*, 66, 95–102. DOI: 10.1016/j.image.2018.05.008

Garg, S., Kaur, K., Aujla, G. S., Kaddoum, G., Garigipati, P., & Guizani, M. (2023). Trusted explainable AI for 6G-Enabled edge cloud ecosystem. *IEEE Wireless Communications*, 30(3), 163–170. DOI: 10.1109/MWC.016.220047

Geng, S., Kivinen, J., Zhao, X., & Vainikainen, P. (2008). Millimeter-wave propagation channel characterization for short-range wireless communications. *IEEE Transactions on Vehicular Technology*, 58(1), 3–13.

Ghassemlooy, Z., Arnon, S., Uysal, M., Xu, Z., & Cheng, J. (2015). Emerging Optical Wireless Communications-Advances and Challenges. *IEEE Journal on Selected Areas in Communications*, 33(9), 1738–1749. DOI: 10.1109/JSAC.2015.2458511

Ghesla, C. (2017). Defaults in green electricity markets: Preference match not guaranteed. *Journal of the Association of Environmental and Resource Economists*, 4(S1), S37–S84. DOI: 10.1086/691977

Gnanalakshmi, V., Raaj, R., & Suresh Kumar, V. (2021). Design of Quad feed end-fire microstrip patch antenna for Airborne Systems. *Journal of Physics: Conference Series*, 2062(1), 1–10. DOI: 10.1088/1742-6596/2062/1/012003

Gohil, J. H., & Nb, V. (2021). Design and simulation of Sierpmski Triangular Fractal Antenna for S-Band. *IOP Conference Series: Materials Science and Engineering*.p.1–10.

Goścień, R., Ellinas, G., Furdek, M., Kolios, P., Lechowicz, P., Manousakis, K., Savva, G., Skorin-Kapov, N., & Walkowiak, K. (2020). *Security-aware carrier network planning*. Guide to Disaster-Resilient Communication Networks.

Goyal, R., & Vishwakarma, D. K. (2018). Design of a graphene-based patch antenna on glass substrate for high-speed terahertz communications. *Microwave and Optical Technology Letters*, 60(7), 1594–1600. DOI: 10.1002/mop.31216

Granström, K., Renter, S., Fatemi, M., & Svensson, L. (2017). *Pedestrian tracking using Velodyne data—Stochastic optimization for extended object tracking. In 2017 IEEE intelligent vehicles symposium (Iv)*. IEEE.

Grillo, M., Dowhuszko, A. A., Khalighi, M.-A., & Hamalainen, J. (2021). Resource allocation in a Quantum Key Distribution Network with LEO and GEO trusted-repeaters. *2021 17th International Symposium on Wireless Communication Systems (ISWCS)*. DOI: 10.1109/ISWCS49558.2021.9562139

Gubbi, J., Buyya, R., Marusic, S., & Palaniswami, M. (2013). Internet of Things (IoT): A vision, architectural elements, and future directions. *Future Generation Computer Systems*, 29(7), 1645–1660.

Gu, L., & Dong, Y. (2023). Compact dual-band bandpass filter with high selectivity based on spiral-stub-loaded dual-mode QMSIW resonators. *Microwave and Optical Technology Letters*, 65(2), 559–566. DOI: 10.1002/mop.33541

Gulati, D., Sharif, U., Zhang, F., Clarke, D., & Knoll, A. (2017). *Data association solution or avoidance: Evaluation of a filter based on RFS framework and factor graphs with SME. In 2017 IEEE international conference on multisensor fusion and integration for intelligent systems (MFI)*. IEEE.

Guo, H., & Ben, B. (2019). *Reinforcement Learning-Enabled Reliable Wireless Sensor Networks in Dynamic Underground Environments. ArXiv*. Cornell University., DOI: 10.1109/MILCOM47813.2019.9021051

Guo, L., Huang, F., & Tang, X. (2014). A novel integrated MEMS helix antenna for terahertz applications. *Optik (Stuttgart)*, 125(1), 101–103. DOI: 10.1016/j.ijleo.2013.06.016

Guo, S., Wang, J., Wang, X., & Tao, D. (2021). Online multiple object tracking with cross-task synergy. In *Proceedings of the IEEE/CVF conference on computer vision and pattern recognition* (pp. 8136–8145). DOI: 10.1109/CVPR46437.2021.00804

Gupta, A. (2022, May). Cellulose-based Planar Antennas for Ingestible Electronics. *IEEE Transactions on Antennas and Propagation*, 70(5), 1122–1130.

Gupta, M., & Mathur, V. (2017). Wheel shaped modified fractal antenna realization for wireless communications. *AEÜ. International Journal of Electronics and Communications*, 79, 257–266. DOI: 10.1016/j.aeue.2017.06.017

Gupta, U., & Dwivedi, S. "Frequency Reconfigurable Antenna Using Metamaterial Split Ring Resonators for Smart Applications", *2021 International Conference on Electrical, Communication, and Computer Engineering (ICECCE)*, 2021. DOI: 10.1109/ICECCE52056.2021.9514067

Hall, D. L., & Llinas, J. (1997). An introduction to multisensor data fusion. *Proceedings of the IEEE*, 85(1), 6–23. DOI: 10.1109/5.554205

Hamdan, S., Ayyash, M., & Almajali, S. (2020). Edge-computing architectures for internet of things applications: A survey. *Sensors (Basel)*, 20(22), 6441. DOI: 10.3390/s20226441 PMID: 33187267

Hamid Rezatofighi, S., Milan, A., Zhang, Z., Shi, Q., Dick, A., & Reid, I. (2015). Joint probabilistic data association revisited. In *Proceedings of the IEEE international conference on computer vision* (pp. 3047–3055).

Hamid, Y., & Sugumaran, M. (2020). A t-SNE based non linear dimension reduction for network intrusion detection. *International Journal of Information Technology : an Official Journal of Bharati Vidyapeeth's Institute of Computer Applications and Management*, 12(1), 125–134. DOI: 10.1007/s41870-019-00323-9

Hasan, B. M. S., & Abdulazeez, A. M. (2021). A review of principal component analysis algorithm for dimensionality reduction. *Journal of Soft Computing and Data Mining*, 2(1), 20–30.

Hassan, K. (2020, July). Folded Dipole Edible Antennas for Smart Packaging Applications. *Journal of Food Engineering*, 292, 110310.

Hassan, M. R., Karmakar, G. C., Kamruzzaman, J., & Srinivasan, B. (2017). Exclusive use spectrum access trading models in cognitive radio networks: A survey. *IEEE Communications Surveys and Tutorials*, 19(4), 2192–2231.

Heath Jr, R. W. (2013). What is the role of MIMO in future cellular networks: Massive? coordinated? mmwave?. ICC Wksp. Plenary: Beyond LTE-A.

Heikkilä, H., Maalouf, W., & Campello, G. (2021). The United Nations Office on Drugs and Crime's efforts to strengthen a culture of prevention in low-and middle-income countries. *Prevention Science*, 22(1), 18–28. DOI: 10.1007/s11121-020-01088-5 PMID: 31953793

Hemadeh, I. A., Satyanarayana, K., El-Hajjar, M., & Hanzo, L. (2017). Millimeter-wave communications: Physical channel models, design considerations, antenna constructions, and link-budget. *IEEE Communications Surveys and Tutorials*, 20(2), 870–913.

Henrique, P. S. R., & Prasad, R. (2020, October). The road for 6G multimedia applications. In 2020 23rd International Symposium on Wireless Personal Multimedia Communications (WPMC) (pp. 1-6). IEEE.

Hesham, H., Ismail, T., & Darweesh, M. S. (2020, July). Indoor localization and movement prediction algorithms with light-fidelity. In 2020 22nd international conference on transparent optical networks (ICTON) (pp. 1-4). IEEE. DOI: 10.1109/ICTON51198.2020.9203505

He, Y., Han, J., Yu, W., Hong, X., Wei, X., & Gong, Y. (2020). City-scale multicamera vehicle tracking by semantic attribute parsing and cross-camera tracklet matching. In *Proceedings of the IEEE/CVF conference on computer vision and pattern recognition workshops* (pp. 576–577).

Hocini, A., Temmar, M. N., Khedrouche, D., & Zamani, M. (2019). Novel approach for the design and analysis of a terahertz microstrip antenna based on photonic crystals. *Photonics and Nanostructures*, 36, 100723. DOI: 10.1016/j.photonics.2019.100723

Holloway, D., Dienstfrey, A., Kuester, E. F., O'Hara, J. F., Azad, A. K., & Taylor, A. J. (2009). A Discussion on the Interpretation and Characterization of Metafilms/Metasurfaces: The Two Dimensional Equivalent of Metamaterials. *Metamaterials (Amsterdam)*, 3(2), 100–112. DOI: 10.1016/j.metmat.2009.08.001

Holloway, K., Kuester, E. F., Baker-Jarvis, J., & Kabos, P. (2003). A Double Negative (DNG) Composite Medium Composed of Magneto-Dielectric Spherical Particles Embedded in a Matrix. *IEEE Transactions on Antennas and Propagation*, 51(10), 2596–2603. DOI: 10.1109/TAP.2003.817563

Hong, J. S. G., & Lancaster, M. J. (2004). *Microstrip filters for RF/microwave applications*. John Wiley & Sons.

Ho, P. W., Cheng, C. H., Liao, Y. S., Chi, Y. C., Sadhu, A. S., Matsumoto, A., Akahane, K., Chen, L.-Y., Kuo, H.-C., & Lin, G. R. (2024). Blue micro-LED with a red/green blended polymer film for 3.5-Gbps visible light communication employing adaptive SNR-Flattening Algorithm. *Next Nanotechnology*, 5, 100043. DOI: 10.1016/j.nxnano.2024.100043

Hosamo, H. H., Svennevig, P. R., Svidt, K., Han, D., & Nielsen, H. K. (2022). A Digital Twin predictive maintenance framework of air handling units based on automatic fault detection and diagnostics. *Energy and Building*, 261, 111988. DOI: 10.1016/j.enbuild.2022.111988

Hossain, M. A., Hossain, A. R., & Ansari, N. (2022). AI in 6G: Energy-efficient distributed machine learning for multilayer heterogeneous networks. *IEEE Network*, 36(6), 84–91. DOI: 10.1109/MNET.104.2100422

Huang, H. C. (2018, March). Overview of antenna designs and considerations in 5G cellular phones. In 2018 International Workshop on Antenna Technology (iWAT) (pp. 1-4). IEEE.

Huang, C., Zappone, A., Alexandropoulos, G. C., Debbah, M., & Yuen, C. (2019). Reconfigurable intelligent surfaces for energy efficiency in wireless communication. *IEEE Transactions on Wireless Communications*, 18(8), 4157–4170. DOI: 10.1109/TWC.2019.2922609

Huang, K. C., & Edwards, D. J. (2008). *Millimetre wave antennas for gigabit wireless communications: a practical guide to design and analysis in a system context*. John Wiley & Sons.

Huang, Y., & Clerckx, B. (2017). Large-Scale Multiantenna Multisine Wireless Power Transfer. *IEEE Transactions on Signal Processing*, 65(21), 5812–5827. DOI: 10.1109/TSP.2017.2739112

Hu, E., Deng, Z., Jiang, K., & Wu, C. (2020). Kullback-Leibler differential entropy equation based CIMM-PDA for reliable positioning. *Alexandria Engineering Journal*, 59(4), 2607–2615. DOI: 10.1016/j.aej.2020.04.024

Hugues-Salas, E., Ntavou, F., Gkounis, D., Kanellos, G. T., Nejabati, R., & Simeonidou, D. (2019). Monitoring and physical-layer attack mitigation in SDN-controlled quantum key distribution networks. *Journal of Optical Communications and Networking*, 11(2), A209–A218. DOI: 10.1364/JOCN.11.00A209

Hung, W.-C., Kretzschmar, H., Lin, T.-Y., Chai, Y., Yu, R., Yang, M.-H., & Anguelov, D. (2020). SoDA: Multi-object tracking with soft data association. arXiv preprint arXiv:2008.07725.

Huo, Y., Lin, X., Di, B., Zhang, H., Hernando, F. J. L., Tan, A. S., Mumtaz, S., Demir, Ö. T., & Chen-Hu, K. (2023). Technology trends for massive MIMO towards 6G. *Sensors (Basel)*, 23(13), 6062. DOI: 10.3390/s23136062 PMID: 37447911

Hussain, F., Khupse, S., & Khan, M. R. S. (2019). An Empirical Study on Li-Fi Technology. *International Research Journal of Innovations in Engineering and Technology*, 3(11), 58.

Hussein, H. S., Elsayed, M., Mohamed, U. S., Esmaiel, H., & Mohamed, E. M. (2019). Spectral Efficient Spatial Modulation Techniques. *IEEE Access : Practical Innovations, Open Solutions*, 7, 1454–1469. DOI: 10.1109/ACCESS.2018.2885826

Hu, Y. H., Yuan, X., Zhang, G., & Schmeink, A. (2021). Sustainable Wireless Sensor Networks With UAV-Enabled Wireless Power Transfer. *IEEE Transactions on Vehicular Technology*, 70(8), 8050–8064. DOI: 10.1109/TVT.2021.3090849

Huynh, L. N., & Huh, E. (2021). Envisioning edge computing in future 6G wireless networks. *2021 Fifth World Conference on Smart Trends in Systems Security and Sustainability (WorldS4)*, 307-311. https://doi.org/DOI: 10.1109/WorldS451998.2021.9514026

IEA (2023). “Global EV Outlook 2023: Scaling up the transition to electric mobility.”

Iffat Naqvi, S., Hussain, N., Iqbal, A., Rahman, M., Forsat, M., Mirjavadi, S. S., & Amin, Y. (2020). Integrated LTE and millimeter-wave 5G MIMO antenna system for 4G/5G wireless terminals. *Sensors (Basel)*, 20(14), 3926.

Ijeh, I. C., Khalighi, M. A., Elamassie, M., Hranilovic, S., & Uysal, M. (2022). Outage probability analysis of a vertical underwater wireless optical link subject to oceanic turbulence and pointing errors. *Journal of Optical Communications and Networking*, 14(6), 439. DOI: 10.1364/JOCN.454191

Imasuen, I. I., & Nwajana, A. O. (2022). Balancing WTO-TRIPS standard against Nigeria counterfeit regulatory efforts: A critique. [IJPPPHCE]. *International Journal of Public and Private Perspectives on Healthcare, Culture, and the Environment*, 6(1), 1–19. DOI: 10.4018/IJPPPHCE.301575

Imran, A. Z. M., Thomas, M., Ogbodo, E., & Ukala, A. (2024, April). ITS-G5 Antennas: Powering Future Urban Transports. In *2024 Photonics & Electromagnetics Research Symposium (PIERS)* (pp. 1-6). IEEE.

Inomata, M., Yamada, W., Kuno, N., Sasaki, M., Kitao, K., Nakamura, M., Ishikawa, H., & Oda, Y. (2021). Terahertz propagation characteristics for 6G mobile communication systems. *2021 15th European Conference on Antennas and Propagation (EuCAP)*. https://doi.org/DOI: 10.23919/EuCAP51087.2021.9411143

ITU-R. Report ITU-R, SM.2303-2. (2017). "Wireless power transmission using technologies other than radio frequency beam," 2017.

Iyampalam, P., & Ganesan, I. (2020). Design of Sierpinski Knopp Inspired Fractal Antenna for Public Safety Applications. *Wireless Personal Communications*, 114(1), 227–239. DOI: 10.1007/s11277-020-07360-9

Iyer, S., Patil, A., Bhairanatti, S., Halagatti, S., & Pandya, R. J. (2022). A survey on technological trends to enhance spectrum-efficiency in 6g communications. *Transactions of the Indian National Academy of Engineering : an International Journal of Engineering and Technology*, 7(4), 1093–1120. DOI: 10.1007/s41403-022-00372-w

Jain, J. K., & Chauhan, D. (2021). An Energy-Efficient Model for Internet of Things Using Compressive Sensing. *Journal of Management Information and Decision Sciences*, 24, 1–7.

James, G., Witten, D., Hastie, T., Tibshirani, R., & Taylor, J. (2023). Unsupervised learning. In *An Introduction to Statistical Learning: with Applications in Python* (pp. 503–556). Springer International Publishing. DOI: 10.1007/978-3-031-38747-0_12

Jamuna, R., Abarna, B., Kumar, A. S., Ganesh, L., & Pranesh, H. (2023). LIFI for Smart Transportation: Enabling Secure and Safe Vehicular Communication. Journal of Xi'an Shiyou University. *Natural Sciences Education*, 19(5), 1–8.

Janjua, J. I., Khan, T. A., Khan, M. S., & Nadeem, M. (2021, October). Li-Fi communications in smart cities for truly connected vehicles. In 2021 2nd International Conference On Smart Cities, Automation & Intelligent Computing Systems (ICON-SONICS) (pp. 1-6). IEEE. DOI: 10.1109/ICON-SONICS53103.2021.9617200

Jantree, P., Lerdwanittip, R., & Namsang, A. Pakkaraekthalin, Sirawattanachai N.(2021).A Dual-Band Fractal Antenna using ICPW Technique for LTE and WiMAX Systems. 9th International Electrical Engineering Congress (iEECON).p. 547–550.

Játiva, P. P., Azurdia-Meza, C. A., Cañizares, M. R., Sánchez, I., Seguel, F., Zabala-Blanco, D., & Carrera, D. F. (2022). Performance analysis of IEEE 802.15. 7-based visible light communication systems in underground mine environments. *Photonic Network Communications*, 43(1), 23–33. DOI: 10.1007/s11107-022-00965-z

Jayasinghe, J., Andújar, A., & Anguera, J. (2018). On the properties of Sierpinski gasket fractal microstrip antennas. *Microwave and Optical Technology Letters*, 61(3), 772–776. DOI: 10.1002/mop.31605

Jeon, H.-B., Kim, S.-M., Moon, H.-J., Kwon, D.-H., Lee, J.-W., Chung, J.-M., Han, S.-K., Chae, C.-B., & Alouini, M.-S. (2023). Free-space optical communications for 6G wireless networks: Challenges, opportunities, and prototype validation. *IEEE Communications Magazine*, 61(4), 116–121. DOI: 10.1109/MCOM.001.2200220

Jiang, T., Gradus, J. L., & Rosellini, A. J. (2020). Supervised machine learning: A brief primer. *Behavior Therapy*, 51(5), 675–687. DOI: 10.1016/j.beth.2020.05.002 PMID: 32800297

Jiang, W., Zhou, Q., He, J., Habibi, M. A., Melnyk, S., El-Absi, M., Han, B., Renzo, M. D., Schotten, H. D., Luo, F.-L., El-Bawab, T. S., Juntti, M., Debbah, M., & Leung, V. C. (2024). Terahertz communications and sensing for 6G and beyond: A comprehensive review. *IEEE Communications Surveys and Tutorials*, 26(4), 2326–2381. DOI: 10.1109/COMST.2024.3385908

Jiang, Z., Huynh, D. Q., Zhang, J., & Wu, Q. (2017). *Part-based data association for visual tracking. In 2017 international conference on digital image computing: techniques and applications (DICTA)*. IEEE.

Jin, Z., Lin, R., Liu, X., Sun, D., Liu, B., Tao, T., & Tian, P. (2024). Green micro-LED With a Bandwidth Exceeding 2 GHz for 9-Gbps Visible Light Communication Based on SNR Gap-Dependent Bit and Power Loading. Journal of Lightwave Technology.

Joerger, M., & Hassani, A. (2020). *A new data association method using Kalman filter innovation vector projections. In 2020 IEEE/ION Position, location and navigation symposium (PLANS)*. IEEE.

Johnson, M. (2020, November). Comparative Study of Edible Materials for Antenna Applications. *IEEE Transactions on Antennas and Propagation*, 68(11), 5314–5323.

Joyce Jacob, J., Abinaya S, Divya Priya R, Poonam Khatarkar, Abdullaeva Barno and G. Sathish kumar, (2023) E3S Web Conf., 399 01010 DOI: https://doi.org/.DOI: 10.1051/e3sconf/202339901010

Judge, M. A., Khan, A., Manzoor, A., & Khattak, H. A. (2022, May). Overview of smart grid implementation: Frameworks, impact, performance and challenges. *Journal of Energy Storage*, 49, 104056. DOI: 10.1016/j.est.2022.104056

Kakkars, R. S. (2021). Modified Koch Fractal Antenna for Multi and Wideband Wireless Applications. *ECTI Transactions on Electrical Engineering, Electronics, and Communications*, 19(2), 182–189. DOI: 10.37936/ecti-eec.2021192.242065

Kamalinejad, P., Mahapatra, C., Sheng, Z., Mirabbasi, S., Leung, V. C. M., & Guan, Y. L. (2015). Wireless energy harvesting for the Internet of Things. *IEEE Communications Magazine*, 53(6), 102–108. DOI: 10.1109/MCOM.2015.7120024

Kamga, G. N., & Aissa, S. (2019). Wireless Power Transfer in mmWave Massive MIMO Systems With/Without Rain Attenuation. *IEEE Transactions on Communications*, 67(1), 176–189. DOI: 10.1109/TCOMM.2018.2799217

Karthika, R., & Balakrishnan, S. (2015). Wireless communication using Li-Fi technology. SSRG International Journal of Electronics and Communication Engineering (SSRG-IJECE), 2(3), 32-40.

Karthikeyan, S., Bragruthshibu, H., Logesh, R., Srinivasan, K., & Tarjanbabu, S. "Solar Based Fast Tag Charger for Electrical Vehicle," 2021 7th International Conference on Advanced Computing and Communication Systems (ICACCS), Coimbatore, India, 2021, pp. 776-779. DOI: 10.1109/ICACCS51430.2021.9441940

Kashyap, S., Bjornson, E., & Larsson, E. G. (2015). *On the feasibility of wireless energy transfer using massive antenna arrays in Rician channels. KTH Publication Database DiVA*. KTH Royal Institute of Technology., DOI: 10.1109/SPAWC.2015.7226997

Katz, M., & Ahmed, I. (2020). Opportunities and challenges for visible light communications in 6G. 2020 2nd 6G wireless summit (6G SUMMIT), 1-5.

Katz, M., & O'Brien, D. (2019). Exploiting novel concepts for visible light communications: From light-based IoT to living surfaces. *Optik (Stuttgart)*, 195, 163176. DOI: 10.1016/j.ijleo.2019.163176

Kaur, T., Sharma, Y., & Sikand, R. (2024, July). Futuristic aspects of Li-Fi technology. In AIP conference proceedings (Vol. 3121, No. 1). AIP Publishing. DOI: 10.1063/5.0221807

Kaur, K., Mangat, V., & Kumar, K. (2020). A comprehensive survey of service function chain provisioning approaches in SDN and NFV architecture. *Computer Science Review*, 38, 100298. DOI: 10.1016/j.cosrev.2020.100298

Keerthi, R. S., Dhabliya, D., Elangovan, P., Borodin, K., Parmar, J., & Patel, S. K. (2021). Tunable high-gain and multiband microstrip antenna based on liquid/copper split-ring resonator superstrates for C/X band communication. *Physica B, Condensed Matter*, 618, 413203.

Kemmerer, R. A. (1985). Testing formal specifications to detect design errors. *IEEE Transactions on Software Engineering*, SE-11(1), 32–43. DOI: 10.1109/TSE.1985.231535

Kenner, A., & Oladele, O. (2024). Software-Defined Networking (SDN) and Network Function Virtualization (NFV) in Telecommunication: A Computer Science Perspective.

Khadersab, A., & Shivakumar, S. (2018). Vibration Analysis Techniques for Rotating Machinery and its effect on Bearing Faults. *Procedia Manufacturing*, 20, 247–252. DOI: 10.1016/j.promfg.2018.02.036

Khaleel, S. A., Hamad, E. K., Parchin, N. O., & Saleh, M. B. (2022). MTM-inspired graphene-based THz MIMO antenna configurations using characteristic mode analysis for 6G/IoT applications. *Electronics (Basel)*, 11(14), 2152.

Khalid, R., Naeem, M., & Ejaz, W. (2022). Autonomous Aerial Networks with Wireless PowerTransfer: Resource Optimization, Standardization, and Challenges. *IEEE Communications Standards Magazine*, 6(3), 24–31. DOI: 10.1109/MCOMSTD.0001.2200022

Khalil, R. A., Safelnasr, Z., Yemane, N., Kedir, M., Shafiqurrahman, A., & Saeed, N. (2024). Advanced learning technologies for intelligent transportation systems: Prospects and challenges. *IEEE Open Journal of Vehicular Technology*, 5, 397–427. DOI: 10.1109/OJVT.2024.3369691

Khaloopour, L., Su, Y., Raskob, F., Meuser, T., Bless, R., Würsching, L., . . . Jamali, V. (2024). Resilience-by-Design Concepts for 6G Communication Networks. arXiv preprint arXiv:2405.17480.

Khan, L. U., Yaqoob, I., Imran, M., Han, Z., & Hong, C. S. (2020). 6G wireless systems: A vision, architectural elements, and future directions. *IEEE Access : Practical Innovations, Open Solutions*, 8, 147029–147044. DOI: 10.1109/ACCESS.2020.3015289

Kharche, S., & Kharche, J. (2023). 6G Intelligent Healthcare Framework: A Review on Role of Technologies, Challenges and Future Directions. J. *Mobile Multimedia*, 19(3), 603–644. DOI: 10.13052/jmm1550-4646.1931

Khattak, M. K., Kahng, S., Khattak, M. S., Rehman, A., Lee, C., & Han, D. (2017, July). A low profile, wideband and high gain beam-steering antenna for 5G mobile communication. In 2017 IEEE international symposium on antennas and propagation & USNC/URSI national radio science meeting (pp. 2575-2576). IEEE.

Khezzar, D., Khedrouche, D., & Denidni, T. A. (2021). New design of a broadband PBG-based antenna for THz band applications. *Photonics and Nanostructures*, 46, 100947. DOI: 10.1016/j.photonics.2021.100947

Kim, C., Fuxin, L., Alotaibi, M., & Rehg, J. M. (2021). Discriminative appearance modelling with multi-track pooling for real-time multi-object tracking. In *Proceedings of the IEEE/CVF conference on computer vision and pattern recognition* (pp.9553–9562).

Kim, J., Park, S., & Lee, H. (2021). Blockchain-based standardization for RFID wireless charging. *IEEE Transactions on Industrial Electronics*.

Km, A., & Duttagupta, S. (2023). Hdl-ready mac layer implementation for multi-node Li-Fi communications. *International Journal of Information Technology : an Official Journal of Bharati Vidyapeeth's Institute of Computer Applications and Management*, 15(4), 2039–2051. DOI: 10.1007/s41870-023-01255-1

Koutsoupidou, M., Karanasiou, I. S., & Uzunoglu, N. (2014). Substrate constructed by an array of split ring resonators for a THz planar antenna. *Journal of Computational Electronics*, 13(3), 593–598. Advance online publication. DOI: 10.1007/s10825-014-0575-y

Krishna, C. M., Das, S., Nella, A., Lakrit, S., & Madhav, B. T. P. (2021). A micro-sized rhombus shaped THz antenna for high-speed short-range wireless communication applications. *Plasmonics*, 16(6), 2167–2177. DOI: 10.1007/s11468-021-01472-z

Krishna, ChM., Das, S., Lakrit, S., Lavadiya, S., Madhav, B. T. P., & Sorathiya, V. (2021). Design and analysis of a super wideband (0.09−30.14 THz) graphene-based log periodic dipole array antenna for terahertz applications. *Optik (Stuttgart)*, 247, 167991. DOI: 10.1016/j.ijleo.2021.167991

Kumar Singh, S., Pan, Y., & Hyuk Park, J. (2021). OTS scheme based secure architecture for energy-efficient IoT in edge infrastructure. *Computers, Materials & Continua*, 66(3), 2905–2922. DOI: 10.32604/cmc.2021.014151

Kumar, A. (2021, March). Biodegradable Materials for Edible Electronics. *Nature Sustainability*, 4(3), 166–175.

Kumar, A., "Examining healthcare services utilizing cloud technology in intelligent urban environments," *In Advances in healthcare information systems and administration book series,* pp. 77–98, Oct 2024.

Kumari, R., Tomar, V. K., & Sharma, A. (2022). Miniaturization and performance enhancement of super wide band four element MIMO antenna using DNG metamaterial for THz applications. *Optical and Quantum Electronics*, 54(9), 577. Advance online publication. DOI: 10.1007/s11082-022-04011-0

Kumar, K. V. P., Velidi, V. K., Althuwayb, A. A., & Rao, T. R. (2021). Microstrip dual-band bandpass filter with wide bandwidth using paper substrate. *IEEE Microwave and Wireless Components Letters*, 31(7), 833–836. DOI: 10.1109/LMWC.2021.3077879

Kumar, R. (2020, June). Edible Polymers in Antenna Design: Challenges and Solutions. *Advanced Functional Materials*, 30(23), 2002201.

Kumar, S. (2021, February). Novel Materials for Edible Antenna Applications. *Journal of Materials Chemistry. C, Materials for Optical and Electronic Devices*, 9(6), 1470–1481.

Kumar, S., Dixit, A. S., Malekar, R. R., Raut, H. D., & Shevada, L. K. (2020). Fifth generation antennas: A comprehensive review of design and performance enhancement techniques. *IEEE Access : Practical Innovations, Open Solutions*, 8, 163568–163593.

Kure, H. I., & Nwajana, A. O. (2022). *Protection of critical infrastructure using an Integrated Cyber-security Risk Management (i-CSRM) framework. In 5G Internet of Things and Changing Standards for Computing and Electronic Systems*. IGI Global.

Kushwaha, R. K., & Karuppanan, P. (2021). Investigation and design of microstrip patch antenna employed on PCs substrates in THz regime. *Aust J Electr Electron Eng*, 18(2), 118–125. DOI: 10.1080/1448837X.2021.1936779

Kushwaha, R. K., Karuppanan, P., & Malviya, L. D. (2018). Design and analysis of novel microstrip patch antenna on photonic crystal in THz. *Physica B, Condensed Matter*, 545, 107–112. DOI: 10.1016/j.physb.2018.05.045

Kushwaha, R. K., Karuppanan, P., & Srivastava, Y. (2018). Proximity feed multiband patch antenna array with SRR and PBG for THz applications. *Optik (Stuttgart)*, 175, 78–86. DOI: 10.1016/j.ijleo.2018.08.139

Lambrechts, W., Sinha, S., Lambrechts, W., & Sinha, S. (2019). Bridging the Digital Divide: Innovative Last Mile Internet Connectivity Solutions. Last Mile Internet Access for Emerging Economies, 1-40. Chauhan, D., & Jain, J. K. (2023, January). Profiling Network Traffic by Using Classification Techniques in Machine Learning. In International Conference on Smart Trends in Computing and Communications (pp. 113-123). Singapore: Springer Nature Singapore.

Lambrechts, J., & Sinha, S. (2016). *Microsensing Networks for Sustainable Cities* (Vol. 18). Springer., DOI: 10.1007/978-3-319-28358-6

Lan, H., Pan, Q., Yang, F., Sun, S., & Li, L. (2016). Variational Bayesian approach for joint multitarget tracking of multiple detection systems. In 2016 19th international conference on information fusion (FUSION), (pp. 1260–1267). IEEE.

Lang, H. D. (2018). Optimization of wireless power transfer systems with multiple transmitters and receivers. *Ph.D. dissertation, Dept. Elect. Eng., Toronto Univ., Toronto, Canada.*

Le, T. D. Georges Kaddoum, Tran, H.V., & Chadi Abou-Rjeily. (2021). Evolution Strategies for Lightwave Power Transfer Networks. *IEEE Wireless Communications Letters, 10*(11), 2572–2576. DOI: 10.1109/LWC.2021.3107731

Lee, S. Bin, Hyun, D., Kang, T. J., Yang, C., Shin, S., Kim, H., Park, S., Kong, T. S., & Kim, H. D. (2015). Identification of false rotor fault indications produced by on-line MCSA for medium voltage induction machines. IEEE Conference Record of Annual Pulp and Paper Industry Technical Conference, 2015-July (1), 729–739.

Lee, G., Sung, Y., & Seo, J. (2016). Randomly-Directional Beamforming in Millimeter-Wave Multiuser MISO Downlink. *IEEE Transactions on Wireless Communications*, 15(2), 1086–1100. DOI: 10.1109/TWC.2015.2483493

Lee, J. (2020, September). Edible Dipole Antennas for Digestive Monitoring. *Applied Physics Letters*, 117(12), 123502.

Lee, K.-H., Kanzawa, Y., Derry, M., & James, M. R. (2018). Multi-target track-to- track fusion based on permutation matrix track association. In 2018 [IEEE.]. *IEEE Intelligent Vehicles Symposium*, (IV), 465–470.

Leema, R. G., Rajesh, R., Rajeswari, M., Akshaya, V., Saravanan, D., & Sangeetha, N. (2021, July). Women Safety Android Application with Hardware Device. In *2021 International Conference on System, Computation, Automation and Networking (ICSCAN)* (pp. 1-5). IEEE. DOI: 10.1109/ICSCAN53069.2021.9526474

Lee, R. H., & Leahy, R. Segmentation of multisensor images, *Proceedings of the 6th Multi-dimensional Signal Processing Workshop*, Sept. 1989, Pacific Grove, USA, 23. DOI: 10.1109/MDSP.1989.96998

Lee, S. (2023, August). Edible Antennas Using Carbohydrate Materials for Biodegradable Wireless Communication Systems. *IEEE Sensors Journal*, 22(8), 6745–6752.

Lee, T. H., Yoon, K. C., & Kim, K. G. (2022). Miniaturized Dual-Band Bandpass Filter Using T-Shaped Line Based on Stepped Impedance Resonator with Meander Line and Folded Structure. *Electronics (Basel)*, 11(2), 219. DOI: 10.3390/electronics11020219

Lee, W. C. (2015). *Integrated wireless propagation models*. McGraw-Hill Education.

Lei, L., Zhang, Y., Shen, X. S., Lin, C., & Zhong, Z. (2013). Performance analysis of device-to-device communications with dynamic interference using stochastic Petri nets. *IEEE Transactions on Wireless Communications*, 12(12), 6121–6141.

Leitenstorfer, A., Moskalenko, A. S., Kampfrath, T., Kono, J., Castro-Camus, E., Peng, K., Qureshi, N., Turchinovich, D., Tanaka, K., Markelz, A. G., Havenith, M., Hough, C., Joyce, H. J., Padilla, W. J., Zhou, B., Kim, K.-Y., Zhang, X.-C., Jepsen, P. U., Dhillon, S., & Cunningham, J. (2023, June). The 2023 terahertz science and technology roadmap. *Journal of Physics. D, Applied Physics*, 56(22), 223001. Advance online publication. DOI: 10.1088/1361-6463/acbe4c

Lei, Y.. (2015). A deep learning-based method for machinery health monitoring with big data. *Jixie Gongcheng Xuebao*, 51(21), 49–56. DOI: 10.3901/JME.2015.21.049

Lei, Y.. (2017). Applying machine learning to machine fault diagnosis: A review and roadmap. Mechanical Systems and Signal Processing, 2020. 138: p. 106587, Li, Z.; Wang, Y.; Wang, K.-S. Intelligent predictive maintenance for fault diagnosis and prognosis in machine centers: Industry 4.0 scenario. *Advances in Manufacturing*, 5, 377–387.

Letaief, K. B., Chen, W., Shi, Y., Zhang, J., & Zhang, Y. J. A. (2019, August). The Roadmap to 6G: AI Empowered Wireless Networks. *IEEE Communications Magazine*, 57(8), 84–90. DOI: 10.1109/MCOM.2019.1900271

Li, C., Li, J., & Chen, W. (2022). U.S. Patent No. 11,284,416. Washington, DC: U.S. Patent and Trademark Office.

Li, X. F., & Xiao, J. K. (2023). Dual-Band Bandpass Filter Based on Suspended Coplanar Waveguide-Microstrip Hybrid. IEEE Transactions on Circuits and Systems I: Regular Papers.

Li, Y. C., Wu, D. S., Zhang, T. Z., & Xue, Q. (2023). Dual-band dual-channel bandpass filters using high quality factor dielectric resonators. IEEE Transactions on Circuits and Systems II: Express Briefs.

Liang, T., Lan, L., & Luo, Z. (2020). Enhancing the association in multi-object tracking via neighbor graph. arXiv preprint arXiv:2007.00265.

Liang, Y., & Zeng, J. (2021, October). Ai application in english vocational education through 6g revolution. In International Conference on 5G for Future Wireless Networks (pp. 532-545). Cham: Springer International Publishing.

Liao, L., & Köttig, F. (2014). Review of hybrid prognostics approaches for remaining useful life prediction of engineered systems, and an application to battery life prediction. *IEEE Transactions on Reliability*, 63(1), 191–207. DOI: 10.1109/TR.2014.2299152

Li, D. (2020). Ergodic Capacity of Intelligent Reflecting Surface-Assisted Communication Systems With Phase Errors. *IEEE Communications Letters*, 24(8), 1646–1650. DOI: 10.1109/LCOMM.2020.2997027

Li, F. (2022, March). Hybrid Edible Antennas: Combining Materials for Enhanced Performance. *Sensors and Actuators. B, Chemical*, 352, 130774.

Li, H. (2022, September). Edible Antennas for Food Safety and Environmental Monitoring. *IEEE Transactions on Antennas and Propagation*, 70(9), 2905–2915.

Li, J. (2022). Hybrid Edible Antenna Design Using Biodegradable Substrates and Conductors. *IEEE Access : Practical Innovations, Open Solutions*, 10, 30432–30443.

Li, J., Dang, S., Wen, M., Li, Q., Chen, Y., Huang, Y., & Shang, W. (2023). Index modulation multiple access for 6G communications: Principles, applications, and challenges. *IEEE Network*, 37(1), 52–60. DOI: 10.1109/MNET.002.2200433

Li, J., Della Penda, D., Sahlin, H., Schliwa-Bertling, P., Folke, M., & Stattin, M. (2021). An overview of 5g system accessibility differentiation and control. *IEEE Communications Standards Magazine*, 5(4), 104–111. DOI: 10.1109/MCOMSTD.0001.2000067

Li, J., Gao, T., He, B., Zheng, W., & Lin, F. (2023). Power Allocation and User Grouping for NOMA Downlink Systems. *Applied Sciences (Basel, Switzerland)*, 13(4), 2452. DOI: 10.3390/app13042452 PMID: 39354955

Lin, H. L., Liao, P. K., & Wu, W. D. (2018). U.S. Patent Application No. 15/835,768.

Linnartz, J. P., Corrêa, C. R. B., Cunha, T. B., Tangdiongga, E., Koonen, T., Deng, X., & Jungnickel, V. (2021, June). ELIoT: New features in LiFi for next-generation IoT. In 2021 Joint European Conference on Networks and Communications & 6G Summit (EuCNC/6G Summit) (pp. 148-153). IEEE.

Lips, S., Amola, O., & Draheim, D. (2021, November). Designing a Crisis Management Mobile Application Solution in Nigeria. In *The 23rd International Conference on Information Integration and Web Intelligence* (pp. 571-579). DOI: 10.1145/3487664.3487807

Li, Q., Wen, M., Basar, E., Poor, H. V., & Chen, F. (2019). Spatial Modulation-Aided Cooperative NOMA: Performance Analysis and Comparative Study. *IEEE Journal of Selected Topics in Signal Processing*, 13(3), 715–728. DOI: 10.1109/JSTSP.2019.2898099

Li, R., Zhao, Z., Chen, X., Palicot, J., & Zhang, H. (2014). TACT: A Transfer Actor-Critic Learning Framework for Energy Saving in Cellular Radio Access Networks. *IEEE Transactions on Wireless Communications*, 13(4), 2000–2011. DOI: 10.1109/TWC.2014.022014.130840

Li, S., & Zheng, X. (2021).Antenna Design Based on Minkowski Like Fractal Structure.*19th International Conference on Optical Communications and Networks (ICOCN)*.p.1–3.

Liu, H., Zong, J., Wang, Q., Liu, Y., & Yang, F. (2022). Cloud native based intelligent RAN architecture towards 6G programmable networking. *2022 7th International Conference on Computer and Communication Systems (ICCCS)*. https://doi.org/DOI: 10.1109/ICCCS55155.2022.9846266

Liu, Q., Chu, Q., Liu, B., & Yu, N. (2020). GSM: Graph similarity model for multi-object tracking. In IJCAI (pp. 530–536).

Liu, F., Yuan, W., Masouros, C., & Yuan, J. (2020). Radar-Assisted Predictive Beamforming for Vehicular Links: Communication Served by Sensing. *IEEE Transactions on Wireless Communications*, 19(11), 7704–7719. DOI: 10.1109/TWC.2020.3015735

Liu, G., Sun, Z., & Jiang, T. (2019). Joint time and energy allocation for QoS-aware throughput maximization in MIMO-based wireless powered underground sensor networks. *IEEE Transactions on Communications*, 67(2), 1400–1412. DOI: 10.1109/TCOMM.2018.2874990

Liu, H., Kuang, J., Cao, Y., & Zhang, Y. (2023). Flexible Design of Dual-Band Common-Mode Filters Using Hairpin Ring Resonators and Defected Ground Slots. *IEEE Transactions on Electromagnetic Compatibility*, 65(5), 1360–1370. DOI: 10.1109/TEMC.2023.3304532

Liu, H., Wang, Z., Hu, S., Xu, H. X., & Ren, B. (2020). Design of tri-band balanced filter with wideband common-mode suppression and upper stopband using square ring loaded resonator. *IEEE Transactions on Circuits and Wystems. II, Express Briefs*, 67(10), 1760–1764. DOI: 10.1109/TCSII.2019.2949772

Liu, J., Xiong, K., Lu, Y., Wing, D., Zhong, Z., & Han, Z. (2020). Energy Efficiency in Secure IRS-Aided SWIPT. *IEEE Wireless Communications Letters*, 9(11), 1884–1888. DOI: 10.1109/LWC.2020.3006837

Liu, Q., Lv, T., & Lin, Z. (2018). Energy-Efficient Transmission Design in Cooperative Relaying Systems Using NOMA. *IEEE Communications Letters*, 22(3), 594–597. DOI: 10.1109/LCOMM.2018.2790379

Liu, Q., Zhang, D. W., Gong, K., Qian, H., Qu, W. Z., & An, N. (2023). Single-and dual-band bandpass filters based on multiple-mode folded substrate-integrated waveguide cavities. *IEEE Transactions on Microwave Theory and Techniques*, 71(12), 5335–5345. DOI: 10.1109/TMTT.2023.3276058

Liu, R., Yang, B., Zio, E., & Chen, X. (2018). Artificial intelligence for fault diagnosis of rotating machinery: A review. *Mechanical Systems and Signal Processing*, 108, 33–47. DOI: 10.1016/j.ymssp.2018.02.016

Liu, W. (2022, January). Sustainable Design Strategies for Edible Electronics. *Environmental Science & Technology*, 56(2), 1134–1143.

Liu, X. (2022, July). High Efficiency Edible Antennas for Biomedical Applications. *IEEE Transactions on Antennas and Propagation*, 70(7), 2486–2495.

Liu, X., & Ansari, N. (2019). Toward Green IoT: Energy Solutions and Key Challenges. *IEEE Communications Magazine*, 57(3), 104–110. DOI: 10.1109/MCOM.2019.1800175

Liu, Y., Ding, Z., Elkashlan, M., & Poor, H. V. (2016). Cooperative Non-orthogonal Multiple Access With Simultaneous Wireless Information and Power Transfer. *IEEE Journal on Selected Areas in Communications*, 34(4), 938–953. DOI: 10.1109/JSAC.2016.2549378

Liu, Y., Yao, L., Xiong, W., & Zhou, Z. (2018). Joint kinematic and feature tracking of ships with satellite electronic information. *Journal of Navigation*, 71(5), 1178–1194. DOI: 10.1017/S0373463318000152

Liu, Z., Li, Q., & Mu, C. (2013). A hybrid LSSVR/HMM based prognostics approach. *Sensors (Basel)*, 13(5), 5542–5560. DOI: 10.3390/s130505542 PMID: 23624688

Li, X. (2020, November). Advanced Fabrication Methods for Edible Electronics. *Journal of Micromechanics and Microengineering*, 30(11), 115002.

Li, X., Zhang, L., & Wu, D. (2021). Intrusion detection for RFID-enabled EV systems: A deep learning approach. *Computers & Security*.

Li, Y., Huang, J., Sun, Q., Sun, T., & Wang, S. (2021). Cognitive service architecture for 6G core network. *IEEE Transactions on Industrial Informatics*, 17(10), 7193–7203. DOI: 10.1109/TII.2021.3063697

Lopez, A., Alves, H., Souza, R. D., & Montejo-Sanchez, S. (2019). Statistical Analysis of Multiple Antenna Strategies for Wireless Energy Transfer. *IEEE Transactions on Communications*, 67(10), 7245–7262. DOI: 10.1109/TCOMM.2019.2928542

Lopez, O. L. A., Mahmood, N. H., Alves, H., Lima, C. M., & Latva-aho, M. (2020). Ultra-Low Latency, Low Energy, and Massiveness in the 6G Era via Efficient CSIT-Limited Scheme. *IEEE Communications Magazine*, 58(11), 56–61. DOI: 10.1109/MCOM.001.2000425

Lu, L., Liu, C., Zhang, C., Hu, Z., Lin, S., Liu, Z., Zhang, M., Liu, X., & Chen, J. (2023). Architecture for self-evolution of 6g core network based on intelligent decision making. *Electronics (Basel)*, 12(15), 3255. DOI: 10.3390/electronics12153255

Luo, RC, & Kay, MG. Multisensor integration and fusion in intelligent systems. (. (1989, September/October). Institute of Electrical and Electronics Engineers Transactions on Systems. *Man and Cybernetics*, 19(5), 901–931. DOI: 10.1109/21.44007

Luo, Z., Attari, M., Habibi, S., & Von Mohrenschildt, M. (2019). Online multiple manoeuvring vehicle tracking system based on multi-model smooth variable structure filter. *IEEE Transactions on Intelligent Transportation Systems*.

Lu, X., Wang, P., Niyato, D., Kim, D. I., & Han, Z. (2015). Wireless Networks With RF Energy Harvesting: A Contemporary Survey. *IEEE Communications Surveys and Tutorials*, 17(2), 757–789. DOI: 10.1109/COMST.2014.2368999

M. Kokkonen, S. Myllymaki, and H. Jantunen, "3×3 dipole lens antenna at 300 GHz with different permittivity lenses," in 2nd 6G Wireless Summit 2020: Gain Edge for the 6G Era, 6G SUMMIT 2020, 2020, pp. 4–7, .DOI: 10.1109/6GSUMMIT49458.2020.9083825

Ma, G., Khalili, M., Parthiban, R., & Katz, M. (2023). A low-complexity handover scheme using unsupervised learning techniques for 6G multi-networking. *2023 2nd International Conference on 6G Networking (6GNet), 101*, 1-5. https://doi.org/DOI: 10.1109/6GNet58894.2023.10317655

Macon, C. A., Trott, K. D., & Kempel, L. C. (2003). A practical approach to modeling doubly curved conformal microstrip antennas. *Electromagnetic Waves*, 40, 295–314. DOI: 10.2528/PIER02122903

Mahadevaswamy, C. H., Vasistha, A. R., & Nwajana, A. O. (2024, July). Dual-band bandpass filter derived from the transformation of a single-band bandpass filter. In 2024 24th International Microwave and Radar Conference (MIKON) (pp. 200-203). IEEE. DOI: 10.23919/MIKON60251.2024.10633986

Mahemuti, B., Inoue, D., Kakugo, A., & Konagaya, A. (2016). Investigation of the microtubule dynamics with probabilistic data association filter. In 2016 IEEE 11th annual international conference on nano/micro engineered and molecular systems (NEMS), (pp. 101–106). IEEE. DOI: 10.1109/NEMS.2016.7758209

Mahmood, N. H., L´opez, O. L. A., Park, O. S., Moerman, I., Mikhaylov, K., Mercier, E., Munari, A., Clazzer, F., B¨ocker, S., & Bartz (Eds.), H. (2020). White paper on critical and massive machine type communication towards 6G. *6G Research Vision.*

Mahmood, N. H., Alves, H., Lopez, O. A., Shehab, M., Osorio, D. P., & Latva-Aho, M. (2020). Six key features of machine type communication in 6G. *2020 2nd 6G Wireless Summit (6G SUMMIT).* https://doi.org/DOI: 10.1109/JIOT.2022.3216402

Mahmud, I., & Medha, M. B. M. Hasanuzzaman, Global challenges of electric vehicle charging systems and its future prospects: A review, Research in Transportation Business & Management, Volume 49, 2023,101011, ISSN 2210-5395, https://doi.org/DOI: 10.1016/j.rtbm.2023.101011

Mahmud, R. H. (2020). Terahertz microstrip patch antennas for the surveillance applications. *Kurdistan Journal of Applied Research*, 5(1), 16–27.

Maktoubian, J., Taskhiri, M. S., & Turner, P. (2021). Intelligent Predictive Maintenance (IPdM) in Forestry: A Review of Challenges and Opportunities. *Forests*, 12(11), 1495. DOI: 10.3390/f12111495

Ma, M., You, F., Qian, T., Shen, C., Qin, R., Wu, T., & He, S. (2022). A wide stopband dual-band bandpass filter based on asymmetrical parallel-coupled transmission line resonator. *IEEE Transactions on Microwave Theory and Techniques*, 70(6), 3213–3223. DOI: 10.1109/TMTT.2022.3168024

Mansouri, Y., & Babar, M. A. (2021). A review of edge computing: Features and resource virtualization. *Journal of Parallel and Distributed Computing*, 150, 155–183. DOI: 10.1016/j.jpdc.2020.12.015

Manvi Chopra, T. (2022). A Review Paper on on Light Fidility Technology and its Applications. *Journal of Algebraic Statistics*, 13(2), 1287–1293.

Mao, B., Kawamoto, Y., & Kato, N. (2020). AI-based joint optimization of QoS and security for 6G energy harvesting Internet of Things. *IEEE Internet of Things Journal*, 7(8), 7032–7042. DOI: 10.1109/JIOT.2020.2982417

Martínez, V. M., Ribeiro, M. R., & Mota, V. F. (2024). Wi-Fi faces the new wireless ecosystem: A critical review. *Annales des Télécommunications*, 79(5), 397–413. DOI: 10.1007/s12243-023-00995-2

Ma, S., Zhang, F., Li, H., Zhou, F., Wang, Y., & Li, S. (2019). Simultaneous Lightwave Information and Power Transfer in Visible Light Communication Systems. *IEEE Transactions on Wireless Communications*, 18(12), 5818–5830. DOI: 10.1109/TWC.2019.2939242

Maslo, A., Sarajlić, N., Hodžić, M., & Mujčić, A. (2021). "Optical Network Security Attacks by Tapping and Encrypting Optical Signals." Advanced Technologies, Systems, and Applications V. *Papers Selected by the Technical Sciences Division of the Bosnian-Herzegovinian American Academy of Arts and Sciences*, 2020, 321–332.

Mathijssen, M., & Tanner, M. (2022, May). The use of a user-centric smart mobile application prototype for supporting safety and security in a city: a design science method. In *International Conference on Social Implications of Computers in Developing Countries* (pp. 270-294). Cham: Springer International Publishing. DOI: 10.1007/978-3-031-19429-0_17

Mazher, A., Chaudary, E., Limiti, E., & Falcone, F. Gul E Tayyaba Zehra, Mohammad Alibakhshikenari, Syed Naheel Raza Rizvi. (2021). Wideband Small Fractal Antenna with Simple Design Strategy for 5G Sub-6-GHz Wireless Communications.XXXIVth General Assembly and Scientific Symposium of the International Union of Radio Science (URSI GASS).p. 1–4.

Meinhardt, T., Kirillov, A., Leal-Taixe, L., & Feichtenhofer, C. (2021). Trackformer: Multi-object tracking with transformers. arXiv preprint arXiv:2101.02702.

Mekala, M. S., Viswanathan, P., Srinivasu, N., & Varma, G. P. S. (2019, December). Accurate decision-making system for mining environment using Li-Fi 5G technology over IoT framework. In 2019 International Conference on contemporary Computing and Informatics (IC3I) (pp. 74-79). IEEE.

Meneses, M., Matos, L., Prado, B., de Carvalho, A., & Macedo, H. (2020). Learning to associate detections for real-time multiple object tracking. arXiv preprint arXiv: 2007.06041.

Miao, Z.-W., Hao, Z.-C., Wang, Y., Jin, B.-B., Wu, J.-B., & Hong, W. (2018). A 400-GHz High-Gain Quartz-Based Single Layered Folded Reflectarray Antenna for Terahertz Applications. *IEEE Transactions on Terahertz Science and Technology*, 9(1), 78–88. DOI: 10.1109/TTHZ.2018.2883215

Mishra, P., & Singh, G. (2023). Sustainable Smart City to Society 5.0. In *Sustainable Smart Cities: Enabling Technologies, Energy Trends and Potential Applications* (pp. 39–57). Springer International Publishing. DOI: 10.1007/978-3-031-33354-5_2

Mitev, M., Chorti, A., Poor, H. V., & Fettweis, G. P. (2023). What physical layer security can do for 6G security. *IEEE Open Journal of Vehicular Technology*, 4, 375–388. DOI: 10.1109/OJVT.2023.3245071

Moghadasi, N., Collier, Z. A., Koch, A., Slutzky, D. L., Polmateer, T. L., Manasco, M. C., & Lambert, J. H. (2022, September). Trust and security of electric vehicle-to-grid systems and hardware supply chains. *Reliability Engineering & System Safety*, 225, 108565. DOI: 10.1016/j.ress.2022.108565

Moghaddam, E. E., Beyranvand, H., & Salehi, J. A. (2020). Crosstalk-aware resource allocation in survivable space-division-multiplexed elastic optical networks supporting hybrid dedicated and shared path protection. *Journal of Lightwave Technology*, 38(6), 1095–1102. DOI: 10.1109/JLT.2019.2954177

Mohanty, A. R. (2014). Machinery condition monitoring: Principles and practices. In *Machinery Condition Monitoring*. Principles and Practices. DOI: 10.1201/9781351228626-17

Mohapatra, H., & Dalai, A. K. (2022). IoT based V2I framework for accident prevention. *2022 2nd International Conference on Artificial Intelligence and Signal Processing (AISP), 0*, 1-4. https://doi.org/ DOI: 10.1109/AISP53593.2022.9760623

Moharrami, M., Fallahpour, A., Beyranvand, H., & Salehi, J. A. (2017). Resource allocation and multicast routing in elastic optical networks. *IEEE Transactions on Communications*, 65(5), 2101–2113. DOI: 10.1109/TCOMM.2017.2667664

Mohjazi, L., Muhaidat, S., Abbasi, Q. H., Imran, M. A., Dobre, O. A., & Renzo, M. D. (2022). Battery Recharging Time Models for Reconfigurable Intelligent Surfaces-Assisted Wireless Power Transfer Systems. *IEEE Transactions on Green Communications and Networking*, 6(2), 1173–1185. DOI: 10.1109/TGCN.2021.3120834

Mohsin, M. J., & Murdas, I. A. (2023). Performance analysis of an outdoor Li-Fi system-based AO-OFDM architecture under different FSO turbulence and weather conditions. *Optik (Stuttgart)*, 273, 170427. DOI: 10.1016/j.ijleo.2022.170427

Moro-Visconti, R., Cruz Rambaud, S., & López Pascual, J. (2023). Artificial intelligence-driven scalability and its impact on the sustainability and valuation of traditional firms. *Humanities & Social Sciences Communications*, 10(1), 1–14. DOI: 10.1057/s41599-023-02214-8

Mounir, M., El_Mashade, M. B., & Mohamed Aboshosha, A. (2022). On the selection of power allocation strategy in power domain non-orthogonal multiple access (PD-NOMA) for 6G and beyond. *Transactions on Emerging Telecommunications Technologies*, 33(6), e4289.

Mou, X., Gladwin, D. T., Zhao, R., & Sun, H. (2019). Survey on magnetic resonant coupling wireless power transfer technology for electric vehicle charging. *IET Power Electronics*, 12(12), 3005–3020. DOI: 10.1049/iet-pel.2019.0529

Mubashar, R., Siddique, M. A. B., Rehman, A. U., Asad, A., & Rasool, A. (2021). Comparative performance analysis of short-range wireless protocols for wireless personal area network. *Iran Journal of Computer Science*, 4(3), 201–210. DOI: 10.1007/s42044-021-00087-1

Mudumbai, R., Singh, S. K., & Madhow, U. (2009, April). Medium access control for 60 GHz outdoor mesh networks with highly directional links. In *IEEE INFOCOM 2009* (pp. 2871–2875). IEEE.

Mukherjee, A. N. (2022). Application of artificial intelligence: Benefits and limitations for human potential and labor-intensive economy–an empirical investigation into pandemic ridden Indian industry. *Management Matters*, 19(2), 149–166. DOI: 10.1108/MANM-02-2022-0034

Mu, R., Wu, Y., Pan, L., Zhao, W., & Wang, W. (2023). A Miniaturized Low-Loss Switchable Single-and Dual-Band Bandpass Filter. *International Journal of RF and Microwave Computer-Aided Engineering*, 2023, 2023. DOI: 10.1155/2023/9025980

Muralikrishnan, P., & Kalaivani, M. (2020). IOT Based Electric Vehicle Charging Station Using Arduino Uno. *International Journal of Advanced Science and Technology*, 29, 4101–4106.

Musumeci, F., Rottondi, C., Nag, A., Macaluso, I., Zibar, D., Ruffini, M., & Tornatore, M. (2018). An overview on the application of machine learning techniques in optical networks. *IEEE Communications Surveys and Tutorials*, 21(2), 1383–1408. DOI: 10.1109/COMST.2018.2880039

Muzaffar, K., Magray, M. I., Karthikeya, G. S., & Koul, S. K. (2019, September). High gain broadband Vivaldi antenna for 5G applications. In *2019 International Conference on Electromagnetics in Advanced Applications (ICEAA)* (pp. 496-497). IEEE.

Nacchia, M., Fruggiero, F., Lambiase, A., & Bruton, K. (2021). A Systematic Mapping of the Advancing Use of Machine Learning Techniques for Predictive Maintenance in the Manufacturing Sector. *Applied Sciences (Basel, Switzerland)*, 11(6), 2546. DOI: 10.3390/app11062546

Naik, K. K., Suman, M., & Rao, E. K. (2021). Design of complementary split ring resonators on elliptical patch antenna with enhanced gain for terahertz applications. *Optik (Stuttgart)*, 243, 167434.

Najumunnisa, M., Sastry, A. S. C., Madhav, B. T. P., Das, S., Hussain, N., Ali, S. S., & Aslam, M. (2022). A metamaterial inspired AMC backed dual band antenna for ISM and RFID applications. *Sensors (Basel)*, 22(20), 8065. Advance online publication. DOI: 10.3390/s22208065 PMID: 36298414

Naser, S., Bariah, L., Muhaidat, S., & Basar, E. (2023). Zero-energy devices empowered 6G networks: Opportunities, key technologies, and challenges. *IEEE Internet of Things Magazine*, 6(3), 44–50. DOI: 10.1109/IOTM.001.2200235

Nassif, A. B., Shahin, I., Attili, I., Azzeh, M., & Shaalan, K. (2019). Speech Recognition Using Deep Neural Networks: A Systematic Review. *IEEE Access : Practical Innovations, Open Solutions*, 7, 19143–19165. DOI: 10.1109/ACCESS.2019.2896880

Nayak, S., & Patgiri, R. (2022, April). 6G communication: A vision on the potential applications. In Edge Analytics: Select Proceedings of 26th International Conference—ADCOM 2020 (pp. 203-218). Singapore: Springer Singapore.

Nayak, S., & Patgiri, R. (2022). 6G Communication: A Vision on the Potential Applications. In Patgiri, R., Bandyopadhyay, S., Borah, M. D., & Emilia Balas, V. (Eds.), *Edge Analytics. Lecture Notes in Electrical Engineering* (Vol. 869). Springer. DOI: 10.1007/978-981-19-0019-8_16

Nejati, A., Sadeghzadeh, R. A., & Geran, F. (2014). Effect of photonic crystal and frequency selective surface implementation on gain enhancement in the microstrip patch antenna at terahertz frequency. *Physica B, Condensed Matter*, 449, 113–120. DOI: 10.1016/j.physb.2014.05.014

Nguyen, D. C., Ding, M., Pathirana, P. N., Seneviratne, A., Li, J., Niyato, D., Dobre, O., & Poor, H. V. (2022). 6G Internet of things: A comprehensive survey. *IEEE Internet of Things Journal*, 9(1), 359–383. DOI: 10.1109/JIOT.2021.3103320

Nguyen, D. H., Matsushima, T., Qin, C., & Adachi, C. (2021). Toward Thing-to-Thing Optical Wireless Power Transfer: Metal Halide Perovskite Transceiver as an Enabler. *Frontiers in Energy Research*, 9, 679125. DOI: 10.3389/fenrg.2021.679125

Nguyen, V. L., Lin, P. C., Cheng, B. C., Hwang, R. H., & Lin, Y. D. (2021). Security and privacy for 6G: A survey on prospective technologies and challenges. *IEEE Communications Surveys and Tutorials*, 23(4), 2384–2428. DOI: 10.1109/COMST.2021.3108618

Niyato, D., Kim, D. I., Maso, M., & Han, Z. (2017). Wireless Powered Communication Networks: Research Directions and Technological Approaches. *IEEE Wireless Communications*, 24(6), 88–97. DOI: 10.1109/MWC.2017.1600116

None Ruisi Ge, None Hong Pan, None Zhibin Lin, None Na Gong, None Jinhui Wang, & None Xiaowei Chen. (2016). *RF-powered battery-less Wireless Sensor Network in structural monitoring*. DOI: 10.1109/EIT.2016.7535298

Norp, T., Sachdeva, P., Anton, C., Trichias, K., & Mohnani, P. (2024). 6G research & Innovation activities in Europe: An overview of EU & Nationally funded programmes. *2024 Joint European Conference on Networks and Communications & 6G Summit (EuCNC/6G Summit)*, 1109-1114. https://doi.org/ DOI: 10.1109/EuCNC/6GSummit60053.2024.10597076

Nwajana, A. O., & Paul, R. (2024, July). 3-way equal filtering power divider using compact folded-arms square open-Loop resonator. In 2024 25th International Microwave and Radar Conference (MIKON) (pp. 191-195). IEEE.

Nwajana, A. O. (2020). Circuit modelling of bandpass/channel filter with microstrip implementation. *Indonesian Journal of Electrical Engineering, and Informatics*, 8(4), 696–705.

Nwajana, A. O. (2021). A step-by-step approach to bandpass/channel filter design. [IJECME]. *International Journal of Electronics, Communications, and Measurement Engineering*, 10(2), 1–14. DOI: 10.4018/IJECME.2021070101

Nwajana, A. O., Dainkeh, A., & Yeo, K. S. (2017). Substrate integrated waveguide (SIW) bandpass filter with novel microstrip-CPW-SIW input coupling. *Journal of Microwaves, Optoelectronics and Electromagnetic Applications*, 16(2), 393–402. DOI: 10.1590/2179-10742017v16i2793

Nwajana, A. O., & Obi, E. R. (2022). A review on SIW and its applications to microwave components. *Electronics (Basel)*, 11(7), 1160. DOI: 10.3390/electronics11071160

Nwajana, A. O., & Obi, E. R. (2023). Application of compact folded-arms square open-loop resonator to bandpass filter design. *Micromachines*, 14(2), 320. DOI: 10.3390/mi14020320 PMID: 36838019

Nwajana, A. O., Obi, E. R., Ijemaru, G. K., Oleka, E. U., & Anthony, D. C. (2021). Fundamentals of RF/Microwave bandpass filter design. In *Handbook of research on 5G networks and advancements in computing, electronics, and electrical engineering* (pp. 149–164). IGI Global. DOI: 10.4018/978-1-7998-6992-4.ch005

Nwajana, A. O., & Paul, R. (2024). Three-Way Equal Filtering Power Divider for Modern Communication Systems. *Engineering Reports*, •••, 13036. DOI: 10.1002/eng2.13036

O'malley, P. (1992). Risk, power and crime prevention. *Economy and Society*, 21(3), 252–275. DOI: 10.1080/03085149200000013

Ogunyombo, O. E., Azeez, O. Y., & Odunlami, D. (2018). Exploring the awareness of emergency helplines among taxi drivers in Ikeja administrative division of Lagos state.

Ojha, V., Sharma, A., & Gupta, S. (2022). Secure sound and data communication via Li-Fi. In *An Interdisciplinary Approach to Modern Network Security* (pp. 153–165). CRC Press. DOI: 10.1201/9781003147176-9

Okano, M. T. IOT and Industry 4.0: The Industrial New Revolution. In Proceedings of the ICMIS-17—International Conference on Management and Information Systems, Bangkok, Thailand, 25–26 September 2017; pp. 75–82.

Okan, T. (2021). High efficiency unslotted ultra-wideband microstrip antenna for sub-terahertz short range wireless communication systems. *Optik (Stuttgart)*, 242, 166859. DOI: 10.1016/j.ijleo.2021.166859

Oleiwi, H. W., & Al-Raweshidy, H. (2022). Cooperative SWIPT THz-NOMA/6G Performance Analysis. *Electronics (Basel)*, 11(6), 873. DOI: 10.3390/electronics11060873

Omede, A. J. (2011). Nigeria: Analysing the security challenges of the Goodluck Jonathan administration. *Canadian Social Science*, 7(5), 90–102.

Omri, N., Al Masry, Z., Mairot, N., Giampiccolo, S., & Zerhouni, N. (2021). Towards an adapted PHM approach: Data quality requirements methodology for fault detection applications. *Computers in Industry*, 127, 103414. DOI: 10.1016/j.compind.2021.103414

Paik, H., Lankecharanteja, K. R., & Reddy, M. (2021). A Miniaturized Fractal Antenna with Square Ring Slots for Ultrawideband Applications. *Progress in Electromagnetics Research Letters*, 99(1), 169–177. DOI: 10.2528/PIERL21070801

Palitharathna, K. W. S., Suraweera, H. A., Godaliyadda, R. I., Herath, V. R., & Ding, Z. (2022). Lightwave Power Transfer in Full-Duplex NOMA Underwater Optical Wireless Communication Systems. *IEEE Communications Letters*, 26(3), 622–626. DOI: 10.1109/LCOMM.2021.3140117

Pana, V. S., Babalola, O. P., & Balyan, V. (2022). 5G radio access networks: A survey. *Array (New York, N.Y.)*, 14, 100170. DOI: 10.1016/j.array.2022.100170

Pan, C., Ren, H., Wang, K., Elkashlan, M., Nallanathan, A., Wang, J., & Hanzo, L. (2020). Intelligent Reflecting Surface Aided MIMO Broadcasting for Simultaneous Wireless Information and Power Transfer. *IEEE Journal on Selected Areas in Communications*, 38(8), 1719–1734. DOI: 10.1109/JSAC.2020.3000802

Pan, S. J., Tsang, I. W., Kwok, J. T., & Yang, Q. (2011). Domain Adaptation via Transfer Component Analysis. *IEEE Transactions on Neural Networks*, 22(2), 199–210. DOI: 10.1109/TNN.2010.2091281 PMID: 21095864

Paolini, E., Valcarenghi, L., Maggiani, L., & Andriolli, N. (2023). Real-time clustering based on deep embeddings for threat detection in 6G networks. *IEEE Access : Practical Innovations, Open Solutions*, 11, 115827–115835. DOI: 10.1109/ACCESS.2023.3325721

Park, C., Kim, J., & Shin, S. (2022). Urban planning strategies for RFID-enabled wireless EV charging. *Transportation Research Part D, Transport and Environment.*

Park, J., & Kim, S. (2022). Standardization and policy frameworks for RFID in EV infrastructure. *IEEE Access : Practical Innovations, Open Solutions.*

Park, S., & Choi, W. (2020). Transmitter Current Control and Receiver Coil Selection in Magnetic MIMO Power Transfer Systems. *IEEE Wireless Communications Letters*, 9(10), 1782–1785. DOI: 10.1109/LWC.2020.3004815

Parra-Loera, R., Thompson, W. E., & Salvi, A. P. Adaptive selection of sensors based on individual performances in a multisensor environment, Proceedings of the SPIE, 1470, Data Structures and Target Classification, April 1991, Orlando, FL, 30-6.

Parvin, K., Hannan, M. A., Hui Mun, L., Hossain Lipu, M. S., Abdolrasol, M. G. M., Jern Ker, P., Muttaqi, K. M., & Dong, Z. Y. (2022, October). The future energy internet for utility energy service and demand- side management in smart grid: Current practices, challenges and future directions. *Sustainable Energy Technologies and Assessments*, 53, 102648. DOI: 10.1016/j.seta.2022.102648

Patel, M. (2023). Advanced Ingestible Antenna Designs for Smart Healthcare Applications. *IEEE Access : Practical Innovations, Open Solutions*, 11, 20657–20667.

Patel, R. (2020, September). Fabrication Techniques for Edible Antennas. *IEEE Transactions on Microwave Theory and Techniques*, 68(9), 3525–3534.

Patel, S. K., & Kosta, Y. P. (2012). Meandered multiband metamaterial square microstrip patch antenna design. Waves Rand. *Complex Media*, 2012, 723837. DOI: 10. 1080/ 17455 030

Patti, F. P. (2019). The European road to autonomous vehicles. *Fordham Int'l LJ*, 43, 125.

Paul, R., Sebastian, N., Yadukrishnan, P. S., & Vinod, P. (2021). Study on data transmission using Li-Fi in vehicle to vehicle anti-collision system. *Computer Networks, Big Data and IoTProceedings of ICCBI*, 2020, 519–540.

Pech, M., & Vrchota, J. (2021). Bednárˇ, J. Predictive Maintenance and Intelligent Sensors in Smart Factory [Review]. *Sensors (Basel)*, 21, 1470. DOI: 10.3390/s21041470 PMID: 33672479

Pérez-López, C. D., Soto-Bovó, L. M., Colbes, J., & Pinto-Roa, D. P. (2021). Semi-dynamic Routing and Spectrum Assignment with variable bandwidth in Elastic Optical Networks. Bee-inspired Algorithms approach. In *2021 XLVII Latin American Computing Conference (CLEI)*. IEEE. DOI: 10.1109/CLEI53233.2021.9639900

Perwej, Y. (2017). The next generation of wireless communication using Li-Fi (Light Fidelity) technology. *Journal of Computer Networks*, 4(1), 20–29. DOI: 10.12691/jcn-4-1-3

Pexyean, T., Saraubon, K., & Nilsook, P. (2023, August). IoT, 6G and Digital Twin For Smart Campus. In 2023 Research, Invention, and Innovation Congress: Innovative Electricals and Electronics (RI2C) (pp. 46-50). IEEE.

Pires, R., Blue, H., Malinen, J., De Angelis, M., Giardina, P. G., Landi, G., Laukkanen, M., Ahola, K., & Porambage, P. (2024). Closed-loop automation in 6G for minimum downtime task continuity in surveillance Cobots. *2024 Joint European Conference on Networks and Communications & 6G Summit (EuCNC/6G Summit)*, 860-865. https://doi.org/DOI: 10.1109/EuCNC/6GSummit60053.2024.10597048

Polymeni, S., Skoutas, D. N., Sarigiannidis, P., Kormentzas, G., & Skianis, C. (2024). Smart Agriculture and Greenhouse Gas Emission Mitigation: A 6G-IoT Perspective. *Electronics (Basel)*, 13(8), 1480. DOI: 10.3390/electronics13081480

Porambage, P., Gur, G., Moya Osorio, D. P., Livanage, M., & Ylianttila, M. (2021). 6G security challenges and potential solutions. *2021 Joint European Conference on Networks and Communications & 6G Summit (EuCNC/6G Summit)*. https://doi.org/DOI: 10.1109/EuCNC/6GSummit51104.2021.9482609

Pradhan, N. C., Koziel, S., Barik, R. K., Pietrenko-Dabrowska, A., & Karthikeyan, S. S. (2023). Miniaturized dual-band SIW-based bandpass filters using open-loop ring resonators. *Electronics (Basel)*, 12(18), 3974. DOI: 10.3390/electronics12183974

Press, C. (2017). *IoT Fundamentals: Networking Technologies*. Protocols, and Use Cases for the Internet of Things.

Psomas, C., & Krikidis, I. (2019). Energy beamforming in wireless powered mm-Wave sensor networks. *IEEE Journal on Selected Areas in Communications*, 37(2), 424–438. DOI: 10.1109/JSAC.2018.2872381

Punchihewa, Y. G., Vo, B.-T., Vo, B.-N., & Kim, D. Y. (2018). Multiple object tracking in unknown backgrounds with labeled random finite sets. *IEEE Transactions on Signal Processing*, 66(11), 3040–3055. DOI: 10.1109/TSP.2018.2821650

Qamar, F., Hindia, M. N., Rahman, T. A., Hassan, R., Dimyati, K., & Nguyen, Q. N. (2021). Propagation characterization and analysis for 5G mmWave through field experiments. *Computers, Materials & Continua*, 68(2), 2249–2264.

Qamar, F., Siddiqui, M. U. A., Hindia, M. N., Hassan, R., & Nguyen, Q. N. (2020). Issues, challenges, and research trends in spectrum management: A comprehensive overview and new vision for designing 6G networks. *Electronics (Basel)*, 9(9), 1416.

Qaraqe, M., Usman, M., Serbes, A., Ansari, I. S., & Alouini, M.-S. (2022). Power Hotspots in Space: Powering CubeSats via Inter-Satellite Optical Wireless Power Transfer. *IEEE Internet of Things Magazine*, 5(3), 180–185. DOI: 10.1109/IOTM.001.001.2200071

Qu, C., Zhang, Y., Zhang, X., & Yang, Y. (2020). Reinforcement learning-based data association for multiple target tracking in clutter. *Sensors (Basel)*, 20(22), 6595. DOI: 10.3390/s20226595 PMID: 33218053

R, Deepa., A., Sivasamy., S., Selvam. (2024). Intelligent networking unleashed: exploring ai and ml for enhanced 6g wireless communication systems. 228-242. DOI: 10.58532/V3BINC3P8CH3

Rabia, N. A., Ali, S., Sajid, A., & Zafar, A. (2020). A security review over wi-fi and li-fi. Information Management and Computer Science, 3(1), 1-9.

Rahimian, A., Alomainy, A., & Alfadhl, Y. (2016, November). A flexible printed millimetre-wave beamforming network for WiGig and 5G wireless subsystems. In 2016 Loughborough Antennas & Propagation Conference (LAPC) (pp. 1-5). IEEE.

Rajagopal, S., Abu-Surra, S., Pi, Z., & Khan, F. (2011, December). Antenna array design for multi-gbps mmwave mobile broadband communication. In 2011 IEEE Global Telecommunications Conference-GLOBECOM 2011 (pp. 1-6). IEEE.

Rajagopal, C., Suseela, S. B., Noorullakhan, N., & Sankararajan, R. (2016). Compact modified 'T' slot circular patch quad band antenna for MIMO applications. *International Journal of Microwave and Wireless Technologies*, 9(4), 865–873. DOI: 10.1017/S1759078716000696

Ramesh, B., Changala, R., Kalangi, P. K., & Bhavsingh, M. (2024). Optimizing 6G Network Slicing with the EvoNetSlice Model for Dynamic Resource Allocation and Real-Time QoS Management. *International Research Journal of Multidisciplinary Technovation*, 6(3), 325–340.

Rangan, S., Rappaport, T. S., & Erkip, E. (2014). Millimeter-wave cellular wireless networks: Potentials and challenges. *Proceedings of the IEEE*, 102(3), 366–385.

Rangesh, A., Maheshwari, P., Gebre, M., Mhatre, S., Ramezani, V., & Trivedi, M. M. (2021). TrackMPNN: A message passing graph neural architecture for multi-object tracking. arXiv preprint arXiv:2101.04206.

Rao, C. M., Prasuna, G., Chapala, H. K., Jeebaratnam, N., Navulla, D., & Verma, A. (2023, February). Designing a reliable and cost-effective Internet of Medical Things (IoMT) topology to minimize the maintenance and deployment cost. In 2023 Fifth International Conference on Electrical, Computer and Communication Technologies (ICECCT) (pp. 1-7). IEEE.

Rappaport, T. S. (2016, March 10). Spectrum Frontiers: The new world of millimeter-wave mobile communication. Keynote presentation at the Federal Communications Commission (FCC) Headquarters.

Rappaport, T. S., Eshar Ben-Dor, Murdock, J., & Qiao, Y. (2012). 38 GHz and 60 GHz angle-dependent propagation for cellular & peer-to-peer wireless communications. *International Conference on Communications*.DOI: 10.1109/ICC.2012.6363891

Rappaport, T. S., Heath, R. W., Daniels, R. C., & Murdock, J. N. (2013). Broadband millimeter-wave propagation measurements and models using adaptive-beam antennas for outdoor urban cellular communications. IEEE Transactions on Antennas and Propagation, 61(4), 1850–1859. DOI: 10.1109/TAP.2013.2245062

Rappaport, T. S., Heath, R. W., Daniels, R. C., & Murdock, J. N. (2013). Millimeter wave mobile communications for 5G cellular: It will work! IEEE Access, 1, 335–349. DOI: 10.1109/ACCESS.2013.2260813

Rappaport, T. S., Heath, R. W., Daniels, R. C., & Murdock, J. N. (2014). Millimeter wave wireless communication. Prentice Hall.

Rappaport, T., Heath, R., Daniels, R., & Murdock, J. (2014). Millimeter wave wireless communications (Prentice Hall Communications Engineering and Emerging Technologies Series). Pearson Education. Retrieved from https://books.google.co.uk/books?id=0Jh6BAAAQBAJ

Rappaport, T. S., Xing, Y., MacCartney, G. R., Molisch, A. F., Mellios, E., & Zhang, J. (2017, December). Overview of millimeter wave communications for fifth-generation (5G) wireless networks—With a focus on propagation models. *IEEE Transactions on Antennas and Propagation*, 65(12), 6213–6230. Advance online publication. DOI: 10.1109/TAP.2017.2734243

Ravikumar, S. S., Betgeri, S., & Singh, S. K. (2024). Strategies for mitigating security concerns in IoT-enabled smart cities. *Advances in Computational Intelligence and Robotics*, 239-273. https://doi.org/ DOI: 10.4018/979-8-3693-2373-1.ch012

Reddy, G. T., Kumar Reddy, M. P., Lakshmana, K., Kaluri, R., Singh Rajput, D., Srivastava, G., & Baker, T. (2020). Analysis of dimensionality reduction techniques on big data. *IEEE Access : Practical Innovations, Open Solutions*, 8, 54776–54788. DOI: 10.1109/ACCESS.2020.2980942

Rehman, A. U., Alblushi, I. G. M., Khalid, H. M., Said, Z., Iqbal, A., & Muyeen, S. M. (2024). Techno economics and energy dynamics of a solar powered smart charging infrastructure for electric vehicles with advanced IoT based monitoring and RFID based security. *Sustainable Futures : An Applied Journal of Technology, Environment and Society*, 8, 100376.

Rene, E. A., Tounsi Fokui, W. S., & Nembou Kouonchie, P. K. (2023, June). Optimal allocation of plug-in electric vehicle charging stations in the distribution network with distributed generation. *Green Energy Intell. Transport.*, 100094(3), 100094. Advance online publication. DOI: 10.1016/j.geits.2023.100094

Ren, Z., Wang, Z., Sun, Z., Guo, Y., Song, W., Zhang, H., Chen, C., Guo, B., Yu, Z., Zhou, X., & Zhang, D. (2024). Characterizing the Through-Wall Sensing Mechanism of Wi-Fi Signals with a Refraction-Aware Fresnel Zone Model. *IEEE Transactions on Mobile Computing*, 23(12), 13438–13454. DOI: 10.1109/TMC.2024.3425847

Renzo, M., Zappone, A., Debbah, M., Alouini, M.-S., Yuen, C., Rosny, J., & Tretyakov, S. (2020). Smart Radio Environments Empowered by Reconfigurable Intelligent Surfaces: How It Works, State of Research, and The Road Ahead. *IEEE Journal on Selected Areas in Communications*, 38(11), 2450–2525. Advance online publication. DOI: 10.1109/JSAC.2020.3007211

Rigogiannis, N., Kotsidimou, A., Arzanas, I., Kyritsi, C., & Papanikolaou, N. (2022, August). Comparative Performance Study of Hybrid Si/SiC Insulated-Gate Bipolar Transistors. In 2022 IEEE 7th Forum on Research and Technologies for Society and Industry Innovation (RTSI) (pp. 191-197). IEEE.

Ritter, C., Imle, A., Lee, J. Y., Müller, B., Fackler, O. T., Bartenschlager, R., & Rohr, K. (2018). Two-filter probabilistic data association for tracking of virus particles in fluorescence microscopy images. In 2018 IEEE 15th international symposium on biomedical imaging (ISBI 2018), (pp. 957–960). IEEE. DOI: 10.1109/ISBI.2018.8363729

Rocher, G., Tigli, J. Y., Lavirotte, S., & Ferry, N. (2024, July). A Framework Towards Assessing the Resilience of Urban Transport Systems. In *Proceedings of the 19th International Conference on Availability, Reliability and Security* (pp. 1-10). DOI: 10.1145/3664476.3670435

Roh, W. (2014). Millimeter-wave beamforming as an enabling technology for 5G cellular communications: Theoretical feasibility and prototype results. IEEE Communications Magazine, 52(2), 106–113. DOI: 10.1109/MCOM.2014.6736752

Roslan, N. H., Awang, A. H., & Hizan, H. M. (2008) The effect of photonic crystal parameters on the terahertz photonic crystal cavities microstrip antenna performances. In: *Proceedings of the 2018 IEEE International RF and Microwave Conference (RFM)*, Penang, Malaysia, 17–19 December 2018, pp 147–150.

Rowshan, M., Qiu, M., Xie, Y., Gu, X., & Yuan, J. (2024). *Channel Coding Towards 6G: Technical Overview and Outlook*. IEEE Open Journal of the Communications Society.

Rufino Henrique, P. S., & Prasad, R. (2021). 6G networks for the next generation of digital TV beyond 2030. *Wireless Personal Communications*, 121(2), 1363–1378. DOI: 10.1007/s11277-021-09070-2 PMID: 34566262

Sadman, A. S., Kar, A., Emon, A. R., & Mubin, M. A. (2024, April). An Overview of the Architecture and Challenges of RTM System by Integrating 6G Technology with UAV Network. In 2024 3rd International Conference on Advancement in Electrical and Electronic Engineering (ICAEEE) (pp. 1-6). IEEE.

Saeed, M. A., & Nwajana, A. (2023, September). U-Shaped terahertz microstrip patch antenna for 6G future communications. In 2023 7th International Electromagnetic Compatibility Conference (EMC Turkiye) (pp. 1-4). IEEE.

Saeed, N., Almorad, H., Dahrouj, H., Al-Naffouri, T. Y., Shamma, J. S., & Alouini, M. S. (2021). Point-to-point communication in integrated satellite-aerial 6G networks: State-of-the-art and future challenges. *IEEE Open Journal of the Communications Society*, 2, 1505–1525. DOI: 10.1109/OJCOMS.2021.3093110

Safitri, C., Galina, M., & Simatupang, J. W. (2022). Light-fidelity as next generation network technology: a bibliometric survey and analysis.

Safitri, C., Harjono, M. S., Hasrito, E. S., & Roestam, R. (2024). Comprehensive Survey: Quality of Service in Railway Communication Using Information-Centric Networking and Light Fidelity. *IEEE Transactions on Intelligent Transportation Systems*, 25(12), 19218–19251. DOI: 10.1109/TITS.2024.3472698

Saha, R. K. (2020a). Interweave shared-use model for dynamic spectrum access in millimeter-wave mobile systems for 6G. Proceedings of the 2020 IEEE 92nd Vehicular Technology Conference (VTC2020-Fall), 1–6.

Saha, R. K. (2020b). A flexible licensed spectrum allocation technique for millimeter-wave mobile systems toward 6G. Proceedings of the 2020 23rd International Symposium on Wireless Personal Multimedia Communications (WPMC), 1–6.

Sakarellos, V. K. (2012). Cooperative diversity performance in millimeter wave radio systems. IEEE Transactions on Communications, 60(12), 3592–3603. DOI: 10.1109/TCOMM.2012.092612.110885

Salama, M. (2021, April). Starch-based Edible Patch Antennas for Smart Packaging. *Sensors and Actuators. B, Chemical*, 331, 129–138.

Salami, G., Faruk, N., Surajudeen-Bakinde, N., & Ngobigha, F. (2019, October). Challenges and Trends in 5G Deployment: A Nigerian Case Study. In *2019 2nd International Conference of the IEEE Nigeria Computer Chapter (NigeriaComputConf)* (pp. 1-6). IEEE.

Salman, K. D., & Hamza, E. K. (2021). Visible light fidelity technology: Survey. *Iraqi J. Comput. Commun. Control Syst. Eng*, 21(2), 1–15.

Salscheider, N. O. (2021). Object tracking by detection with visual and motion cues. arXiv preprint arXiv:2101.07549

Samimi, M. K., & Rappaport, T. S. (2015). 3-D statistical channel model for millimeter-wave outdoor mobile broadband communications. In Proceedings of the IEEE International Conference on Communications (ICC) (pp. 2430–2436). DOI: 10.1109/ICC.2015.7248615

Sarkar, A., Agarwal, S., & Nath, A. (2015). Li-Fi technology: data transmission through visible light. International Journal of Advance Research in Computer Science and Management Studies, 3(6).

Sarwar, U., Muhammad, M., Mokhtar, A. A., Khan, R., Behrani, P., & Kaka, S. (2024). Hybrid intelligence for enhanced fault detection and diagnosis for industrial gas turbine engine. *Results in Engineering*, 21, 101841. DOI: 10.1016/j.rineng.2024.101841

Sayyad. S, S. Kumar, A. Bongale, P. Kamat, S. Patil and K. Kotecha, (2021) "Data-Driven Remaining Useful Life Estimation for Milling Process: Sensors, Algorithms, Datasets, and Future Directions," in IEEE Access, . DOI: 10.1109/ACCESS.2021.3101284

Sayyad, S., Kumar, S., Bongale, A., Bongale, A. M., & Patil, S. (2021). Estimating Remaining Useful Life in Machines Using Artificial Intelligence: A Scoping Review. *Library Philosophy and Practice*, 2021(January), 1–26.

Selvi, S. A., Rajesh, R. S., & Ajisha, M. A. T. (2019, December). An efficient communication scheme for Wi-Li-Fi network framework. In 2019 Third International conference on I-SMAC (IoT in Social, Mobile, Analytics and Cloud)(I-SMAC) (pp. 697-701). IEEE. DOI: 10.1109/I-SMAC47947.2019.9032650

Serôdio, C., Cunha, J., Candela, G., Rodriguez, S., Sousa, X. R., & Branco, F. (2023). The 6G ecosystem as support for IoE and private networks: Vision, requirements, and challenges. *Future Internet*, 15(11), 348. DOI: 10.3390/fi15110348

Shah, A. S. (2022, January). A survey from 1G to 5G including the advent of 6G: Architectures, multiple access techniques, and emerging technologies. In 2022 IEEE 12th Annual Computing and Communication Workshop and Conference (CCWC) (pp. 1117-1123). IEEE.

Shalini, M., & Madhan, M. (2019). Design and analysis of a dualpolarized graphene based microstrip patch antenna for terahertz applications. *Optik (Stuttgart)*, 194, 163050. DOI: 10.1016/j.ijleo.2019.163050

Shamim, S. M., Das, S., Hossain, M. A., & Madhav, B. T. P. (2021). Investigations on Graphene-Based Ultra-Wideband (UWB) microstrip patch antennas for Terahertz (THz) Applications. *Plasmonics*, 16(5), 1623–1631. DOI: 10.1007/s11468-021-01423-8

Shamim, S. M., Uddin, M. S., Hasan, M., & Samad, M. (2021). Design and implementation of miniaturized wideband microstrip patch antenna for high-speed terahertz applications. *Journal of Computational Electronics*, 20(1), 604–610. DOI: 10.1007/s10825-020-01587-2

Shankar, E., Kumar, K. V. P., & Velidi, V. K. (2023, April). A miniaturized dual narrow-band bandpass filter using a hairpin resonator and coupled lines. In *2023 International Conference on Recent Advances in Electrical, Electronics, Ubiquitous Communication, and Computational Intelligence (RAEEUCCI)* (pp. 1-4). IEEE. DOI: 10.1109/RAEEUCCI57140.2023.10134052

Shao, S. Y., Sun, W. J., Yan, R. Q., Wang, P., & Gao, R. X. (2017). A Deep Learning Approach for Fault Diagnosis of Induction Motors in Manufacturing. [English Edition]. *Chinese Journal of Mechanical Engineering*, 30(6), 1347–1356. DOI: 10.1007/s10033-017-0189-y

Sharma, N., Singh, H. S., & Khanna, R. (2021).Design and Analysis of Multiband Fractal Antenna for MIMO/Diversity Applications. Wireless Personal Communication.p.1–16.

Sharma, S., Darak, S. J., & Srivastava, A. (2019). Transfer reinforcement learning based framework for energy savings in cellular base station network. *Proc. Asia Pacific Radio Sci. Conf.*, 1–4. G., E. B., Bennis, M., & Debbah, M. (2014). Anticipatory caching in small cell networks: A transfer learning approach. *Proc. Kuvs Workshop Anticipatory Netw.*, 1–3. Ba¸ E., g, stu., Bennis, M., & Debbah, M. (2015). A transfer learning approach for cache-enabled wireless networks. *Proc. Int. Symp. Model. Optim. Mobile, Ad-Hoc, Wireless Netw.*, 161–166. DOI: 10.23919/URSIAP-RASC.2019.8738418

Sharma, G., Sharma, H., Pareek, R., Gour, N., Sharma, R. S., & Kumar, A. (2021). Self-healing topology for DDoS attack identification & discovery protocol in software-defined networks. *Journal of Discrete Mathematical Sciences and Cryptography*, 24(8), 2221–2232. DOI: 10.1080/09720529.2021.2009192

Sharma, M., Sharma, K., Dwivedi, S., Singh, D. K., Gill, R., & Kumar, N. (2023, December). A Multiple Millimeter-Wave 5G MIMO Antenna Including n257/n258/n259/n260/n261 FR2-Bands with High Diversity Performance. In *2023 International Conference on Power Energy, Environment & Intelligent Control (PEEIC)* (pp. 279-284). IEEE.

Sharma, P. K., Ryu, J. H., Park, K. Y., Park, J. H., & Park, J. H. (2018). Li-Fi based on security cloud framework for future IT environment. *Human-centric Computing and Information Sciences*, 8(1), 1–13. DOI: 10.1186/s13673-018-0146-5

Sharma, S., Darak, S. J., & Srivastava, A. (2017). Energy saving in heterogeneous cellular network via transfer reinforcement learning based policy. *Proc. Int. Conf. Commun. Syst. Netw.*, 397–398. DOI: 10.1109/COMSNETS.2017.7945411

Sharma, S., Darak, S. J., Srivastava, A., & Zhang, H. (2016). A transfer learning framework for energy efficient Wi-Fi networks and performance analysis using real data. *IEEE International Conference on Advanced Networks and Telecommunications Systems (ANTS)*. DOI: 10.1109/ANTS.2016.7947853

Sheng, H., Chen, J., Zhang, Y., Ke, W., Xiong, Z., & Yu, J. (2018). Iterative multiple hypothesis tracking with tracklet-level association. *IEEE Transactions on Circuits and Systems for Video Technology*.

Shen, X., Liao, W., & Yin, Q. (2022). A novel wireless resource management for the 6G-enabled high-density Internet of Things. *IEEE Wireless Communications*, 29(1), 32–39. DOI: 10.1109/MWC.003.00311

Shi, E., Zhang, J., Du, H., Ai, B., Yuen, C., Niyato, D., Letaief, K. B., & Shen, X. (2024). RIS-aided cell-free massive MIMO systems for 6G: Fundamentals, system design, and applications. *Proceedings of the IEEE*, 112(4), 331–364. DOI: 10.1109/JPROC.2024.3404491

Shi, Y., An, K., & Li, Y. (2021). Index modulation based frequency hopping: Anti-jamming design and analysis. *IEEE Transactions on Vehicular Technology*, 70(7), 6930–6942. DOI: 10.1109/TVT.2021.3087640

Shi, Y., Lian, L., Shi, Y., Wang, Z., Zhou, Y., Fu, L., Bai, L., Zhang, J., & Zhang, W. (2023). Machine learning for large-scale optimization in 6g wireless networks. *IEEE Communications Surveys and Tutorials*, 25(4), 2088–2132. DOI: 10.1109/COMST.2023.3300664

Silvestri, L., Forcina, A., Introna, V., Santolamazza, A., & Cesarotti, V. (2020). Maintenance transformation through Industry 4.0 technologies: A systematic literature review. *Computers in Industry*, 123, 103335. DOI: 10.1016/j.compind.2020.103335

Singh, G., Rajan, S., & Majumdar, S. (2017). A greedy data association technique for multiple object tracking. In 2017 IEEE third international conference on multimedia big data (BigMM), (pp. 177–184). IEEE. DOI: 10.1109/BigMM.2017.53

Singh, R., Kaushik, A., Shin, W., Di Renzo, M., Sciancalepore, V., Lee, D., . . . Dobre, O. A. (2024). Towards 6G Evolution: Three Enhancements, Three Innovations, and Three Major Challenges. arXiv preprint arXiv:2402.10781.

Singh, A., & Singh, S. (2015). A trapezoidal microstrip patch antenna on photonic crystal substrate for high-speed THz applications. *Photonics and Nanostructures*, 14, 52–62. DOI: 10.1016/j.photonics.2015.01.003

Singhal, S. (2019). Elliptical ring terahertz fractal antenna. *Optik (Stuttgart)*, 194, 163129. Advance online publication. DOI: 10.1016/j.ijleo.2019.163129

Singhal, S., & Singh, A. K. (2017). Modified star-star fractal (MSSF) super-wideband antenna. *Microwave and Optical Technology Letters*, 59(3), 624–630. DOI: 10.1002/mop.30357

Singh, M., Singh, S., & Islam, M. T. (2021). *Highly efficient ultra-wideband MIMO patch antenna array for short range THz applications, emerging trends in terahertz engineering and system technologies devices, materials, imaging, data acquisition and processing.*, DOI: 10.1007/978-981-15-9766-4

Singh, S. K., Yang, L. T., & Park, J. H. (2023). FusionFedBlock: Fusion of blockchain and federated learning to preserve privacy in industry 5.0. *Information Fusion*, 90, 233–240. DOI: 10.1016/j.inffus.2022.09.027

Si, Q., Jin, M., Tsiftsis, T. A., Zhao, N., & Wang, X. (2021). Cooperative SM-Based NOMA Scheme With SWIPT. *IEEE Transactions on Vehicular Technology*, 70(6), 6195–6199. DOI: 10.1109/TVT.2021.3076565

Siriwardhana, Y., Porambage, P., Liyanage, M., & Ylianttila, M. (2021, June). AI and 6G security: Opportunities and challenges. In 2021 Joint European Conference on Networks and Communications & 6G Summit (EuCNC/6G Summit) (pp. 616-621). IEEE.

Siriwardhana, Y., Porambage, P., Liyanage, M., & Ylianttila, M. (2021). AI and 6G security: Opportunities and challenges. *2021 Joint European Conference on Networks and Communications & 6G Summit (EuCNC/6G Summit)*. https://doi.org/DOI: 10.1109/EuCNC/6GSummit51104.2021.9482503

Sirohi, D., Kumar, N., Rana, P. S., Tanwar, S., Iqbal, R., & Hijjii, M. (2023). Federated learning for 6G-enabled secure communication systems: A comprehensive survey. *Artificial Intelligence Review*, 56(10), 11297–11389. DOI: 10.1007/s10462-023-10417-3 PMID: 37362891

Sivukhin, D. V. (1957). The Energy of Electromagnetic Fields in Dispersive Media. *Opt. Spektrosk.*, 3, 308–312.

Skorin-Kapov, N., Furdek, M., Zsigmond, S., & Wosinska, L. (2016). Physical-layer security in evolving optical networks. *IEEE Communications Magazine*, 54(8), 110–117. DOI: 10.1109/MCOM.2016.7537185

Smith, D. R., Padilla, W. J., Vier, D. C., Nemat-Nasser, S. C., & Schultz, S. (2000). Composite Medium with Simultaneously Negative Permeability and Permittivity. *Physical Review Letters*, 84(18), 4184–4186. DOI: 10.1103/PhysRevLett.84.4184 PMID: 10990641

Smrity Dwivedi, "Metamaterials-Based Antenna for 5G and Beyond, "DOI: , Metamaterial Technology and Intelligent Metasurfaces for Wireless Communication Systems, IGI global publishers of timely knowledge, 2023.DOI: 10.4018/978-1-6684-8287-2.ch001

Sodhro, A. H., Pirbhulal, S., Luo, Z., Muhammad, K., & Zahid, N. Z. (2020). Toward 6G architecture for energy-efficient communication in IoT-enabled smart automation systems. *IEEE Internet of Things Journal*, 8(7), 5141–5148. DOI: 10.1109/JIOT.2020.3024715

Soyata, T., Copeland, L., & Heinzelman, W. (2016). RF Energy Harvesting for Embedded Systems: A Survey of Tradeoffs and Methodology. *IEEE Circuits and Systems Magazine*, 16(1), 22–57. DOI: 10.1109/MCAS.2015.2510198

Spagnolo, G. S., Cozzella, L., Leccese, F., Sangiovanni, S., Podestà, L., & Piuzzi, E. (2020, June). Optical wireless communication and Li-Fi: a new infrastructure for wireless communication in saving energy era. In 2020 IEEE International Workshop on Metrology for Industry 4.0 & IoT (pp. 674-678). IEEE. DOI: 10.1109/MetroInd4.0IoT48571.2020.9138180

Srikamu, C., & Jayabharathy, R. (2022). Comparative Analysis of Ergodic Sum Capacity of Cooperative NOMA Aided with Spatial Modulation. *Wireless Personal Communications*, 123(4), 3771–3786. DOI: 10.1007/s11277-021-09313-2

Srikamu, C., Meera, S., & Jayabharathy, R. (2021). Analysis of Bit-Error-Rate for D2D AF Cooperative System Using Various Modulation Techniques. *2021 Sixth International Conference on Wireless Communications, Signal Processing and Networking (WiSPNET)*, 204–208. https://doi.org/DOI: 10.1109/WiSPNET51692.2021.9419452

Srikamu, C., Sahithi, S., Gayathri, M. S. S., Spruha, K. K., Karthikeyan, S., & Jayabharathy, R. (2022). Fairness Index Analysis of NOMA Over OMA System for B5G Applications. *2022 International Conference on Wireless Communications Signal Processing and Networking (WiSPNET)*, 341–344. https://doi.org/DOI: 10.1109/WiSPNET54241.2022.9767104

Srivastava, K., Mishra, B., & Singh, R. (2021). Microstrip-line-fed inverted L-shaped circularly polarized antenna for C-band applications. *International Journal of Microwave and Wireless Technologies*, 14(4), 502–510. DOI: 10.1017/S1759078721000635

Starry, C. (2001). The emerging in vehicle intelligent transportation systems market. *Business Economics (Cleveland, Ohio)*, 36(2), 49–49.

Stauch, J., Bessell, T., Rutten, M., Baldwin, J., Jah, M., & Hill, K. (2017). Joint probabilistic data association and smoothing applied to multiple space object tracking. *Journal of Guidance, Control, and Dynamics*, 41(1), 19–33. DOI: 10.2514/1.G002230

Stork, C., Calandro, E., & Gillwald, A. (2013). Internet going mobile: Internet access and use in 11 African countries. *info, 15*(5), 34-51.

Strinati, E. C., Di Lorenzo, P., Sciancalepore, V., Aijaz, A., Kountouris, M., Gündüz, D., Popovski, P., Sana, M., Stavrou, P. A., Soret, B., Cordeschi, N., Scardapane, S., Merluzzi, M., Zanzi, L., Renato, M. B., Quek, T., Pietro, N. D., Forceville, O., Costanzo, F., & Li, P. (2024). Goal-oriented and semantic communication in 6G AI-native networks: The 6G-GOALS approach. *2024 Joint European Conference on Networks and Communications & 6G Summit (EuCNC/6G Summit), 5*, 1-6. https://doi.org/DOI: 10.1109/EuCNC/6GSummit60053.2024.10597087

Subha, T. D., Subash, T. D., Rani, N. E., & Janani, P. (2020). Li-Fi: A revolution in wireless networking. *Materials Today: Proceedings*, 24, 2403–2413. DOI: 10.1016/j.matpr.2020.03.770

Sujin, K. (2024, August). Won-Yong, Shin., Seong-Lyun, Kim., Seung-Woo, Ko. (2024). Energy-Efficient Edge Learning via Joint Data Deepening-and-Prefetching. *IEEE Transactions on Wireless Communications*, 23(8), 9927–9942. Advance online publication. DOI: 10.1109/TWC.2024.3367352

Sun, P., Jiang, Y., Zhang, R., Xie, E., Cao, J., Hu, X., Kong, T., Yuan, Z., Wang, C., & Luo, P. (2020). Transtrack: Multiple-object tracking with transformer. arXiv preprint arXiv:2012.15460.

Sun, H., Zhu, F., Lin, H., & Gao, F. (2018). Robust Magnetic Resonant Beamforming for Secured Wireless Power Transfer. *IEEE Signal Processing Letters*, 25(8), 1226–1230. DOI: 10.1109/LSP.2018.2849586

Sun, X. (2022, July). Self-healing Edible Antennas with Metamaterial Integration. *Advanced Functional Materials*, 32(18), 2110493.

Sun, Y. (2022, March). Nanotechnology in Edible Antenna Design: Recent Advances. *Nano Energy*, 85, 106018.

Suriyan, G. R., Rahul, K., Rajesh, S., Dhanalakshmi, C., & Udhayakumar, G. (2023, June). Prevention of Road Accidents by Interconnecting Vehicles using LiFi and LoRaWAN Technologies. In *2023 International Conference on Sustainable Computing and Smart Systems (ICSCSS)* (pp. 1383-1387). IEEE. DOI: 10.1109/ICSCSS57650.2023.10169475

Suyama, S., Okuyama, T., Nonaka, N., & Asai, T. (2022). Recent studies on massive MIMO technologies for 5G evolution and 6G. *2022 IEEE Radio and Wireless Symposium (RWS)*. https://doi.org/DOI: 10.1109/RWS53089.2022.9719949

Syed, J. N., Sharma, S. K., Patwary, M. N., & Asaduzzaman, M. (2023). Enhanced URLLC-Enabled Edge Computing Framework for Device-Level Innovation in 6G. *Authorea Preprints*.

Systems, M. I. (2023). Retracted: Assessment and Test-Case Study of Wi-Fi Security through the Wardriving Technique.

Szafraniec, B., Marshall, T. S., & Nebendahl, B. (2012). Performance monitoring and measurement techniques for coherent optical systems. *Journal of Lightwave Technology*, 31(4), 648–663. DOI: 10.1109/JLT.2012.2212234

Tan, L., Dong, X., Ma, Y., & Yu, C. (2018). A multiple object tracking algorithm based on YOLO detection. In 2018 11th international congress on image and signal processing, biomedical engineering and informatics (CISP-BMEI), (pp. 1–5). IEEE. DOI: 10.1109/CISP-BMEI.2018.8633009

Tan, B., Zhang, Y., Pan, S. J., & Yang, Q. (2017). Distant domain transfer learning. *Proc. Nat. Conf. Artif. Intell.*, 2604–2610.

Tang, Z., Gu, R., & Hwang, J.-N. (2018). Joint multi-view people tracking and pose estimation for 3D scene reconstruction. In 2018 IEEE international conference on multimedia and expo (ICME), (pp. 1–6). IEEE.

Tang, F., Mao, B., Kawamoto, Y., & Kato, N. (2021). Survey on machine learning for intelligent end-to-end communication toward 6G: From network access, routing to traffic control and streaming adaption. *IEEE Communications Surveys and Tutorials*, 23(3), 1578–1598. DOI: 10.1109/COMST.2021.3073009

Tang, W., Chen, M. Z., Chen, X., Dai, J. Y., Han, Y., Di Renzo, M., Zeng, Y., Jin, S., Cheng, Q., & Cui, T. J. (2021). Wireless Communications With Reconfigurable Intelligent Surface: Path Loss Modeling and Experimental Measurement. *IEEE Transactions on Wireless Communications*, 20(1), 421–439. DOI: 10.1109/TWC.2020.3024887

Tan, L., Guo, J. X., Huang, X., & Wen, F. (2016). Output power stabilisation of wireless power transfer system with multiple transmitters. *IET Power Electronics*, 9(7), 1374–1380. DOI: 10.1049/iet-pel.2015.0577

Taylor, G. (2019, June). Self-powered Edible Electronics: Toward Sustainable Ingestible Devices. *IEEE Sensors Journal*, 19(12), 4180–4189.

Tehrani, R. H., Vahid, S., Triantafyllopoulou, D., Lee, H., & Moessner, K. (2016). Licensed spectrum sharing schemes for mobile operators: A survey and outlook. IEEE Communications Surveys & Tutorials, 18(4), 2591–2623. DOI: 10.1109/COMST.2016.2596343

Temmar, M. N. E., Hocini, A., Khedrouche, D., & Denidni, T. A. (2021). Analysis and design of MIMO indoor communication system using terahertz patch antenna based on photonic crystal with graphene. *Photonics and Nanostructures*, 43, 100867. DOI: 10.1016/j.photonics.2020.100867

Temmar, M. N. E., Hocini, A., Khedrouche, D., & Zamani, M. (2019). Analysis and design of a terahertz microstrip antenna based on a synthesized photonic bandgap substrate using BPSO. *Journal of Computational Electronics*, 18(1), 231–240. DOI: 10.1007/s10825-019-01301-x

Temmar, M. N., Hocini, A., Khedrouche, D., & Denidni, T. A. (2020). Enhanced flexible terahertz microstrip antenna based on modified silicon-air photonic crystal. *Optik (Stuttgart)*, 217, 164897. DOI: 10.1016/j.ijleo.2020.164897

Thomas, M. T. C. A. J., & Joy, A. T. (2006). *Elements of information theory*. Wiley-Interscience.

Thrane, Wass, Piels, Diniz, Jones, & Zibar. (2017). Machine Learning Techniques for Optical Performance Monitoring from Directly Detected PDM-QAM Signals. Journal of Lightwave Technology, 35(4), 868-875. .DOI: 10.1109/JLT.2016.2590989

Tiddens, W., Braaksma, J., & Tinga, T. (2020). Exploring predictive maintenance applications in industry. *Journal of Quality in Maintenance Engineering*, 28(1), 68–85. DOI: 10.1108/JQME-05-2020-0029

Timothy, M. (2020). An android location-based crime reporting system using the Google Map API. *University of Pitesti Scientific Bulletin: Electronics and Computers Science, 20*(1).

Tizikara, D. K., Serugunda, J., & Katumba, A. (2022). Machine learning-aided optical performance monitoring techniques: A review. *Frontiers in Communications and Networks*, 2, 756513. DOI: 10.3389/frcmn.2021.756513

Tomaszewski, L., & Kołakowski, R. (2023, January). Mobile services for smart agriculture and forestry, biodiversity monitoring, and water management: Challenges for 5G/6G networks. In Telecom (Vol. 4, No. 1, pp. 67-99). MDPI.

Toumi, N., & Dimitrovski, T. (2024). AI-native architecture for 6G networks and services with model dependencies. *2024 Joint European Conference on Networks and Communications & 6G Summit (EuCNC/6G Summit)*, 901-906. https://doi.org/DOI: 10.1109/EuCNC/6GSummit60053.2024.10597080

Tran, H. Q., & Vien, Q.-T. (2023). SWIPT-Based Cooperative NOMA for Two-Way Relay Communications: PSR versus TSR. *Wireless Communications and Mobile Computing*, 2023, 1–14. DOI: 10.1155/2023/6785737

Tran, H. V., Kaddoum, G., Elgala, H., Abou-Rjeily, C., & Kaushal, H. (2020). Lightwave Power Transfer for Federated Learning-Based Wireless Networks. *IEEE Communications Letters*, 24(7), 1472–1476. DOI: 10.1109/LCOMM.2020.2985698

Tran, H.-V., & Kaddoum, G. (2018). RF Wireless Power Transfer: Regreening Future Networks. *IEEE Potentials*, 37(2), 35–41. DOI: 10.1109/MPOT.2017.2749118

Trichias, K., Kaloxylos, A., & Willcock, C. (2024). 6G global landscape: A comparative analysis of 6G targets and technological trends. *2024 Joint European Conference on Networks and Communications & 6G Summit (EuCNC/6G Summit)*, 1-6. https://doi.org/DOI: 10.1109/EuCNC/6GSummit60053.2024.10597064

Tripathi, S. K., & Kumar, A. (2019). High gain miniaturised photonic band gap terahertz antenna for size-limited applications. *Aust J Electr Electron Eng*, 16(2), 74–80. DOI: 10.1080/1448837X.2019.1602944

Udmale, S. S., & Singh, S. K. (2019). Application of spectral kurtosis and improved extreme learning machine for bearing fault classification. *IEEE Transactions on Instrumentation and Measurement*, 68(11), 4222–4233. DOI: 10.1109/TIM.2018.2890329

Uguroglu, S., & Carbonell, J. (2011). Feature selection for transfer learning. *Proc. Joint Eur. Conf. Mach. Learn. Knowl. Discovery Databases,* 430–442.

Ullah S, Ruan C, Haq TU, Zhang X (2019) High performance THz patch antenna using photonic band gap and defected ground structure. J Electromagn Waves Appl 33(15):1943–1954. . 2019. 16549 29.DOI: 10. 1080/ 09205 071

Umoga, U. J., Sodiya, E. O., Ugwuanyi, E. D., Jacks, B. S., Lottu, O. A., Daraojimba, O. D., & Obaigbena, A. (2024). Exploring the potential of AI-driven optimization in enhancing network performance and efficiency. *Magna Scientia Advanced Research and Reviews*, 10(1), 368–378. DOI: 10.30574/msarr.2024.10.1.0028

Union, I. (2015). IMT traffic estimates for the years 2020 to 2030. *ITU-R Radiocommun. Sector ITU, Geneva, Switzerland, Tech. Rep. ITU-R M*, 2370.

Uusitalo, M. A., Rugeland, P., Boldi, M. R., Strinati, E. C., Demestichas, P., Ericson, M., Fettweis, G. P., Filippou, M. C., Gati, A., Hamon, M., Hoffmann, M., Latva-Aho, M., Parssinen, A., Richerzhagen, B., Schotten, H., Svensson, T., Wikstrom, G., Wymeersch, H., Ziegler, V., & Zou, Y. (2021). 6G vision, value, use cases and technologies from European 6G flagship project Hexa-X. *IEEE Access : Practical Innovations, Open Solutions*, 9, 160004–160020. DOI: 10.1109/ACCESS.2021.3130030

Uzair, M. (2022). Vehicular wireless communication standards: challenges and comparison. International Journal of Electrical and Computer Engineering Systems, 13(5), 379-397.

Vaezi, M., Azari, A., Khosravirad, S. R., Shirvanimoghaddam, M., Azari, M. M., Chasaki, D., & Popovski, P. (2022). Cellular, Wide-Area, and Non-Terrestrial IoT: A Survey on 5G Advances and the Road Toward 6G. *IEEE Communications Surveys and Tutorials*, 24(2), 1117–1174. DOI: 10.1109/COMST.2022.3151028

Varshney, A., & Gençoğlan, D. N. (2024). High-Gain Multi-Band Koch Fractal FSS Antenna for Sub-6 GHz Applications. *Applied Sciences (Basel, Switzerland)*, 14(19), 9022. DOI: 10.3390/app14199022

Vasu Babu, K., Das, S., Varshney, G., Sree, G. N. J., & Madhav, B. T. P. (2022). A micro-scaled graphene-based treeshaped wideband printed MIMO antenna for terahertz applications. *Journal of Computational Electronics*, 21(1), 289–303. DOI: 10.1007/s10825-021-01831-3

Vasu Babu, K., Sree, G. N. J., Kumari, S. V., & Das, S. (2022). Design and analysis of a CPW-fed fractal MIMO THz antenna using an array of parasitic elements. In Das, S., Nella, A., & Patel, S. K. (Eds.), *Terahertz devices, circuits and systems*. Springer., DOI: 10.1007/978-981-19-4105-4_4

Venkatesh, S., Lu, X., Saeidi, H., & Sengupta, K. (2022, August). A programmable terahertz metasurface with circuit-coupled meta-elements in silicon chips: Creating low-cost, large-scale, reconfigurable terahertz metasurfaces. *IEEE Antennas & Propagation Magazine*, 64(4), 110–122. DOI: 10.1109/MAP.2022.3176588

Venkatesh, S., Sturm, D., Lu, X., Lang, R. J., & Sengupta, K. (2022, October). Origami microwave imaging array: Metasurface tiles on a shape-morphing surface for reconfigurable computational imaging. *Advancement of Science*, 9(28), 2105016. Advance online publication. DOI: 10.1002/advs.202105016 PMID: 35896946

Vettikalladi H, Sethi WT, Abas AFB, Ko W, Alkanhal MA, Himdi M (2019) Sub-THz antenna for high-speed wireless communication systems. Int J Antennas Propag 2019, Article ID 9573647, 9 pages. .DOI: 10.1155/2019/9573647

Viswanathan, R., Thomopoulos, S. C. A., & Tumuluri, R. (1988, July). Optimal serial distributed decision fusion. *Institute of Electrical and Electronics Engineers Transactions on Aerospace and Electronic Systems*, AES-24(4), 366–375.

Vivone, G., Braca, P., & Errasti-Alcala, B. (2015). Extended target tracking applied to X-band marine radar data. In *OCEANS 2015-Genova* (pp. 1–6). IEEE. DOI: 10.1109/OCEANS-Genova.2015.7271630

Volkov, A. S., Baskakov, A. E., Bakhtin, A. A., & Gorelik, A. V. (2021). Method for Managing Transport SDN Resources. In 2021 Systems of Signal Synchronization, Generating and Processing in Telecommunications (SYNCHROINFO. IEEE. DOI: 10.1109/SYNCHROINFO51390.2021.9488419

Vu, A., Ramanandan, A., Chen, A., Farrell, J. A., & Barth, M. (2012). Real-time computer vision/DGPS-aided inertial navigation system for lane-level vehicle navigation. *IEEE Transactions on Intelligent Transportation Systems*, 13(2), 899–913. DOI: 10.1109/TITS.2012.2187641

Vujović, Ž. (2021). Classification model evaluation metrics. *International Journal of Advanced Computer Science and Applications*, 12(6), 599–606. DOI: 10.14569/IJACSA.2021.0120670

Wan, X., Wang, J., Kong, Z., Zhao, Q., & Deng, S. (2018). Multi-object tracking using online metric learning with long short-term memory. In 2018 25th IEEE international conference on image processing (ICIP), (pp. 788–792). IEEE. DOI: 10.1109/ICIP.2018.8451174

Wang, C. X., Gao, X., & Kadoch, M. (2014). Cellular architecture and key technologies for 5G wireless communication networks. IEEE Communications Magazine, 52(2), 122–130. DOI: 10.1109/MCOM.2014.6736751

Wang, C., Xie, X., & Liao, C. (2021). An adaptive fusion model based on Kalman filtering and LSTM for fast tracking of road signs. In 2020 25th international conference on pattern recognition (ICPR), (pp. 1414–1421). IEEE. DOI: 10.1109/ICPR48806.2021.9412296

Wang, J., Ancha, S., Chen, Y.-T., & Held, D. (2020). Uncertainty-aware self-supervised 3D data association. In 2020 IEEE/RSJ international conference on intelligent robots and systems (IROS), (pp. 8125–8132). IEEE.

Wang, Y., Chen, Y., Dai, H., Huang, Y., & Yang, L. (2017). *A learning-based approach for proactive caching in wireless communication networks*. DOI: 10.1109/WCSP.2017.8170984

Wang, C. X., You, X., Gao, X., Zhu, X., Li, Z., Zhang, C., Wang, H., Huang, Y., Chen, Y., Haas, H., Thompson, J. S., Larsson, E. G., Renzo, M. D., Tong, W., Zhu, P., Shen, X., Poor, H. V., & Hanzo, L. (2023). On the road to 6G: Visions, requirements, key technologies, and testbeds. *IEEE Communications Surveys and Tutorials*, 25(2), 905–974. DOI: 10.1109/COMST.2023.3249835

Wang, C., Chen, Y., & Yang, S. (2018, November). Dual-band dual-polarized antenna array with flat-top and sharp cutoff radiation patterns for 2G/3G/LTE cellular bands. *IEEE Transactions on Antennas and Propagation*, 66(11), 5907–5917. DOI: 10.1109/TAP.2018.2866596

Wang, D., Jiang, H., Liang, G., Zhan, Q., Mo, Y., Sui, Q., & Li, Z. (2020). Optical performance monitoring of multiple parameters in future optical networks. *Journal of Lightwave Technology*, 39(12), 3792–3800. DOI: 10.1109/JLT.2020.3043012

Wang, H. (2021, August). Performance Metrics of Edible Antennas in Real-world Applications. *IEEE Transactions on Antennas and Propagation*, 70(8), 3769–3780.

Wang, H., & Chen, Z. (2020). Threat modeling and defense mechanisms for RFID-based IoT systems. *Journal of Cybersecurity*.

Wang, H., Zhang, Z., Zhu, B., Dang, J., Wu, L., Wang, L., Zhang, K., Zhang, Y., & Li, G. Y. (2021). Performance Analysis of Multi-Branch Reconfigurable Intelligent Surfaces-Assisted Optical Wireless Communication System in Environment With Obstacles. *IEEE Transactions on Vehicular Technology*, 70(10), 9986–10001. DOI: 10.1109/TVT.2021.3092722

Wang, J. (2021, February). High Performance Edible Antennas for Medical Monitoring. *Journal of Biomedical Materials Research*, 109(2), 345–355.

Wang, J., Xu, C., Zhang, J., & Zhong, R. (2021). Big data analytics for intelligent manufacturing systems: A review. *Journal of Manufacturing Systems*, 62, 738–752. DOI: 10.1016/j.jmsy.2021.03.005

Wang, M., Lin, Y., Tian, Q., & Si, G. (2021). Transfer Learning Promotes 6G Wireless Communications: Recent Advances and Future Challenges. *IEEE Transactions on Reliability*, 70(2), 790–807. DOI: 10.1109/TR.2021.3062045

Wang, Q., Liu, Y., Wang, Y., Xing, Y., Zong, J., & Liu, H. (2023). Deep edge node enables AI native toward 6G*. *2023 IEEE International Symposium on Broadband Multimedia Systems and Broadcasting (BMSB)*, 1-6. https://doi.org/DOI: 10.1109/BMSB58369.2023.10211613

Wang, T. (2023, April). Flexible and Edible Antennas for Nutritional Tracking. *IEEE Transactions on Biomedical Circuits and Systems*, 17(4), 542–552. PMID: 37318976

Wang, X., Wang, J., He, L., & Song, J. (2017). Spectral Efficiency Analysis for Downlink NOMA Aided Spatial Modulation With Finite Alphabet Inputs. *IEEE Transactions on Vehicular Technology*, 66(11), 10562–10566. DOI: 10.1109/TVT.2017.2737327

Wang, Y. (2022, June). Advanced Edible Antenna Designs for Smart Food Packaging. *Sensors and Actuators. A, Physical*, 319, 112447.

Wang, Y. (2023, August). Loop Antennas Made from Rice Paper and Edible Gold for Ingestible Sensing. *IEEE Antennas & Propagation Magazine*, 65(4), 54–63.

Wang, Y., Martinsen, K., Yu, T., & Wang, K. (2022). *Advanced Manufacturing and Automation XI*. Springer. DOI: 10.1007/978-981-19-0572-8

Wang, Y., Zhang, W., Liu, C., Sun, J., & Wang, C.-X. ''Reconfigurable intelligent surface for NLOS integrated sensing and communications,'' in *Proc. IEEE/CIC Int. Conf. Commun. China (ICCC)*, IEEE, Aug. 2022, pp. 708–712, DOI: 10.1109/ICCC55456.2022.9880639

Wang, Z. (2023, May). Optimization Techniques for Edible Antenna Designs. *IEEE Transactions on Antennas and Propagation*, 71(5), 2231–2240.

Wei, G. Y., Wang, Y. X., Liu, J., & Li, H. P. (2023). Design of a Planar Compact Dual-Band Bandpass Filter with Multiple Transmission Zeros Using a Stub-Loaded Structure. *Progress in Electromagnetics Research Letters*, 109, 109. DOI: 10.2528/PIERL22122603

Wei, Z., Wang, L., Li, Z., Chen, C. J., Wu, M. C., Wang, L., & Fu, H. Y. (2022). Micro-LEDs illuminate visible light communication. *IEEE Communications Magazine*, 61(4), 108–114. DOI: 10.1109/MCOM.002.2200109

Weng, X., Wang, Y., Man, Y., & Kitani, K. (2020). Graph neural networks for 3Dmulti-object tracking. arXiv preprint arXiv:2008.09506.

Wen, L.-H., Gao, S., Mao, C.-X., Luo, Q., Hu, W., Yin, Y., & Yang, X. (2018). A wideband dual-polarized antenna using shorted dipoles. *IEEE Access : Practical Innovations, Open Solutions*, 6, 39725–39733. DOI: 10.1109/ACCESS.2018.2855425

Wen, M., Zheng, B., Kim, K. J., di Renzo, M., Tsiftsis, T. A., Chen, K.-C., & Al-Dhahir, N. (2019). A Survey on Spatial Modulation in Emerging Wireless Systems: Research Progresses and Applications. *IEEE Journal on Selected Areas in Communications*, 37(9), 1949–1972. DOI: 10.1109/JSAC.2019.2929453

Wu, H., & Li, W. (2016). *Robust online multi-object tracking based on KCF trackers and reassignment. In 2016 IEEE global conference on signal and information processing (GlobalSIP)*. IEEE.

Wu, J. (2022). Edible Antenna Technologies: State-of-the-Art and Future Perspectives. *IEEE Access : Practical Innovations, Open Solutions*, 10, 43589–43601.

Wu, Q., & Zhang, R. (2020). Joint Active and Passive Beamforming Optimization for Intelligent Reflecting Surface Assisted SWIPT Under QoS Constraints. *IEEE Journal on Selected Areas in Communications*, 38(8), 1735–1748. DOI: 10.1109/JSAC.2020.3000807

Wu, Q., & Zhang, R. (2020a). Weighted Sum Power Maximization for Intelligent Reflecting Surface Aided SWIPT. *IEEE Wireless Communications Letters*, 9(5), 586–590. DOI: 10.1109/LWC.2019.2961656

Wu, S. (2021, October). Metamaterial-enhanced Edible Antennas: Performance and Applications. *IEEE Transactions on Antennas and Propagation*, 69(10), 7420–7429.

Wu, T., & Zhang, Y. (2020). Dynamic wireless EV charging: Technical and economic perspectives. *Journal of Advanced Transportation*.

Wu, X., Jargon, J. A., Skoog, R. A., Paraschis, L., & Willner, A. E. (2009). Applications of artificial neural networks in optical performance monitoring. *Journal of Lightwave Technology*, 27(16), 3580–3589. DOI: 10.1109/JLT.2009.2024435

Xie, T. (2021, November). Metamaterial-enhanced Antennas for Food Monitoring Applications. *IEEE Transactions on Microwave Theory and Techniques*, 69(11), 4555–4565.

Xie, X., & Ding, Z. (2023). Backscatter-assisted Non-orthogonal multiple access network for next generation communication. *IET Signal Processing*, 17(4), e12211. Advance online publication. DOI: 10.1049/sil2.12211

Xie, X., Li, M., Shi, Z., Tang, H., & Huang, Q. (2021). User selection and dynamic power allocation in the SWIPT-NOMA relay system. *EURASIP Journal on Wireless Communications and Networking*, 2021(1), 124. DOI: 10.1186/s13638-021-01998-0

Xiong, S., Guan, X., Ren, B., Wan, S., & Zhang, X. (2023, November). Dual-Band Bandpass Filter Based on Two Pairs of Degenerate Modes of a Regular Hexagonal Prism Dielectric Waveguide Resonator. In 2023 Cross Strait Radio Science and Wireless Technology Conference (CSRSWTC) (pp. 01-03). IEEE. DOI: 10.1109/CSRSWTC60855.2023.10426903

Xi-yang, Z., Xiao-li, W., & Liangqun, L. (2018). Online multi-object tracking via maximum entropy intuitionistic fuzzy data association. In 2018 14th IEEE international conference on signal processing (ICSP), (pp. 803–806). IEEE. DOI: 10.1109/ICSP.2018.8652401

Xu, C. (2024, May 07). Zhiyong, Feng., Andrew, Zhang., Zhaohui, Yang., Xin, Yuan., Xuhua, He., Ping, Zhang. (2024). Integrated Communication, Sensing, and Computation Framework for 6G Networks. *Sensors (Basel)*, 24(10), 2968. Advance online publication. DOI: 10.3390/s24102968

Xu, C., & Zhou, Y. (2018). Hierarchical online multi-person pose tracking with multiple cues. In *International conference on neural information processing* (pp. 318–328). Springer. DOI: 10.1007/978-3-030-04224-0_27

Xu, H. (2021, August). The Role of Metamaterials in Edible Antennas. *IEEE Transactions on Microwave Theory and Techniques*, 69(8), 3395–3404.

Xu, L. (2021, October). Advanced Nanomaterials for High-performance Edible Antennas. *Nano Letters*, 21(10), 4195–4203.

Xu, Q. (2021, April). Nanomaterials in Edible Antenna Design: Enhancements and Challenges. *Advanced Materials*, 33(15), 2007372. PMID: 33694214

Xu, Z. Q., & Gao, Y. Y. (2022). NEXT-GEN MAINTENANCE FOR URBAN RAIL TRANSIT: AN IOT-5G SYSTEM DESIGN. *Academic Journal of Science. Engineering and Technology*, 7(2), 8–18.

Xu, Z., Wu, Y., Dong, Q., & Wang, W. (2022). Miniaturized dual-band filter using dual-mode dielectric waveguide resonator. *IEEE Microwave and Wireless Components Letters*, 32(12), 1411–1414. DOI: 10.1109/LMWC.2022.3193424

Yajun, Zhao., Guanghui, Yu., Hanqing, Xu. (2019). 6G mobile communication networks: vision, challenges, and key technologies. DOI: 10.1360/N112019-00033

Yang, S., Thormann, K., & Baum, M. (2018). Linear-time joint probabilistic data association for multiple extended object tracking. In 2018 IEEE 10th sensor array and multichannel signal processing workshop (SAM), (pp. 6–10). IEEE. DOI: 10.1109/SAM.2018.8448430

Yang, X. L., Zhu, X. W., & Wang, X. (2023). Dual-band substrate integrated waveguide filters based on multi-mode resonator overlapping mode control. IEEE Transactions on Circuits and Systems II: Express Briefs.

Yang, Zhong, Chen, Alphones, & Xie. (2020). Deep-reinforcement-learning-based energy-efficient resource management for social and cognitive Internet of Things. Internet of Things Journal, 7(6), 5677-5689.

Yang, D., Liu, S., & Zhao, Z. (2017, May). A broadband dual-polarized printed dipole antenna with low cross-polarization and high isolation for base station applications. *Microwave and Optical Technology Letters*, 59(5), 1107–1111. DOI: 10.1002/mop.30467

Yang, G., Reza, M., & Zhang, R. (2017). Magnetic MIMO Signal Processing and Optimization for Wireless Power Transfer. *IEEE Transactions on Signal Processing*, 65(11), 2860–2874. DOI: 10.1109/TSP.2017.2673816

Yang, G., Vedady, M. R., & Zhang, R. (2016). *Magnetic beamforming for wireless power transfer. ArXiv*. Cornell University., DOI: 10.1109/ICASSP.2016.7472415

Yang, H. (2021, August). Edible Slot Antennas for Medical Applications. *Nature Biomedical Engineering*, 5(8), 723–732. PMID: 34131325

Yang, H., Alphones, A., Xiong, Z., Niyato, D., Zhao, J., & Wu, K. (2020). Artificial-intelligence-enabled intelligent 6G networks. *IEEE Network*, 34(6), 272–280. DOI: 10.1109/MNET.011.2000195

Yang, J. (2022, November). Challenges and Solutions in Edible Electronics. *IEEE Transactions on Electron Devices*, 69(11), 4960–4970.

Yang, L. (2022). Future Directions in Edible Antenna Technology: Materials and Manufacturing. *IEEE Reviews in Biomedical Engineering*, 15, 235–248.

Yang, L., Zeng, Y., & Zhang, R. (2018). Wireless Power Transfer With Hybrid Beamforming: How Many RF Chains Do We Need? *IEEE Transactions on Wireless Communications*, 17(10), 6972–6984. DOI: 10.1109/TWC.2018.2865313

Yang, Q., Liu, Y., Chen, T., & Tong, Y. (2019). Federated machine learning: Concept and applications. *ACM Transactions on Intelligent Systems and Technology*, 10(2), 1–12. DOI: 10.1145/3298981

Yang, S.-N., Cheng, W.-S., Hsu, Y.-C., Gan, C.-H., & Lin, Y.-B. (2011, December). Charge scheduling of electric vehicles in highways. *Mathematical and Computer Modelling*, •••, 6.

Yang, W. (2021). Innovative Designs for Edible Antennas: A Review. *IEEE Reviews in Biomedical Engineering*, 14, 273–287.

Yang, X. S., Sherratt, S., Dey, N., & Joshi, A. (2024). *Proceedings of Ninth International Congress on Information and Communication Technology: ICICT 2024*, London, Volume 1.

Yang, Y., Ma, M., Wu, H., Yu, Q., You, X., Wu, J., Peng, C., Yum, T.-S. P., Aghvami, A. H., Li, G. Y., Wang, J., Liu, G., Gao, P., Tang, X., Cao, C., Thompson, J., Wong, K.-K., Chen, S., Wang, Z., & Shu, H. (2022). 6G network AI architecture for everyone-centric customized services. *IEEE Network*, 37(5), 71–80. DOI: 10.1109/MNET.124.2200241

Yan, H., Chen, Y., & Yang, S.-H. (2020). UAV-Enabled Wireless Power Transfer With Base Station Charging and UAV Power Consumption. *IEEE Transactions on Vehicular Technology*, 69(11), 12883–12896. DOI: 10.1109/TVT.2020.3015246

Yan, J. M., Kang, B. J., Yang, Y., & Cao, L. (2023). Single-and Dual-Band Bandpass Filters Based on a Novel Microstrip Loop-Type Resonators Loaded with Shorted Stubs. *Progress in Electromagnetics Research Letters*, 113, 113. DOI: 10.2528/PIERL23083105

Yan, Z., Zhang, X., He, X., Ye, Y., Hou, B., Hu, C., & Wen, W. (2023). A high performance dual-band switchable bandpass filter and its application in 5G signal detector. *IEEE Transactions on Components, Packaging, and Manufacturing Technology*, 13(8), 1242–1253. DOI: 10.1109/TCPMT.2023.3302729

Yeo, K. S. K., & Nwajana, A. O. (2013). A novel microstrip dual-band bandpass filter using dual-mode square patch resonators. *Progress In Electromagnetics Research*, C, 36, 233–247.

Yigitcanlar, T., Downie, A. T., Mathews, S., Fatima, S., MacPherson, J., Behara, K. N., & Paz, A. (2024). Digital technologies of transportation-related communication: Review and the state-of-the-art. *Transportation Research Interdisciplinary Perspectives*, 23, 100987. DOI: 10.1016/j.trip.2023.100987

Ying, W., Zhang, Y., Huang, J., & Yang, Q. (2018). Transfer learning via learning to transfer. *Proc. Int. Conf. Mach. Learn.*, 5072–5081.

Yoon, J. H., Lee, C.-R., Yang, M.-H., & Yoon, K.-J. (2019). Structural constraint data association for online multi-object tracking. *International Journal of Computer Vision*, 127(1), 1–21. DOI: 10.1007/s11263-018-1087-1 PMID: 33664553

Younssi, M., Jaoujal, A., Yaccoub, M. D., El Moussaoui, A., & Aknin, N. (2013). Study of a microstrip antenna with and without superstrate for terahertz frequency. *International Journal of Innovation and Applied Studies*, 2(4), 369–371.

Yrjölä, S., Matinmikko-Blue, M., & Ahokangas, P. (2023). Developing 6G visions with stakeholder analysis of 6G ecosystem. *2023 Joint European Conference on Networks and Communications & 6G Summit (EuCNC/6G Summit)*. https://doi.org/DOI: 10.1109/EuCNC/6GSummit58263.2023.10188379

Yu, C. H., Huang, Y. C., & Chen, H. (2011). Resource sharing optimization for device-to-device communication underlying cellular networks. IEEE Transactions on Wireless Communications, 10(8), 2752–2763. DOI: 10.1109/TWC.2011.060911.101290

Yuan, T., Krishnan, K., Chen, Q., Breu, J., Roth, T. B., Duraisamy, B., Weiss, C., Maile, M., & Gern, A. (2017). Object matching for inter-vehicle communication systems—An IMM-based track association approach with sequential multiple hypothesis test. *IEEE Transactions on Intelligent Transportation Systems*, 18(12), 3501–3512. DOI: 10.1109/TITS.2017.2723894

Yuan, W., Li, S., Wei, Z., Cui, Y., Jiang, J., Zhang, H., & Fan, P. (2023). New delay Doppler communication paradigm in 6G era: A survey of orthogonal time frequency space (OTFS). *China Communications*, 20(6), 1–25. DOI: 10.23919/JCC.fa.2022-0578.202306

Yu, H., Li, G., Su, L., Zhong, B., Yao, H., & Huang, Q. (2020). Conditional GAN based individual and global motion fusion for multiple object tracking in UAV videos. *Pattern Recognition Letters*, 131, 219–226. DOI: 10.1016/j.patrec.2019.12.018

Yu, Z., Yu, J., Ran, X., & Zhu, C. (2017). A novel Koch and Sierpinski combined fractal antenna for 2G/3G/4G/5G/WLAN/navigation applications. *Microwave and Optical Technology Letters*, 59(9), 2147–2155. DOI: 10.1002/mop.30698

Zein, S., Salleh, N., & Grundy, J. (2016). A systematic mapping study of mobile application testing techniques. *Journal of Systems and Software*, 117, 334–356. DOI: 10.1016/j.jss.2016.03.065

Zeng, D., Xiong, L., Yu, Z., Chen, Q., Fu, Z., Li, Z., Zhang, P., Xu, P., Qian, Z., Xiao, H., Fang, P., Li, Z., & Leng, B. (2020). A priority data association policy for multitarget tracking on intelligent vehicle risk assessment. *Remote Sensing (Basel)*, 12(19), 3255. DOI: 10.3390/rs12193255

Zeng, Y., Clerckx, B., & Zhang, R. (2017). Communications and signals design for wireless power transmission. *IEEE Transactions on Communications*, 65(5), 2264–2290. DOI: 10.1109/TCOMM.2017.2676103

Zhang, J., Ge, X., Li, Q., Guizani, M., & Zhang, Y. (2017). 5G millimeter-wave antenna array: Design and challenges. IEEE Wireless Communications, 24(2), 106–112. DOI: 10.1109/MWC.2017.1600236

Zhang, C., Huang, Y., Wang, Z., Jiang, H., & Yan, D. (2018). Cross-camera multi-person tracking by leveraging fast graph mining algorithm. *Journal of Visual Communication and Image Representation*, 55, 711–719. DOI: 10.1016/j.jvcir.2018.08.006

Zhang, G., Zhou, X., Hu, J., Tam, K. W., Li, D., Zhang, Z., Ma, T., & Yang, J. (2024). Multilayer Packaged Single-and Dual-Band Dual-Channel Bandpass Filters Based on Substrate Integrated Waveguide Cavities. *IEEE Transactions on Components, Packaging, and Manufacturing Technology*, 14(1), 114–121. DOI: 10.1109/TCPMT.2023.3330170

Zhang, H., Shlezinger, N., Guidi, F., Dardari, D., Imani, M. F., & Eldar, Y. C. (2022). Near-Field Wireless Power Transfer for 6G Internet of Everything Mobile Networks: Opportunities and Challenges. *IEEE Communications Magazine*, 60(3), 12–18. DOI: 10.1109/MCOM.001.2100702

Zhang, L. (2022, October). Compact Microstrip Antennas for Wearable Health Devices. *IEEE Transactions on Microwave Theory and Techniques*, 70(10), 3756–3766.

Zhang, L., Li, X., Tang, Y., Xin, J., & Huang, S. (2022). A survey on QoT prediction using machine learning in optical networks. *Optical Fiber Technology*, 68, 102804. DOI: 10.1016/j.yofte.2021.102804

Zhang, P. (2021, July). Gelatin-based Slot Antennas for Biomedical Monitoring. *IEEE Transactions on Biomedical Engineering*, 68(7), 1820–1830.

Zhang, Q. (2021, May). Performance Analysis of Edible Antennas in Digestive Environments. *IEEE Transactions on Biomedical Engineering*, 68(5), 1301–1311. PMID: 32991277

Zhang, R. (2022, December). Metamaterial-Based Edible Antennas for Biomedical Applications. *IEEE Transactions on Biomedical Engineering*, 69(12), 3456–3464. PMID: 35622786

Zhang, R., & Ho, C. K. (2013). MIMO Broadcasting for Simultaneous Wireless Information and Power Transfer. *IEEE Transactions on Wireless Communications*, 12(5), 1989–2001. DOI: 10.1109/TWC.2013.031813.120224

Zhang, S. (2021, May). High-Performance Edible Antennas for Environmental Sensing. *IEEE Sensors Journal*, 21(10), 10845–10854.

Zhang, T. (2021, March). Edible Conductors and Their Performance in Antenna Design. *Journal of Materials Science*, 56(6), 4556–4568.

Zhang, X.-G., Yang, G.-H., & Ren, X.-X. (2022). Network steganography-based security framework for cyber-physical systems. *Information Sciences*, 609, 963–983. DOI: 10.1016/j.ins.2022.07.081

Zhang, Y. (2021, August). Design Challenges in Edible Antenna Technology. *Advanced Functional Materials*, 31(29), 2104521.

Zhang, Y., Azizui, M. F. A. B., & Yan, C. (2024). The Application of 6G and Augmented Reality Technology in Education and Training. *Wireless Personal Communications*, •••, 1–19. DOI: 10.1007/s11277-024-11038-x

Zhang, Y., Chen, D., & Jiang, T. (2021). Robust Beamforming Design for Magnetic MIMO Wireless Power Transfer Systems. *IEEE Transactions on Signal Processing*, 69, 5066–5077. DOI: 10.1109/TSP.2021.3108686

Zhang, Y., He, S., & Yang, Z. (2022). Security challenges in RFID-based electric vehicle charging systems. *IEEE Transactions on Industrial Informatics*.

Zhang, Y., Xu, S., & Wan, Y. (2020). Performance improvement of centrifugal compressors for fuel cell vehicles using the aerodynamic optimization and data mining methods. *International Journal of Hydrogen Energy*, 45(19), 11276–11286. DOI: 10.1016/j.ijhydene.2020.02.026

Zhao, Q., & Grace, D. (2024). *Agent transfer learning for cognitive resource management on multi-hop backhaul networks*. Ieee.org. https://ieeexplore.ieee.org/document/6633554,1-10

Zhao, H., Zhang, X., Hu, J., Chen, Z. D., & Liang, Y. C. (2020). A hybrid-equivalent surface-edge current model for simulation of V2X communication antennas with arbitrarily shaped contour. *IEEE Internet of Things Journal*, 8(10), 8064–8077. DOI: 10.1109/JIOT.2020.3042336

Zhao, J. (2020, September). Recent Advances in Edible Electronics. *Advanced Functional Materials*, 31(34), 2101180.

Zhao, J., Cai, T., Duan, S., Feng, H., Chen, C., & Zhang, X. (2016). A General Design Method of Primary Compensation Network for Dynamic WPT System Maintaining Stable Transmission Power. *IEEE Transactions on Power Electronics*, 31, 8343–8358. DOI: 10.1109/TPEL.2016.2516023

Zhao, L. (2023, April). Advanced Design Techniques for Edible Antennas. *IEEE Transactions on Microwave Theory and Techniques*, 71(4), 1970–1980.

Zhao, L., Wang, Z., & Wang, X. (2021). Wireless Power Transfer Empowered by Reconfigurable Intelligent Surfaces. *IEEE Systems Journal*, 15(2), 2121–2124. DOI: 10.1109/JSYST.2020.2985121

Zhao, P., Hoi, S. C., Wang, J., & Li, B. (2014). Online transfer learning. *Artificial Intelligence*, 216, 76–102. DOI: 10.1016/j.artint.2014.06.003

Zhao, Q., Grace, D., & Clarke, T. (2015). Transfer learning and cooperation management: Balancing the quality of service and information exchange overhead in cognitive radio networks. *Transactions on Emerging Telecommunications Technologies*, 26(2), 290–301. DOI: 10.1002/ett.2913

Zhao, Q., & Ma, Y. (2019). Interoperability frameworks for next-generation EVs. *Energy Conversion and Management.*

Zhao, R. (2022, June). Optimization Techniques for Performance Improvement of Edible Antennas. *IEEE Transactions on Electron Devices*, 69(6), 2345–2353.

Zhao, X. (2020, January). Self-Powered Edible Antennas: A Review. *IEEE Sensors Journal*, 20(2), 567–577.

Zhao, Y. (2020, December). High-performance Edible Antennas: Design and Application. *IEEE Transactions on Microwave Theory and Techniques*, 68(12), 5256–5267.

Zhao, Y. (2023, March). Advanced Metamaterials for Edible Electronics. *IEEE Transactions on Electron Devices*, 70(3), 1234–1242.

Zhao, Y., Huang, J., Li, X., & Xu, C. (2020). Optimal node placement for magnetic relay and MIMO wireless power transfer network. *Proc. 15th Int. Conf. Wireless Algorithms Syst. Appl.*, 743–754. DOI: 10.1007/978-3-030-59016-1_61

Zhao, Y., Li, X., Ji, Y., & Xu, C.-Z. (2019). Random Energy Beamforming for Magnetic MIMO Wireless Power Transfer System. *IEEE Internet of Things Journal*, 7(3), 1773–1787. DOI: 10.1109/JIOT.2019.2962699

Zheng, K., Jiao, Y., & Chen, L. (2013). Stochastic performance analysis of a wireless finite-state Markov channel. IEEE Transactions on Wireless Communications, 12(2), 782–793. DOI: 10.1109/TWC.2012.110212.121883

Zhong, C., Hu, X., Chen, X., Ng, D. W. K., & Zhang, Z. (2018). Spatial Modulation Assisted Multi-Antenna Non-Orthogonal Multiple Access. *IEEE Wireless Communications*, 25(2), 61–67. DOI: 10.1109/MWC.2018.1700062

Zhong, W., Yu, R., Xie, S., Zhang, Y., & Yau, D. K. Y. (2018, July). On stability and robustness of demand response in V2G mobile energy networks. *IEEE Transactions on Smart Grid*, 9(4), 3203–3212. DOI: 10.1109/TSG.2016.2629378

Zhou, Z., Chen, D., & Xie, S. (Shengquan). (2007). Springer Series in Advanced Manufacturing. In Thermoplastics and Thermoplastic Composites.

Zhou, L. (2022, August). Metamaterial-based Edible Antennas: A New Frontier in Ingestible Wireless Systems. *Nature Communications*, 13(1), 3498. PMID: 35715397

Zhu, X., & Goldberg, A. B. (2022). *Introduction to semi-supervised learning*. Springer Nature.

Zhu, X., Wang, Z., & Cao, J. (2017). NOMA-Based Spatial Modulation. *IEEE Access : Practical Innovations, Open Solutions*, 5, 3790–3800. DOI: 10.1109/ACCESS.2017.2688019

Ziegler, V., Viswanathan, H., Flinck, H., Hoffmann, M., Räisänen, V., & Hätönen, K. (2020). 6G architecture to connect the worlds. *IEEE Access : Practical Innovations, Open Solutions*, 8, 173508–173520. DOI: 10.1109/ACCESS.2020.3025032

Zong, J., Liu, Y., Liu, H., Wang, Q., & Chen, P. (2022). 6G cell-free network architecture. *2022 IEEE 2nd International Conference on Electronic Technology, Communication and Information (ICETCI)*. https://doi.org/DOI: 10.1109/ICETCI55101.2022.9832308

Zong, B., Fan, C., Wang, X., Duan, X., Wang, B., & Wang, J. (2019). 6G Technologies: Key Drivers, Core Requirements, System Architectures, and Enabling Technologies. *IEEE Vehicular Technology Magazine*, 14(3), 18–27. DOI: 10.1109/MVT.2019.2921398

Zonta, T., da Costa, C. A., Righi, R. D. R., de Lima, M. J., da Trindade, E. S., & Li, G. P. (2020). Predictive maintenance in the Industry 4.0: A systematic literature review. *Computers & Industrial Engineering*, 150, 106889. DOI: 10.1016/j.cie.2020.106889

About the Contributors

Augustine O. Nwajana (Senior Member, IEEE) received the M.Sc. (distinction) and the Ph.D. degrees in electrical and electronic engineering from the University of East London, U.K., in 2010 and 2017, respectively. He also received the research CPD certificate in Practical Antenna Design from the University of Oxford, U.K., in 2019. He was with Siemens AG as telecommunications engineer, with Coventry University as associate lecturer, and with the University of East London as part-time lecturer. He is currently a senior lecturer (assistant professor) at the University of Greenwich where his current research interests include radio frequency (RF) energy harvesting/scavenging, battery-less RF/microwave systems, passive RF/microwave devices, and more. He has authored/co-authored five books, several journal articles, book chapters, and conference papers. Dr Nwajana has served as keynote speaker and TPC member to several conferences and is currently the Editor-in-Chief for the International Journal of Electronics, Communications, and Measurement Engineering (IJECME) published by IGI Global, USA. He is a fellow of the Higher Education Academy.

* * *

Stella N. Arinze received her Master of Engineering and Doctor of Philosophy degrees in Electrical and Electronic Engineering from Enugu State University of Science and Technology, Nigeria. Currently, she serves a lecturer in the Department of Electrical and Electronic Engineering at the same University. Her research interests include Wireless Communication and Networks, Light Fidelity Technology, Internet of things and Antenna. She has authored/co-authored several journal articles, book chapters and conference papers.

Srikamu C. received her B.E. degree in Electronics and Communication Engineering from Anna University in 2006, her Master's Degree in Applied Electronics from PRIST University, Thanjavur in 2011 and her Ph.D. degree from SASTRA Deemed to be University, Thanjavur in 2024. She is currently a Post-Doctoral Fellow in the IITTNiF-SASTRA project. Her research interests include Signal Processing, Beamforming, Cooperative Relaying, Spatial Modulation, Power allocation in NOMA and Optimization.

Lakshmi D. is presently designated as a Senior Associate Professor in the School of Computing Science and Engineering (SCSE) & Assistant Director, at the Centre for Innovation in Teaching & Learning (CITL) at VIT Bhopal. She has 17 international conference presentations, and 21 international journal papers inclusive of SCOPUS & SCI (cumulative impact factor 31). 3 SCOPUS inee book chapters. A total of 24 patents are in various states and 18 patents have been granted at both national and international levels. One Edited book with Taylor & Francis (SCOPUS Indexed). She has won two Best Paper awards at international conferences, one at the IEEE conference and another one at EAMMIS

2021. She received two awards in the year 2022. She received two awards in the year 2022. She has addressed innumerable guest lectures, acted as a session chair, and was invited as a keynote speaker at several international conferences. She has conducted FDPs that cover approximately ~80,000 plus faculty members including JNTU, TEQIP, SERB, SWAYAM, DST, AICTE, MHRD, ATAL, ISTE, Madhya Pradesh Government-sponsored, and self-financed workshops across India on various titles.

Smrity Dwivedi is working as an Associate Professor in the department of Electronics Engg., IIT (BHU), Varanasi. She is Senior Member of IEEE, Fellow of IETE, Member of IET, LMISC, Guest Editor in Elsevier, Life member of VEDA society. She completed her doctoral work from IIT (BHU), VNS, INDIA in RF and Microwaves. She worked as a research fellow (postdoctoral type work) from IIT-BHU, INDIA. She has published many research papers in international journals, international conferences and national conferences. She is also published many chapters in reputed books.

Solomon H. Ebenuwa has a B. Ed in Physic, an MSc in Computing, and a PhD from the University of East London in Computing. He is a Fellow of the U.K. Higher Education Academy (FHEA). Dr Solomon Ebenuwa has worked in various institutions, from Secondary Schools to Higher education (University). Dr. Solomon's research interests are in the area of imbalanced data, algorithm optimization, and deep learning for real-life improvement in predictive modelling. He is currently lecturing at the University of Greenwich in the United Kingdom.

Gregory Omozuhiomwen Egbe holds B.Eng in Electrical and Electronics Engineering, from Enugu State University and Technology, he also holds master's in engineering management (MSc) and PhD. degree in Industrial Engineering both from the University of Benin, Nigeria. A data analysist, with a successful track record with solid background and several years of experience in the oil and gas sector. He is currently the General Manager Drilling and Measurement Vaughn Energy Services Limited, before now he served in different capacity in various organization in Nigeria such a Field Engineer in Kasolute Nigeria Limited, Field Service Engineer in Vigeo Limited, Field Service Manger Vaughn Energy Service Limited, Acting General Manager Operations (west) Vaughn Energy Services Limited. He has authored several journal articles publication and he is very active in the areas of asset integrity management, reliability, availability, maintainability (RAM) and Artificial Intelligence.

Jay Kumar Jain is currently working as an Assistant Professor in Department of Mathematics, Bioinformatics and Computer Applications, MANIT, Bhopal. He did his Ph. D. from Maulana Azad National Institute of Technology, Bhopal in 2015. He was awarded with Research Fellowship by MHRD for completing his Ph.D. He has research as well as teaching experience of about 13 years. He has published around 55+ research papers and book chapters in SCI/SCOPUS/Referred International/ National Journal and Conferences He has also authored 03 books for International Publisher. He has also granted 3 International Patents and published/granted 3 Indian patents and also registered 2 copyrights in IPR, India. He has been a reviewer in many international journals/conferences including Elsevier, IEEE Access, and Springer. He has lifetime membership of various professional societies such as IEEE, CSI, Franklin, IDES, IAENG, SDIWC, and many more. His research interests include Wireless Sensor Networks, Internet of Things, and Mobile Ad hoc Networks.

Pooja Jaiswal received the B.Sc. degree in Science from J.K. Institute of Applied Physics & Technology, University of Allahabad, Allahabad (U.P.) in 2007. She received the B.Tech degree in Electronics and communication from J.K. Institute of Applied Physics & Technology, University of Allahabad, Allahabad (U.P.) in 2010. She received the M.Tech degree in Electronics and communication from J.K. Institute of Applied Physics & Technology, University of Allahabad, Allahabad (U.P.) in 2012. Her research interest includes Antenna and Microwave communication and their applications. Presently she is pursuing PhD and acting as an Assistant professor in United college of Engineering & Research, Prayagraj.

Karthika K. received her B.E. degree in Electronics & Communication Engineering from Maharaja Engineering College, Avinashi, India in 2013 and M.E degree in Communication Systems from Mepco Schlenk Engineering College, Sivakasi, India in 2015. She has obtained her Ph.D Degree from Anna University, India. Currently, she is working as an Assistant Professor in the Department of Electronics and Communication Engineering at Kumaraguru College of Technology, Coimbatore, India. Her research interest includes Ultra-Wideband (UWB) Antennas, Reconfgurable Antennas, Rectennas and Fractal Antennas.

Kavitha K. received her Bachelor's degree in Electronics and Communication Engineering from Madurai Kamaraj University, Madurai, India and her Master's degree in Microwave & Optical Engineering from Madurai Kamaraj University, Madurai, India in She completed her doctoral degree at Anna University, Chennai, specializing in Hyperspectral Image Classification. Currently, she holds the position of Professor and Head in the Department of Electronics and Communication Engineering at Velammal College of Engineering and Technology in Madurai, Tamil Nadu. Her career spans over 23 years, during which she has made significant contributions in the realms of teaching and research. Her research interests encompass diverse areas, including Antenna Design, Hyperspectral Image Classification and Optical System Design. She has published and presented over 100 research articles in renowned international journals and international and national conferences. Presently, she is supervising the research works of ten Ph.D. scholars in various areas of her expertise. In recognition of her outstanding contributions to education, Dr. K. Kavitha was honored with Dr. A.P.K. Abdul Kalam Teachers Excellence Award.

Sujanth Narayan K. G., born in Chennai, Tamil Nadu, is an Indian researcher and Technical Lead at the Technology Innovation Hub in Positioning and Precision Technologies at IIT Tirupati since May 2024. He holds a Bachelor of Engineering (B.E.) in Electronics and Communication Engineering from R.M.K. Engineering College (2017) and a Master of Technology (M.Tech.) in Communication Systems from SASTRA Deemed University (2019). He is completed his Ph.D. in Antennas and RF System Design from SASTRA Deemed University (2024). Sujanth has made significant contributions to his field, with one published patent titled "Fully Generalized Spatial Modulation Aided CNOMA System" (2024) and numerous publications, achieving an h-index of 5 and 76 citations. His research areas include antenna design, RF system design, V2X communication, wireless communication, ionospheric studies, and information theory.

Chandramouli H Mahadevaswamy received a B.Eng. degree in 2019 from Visvesvaraya Technological University in India, and an M.Sc. (Distinction) degree from the University of Greenwich, U.K., in 2023. He completed a master's thesis on Advanced Circuit Design for a Compact 9-pole Triple-

Band bandpass filter for satellite communication systems. He worked as an Electrical Design Engineer at JVS Electronics Pvt.ltd from 2019 to 2020. He was responsible for providing DC-DC converter design and quality control operations for Relay manufacturing. He was with Lecon Energetics Pvt.ltd as an Electrical Application Engineer from 2020 to 2022 and has expertise in power system design, SCADA communications, and PLC programming. He is currently an Application Engineer for Automation at Schweitzer Engineering Laboratories working on building a smart substation that includes SCADA communications with IEC 61850 and cyber security designs. His major project was titled "Wireless Power Theft Monitoring and Detection System". This proposed system design avoids human intervention in Power Management.

Palak Mangal is a second-year student pursuing a Bachelor of Technology in Electronics and Communication Engineering at VIT Bhopal University. With a keen interest in electronics and communications, Palak has demonstrated strong academic performance and a passion for technology innovation. During their time at VIT Bhopal, Palak has engaged in various technical projects and extracurricular activities, further honing their skills and knowledge in the field.

Hitesh Mohapatra received the B.E. degree in Information Technology from Biju Patnaik University of Technology (BPUT), Odisha in 2006, and the MTech. Degree in CSE from Odisha University of Technology and Research (OUTR) University, Odisha in 2009. He received his Ph.D. in Computer Science & Engineering in 2021 from Veer Surendra Sai University of Technology (VSSUT), Burla, India. He has contributed 30 SCI and Scopus indexed research papers, 21 international/national conferences and books on Software Engineering and C Programming respectively. He has 15 years of teaching experience both in industry and academia. He has served the research community with various capacities like session chair, technical chair, keynote speaker etc. His research interests include wireless sensor network, smart city, smart grid and smart water. Currently he is working as Associate professor at School of Computer Engineering, KIIT DU, Bhubaneswar, and Odisha.

Eugene Ogbodo holds a PhD in Electrical Engineering from the University of Greenwich. He specializes in microwave engineering, focusing on the design and development of microwave devices for communication systems, including satellite communication and base station technology. Currently, Dr. Ogbodo is affiliated with the University of Hertfordshire, where he serves as a researcher and academic. He has contributed significantly to the fields of circuit design, system integration, and the miniaturization of passive microwave devices, with numerous peer-reviewed publications showcasing his innovative work in microwave technology and engineering applications.

Nandita Pradhan is the head of the ECE and CRT department. She did B.E. in Electronics and Telecommunication Engineering in 1985, M.E. in Communication Systems from the Govt. Engineering College, Jabalpur in 1991 and Ph.D. in Electronics and Communication Engineering from AKTU, Lucknow. She began her professional journey as a Lecturer in the Department of ECE at Government Engineering College, Jabalpur in 1985 and served in that position until 1997. From 2001 to 2011, she worked as a Professor and Head of the Departments of ECE and EIE at Galgotias College of Engineering and Technology in Greater Noida. Since 2014, she has been associated with United Group of Institutions (UGI), Allahabad, where she currently holds the positions of Professor, Dean of Students Welfare, and Head of the Department of ECE. She also actively promotes value addition programs and serves as a coordinator for the GATE program, Innovation Cell, and Startup Cell at UGI, Allahabad. She has also published several papers in esteemed journals and conferences, with a particular emphasis on areas such as Digital Signal Processing (DSP), Digital Image Processing (DIP), Medical Imaging, and Biomedical Instrumentation.

Jayabharathy R. received the B.E. degree in Electronics and Communication Engineering from Thiagarajar College of Engineering (Madurai Kamaraj Universit)y, the M.E. degree in Power Electronics and Drives from Bharathidasan University and the Ph.D. degree in Wireless Communication from SASTRA Deemed to be University, India. Her research interests include Digital Signal Processors, Network Theory and Indoor Wireless Communication Systems. She has published more than 30 research articles in reputed International Journals and Conferences. She is currently working as Senior Assistant Professor in the School of EEE, SASTRA Deemed to be University. She is a IEEE Member and life member of the Indian Society of Systems for Science and Engineering (ISSE).

Karthikeyan S. completed his Ph.D. from SASTRA Deemed to be University, Thanjavur, Tamil Nadu, India in 2024. He is currently serving as Assistant Professor in the Department of Electronics and Communication Engineering at SRM Institute of Science and Technology, Tiruchirappalli. His research interests include energy-efficient wireless body area network, bio-inspired optimization algorithms and sustainable solutions.

Rajkumar S. received his bachelor's degree in Electronics and Communication Engineering from Anna University, Chennai, and his master's degree in VLSI Design from the same institution. He completed his doctoral degree at Anna University, Chennai, specializing in VLSI device modeling. Currently, he serves as an Assistant Professor in the Department of Electronics Engineering (VLSI Design and Technology) at Velammal College of Engineering and Technology, Madurai, Tamil Nadu, India. His career spans over 11 years, during which he has made significant contributions to teaching and research. His research interests include diverse areas such as modeling and simulation of analog VLSI. He has published over 10 research papers in international journals and presented 20 papers at international conferences. He has also received the award for "Best Outcome in Academic Performance".

Anup Raju Vasistha holds a Bachelor of Engineering (BE) degree in the field of Electrical and Electronics Engineering (EEE), awarded in 2019 with the accolade of "Best Outgoing Student" from his university in India. He further enhanced his academic credentials in 2023 with a Master of Science (MSc) in Electrical and Electronics Engineering from the University of Greenwich, London, UK, where he graduated with merit. His master's studies emphasized filter designing, analysis of electrical power systems, and designing power converters. His MSc thesis focused on using Advanced Design Systems to develop filters for frequency spectrum applications in wireless and satellite communication systems. From 2019 to 2021, Anup served as an Automation Engineer at ABB in India, where he was responsible for designing and commissioning activities in manufacturing and process industries. His role included extensive work in PLC automation, as well as designing HMI and SCADA systems for international clients. Currently, Anup is an Application Engineer in Automation for the Europe region at Schweitzer Engineering Laboratories (SEL) in Reading, UK. His expertise encompasses power system automation and cybersecurity related to SEL products. In this capacity, he assists customers with product applications and contributes to the development of product specifications and literature.

Index

Symbols

A

B

C

D

E

F

G

H

I

L

M

N

O

P

R

S

T

U

V

W

www.ingramcontent.com/pod-product-compliance
Lightning Source LLC
LaVergne TN
LVHW060350300926
849162LV00006B/1236